Agroecology and Integrated Farming System

In most developing countries, agriculture has grown from merely an art to a science, but it does not yet maximize its business potential. In these countries, subsistence farming dominates, and farmers face the increasing impact of climate change and natural disasters. An integrated farming system (IFS) model yields minimum risk and maximum environmental benefit. The latest cutting-edge technologies applicable to each component of IFS and the science behind an agro-ecological approach are discussed at length in this book, which takes a holistic approach towards sustainable agricultural production technologies that result in maximum profit for the farming community. Also, it considers practices that care for natural resource bases and leave behind minimal environmental footprints. To keep prepared for climate change and natural disasters, appropriate contingency measures to tackle these unwanted situations are detailed.

The book offers comprehensive coverage of the most essential topics, including:

- Modern technologies, new concepts and innovations such as 3D farming, Integrated System of Rice Intensification (ISRI), hydroponics, rooftop farming and water budgeting.
- The use of IT for supporting IFS and environmental aspects related to greenhouse gas (GHG) emission.
- Information on organic farming covering all its aspects, present situation, market-related issues and future options.
- *In-situ* input generation procedures that are integral to recycling and their effective reuse.
- Region-specific IFS models based on soil, climate and farmers' requirements for different agroclimatic situations.
- IFS management aspects including water harvesting, conservation, increased productivity and drainage.
- Latest information on the socio-economic factors, impacts, government orientations, policy framework towards agriculture and environmental aspects, and the future road map to make IFS a success.

This book will serve as a handy reference for academics, researchers, students, progressive farmers and policymakers aiming to make agriculture more resilient, sustainable and eco-friendly.

Agroecology and Integrated Farming System

Sukanta Kumar Sarangi, Rajeeb Kumar Mohanty,
Sukham Munilkumar and Jitendra Kumar Sundaray

CRC Press
Taylor & Francis Group
Boca Raton London New York

CRC Press is an imprint of the
Taylor & Francis Group, an **informa** business

Designed cover image: Author's own image

First edition published 2025
by CRC Press
2385 NW Executive Center Drive, Suite 320, Boca Raton FL 33431

and by CRC Press
4 Park Square, Milton Park, Abingdon, Oxon, OX14 4RN

CRC Press is an imprint of Taylor & Francis Group, LLC

ISBN: 9781032145648 (hbk)
ISBN: 9781032145655 (pbk)
ISBN: 9781003239963 (ebk)

DOI: 10.1201/9781003239963

Typeset in Times
by Deanta Global Publishing Services, Chennai, India

Contents

Preface

Agroecology is a holistic and integrated approach that simultaneously applies ecological and social concepts and principles to the design and management of sustainable agriculture and food systems. It seeks to optimize the interactions among plants, animals, humans and the environment while also addressing the need for socially equitable food systems within which people can exercise choice over what they eat and how and where it is produced. Agroecology is concurrently a science, a set of practices and a social movement and has evolved as a concept over recent decades to expand in scope from a focus on fields and farms to encompass the entirety of agriculture and food systems. It now represents a transdisciplinary field that includes the ecological, sociocultural, technological, economic and political dimensions of food systems, from production to consumption. The book *Agroecology and Integrated Farming System* explores the diverse aspects of agroecology and farming system.

The book consists of 15 chapters and focuses on the importance of agroecology and integrated farming system (IFS). The chapters narrate the concept, components and integration of agroecology, sustainable integrated farming systems for food and nutrition security, integration of enterprises and efficient resource use in diverse farming systems, application of information technology for advanced IFS research and development (R&D) and integration of modern tools, including artificial intelligence. The latter part of the book explains the modern tools for greenhouse gas emission estimation and mitigation in IFS, organic farming systems, *in situ* resource generation, nutrient management, ITK and use for IFS, impact of climate change on agriculture and adaptations through IFS, IFS models for different agroclimatic zones, high-value aquaculture, agribusiness options and farmers producer organizations (FPOs), agroecology and biodiversity, rainwater harvesting and management for IFS. The concluding chapters cover the socioeconomic impact of IFS, road map for the future – doubling farmers' income and policy framework requirements for the success of IFS. We stongly believe that the book covers the latest findings and information on different aspects of IFS for stepping up agricultural productivity, profitability, resource use efficiency and environmental benefits.

Overall, *Agroecology and Integrated Farming System* highlights the importance of adopting innovative technologies and policies to ensure the sustainable development of the IFS. We hope that this book will serve as a valuable resource for students, teachers, researchers, policymakers and practitioners working in the fields of agriculture, aquaculture, animal husbandry and allied disciplines and contribute to the advancement of sustainable integrated farming systems.

We take this opportunity to sincerely thank our publisher CRC Press (Taylor and Francis Books India Pvt. Ltd.) for selecting our book for publication and extending strong cooperation during the entire publication process. We are extremely thankful to the anonymous reviewers for giving their positive comments and recommending our book for publication. We gratefully acknowledge the Indian Council of Agricultural Research (ICAR) for providing support for conducting the research and providing an opportunity for serving the farming community of India.

Dr. Sukanta Kumar Sarangi
Dr. Rajeeb Kumar Mohanty
Dr. Sukham Munikumar
Dr. Jitendra Kumar Sundaray

Foreword

Agroecology integrates ecological principles into agriculture, promoting sustainable practices and biodiversity conservation. It represents an overarching and comprehensive systems framework to guide public policies towards sustainable agriculture and food systems. Agroecology actively engages different stakeholders through interdisciplinary mechanisms, favouring responsible and transparent governance of resources. This approach supports the simultaneous achievement of multiple sustainability objectives – economic, environmental, social, nutritional, health and cultural – holistically and in an integrated manner at different levels and scales, while being adapted to various environmental and cultural contexts.

Agroecology emphasizes soil health for sustainable agriculture, promoting nutrient cycling, organic matter accumulation and soil structure. Agroecology helps to protect, restore and improve agriculture and food systems in the face of climate shocks and stressors. Diverse, well-integrated agro-ecological systems can promote greater carbon sequestration, increase resilience of livelihoods and provide climate change mitigation and adaptation solutions. Furthermore, agroecology offers innovative solutions and attractive employment for youth, characterized by knowledge-intensive, environmentally friendly, socially responsible and skilled labour for agricultural production. It applies practices of solidarity through collective actions to reduce gender inequity by addressing underlying power imbalances that perpetuate discrimination and oppression.

Integrated farming systems (IFS) promote sustainable and inclusive agricultural development. Integrated farming systems (IFS) can improve food and nutrition security and increase income and livelihood opportunities for farmers and their communities. IFS establish a conducive environment for farmers and other stakeholders, ensure the effective utilization of available natural resources and contribute to achieving broader development objectives.

Agroecology and Integrated Farming System provides valuable insights into the role of IFS in addressing the complex challenges of food security, nutrition and sustainability. Agro-ecological aspects of IFS related to soil health, pest and disease regulation, pollination, genetic diversity and ecosystem resilience are highlighted. I compliment the authors for their valuable contributions in this book, and I hope that the book will serve as an excellent source of reference material for researchers, farmers, students and other stakeholders.

Dr. Himanshu Pathak
Secretary (DARE) & Director General (ICAR)
Dated 6 August 2024
New Delhi

About the Authors

Sukanta Kumar Sarangi obtained a BSc Agriculture (1995) with a first position and Gold Medal from the Odisha University of Agriculture & Technology (OUAT), Bhubaneswar; MSc Agronomy (1998) with first rank and University Gold Medal from OUAT, Bhubaneswar; and PhD (2003) from Banaras Hindu University, Varanasi. Dr Sarangi worked with Prof. M. S. Swaminathan Research Foundation (1999–2000) on conservation, utilization and enhancement of rice biodiversity. Served as Scientist (2003–2008) at ICAR-Research Complex for the North Eastern Hill Region, Meghalaya; Senior Scientist (2008–2014); and Principal Scientist (2014–2023) at ICAR-Central Soil Salinity Research Institute, Regional Research Station, West Bengal. Currently he heads the Division of Agricultural Technologies for Women at ICAR-Central Institute for Women in Agriculture, Bhubaneswar, Odisha, India.

In his research on farming systems in hill and coastal regions, Dr Sarangi has developed resource conservation technologies for cropping systems intensification, including: a cost-effective, water-productive improved package of practices for cultivation of diverse crops; salt-tolerant varieties; integrated nutrient management practices; conservation agricultural practices; and technology for enhancing income and nutritional security of smallholder farmers. He has established strong research collaborations with several countries including Australia, the Philippines, Vietnam, Bangladesh and Nepal.

He has been made a Fellow of the Indian Society of Agronomy (ISA), the Indian Society of Coastal Agricultural Research (ISCAR), the Association of Rice Research Workers (ARRW) and Honorary Fellow of Pragati International Scientific Research Foundation (PISRF). He currently serves as the editor of the *Indian Journal of Agronomy* and editor-in-chief of the *Journal of the Indian Society of Coastal Agricultural Research.*

Rajeeb Kumar Mohanty was born in 1969 in Balasore district of Odisha. He holds an MSc in Fisheries Management (ICAR-Central Institute of Fisheries Education, Mumbai), a PhD in Aquaculture and a DSc in Environmental Science. He joined the Agricultural Research Service of ICAR and started his professional career as a scientist in 1997; presently, he is Principal Scientist (Aquaculture) at the ICAR-Indian Institute of Water Management, Bhubaneswar.

Dr Mohanty has made an outstanding research contribution to aquaculture water use efficiency, water budgeting, nutritional water productivity, water footprint, EIA, bioresource technology, compensatory growth performance of IMCs, run-off recycling aquaculture, self-reliant integrated aquaculture, multiple use management of various aquaculture production systems, periphyton-based carp-SIS polyculture and semi-intensive shrimp monoculture in improving aquaculture productivity, and the livelihood of small and marginal farmers.

Significant work done by Dr Mohanty is reflected through the implementation of 17 research projects of national importance as project leader, more than 140 publications in reputed high-impact journals, 8 commercialized technologies, recipient of 7 national awards, such as ICAR–Team Research Award (2001–2002 and 2003–2004), Vasantrao Naik Award (2001 and 2007), Indira Gandhi Rajbhasha Puraskar (2005–2006), Dr Hiralal Chaudhuri Young Scientist Award (2003), BKJF-INCSW Team Award 2018, NESA Fellowship 2019 and NABS Fellowship 2020, as well as

5 other society awards. In addition, Dr Mohanty has guided the research work of two PhD scholars and eight MSc students in the fields of aquaculture, water management and wetland ecology.

Sukham Munikumar, born in Imphal, Manipur, graduated from the College of Fisheries, Ratnagiri, Maharashtra, and later went on to complete his master's and doctoral degrees from the Central Institute of Fisheries Education, Mumbai. He did his post-doctoral fellowship research at Curtin University, Perth, Australia, and attended short-term courses at Ben Gurion University, Israel, and the University of Las Palmas, Gran Canaria, Spain. He has diverse experience, having worked in the shrimp hatchery industry in Andhra Pradesh and participated in a deep-sea fishing project with CMLRE (Department of Ocean Development, Ministry of Earth Sciences). He started his teaching, research and extension career as Assistant Professor (Aquaculture) at the College of Fisheries, Agartala, Tripura. He has contributed to eight externally funded projects, including four projects as Principal Investigator. He has also worked as PI on two institutional projects as well as contributing as Co-PI to ten projects. He has served five PhD and 11 MFSc students as a major guide and has served as an Advisory Member for six PhD and 14 MFSc students. He has coordinated more than 80 skill development programmes of seven days' duration for more than 3,000 farmers during the last 14 years. He has published more than 60 research articles and 15 book chapters, including 10 manuals. He is a Fellow of the Zoological Society of India, Bodhgaya, India, and the Applied Zoologists Research Association (AZRA), Cuttack, Odisha. Currently he is working as Principal Scientist in the Aquaculture Division, ICAR- Central Institute of Fisheries Education (Deemed University), Mumbai.

Jitendra Kumar Sundaray completed his PhD (Animal and Marine Bioresources Science) from Kyushu University, Fukuoka, Japan, and is presently working as Head, Fish Genetics and Biotechnology Division, ICAR-Central Institute of Freshwater Aquaculture, Bhubaneswar. He has served as Officer-in-Charge of the Kakdwip Research Centre of ICAR CIBA (2010–2013), Chennai. He completed his tenure (2013–2022) as Head, Division of Fish Genetics Biotechnology, ICAR-CIFA and as Director, ICAR-CIFA during 2017–2018. He is the pioneer team member of the first brackish water fish (seabass) breeding project in India.

He has published more than 132 research publications in both national and international peer-reviewed journals. He has a total citation of 2,527 and H index of 24. He has completed 12 projects as Principal Investigator and headed two externally funded projects. He has completed a Center of Excellence project from DBT on the genetic improvement of *Labeo catla.*

He is a fellow of the Applied Zoologist Research Association, Zoological Society of India, Biotechnology Research Society of India, Inland Fisheries Society of India, Odisha Environmental Society and National Environmental Science Academy. He is the recipient of the young scientist award of AZRA, ZSI, and the Hiralal Choudhary Foundation, ICAR CIFE, Mumbai. He is the recipient of the Prof M. S. Swaminathan Best Fisheries Scientist for the year 2021 and Distinguished Scientist Award from the National Environmental Science Academy and Padma Shri (Dr) Ajaya Parida Life Time Achievement Award for the year 2023.

He has guided two post-doctorate, two PhD and 20 MSc students. He is one of the Vice Presidents of the Applied Zoologist Research Association and he is the associate editor of the *Journal of the Indian Society of Coastal Agricultural Research.* He is the associate editor of *Frontiers in Physiology* and *Frontiers in Molecular and Cellular Reproduction.* He is leading a team on selective breeding and genetic improvement of carp species in India. He is currently involved in selective breeding and halogenomics.

1 Integrated Farming System
Concept, Components and Integration

1.1 INTRODUCTION: CONCEPT OF FARMING SYSTEM

Agriculture is the ancient tradition of human civilization to meet the basic needs of diverse food, fuel for cooking, feed and fodder for domestic animals, shelter for self and animals as well as fibre for clothing. Therefore, to meet all these essential requirements there is a need to grow a variety of crops as well as animals, birds, aquatic fauna, etc. Different components are interrelated in terms of use of resources/inputs and outputs. This relation results in productive use of the byproduct of one component to produce output of greater value and use for human beings. Directly and indirectly, there is benefit to the overall output of the system and system sustainability. The total combination of the individual components results in a farming system with inclusion of the farming family. The individual components also comprise a system within itself. The combination of crops form the cropping system, different animals form the animal system/animal/dairy husbandry; fish and other water-related components form the pisciculture/aquaculture; bird components may be poultry farming/duckery, etc. All these component systems form the farming system. When a particular set of component systems is followed with certain resource flow structure and are integrated, it forms the integrated farming system (IFS). A farm is the piece of land used for agricultural production whereas a firm refers to a business entity, such as an agricultural farm, which is market oriented and involves business.

The farming system is an art, commerce and science involving cultivating plants, rearing animals/livestock to result in products for consumption, marketing and industrial uses. Crop cultivation is an age-old practice of human civilization so also the rearing of animals for use on farms for draft purposes as well as for consumption. Most of the developing countries in Asia, including India, follow an agrarian culture of food production in varied prevailing agroclimatic conditions to meet the regional food requirement as well as to export the surplus production. The crops include cereals, pulses, oilseeds, fruits, forest trees, vegetables, spices, medicinal and ornamental/floricultural plants. Animal husbandry includes dairy cattle for milk, pigs for meat, goats and sheep for wool and meat, honey bees for honey, poultry and duckery for eggs and meat, etc. Aquaculture produces fish of different kinds, pearl culture, ornamental fishes for an aquarium, etc. In a farming system, two or more enterprises are combined, major output is used for consumption and marketing, some of the by-product/wastes are used as input for other enterprise, e.g., cow dung, urine, leftover roughages of animal husbandry are mixed and used as farmyard manure as an input in the cropping system. The straw/stalk/left over from the cropping system is fed to animals as fodder as an input for the dairy system. The farming system is the refined form of traditional mixed farming followed from ancient times, where a combination of enterprises served only for subsistence income and not for generating a marketable surplus. The farming system takes into account the logical and ecological concept of enterprise combination.

Agricultural products are either used directly for consumption, such as fruits, or cooked for consumption. There are products which require processing for further use as well as for value addition. Food processing and preservation add value and price to the original production. After processing, good packaging and preservation are also required for agricultural produce to enhance the shelf life and export the produce to different countries.

DOI: 10.1201/9781003239963-1

The cropping system is generally a land-intensive component of farming system; however, in recent years, hydroponics and other innovations, such as roof top farming and vertical farming, have been developed to reduce the pressure on land. Crop cultivation is dependent upon the prevailing climatic conditions, i.e., related to the agroclimatic zones of the region or country. Field crops are generally season oriented with three seasons, such as *Kharif*, rabi and summer season cultivation. Fruit crops are generally perennial in nature and produce fruits on an annual basis. The cropping system also depends on the land type/elevation of the location. Some crops require higher elevation/uplands as these crops cannot tolerate waterlogging or require drained soils, such as fruits and vegetables. Lowlands are suitable for rice cultivation and aquaculture.

1.2 CROPPING SYSTEM

It refers to the crop production process in the farm by the use of available resources of land, water, labour, other inputs and interaction with fixed farm resources such as land, machinery, variable resources of seed, fertilizer, manures, labour, etc. Most of these resources are owned by a person, who is called the farmer; some resources can also be hired/leased/purchased. The socioeconomic status of a farmer and the growing environment can be better by a sustainable cropping system. It evolves through continuous field testing and designing to meet the farmers needs and the demand of the community. Therefore, cropping system research takes into account the available land with the farmer, its suitability to grow diverse crops by the efficient use of crop rotations and other available resources. Crop rotation encompasses a certain sequence of crops which improve soil health, requires or generates less pest control measures and produces diverse products to fulfil the balanced food demand of the family in addition to providing income to meet other needs. Crop rotation and cropping sequence are the synonyms used to refer recurrent succession of crops on a particular land over a period of time. Cropping pattern designates the yearly sequence and spatial arrangement of crops or fallows in a given land. It is generally misused in place of the cropping system and it should always be given with seasonal area allocated to different crops in a land/area/district/region. For example, the cropping pattern of coastal West Bengal for *Kharif* season is rice 98% and vegetables 2% of the area, whereas cropping pattern during *Rabi* season is fallow 80%, rice 14%, vegetables 4% and other crops 2% of the area. Mono/double/triple cropping refers to one, two and three number of crops, respectively, grown in an area year after year. If only one crop is grown on land year after year, it is called a mono-cropped land, if two, double cropped and so on (Figure 1.1). Monoculture refers to single-crop cultivation in a given field in a crop season, e.g., growing only a maize crop in a sole stand in a field. Similar terminology of growing two or more crops in a piece of land in a season are dual culture, polyculture, etc.

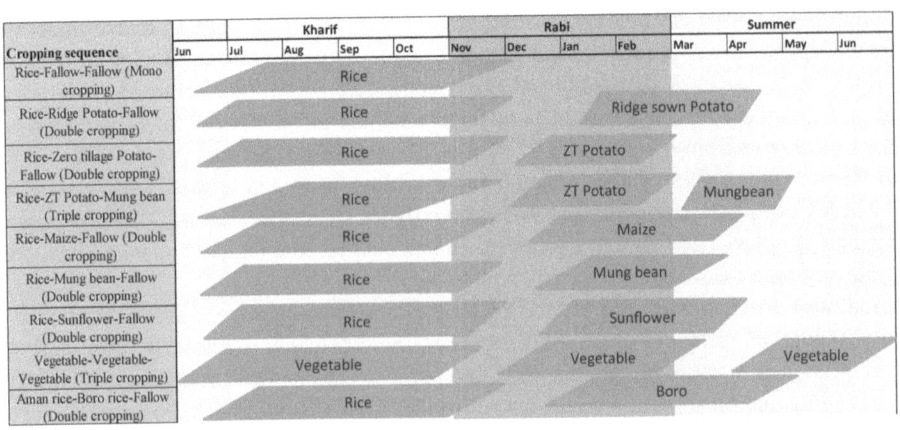

FIGURE 1.1 Mono-, double and triple cropping systems for coastal region of India.

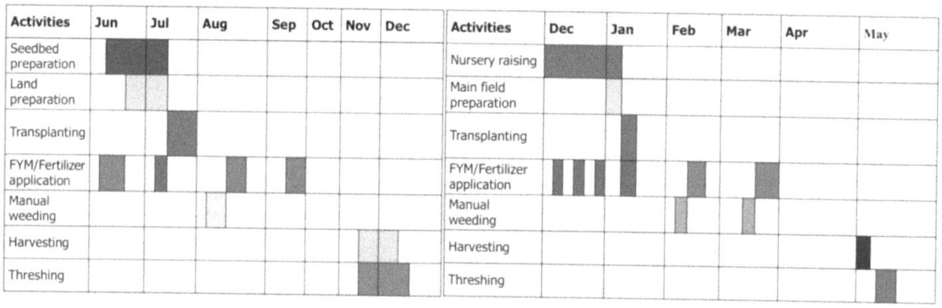

FIGURE 1.2 Crop calendar for *Kharif* (*Aman*) and summer (*Boro*) rice in coastal region.

Cropping intensity refers to the yearly intensity of land use. If the entire area is used once in a year, the cropping intensity becomes 100%. Therefore, cropping intensity is calculated by dividing the gross cropped area (summation of area used for different crops in a particular area) by the net area and multiplied by 100. In intensive cropping systems, the land is utilized intensively, usually high yielding crop varieties are grown, improved management practices are followed and the market demand is fulfilled.

A crop calendar (Figure 1.2) shows the time of activity related to growing of crops in a year. The sequence of activities such as land preparation, sowing/planting, application of inputs, intercultural operations, harvesting, etc. are depicted in different months.

Intercropping and mixed cropping are used to refer to the simultaneous cultivation of two or more crops on a same portion of land, when a definite row pattern or geometry is followed it is called intercropping and when there is a haphazard combination of crops, it refers to mixed cropping. The traditional system of cultivation practice, where two or more crops of different growth habits are broadcasted in the same piece of land to obtain different crop harvests as well as to reduce risk is called mixed cropping. When one crop fails, other crops provide the food to the farming family. For example, rice and maize are grown together in hilly regions as a traditional mixed cropping system. Intercropping is a scientifically designed crop combination with certain row arrangements, taking into account the phenotypic growth habit, root architecture, growth duration, nutrient requirements and withdrawal pattern of component crops. Intercropping system is based on a base crop known as the main crop and the additional crop is called the intercrop. Crop geometry or arrangement of rows are adjusted in an intercropping system to accommodate the intercrop. When the base crop is kept at its optimum density and an additional crop is introduced to this, this is known as additive series. In additive series, the population of the intercrop is less than its pure crop population. In replacement series, the population of the base crop is adjusted to accommodate another crop and both the crops are called component crops. Dual culture is a term used to refer to growing two different components on the same piece of land simultaneously in a particular crop season, e.g., Rice+*Azolla*, Rice+Fish, Rice+Frog, etc.

Relay/*paira*/*uttera* cropping refers to growing of crops in succession without giving a turnaround time or fallowing the land. The current crop relays the land to the subsequent crop as the subsequent crop is sown in the standing present crop before harvest. In coastal regions, such as in Odisha and West Bengal, *Paira*/*uttera* cropping is followed in a Rice-rabi cropping system to grow pulse crops. Before the harvest of a *Kharif* rice crop, the subsequent pulse crops (e.g., grass pea, horse gram, black gram, etc.) are broadcast into the wet soil to utilize the carryover soil moisture as well as the low temperature of the winter season.

Overlapping cropping system is a term used to describe crop combinations in the same piece of land, where a portion of the standing crop is harvested to sow the subsequent crop. In the case of fodder crops, the harvesting is done in instalments to meet the periodic requirement. In the harvested area, some other crop is sown, so that there is overlapping of one crop with the subsequent crop. In multi-storeyed cropping, crops of different canopy height are arranged for optimum utilization of land and solar energy. The trees/tall plants form the upper storey, e.g., coconut, arecanut, other fruit plants, etc.; plants of medium height such as papaya, pegion pea, sugarcane etc. form the second storey; and plants of low height, such as groundnut, ginger, pineapple, sesame, etc., form the third storey. Climbers such as black pepper may be grown along with trees such as arecanut. Crop cafeteria is the modern concept of diverse crop demonstration in the same piece of land to evaluate the performance and to give a scope to the farmers to choose the best performing crop as per individual requirement and suitability. Ratooning or ratoon cropping is followed with crops like sugarcane, where the subsequent crop is grown from the leftover stem/vegetation/stubble of the previous crop. Ratooning is done with the planned harvest of the previous crop so that the cut stalks regenerate and produce shoots, with optimum nutrient management it is grown as a full crop. In recent years, research on ratoon cropping in rice has also been conducted to evaluate the most promising genotypes which perform better under ratooning. Catch crops are the minor crops (in terms of output) grown to utilize the carryover soil moisture, grown after the harvest of the main crop. This is grown to immediately catch the natural resources and produce output. Catch crops are generally of very short duration, e.g., radish/green leafy vegetables crops that produce output in 25–30 days. Recession cropping is a term used to describe the growing of crops in flood-prone areas, where once the flood water recedes, a crop is planned to utilize the saturated and nutrient enriched soil profile.

In the Gangetic plains (middle, upper and trans) and western Himalaya, rice-wheat is the predominant cropping system, whereas a rice-rice system prevails in eastern Himalaya, lower Gangetic plains, eastern and southern plateau and hills, east coast, south and western plains and hills region. Apart from that, maize-wheat is followed in western Himalaya and eastern plateau and hills, pearl millet-wheat in trans-Gangetic plains, central plateau and hills and arid western plains, rice-pulses in eastern plateau and hills, soybean-wheat in central plateau and hills and western plateau and hills, groundnut-wheat in the Gujarat plains and hills region. In iIsland agroclimatic conditions, rice-fallow is the predominant cropping system. Research on sustainable and profitable cropping systems over the years has resulted in efficient alternate systems (Table 1.1).

1.3 ORGANIC FARMING

Organic farming/agriculture relies on inputs of organic origin and avoids the use of substances which are chemical/synthetic in nature (Figure 1.3). The aim is to produce and preserve nature without toxic effects on the environment. Modern agriculture is based on indiscriminate use of synthetic/chemical substances. In recent years, even to make the produce attractive looking or eye catching to the consumers, carcinogenic and other harmful chemicals are used by the middle men selling the produce in retail markets. The agricultural production system not permitting the use of chemical inputs such as fertilizers, pesticides, growth regulators and livestock additives is known as organic farming. This farming system follows crop rotations, crop residue recycling, effective utilization of animal manures, composts, beans, green manures, off and on-farm organic wastes to maintain soil productivity and tilth, to supply plant nutrients and employs biological pest control methods to manage insects, weeds and other pests (Lampkin 1990). In this system, crop nutrition is met through building up of the soil health by Azolla, crop residue incorporation, nitrogen fixation by legumes, application of organic manures, vermicomposts, green/green leaf manures and soil amendments such as gypsum and rock phosphate, sulphur dust, etc. However, the above products are not allowed when these originate from non-organic farms. Use of some of the inputs is restricted, such as organically produced mushroom beds, oil cake and milled products, human excreta, biofertilizers, saw dusts, coir pith, blood meal, meat meal, bone meal brought from other sources with preservatives, fish and fish products, etc.

The aim of organic farming should be to maintain organic production standards, obtain certification, appropriately package and label produce and efficiently market the produce. For the focussed and well-planned development of organic agriculture and quality production, Ministry of Commerce and Industry, Government of India, launched a National Program on Organic Production

TABLE 1.1
Efficient Alternate Cropping Systems For Different Agro-Climatic Zones of India

Agro-Climatic Zone	Alternate Cropping System
Western Himalaya	Rice-potato-onion, Rice-marigold-french bean, Maize+black gram-potato-onion, Rice-radish-potato (mid hills); Rice-potato-potato (low hills)
Eastern Himalaya	Winter rice-onion-cowpea (F), Rice-rice, Rice-rice-rice
Lower Gangetic plains	Rice-potato-jute, Rice (*Kharif*)-rice (*Boro*/summer)-cowpea (F)
Middle Gangetic plains	Rice-potato- mung bean, Rice-onion (eastern UP), Rice-lentil-mung bean (north eastern plains); Rice-potato- mung bean, Rice-mustard-mung bean, Rice-wheat-mung bean (eastern plain sub-humid zone of UP)
Upper Gangetic plains	Green manuring (*Dhaincha*)-potato-okra, Bajra-potato-cluster bean, Bajra-wheat-mung bean (Semi-arid UP); Maize-potato-sunflower, Maize-garlic-mung bean, Maize+black gram-potato-onion, Hybrid rice-wheat- mung bean (central plains of UP)
Trans Gangetic plains	Maize-potato-onion, Groundnut-potato-bajra (F), Maize-potato-mung bean, Maize-wheat- mung bean (Punjab); Pearl millet-potato-mung bean (Haryana)
Eastern plateau and hills	Rice-wheat+potato- mung bean, Rice-potato- mung bean, Rice-potato-cowpea
Central plateau and hills	Cluster bean-onion in irrigated areas, Mung bean-mustard in rainfed areas of Semi-arid eastern plain zone of Rajasthan; Soybean-onion, Black gram-mustard/wheat in sub-humid southern plain zone of Rajasthan; Rice-potato-wheat, Rice-berseem (F+Seed), Rice-pea-wheat for Kaymore plateau and Satpura hill zone of Madhya Pradesh; Soybean-pea-sugarcane, Soybean-potato-okra for Central Narmada valley
Western plateau and hills	Soybean-onion/wheat, Pearl millet-onion/chickpea (dry zone of Maharashtra); Soybean-onion, Turmeric+castor, Soybean-wheat (plateau zone of Maharashtra); Soybean-chickpea/sorghum (Western Vidarbha zone of Maharashtra); Soybean-wheat-potato-late wheat (Malwa plateau and Narmada basin zone of Madhya Pradesh)
Southern plateau and hills	Rice-maize, Maize-onion, Maize-chick pea (Southern Telangana zone of AP); Rice-rice, soybean+pigeon pea-sesamum, Soybean-sunflower, Pigeon pea-sesamum (Northern Telangana zone of AP); Maize-ground nut, Maize-sunnhemp-sunflower (Southern Transitional zone of Karnataka); Rice-rice, Rice/vegetables/marigold-maize-pulses, Rice-ground nut-sesamum, Sugarcane, Maize-ground nut, Rice/maize-pulses (Tamil Nadu)
East coast plains and hills	Rice-maize-cow pea/mung bean, Rice-ground nut-mung bean/sesamum (Coastal inland zone of Odisha)
Western plains and ghat	Rice-rice-okra, Rice-rice-groundnut, Green manure crop (Kerala); Rice-brinjal, Rice-maize (F), Rice-ground nut+sweet corn (Konkan region of Maharashtra)
Gujarat plains and hills	Ground nut-potato-pearl millet, castor-mung bean (North Gujarat); Rice-fenugreek-okra, Rice-onion-cowpea (South Gujarat); Ground nut-wheat/onion- mung bean, Ground nut-pearl millet- mung bean, Ground nut-wheat-sesamum, Cotton-pearl millet/ground nut, Ground nut-castor
Arid western plains	Sole castor, Cluster bean-wheat, Cluster bean-mustard
Islands	Homestead system with coconut/arecanut, jack fruit, guava, sapota, Coconut+ginger, Coconut+tapioca, Rice based system including rice+fish in low-lying areas

Source: After Gangwar and Singh (2011).

(NPOP) during the year 2000 and notified in 2001 under the Foreign Trade and Development Act (FTDR Act). At the national level, the National Centre of Organic Farming (NCOF) located at Ghaziabad, Uttar Pradesh, was established during the year 2004 for promotion of organic farming under the Department of Agriculture, Cooperation and Farmers Welfare, Ministry of Agriculture and Farmers Welfare, Government of India. NCOF implements the National Project on Organic

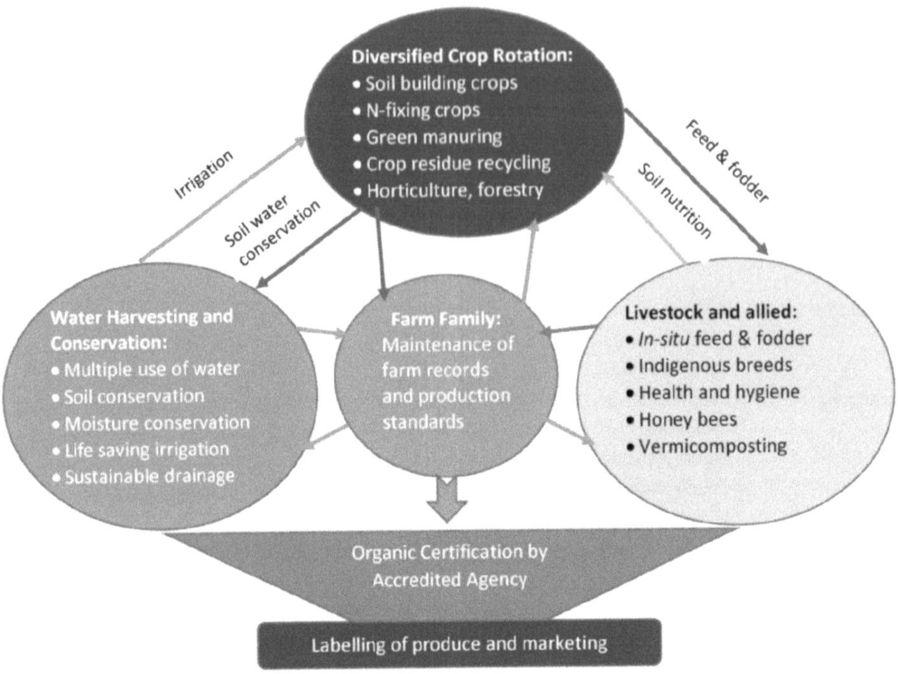

FIGURE 1.3 A schematic diagram to represent the framework of organic agriculture.

Farming (NPOF) through its regional centres located at Bengaluru, Bhubaneswar, Imphal, Jabalpur, Nagpur and Panchkula. The national accreditation body for organic farming comprises of representatives from the Ministry of Agriculture, Ministry of Commerce and Industry, Agricultural and Processed Food Products Export Development Authority (APEDA), Coffee Board, Spices Board and Tea Board. Some of the accredited certification bodies in India are Bureau Veritas, ECOCERT India, IMO Control Pvt., Indian Organic, Lacon Quality, OneCert Asia Agri., SGS India Pvt., Control Union, etc. The Government of India is encouraging organic farming by the implementation of *Paramparagat Krishi Vikas Yojana* (PKVY).

1.4 DAIRY FARMING

Dairy farming consists of cattle and buffalo rearing to produce milk, milk products, manure, etc. Farmers get income from the sale of the produce as well as the sale of animals apart from meeting family needs. Several milk products such as ghee, butter, paneer, cheese, etc. are produced from milk. Cow is traditionally given the status of 'Mother,' the sacred organism that feeds us with wholesome food apart from the use of cow urine for various purposes. Cow urine can be used as a nutrient source for field and horticultural crops by collecting it systematically. An annual urine production of 2000–3000 kg per animal can provide about 20 kg nitrogen and 40 kg potash along with other micronutrients. Cow dung production of 10–20 kg day^{-1} $animal^{-1}$ could be used for biogas production as well as provide the organic manure for enriching the soil nutrient status. There are mainly three types of breeds of cattle and buffalo.

Milch breeds: These are mainly reared for getting milk and not for use in farm operations, e.g., Sahiwal, Red Sindhi, Gir, Deoni, Nimari, etc.

Draught breeds: These breeds are reared for pulling carts, ploughing fields and carrying out other farm operations. Amritmahal, Kangayam, Malvi, Nagori, Krishna Valley, Hallikar, etc. are draught breeds of cattle.

Dual purpose breeds: The cows are good milk yielders and male animals are used for draught purposes, such as Hariana,vOngole, Tharparker, Kankrej, etc.

Apart from the above, exotic breeds such as Jersey, Holstein-Friesian (HF) and cross-bred cows may be included in dairy farming as per the agroclimatic conditions. Jersey cross-bred cows can be reared in coastal regions, as they can tolerate the heat conditions. HF crosses are suitable for hilly and temperate climates as they cannot tolerate heat conditions. Suitable buffalo breeds are Murrah, Ravi, Mehsana, Surti, Jaffarabadi, Nili-Ravi, Nagpuri, Bhadwari, etc.

The dairy and animal husbandry sector contributes about 4.2% to India's gross domestic product (GDP). About seven crores of rural families get their livelihood from this farming system. Milk is the largest crop in India (₹6.5 lakh crores), which is more than the total value of paddy and wheat put together. India is the largest milk producer in the world, contributing 22% of global milk production in 2018.

Proper housing/space (cow: 12–18 m^2 area; buffaloes: 12–15 m^2 area per animal), hygiene (ventilation, drainage, cleanliness, rough and slightly sloping floor), feeding (roughage, concentrates and green fodder) should be provided for profitable dairy farming. Vitamins and mineral mixtures containing salt, calcium, magnesium and phosphorus should also be supplemented with the feed of milch animals. Crop by-products such as maize kernels, stover, dry leaves of pulses, pod covers, ground nut oil cakes, coconut oil cakes, rice gruel, rice bran, etc. form important household animal feed. Forage crops need considerable amount of magnesium (Mg), because cattle consuming grass/fodder with low Mg content suffer from hypomagnesemia (an abnormally low level of blood Mg) which is commonly known as 'Grass Tetany.' This may occur due to a high rate of applying nitrogenous fertilizers, especially ammoniacal fertilizers, or a high level of potassic fertilizers that depress the level of Mg in plant tissues. Mg level of corn plant in young stage is markedly reduced when ammoniacal fertilizers are added. Grass Tetany may occur in the spring season when the N content in soil is quite high but it is in the ammoniacal from, this happens in the winter season as mineralization is affected under low temperature. High protein content of ingested forage crops or other feeds may also depress the absorption of Mg by the animals, leading to hypomagnesemia. To avoid such a problem, balanced fertilization in fodder grasses is considered as a remedial measure. The level of soil Mg may be kept at an optimum level through the use of dolomite limestone, if liming is advisable. Double salt of potassium, magnesium sulphate, adequate amount of single super phosphate, inclusion of legumes in forage crops (as legumes have higher Mg compared to grasses) also alleviate this problem. A dairy farmer should cultivate fodder crops such as maize, bajra, berseem, lucern, etc. (Table 1.2). Hybrid napier bajra (HNB) is a cross product of bajra and napier grass. It possesses the high quality and faster growth characters of bajra and the deep root system and multi-cut habits of napier grass. It is a suitable fodder crop for a year-round supply of green fodder. It provides 8–12% crude protein and 26–28% crude fibre. The biomass comprises of 55–58% total digestible nutrients. The first harvest of fodder is done at about 50–60 days after its planting and subsequent cuttings are done at around a 20–25 days period.

Para grass can be grown in low land areas, rhodes grass under saline soils, guinea grass is shade tolerant and can be grown in a coconut garden as well as between subabul and lebbeck (*Albizzia lebbeck*) trees. Job's tear (*Coix lacryma-jobi*) has been recommended as a prominent fodder crop for the coastal areas due to its dual-stress adaptations of high salt tolerance and also tolerance to seasonal waterlogging (Sarangi and Islam 2019). Studies on biochemical properties of Job's tear showed that due to stigmasterol, this plant tolerates high levels of salts. Other salt-tolerant grasses recommended are Para grass [*Brachiaria mutica* (Forsk) Stapf.] and Karnal grass [*Leptochloa/Diplachne fusca* (Linn.)], these may also be grown as fodder crops for cattle on unutilized land. Para grass is tolerant to waterlogging, therefore suitable in lowland areas. Karnal grass also known as *Kallar* grass (salt grass) is a C_4 species having halophytic characters, it is recommended for sodic soils.

TABLE 1.2
Suitable Fodder Crops To Supplement Fodder Requirement of Dairy Farming

Cereal Fodders		Legume Fodders	
Common name	Botanical name	Common name	Botanical name
Maize	*Zea mays*	Berseem/Egyptian clover	*Trifolium alexandrinum*
Guinea grass	*Panicum maximum*	Lucerne/Alfalfa	*Medicago sativa*
Bajra (pearl millet)	*Pennisetum glaucum*	Senji/sweet clover	*Melilotus indicus*
Jowar (sorghum)	*Sorghum bicolor*	Cowpea/lobia	*Vigna unguiculata*
Napier/elephant grass	*Pennisetum purpureum*	Guar	*Cyamopsis tetragonaloba*
Para grass	*Brachiaria mutica*	Rice bean	*Vigna umbellata*
Dinanath grass	*Pennisetum pedicellatum*	Horse gram/kulthi	*Dolichos biflorus*
Oat	*Avena sativa*	Stylo	*Sylosanthesis hamata/guianensis/humilis*
Rhodes grass	*Chloris gayana*	Shaftal/Persian clover	*Trifolium resupitanum*
Marvel grass	*Dicanthium annulatum*	Butterfly pea	*Clitoria ternatea*
Sabi grass	*Urochloa mosambicensis*	**Non-Leguminous-Non-Graminaceous Fodder**	
Signal grass	*Brachiaria brizantha*	Sunflower	*Helianthus annus*
Congo grass	*Brachiaria ruziziensis*	Mustard	*Brassica campestris*
Dharaf grass	*Crysopogon fulvus*	Turnip	*Brassica rapa*
Anjan/buffel grass	*Cenchrus ciliaris*	Carrot	*Daucus carota*
Yellow anjan	*Cenchrus setigerus*	Beet	*Beta vulgaris*
Sudan grass	*Sorghum sudanese*	Amaranthus	*Amaranthus viridis*
Barley	*Hordeum vulgare*	**Legume Tree Fodder**	
Teosinte	*Euchlaene/Zea mexicana*	Subabul	*Leucaena leucocephala*
Setaria grass	*Setaria onceps*	Mesquite	*Prosopis julliflora*

1.5 DUCK FARMING/DUCKERY

Duck birds require a water body for their growth, which is found in most of the low-lying areas. Ponds, water harvesting structures, lakes, etc. may be used for duckery. It is generally practiced as a supplementary enterprise to get a small but steady income throughout the year. Income is generated through sale of eggs, meat and birds, besides meeting the family requirement. Globally, the highest duck population is recorded in Indonesia, followed by India (about 10 million duck population). In India, there is production and consumption of about 600 million duck eggs in the rural areas with value of ₹180 crores. Andhra Pradesh, Gujarat, Kerala, Odisha and West Bengal are the states where duck farming is practiced. Ducks meet a portion of their feed requirement by foraging in the pond and nearby areas. They consume dropped grains from harvested paddy fields, insects, snails, earthworms, small fishes and other aquatic materials. Duck farming involves less cost compared to poultry rearing as the cost requirement for housing of ducks is significantly less. Ducks are hardier, an auxiliary gender inclusive enterprise and require moderate level of scientific management. Ducks tolerate waterlogging and are well adapted to swampy/marshy, riverside, wet as well as barren land. It is best to integrate ducks with fish farming. Most of the egg laying (95–98%) of ducks happens in the early morning before 9.00 am; therefore the birds are allowed to forage the nearby ponds and water bodies throughout the day. Duck breeds suitable for egg production purpose are Khaki Campbell and Indian Runner. About 240–280 eggs are produced per year per bird. The meat purpose duck breed 'White Pekin' weighs roughly 2.2 to 2.5 kg in a growth period of 42 days, with a feed conversion ratio of 1:2.3 to 2.7. When duckery is combined with fishery, it reduces the cost of feed for ducks, with advantages of fertilizing the fish pond, accelerating the nutrient cycling and oxygenating the pond water due to their dabbling habits. Ducks constitute nearly 10% of the total poultry population in India and add about 7–8% of the total egg production of the country. Egg laying by ducks is more (about 300 eggs bird^{-1} year^{-1}) than chickens and also the size of eggs is larger by about 18–20 g than hen eggs. Floor space of 3–5 ft^2 in shelter and 12–17 ft^2 on outfield is required per bird. The housing should be dry, well-ventilated and rat proof. A slopy pathway made up of bamboo or wood may be provided from the shelter to the waterbody. For higher production, supplementary feed of about 50 kg per annum may be provided for a bird. The feed could be prepared at household level by the use of maize (4 kg), rice polish (3 kg), mustard cake (1 kg), mineral mixture (350 g), salt (250 g), vitamin mixture (250 g), antibiotics (75 g) and coccidiostate (75 g). All ingredients should be free from alfatoxin. Feeding of birds should be done during morning and late afternoon.

Ducks can be released as a scavenging bird in the waterlogged paddy fields to feed upon the larvae and insects of insect pests of rice like plant hoppers, grass hoppers, case worm, stem borers, etc. This helps in maintaining the ecosystem as it reduces the chemical or pesticide pollution along with the benefits of improving feed utilization efficiency, reducing insect problems in the field and reducing the pest management cost in a rice crop.

1.6 POULTRY FARMING

Poultry birds (chickens, turkeys, ducks and geese) are reared for meat, egg or dual purposes and either as a specialized farming or as a component of an integrated farming system. In rural areas, small-scale poultry farming is practiced with local breeds by keeping few birds in the homestead area. These birds forage in the nearby fields and meet most of their feed and water requirements. The household surplus cooked food, broken grains, millets, etc., are fed as supplementary feed. Apart from selling, most of the products of eggs and meat are consumed in the family. Some farmers also generate income by selling feathers. The most common poultry farming is for layer chicken (egg production) and broiler chicken (meat production). This farming does not require high investments; however, initial costs may be met through bank loans or through contract farming. Housing, feed, hygiene require investment for profitable poultry farming. However, farmers can start getting a return in a short period of about three months or less. However, lack of proper knowledge and ideas about the market lead to loss instead of profit. The intensive system requires less space, feed and labour, as the stocking density is very high. However, it involves risk due to incidence of disease, there is abuse to the birds and environmental risks. Free-range poultry farming requires a larger space with the same level of production. In layer poultry farming, hens start laying eggs in 12–20 weeks. Proper ventilation, temperature, lighting, feeding and hygiene are necessary for profitable poultry farming.

For broiler production, a deep litter system, whereas for egg production, a batter cage system is the recommended procedure of farming. Broiler birds need a space of 2.5 ft^2 and egg producing birds need a space of 4 ft^2. In a deep litter system, the floor is covered with bedding or litter material consisting of rice bran, saw dust, groundnut husks, straw, hay, dried leaves, etc. Feeding and drinking trays are provided at an optimum rate for proper growth of the chicken.

There is huge potential for layer poultry farming for egg production in India, as the per capita availability is only 86 eggs $year^{-1}$ during 2019–2020 (Economic Survey 2020–2021) compared to a requirement of 182 eggs $year^{-1}$. Today a variety of poultry breeds are available for poultry farming.

Vanaraja: This is a multi-coloured poultry bird suitable for egg and meat production. This breed can be reared under free range as well as commercial farming. Under commercial farming, Vanaraja can be grown in an intensive/semi-intensive system. This is also suitable for backyard poultry farming for production of 150–200 eggs $year^{-1}$ and meat up to 1.5 kg in 3.5 months. Vanaraja birds are tolerant to diseases, requires less nutrition, growth is faster and they produce more eggs compared to local breeds.

Giriraja: This is also a very good breed of poultry with multi-colour feathers. They are suitable for varied agroclimatic conditions due to their disease resistance and sturdy nature. Meat and egg production is significantly higher than local breeds. Suitable for backyard poultry farming as these birds feed insects, ants, kitchen waste, broken grains, grasses, vegetable wastes, etc. and convert these into productive biomass. Poultry droppings also enrich the soil of the homestead farm.

Gramapriya: It is a coloured layer poultry breed with attractive brown colour eggs. Suitable for all agro-climatic conditions due to their better adaptability to adverse conditions and immunity. Produces about 160 eggs/year and attain body weight of 1.5 kg in three months. Male birds are used for meat.

Kadaknath: Black coloured indigenous poultry breeds of Jhabua district of Madhya Pradesh, India. Females weigh about 1.8 kg; famous for black colour meat with good quality, taste and medicinal values, used for treatment of nervous disorders. Attain body weight of 1.25 kg in about four months. Meat quality is better due to richness in protein (25%), less fat (0.73–1.03%) and low cholesterol (184 mg 100 g^{-1}).

CARI Nirbheek: It is a cross of the Indian native breed Aseel (means pure) with CARI Red. It is a dual-purpose breed for egg as well as meat production, produce about 160 eggs/year and attain a body weight of 1.8 kg in five months. Suitable for backyard poultry farming in semi-arid regions.

CARI Priya: This is a white colour layer poultry breed developed from the White Leghorn breed. It produces about 300 eggs $year^{-1}$.

CARI Shyama (*Kadaknath* cross): It is an indigenous fowl, locally known as '*Kalamasi*' with production of 105 eggs $year^{-1}$ and attain a body weight of 0.92 kg in five months of age. Meat has medicinal properties, rich in protein and iron.

CARI Debendra: Dual-purpose breed developed from a cross of coloured synthetic broiler and Rhode Island Red as male and female, respectively. This poultry breed has attractive plumage colour, attains body weight of about 1.8 kg in three months and produces about 200 eggs $year^{-1}$.

Nicobari: Found in the Andaman and Nicobar Islands as backyard poultry or semi-intensive farming. Adaptable to hot and humid climate with temperature range of 23.3–30.5°C and humidity up to 90%. Produce 148–160 eggs $year^{-1}$ and attain body weight of about 1 kg in nine months.

CARI Sonali: Developed from cross of White Leghorn as male and Rhode Island Red as female. It produce about 280 brown coloured eggs $year^{-1}$ with high feed conversion efficiency.

CARI Upkari: Developed from cross of Indian native chicken with frizzle plumage with CARI Red. These are multi-coloured birds and better adaptable to tropical climate (arid zones). Males attain body weight of 1.69 kg in five months and females produce 220 eggs year^{-1}.

CARI Hitkari: The neck region of this breed is devoid of feathers. It is developed by crossing Indian native naked neck with exotic breed CARI Red. Body weight of males attains 1.76 kg in five months and female produces 200 eggs year^{-1}.

Apart from poultry birds, guinea fowl and emu farming are becoming popular for meat and egg production. Guinea fowl can be reared in the backyard or homestead area having open space and fields for producing about 100 eggs year^{-1} bird^{-1}. They have white dots on feathers of pink, black and yellow colour. It originated from the Guinea coast of West Africa. The most common breed is Peal helmeted, which is reared under a free range scavenging system, having advantages of controlling rats and insects such as grasshoppers, flies and crickets. Emu farming is becoming popular in India (Andhra Pradesh, Delhi, Karnataka, Maharashtra and Tamil Nadu), Bangladesh and other Asian countries. These are large-sized birds of ratite group, which attain about 25 kg body weight in 8.5 months. They have immense economic value in terms of eggs, meat, oil, skin and feathers. Female emu birds, produce large-sized greenish eggs, which weight around 475–650 g per egg. These are hardy birds, adaptable to almost all types of agroecological conditions. Meat is rich in protein and energy, low in fat and cholesterol. Emu birds are compatible with other livestock animals and poultry birds. However, for commercial farming, they require enough land/space, clean water, good and nutritious food, cheap labour, medicines, etc. The location of farming should have good transportation facilities and suitable markets.

1.7 FISH FARMING/PISCICULTURE

Fishes are rich in protein, therefore, they form a nutritious food for human being worldwide. It is estimated that worldwide, fish supplies more than 1.5 billion people with nearly 20 percent of their regular per capita intake of animal protein, approximately 3.0 billion people with 15 percent of such protein, at least half of total animal protein consumption in Bangladesh, Cambodia, Equatorial Guinea, French Guiana, The Gambia, Ghana, Indonesia and Sierra Leone (https://www.greenfacts.org/en/fisheries/l-2/06-fish-consumption.htm, accessed on 5 July 2021). The cultivation of fishes called pisciculture is also a profitable enterprise. It is reported that about 60% of the Indian population consume fish, which varies according to states, with coastal states the major consumers of fish. The per capita average annual fish consumption for all of India is 5–6 kg, while for fish eating people it is 8–9 kg, much below the recommended level of 12 kg capita^{-1} per annum^{-1} (Salim 2016). Therefore, there is a need for increasing fish production in India with the adoption of scientific cultivation practices.

For freshwater pisciculture, there is a need for preparation of the pond which is very much suitable for growth of fishes. Generally, in rural areas, the ponds are silted and covered with aquatic weeds. The most common are water hyacinth (*Eichhornia crassipes*), water lettuce (*Pistia stratiotes*), duck weed (*Lemna minor*), waterthyme (*Hydrilla verticillata*), coontail (*Ceratophyllum demersum*), cattail (*Typha latifolia*), water lily (*Nymphaea* spp.), lotus (*Nelumbo* spp.) water shield (*Brasenia schreberi*), etc. Similarly, deposition of silts at pond bottoms reduces the water depth, alters the pH of pond water beyond 6.5–9.0. Therefore, desilting is required to expose the pond bottom to sunlight before starting the pisciculture. Predatory fishes such as *Puntius* sp., *Oxygaster* sp., *Ambassis* sp., *Amblypharyngodon mala*, *Colisa* sp., *Rasbora* sp., *Aplocheilus* sp., *Laubuca* sp., *Esomus danricus*, etc., should be eradicated by the application of mahua cake @ 2 t ha^{-1}, when pond water depth is 1m. Powdered cake is soaked in water and broadcast over the water body. Alternative piscicides such as tea (*Camellia sinensis*) seed powder @ 1 t ha^{-1} or bleaching powder (30% chlorine) @ 0.35 t ha^{-1}, seed meal of tamarind (*Tamarindus indica*) @ 2 t ha^{-1} may be used. Plant-derived products should be used for better environmental benefit. After 10–15 days of eradication of the predatory and weed fishes, liming should be done @ 225–425 kg ha^{-1} to bring the water pH to suitable level, increase the fertility of the pond and for prevention of diseases. Organic manure application, particularly cow dung manure @ 15–25 t ha^{-1} should be applied to increase the plankton population, which is a natural food for fishes. This dose may be applied in splits, such as 10 t ha^{-1} 15 days before releasing the fingerlings and 5 t ha^{-1} 7 days after releasing the fingerlings. Chemical fertilizers such as Ammonium sulphate 125 kg, single super phosphate 115 kg and potassium nitrate 20 kg ha^{-1} applied 25 days before releasing the fingerlings.

To utilize the nutrient available in different layers of pond water, it is recommended to release about 5000–7500 fingerlings comprising of 20% catla (*Catla catla*), 20% silver carp (*Hypopthalmichthys molitrix*), 30% rohu (*Labeo rohita*), 10% mrigal (*Cirrhinus mrigala*), 10% common carp (*Cyprinus carpio*), 5% Calbasu (*Labeo calbasu*) and 5% Grass carp (*Ctenopharyngodon idella*). The fingerlings of size 7–10 cm length may be used.

1.7.1 Nutrient Management

After releasing the fingerlings, monthly application of fresh cow dung 500 kg, single super phosphate 10 kg and ammonium sulphate 20 kg per hectare of pond area is recommended. The cow dung should be kept in a corner of the pond, whereas SSP and AS are broadcast over the pond water. These applications help in growth of phyto- and zooplankton as natural fish food. Groundnut oil-cake and rice bran may be applied each @ 2.5 kg ha^{-1} day^{-1} for first two months, @ 5 kg ha^{-1} day^{-1} for third and fourth months, @ 7.5 kg ha^{-1} day^{-1} for fifth and sixth months, @ 10 kg ha^{-1} day^{-1} for seventh and eighth months and @ 12.5 kg ha^{-1} day^{-1} for ninth and tenth months. This feed material may be kept in a bamboo basket in a particular place in the pond during daytime.

1.7.2 Freshwater Prawn Culture

Scientific prawn cultivation is a profitable business to get income within five to six months. Suitable fresh water prawn such as *Macrobrachium rosenbergii* and *M. malcohmsonii* may be cultured in freshwater ponds. The pond is prepared as described above. Fingerlings of weight 1–2 g or 6–25 cm length may be collected from authentic sources and released @ 20,000–25,000 numbers ha^{-1}. Prawns need resting places, for which small branches/twigs/coconut fronds @ 1 number 50 m^{-2} area, may be placed at different places in the pond. Basal manure @ 200 kg lime, cow dung 3 t, poultry manure 500 kg, urea 100 kg and SSP 100 kg per hectare should be applied. Additional fertilization of cow dung 1.5 t, poultry manure 250 kg, urea 25 kg and SSP 25 kg may be required during the growth stages. Occasional stirring of pond water is required for good aeration and release of toxic gases. Good hygiene and disease management practices are essential to reduce the risk of mortality.

1.7.3 Brackish Water Prawn/Shrimp Culture

Slightly saline or moderately saline water (TDS of 10–30 ppt ~ EC of 1.5–5.0 dS m^{-1}) may be used for brackish water prawn culture. Specific cultures also sometimes use high saline water beyond 30 ppt. Large-sized and high-priced species such as *Penaeus monodon* (tiger prawn) and *P. indicus* (white prawn) and recently *P. vannamei* are suitable for farming. Water pH should be maintained within 6.8–8.7 and dissolved oxygen of 3.5–4.0 ppm. Areas where the soil is highly acidic such as acid sulphate soils are not suitable for shrimp culture. In coastal areas, stocking density of 40,000 to 60,000 ha^{-1} crop^{-1} may be used and outside coastal areas stocking density may be increased up to 1,00,000 ha^{-1} crop^{-1}. Scientific pond management such as cleaning, drying, liming, elimination of predatory fish, fertilization and aeration should be done. Feeding is one of the most important factors in shrimp farming as it constitutes about 60% of the total investment. Therefore, appropriate feed management on the basis of body weight and scientific principles should be followed to obtain an optimum feed conversion ratio (FCR) of 1.5–1.8. FCR is calculated as weight of feed divided by the net production (final weight minus starting weight).

1.7.4 Crab Farming and Pearl Culture

Apart from pisciculture, crabs and pearl culture also are profitable ventures in coastal areas during the wet season with availability of bountiful rainfall. Ponds with 1.0–1.5 m water depth can be used for crab culture. Mud crab culture is becoming a profitable aquaculture business venture in Asian countries such as China, India, Indonesia, the Philippines Vietnam, etc. Mud crabs can be cultivated in estuaries near mangrove forests in coastal regions. Four species of mud crabs such as *Scylla serrata, S. olivacea, S. tranquebarica* and *S. paramamosain*, are reported with diverse species abundance at the same site in different countries. Bahrain, an island nation in the Persian Gulf, is known as the 'Island of Pearls,' lying between the Qatar Peninsula and the Saudi Arabian eastern coast. The island is well-known for its pearl farming industry. In India pearl oysters are found in the Gulf of Mannar along the Tamil Nadu coast and in the Gulf of Kutch along the Gujarat coast. The Gulf of Mannar is famous for rich oyster resources. Different pearl oyster species are *Pinctada fucata* (=*P. martensii*/*P. radiata*), *P. margaritifera, P. bugillata, P. anomoides, P. chemnitzii* and *P. atropurourea*.

Intensive pisciculture involves high stocking density of about 7500 numbers fingerlings per hectare pond area in the ratio of 4:3:3 (Catla, Rohu and Mrigal). However, in this system the fish pond should be integrated with enterprises such as duckery, dairy, poultry, etc. as per the model of the farming system. This system has the advantage of the use of pond silt directly as manure for fodder and other crops. Integrating fish and dairy diminishes the requirement to buy fertilizers and fish feeds and enhances income generation from the fish farm. Animal manure, particularly cow dung, can be applied directly to the fish ponds. When fresh cow dung is applied to the ponds directly, the energy losses as a result of processing and transportation are avoided. The leftover livestock feeds as well as the animal manure containing partially digested feed can be effectively utilized by the fish; thus, the feed conversion ratio is significantly increased. In this integrated system, the number of farm animals should be harmonious with the available pond area.

1.8 GOATERY

Goatery or goat farming is practiced for the production of milk as well as meat. Goats are multi-functional animals and are important components of the farming system from economics and nutrition aspects for landless, small and marginal farmers in the country. In rural India, small-scale goat farming is an enterprise which is gaining popularity today due to changes in food habits of the urban population. Goats can proficiently thrive on limited available shrubs and trees in stress-prone agro-ecosystems, in less fertile lands where cultivation of high-value crops is limited. Globally (unlike in India), more people drink goat milk compared to cow milk. More people also eat chevon (goat meat) than beef. As per a report of the World Health Organization (WHO), more than 70% of the world's population has some allergy (stomachaches, gas, skin rashes and ear infections) to cow milk. The report about the allergy symptoms on consumption of goat milk is very uncommon. The goat is known as the 'Poor man's cow' in India due to its important contribution to the rural economy and is a very significant enterprise in the dry land farming system. Goats are the best alternative animals for hilly and undulating lands not suitable for other types of animals such as cow or buffalo. There are several breeds of goat suitable for different agro-ecological conditions.

1.8.1 Konkan Kanyal

This breed is native to the Konkan region of Maharashtra, and it is nurtured mostly by the Dhangar and Maratha communities for meat purpose. The colour of these goats is mainly black with white marking in a definite pattern. The ventral surface of the body is white while the legs have white 'stockings.' Konkan Kanyal breeds have two-sided white strips from nostrils to ears; a flat and broad forehead; flat, long drooping ears; backward, straight, pointed, cylindrical horns; white muzzle and long legs, laterally black, medially white from knee to the fetlock joint. The average body weight of adult bucks and does are 35 and 30 kg, respectively. These are regular breeders and produce throughout the year, with a twinning percentage of about 66%.

1.8.2 Sangamneri

The native habitat of the Sangamneri goat breed is located in the semi-arid region of Maharashtra, including Nasik, Ahmednagar and Pune districts. These are medium-stature animals with completely white coat with combinations of black and brown. They have long and drooping ears and both sexes have horns directed backward and upward. The size of litter is largely single still 15–20% goats show twinning, however triplets are uncommon. The normal daily milk production varies from 0.5 to 1.0 kg during the usual lactation period of around 160 days. Though this breed is domesticated mostly for meat purposes, some animals yield significant amounts of milk. Dressing percentage is about 41, 45 and 46% at 6, 9 and 12 months of age, respectively.

1.8.3 Osmanabadi

This is a dual-purpose goat breed yielding both milk and meat. It is from the Osmanabad district of Maharashtra particularly from Latur, Tuljapur and Udgir taluks. Osmanabadi goats are also found in Karnataka state and the Nizamabad district of Andhra Pradesh. These are large in size, with mostly a black colour coat; however, there are also white, brown and spotted Osmanabadi goats. Males are generally horned; however, females may be horned or polled. During a lactation period of 120 days, total milk yield varies from 60 to 180 kg. Under favourable environments they normally breed twice a year and twinning is frequent.

1.8.4 Malabari or Tellicherry

Malabari goats are also known as Cutch breed. These goats are reared for milk, meat and skin (used in the tanning industry) purposes and are found in Kerala. The animals are medium in size, have no uniform colour and the coat varies from completely white to black. Entire males and some females are bearded. The head size is medium with a flat and sometimes a Roman nose, ears are medium in size directed outward and downward. These goats need a semi-intensive management system, including 4 to 6 hours of grazing augmented with stall feeding in the evening. Malabari goats are quite prolific and has a kidding percentage for twinning, triplets and quadruplets is 50, 25 and 5%, respectively. During the lactation period of 178 days, the daily milk yield ranges from 0.5 to 1.5 litres with an average total milk yield of 90 kg.

1.8.5 Jamunapari

This breed of goat is a native of northwestern arid and semi-arid regions of Etawah district in Uttar Pradesh. However, these are found in a number of states such as Assam and West Bengal in the east, Uttarakhand and Jammu and Kashmir in the north, Madhya Pradesh and Jharkhand in central India, and Karnataka in the south. The coat colour is white with tan or black markings on the neck and ears; beard is present in both bucks and does with a tuft of long hair on the hind legs. Jamunapari goats are considered the largest and most graceful breed of the long-legged goats of India. These goats have long and pendulous ears (26–28 cm), tubular with the opening towards the front. They have short horns, which are flat, horizontal and twisting backwards. Kidding frequency is once a year with mostly single births and at times twins. This is a dual-purpose breed with production of good meat and skin. During the average lactation period of 274 days, the milk yield of 280 kg is obtained; however, highest record of 4 kg milk per day and total mild yield of 575 kg in lactation period is reported. The fat content of milk ranges from 5.2 to 7.8 percent.

1.8.6 Ganjam

The Ganjam breed is found in the eastern part of India, mostly in the interior districts (Gajapati, Rayagada and Koraput) of Odisha. These goats are also found in Sikkim. The animals are tall and leggy, the coat may be black, white, brown or spotted, but black goats are predominant. They have short and lustrous hair. Both bucks and does have long, straight horns, directed upward, ears and tail are medium in size. The kidding (82%) takes place once a year and the litter size is mostly single (98.4%). Total milk yield during the average lactation period of 150 days is about 450 kg. As per the 18th Livestock Census (2007), the total number of Ganjam breed goats in India is 148,473.

1.8.7 Black Bengal

This goat breed is found in Bangladesh, West Bengal, Bihar, Assam and Odisha. The coat colour of this breed is typically black but rarely these are also found in brown, white or grey colours. This is reared for milk, meat and skin purpose. The skin is used for shoe making. The animals are dwarf and tight in body structure with small horns, short legs, straight back and both male and females are bearded. The average live weight of buck is about 15 kg and doe is about 12 kg. The speciality of this breed is giving multiple births of two, or more kids with average litter size of 2.1 at a time and kidding is twice a year. Average age at first kidding is nine to ten months. The total average milk yield is 53 kg during the lactation length of 90–120 days.

1.9 SHEEP REARING

Sheep are four-footed, even-toed hoofed mammals typically reared as livestock for multi-purpose utility such as for the production of meat, milk, wool, skin and manure. This animal contributes significantly to rural economy in general and economy of the arid, semi-arid and hill agriculture of the country, in particular. This enterprise forms a reliable source of income to the shepherds by the sale of different produces as well as animals. Rearing of sheep needs low-cost building/infrastructure and less labour compared to other livestock. These animals efficiently and economically convert grass into products such as meat and wool. They generally feed on diverse vegetation under scarce situations. Details of important sheep breeds are given below.

1.9.1 Neelagiri

This breed of sheep is found in the Neelagiri district of Tamil Nadu. It is reared for wool production purposes. The animals are medium in weight, the coat colour of majority of animals is white; however, some have purple spots on the body and face. They have broad and drooping ears. Males are horned while females are hornless. The average body weight of both adult male and female is about 31 kg.

1.9.2 Deccani

This breed of sheep is a combination of the woolly type of breed from Rajasthan and the hairy types from Andhra Pradesh and Tamil Nadu. These are distributed in the Bombay-Deccan area and portions of Karnataka and Andhra Pradesh. These are reared for meat (mutton) and wool purposes. The animals are small, hardy and adapt to poor pastoral conditions. Though they possess a coloured fleece, black and grey colours are more dominant. The annual average yield of wool is 4.54 kg per sheep; however, the quality of wool is generally of a low grade with a mixture of hair and fine fibres. This wool is mostly used for the manufacture of *Kambals* (rough blankets).

1.9.3 Nellore

The Nellore breed of sheep are found in Nellore, Prakasam and Ongole districts of Andhra Pradesh. The animals are tall with little hair except at brisket, withers and breech. Rams are horned, whereas ewes are polled. They have long and drooping ears and the majority of the animals carry wattles. The average body weight of male and female is 36 and 28 kg, respectively. This breed resembles goats in appearance and is the tallest breed of sheep in India. They have a long face, long ears and the entire body is densely covered with short hair. Most of the flocks are of fawn or deep red fawn colour.

1.9.4 Marwari

This breed of sheep is hardy in nature, yield coarser carpet type white wool of a mixed hairy composition. The animals have long legs, a black face, a prominent nose, a short and pointed tail. Under the throat, often,fleshy appendages, known as wattles, are present. This sheep breed is commonly found throughout Jodhpur and parts of Jaipur, Pali and Barmer districts of Rajasthan. However, the animals also migrate to different region of Uttar Pradesh, distant places of Madhya Pradesh and often to the northern parts of Maharashtra. The animals are highly resistant to diseases and worms. The average wool yield per year is 0.90–1.81 kg per animal.

1.9.5 Gaddi

Gaddi animals are small in size, and they are distributed in Kishtwar and Bhadarwah region of Jammu. Large numbers of animals are also found in the Kulu valleys in Himachal Pradesh. During the winter and in summer seasons they graze the highest elevations of the Pir Panjal Mountains, mostly in the Paddar range. Male animals (Rams) are horned, while females (ewes) are hornless; fleece colour is generally white with brown coloured hair on the face. The quality of wool is fine and lustrous with average annual yield of 1.13 kg per sheep, which is clipped thrice a year. The produced wool is used in the Dhariwal mills and Amritsar markets. High-quality Kulu shawls and blankets are manufactured from the undercoat of this sheep breed.

1.9.6 Dorset

This breed of sheep is a native of the United Kingdom. Animals may be polled or horned. Typically their face, ears and legs are white in colour and free from wool. These are reared for wool (average yield 2.75 to 3.25 kg) and production of mutton of superior quality. Male animals (Rams) weigh 80–110 kg, while female animals (Ewes) weigh 50 to 80 kg. Dorset is a hardy breed and performs well under diverse conditions.

1.9.7 Merino

This is the most popular fine wool breed of sheep in the world. These originate from Spain. The animals are white-faced with white feet. Rams are horned whereas the ewes are hornless. Most of the head and legs of the animal are covered by wool. These are extremely hardy animals being able to produce under adverse weather and poor grazing conditions. The ewes live and yield longer than any other breed of sheep.

1.9.8 Rambouillet

This breed of sheep is from France. The animals have a large head with white hair around the nose and ears. Rams are horned whereas the ewes are hornless. The weight of Rams goes up to 125 kg and ewes grow up to a maximum weight of 90 kg. They produce an excellent fine-wool fleece, which is heavy, close, compact and covers most of the body, including the face and legs.

1.10 PIGGERY

Piggery/pig farming/swine husbandry is practiced under diversified as well as intensified agricultural system. Piggery generates income from the sale of meat (pork), live animals (piglets) and manure. Some farmers rear pigs in homestead areas by feeding household food/kitchen waste. Pigs are good converter of inedible feeds, fodders/forages, and some grain by-products produced from mills, meat industries, spoiled feeds and debris into an economic product (pork). The manure obtained from piggery is helpful in sustaining soil fertility. Pigs are non-ruminant animals which are sometimes reared to reduce the large-scale municipal garbage. Pigs are prolific breeders, therefore, due to the rapid increase in number of animals the return to investment is very high in piggery. Important breeds of pigs are given below.

1.10.1 Large White Yorkshire

This is a well-known English bacon breed, which originated from Yorkshire and neighbouring counties in northern England. The sows of Yorkshire breed are good mothers. The body colour of the animal is solid white with intermittent black pigmented spots called 'freckles.' Some of the distinguishing characteristics of Large White Yorkshire breed are large size body, long neck (full to the shoulder), erect ear, medium snout length, slightly dished face and back is slightly arched. They have pink coloured skin, free from wrinkles with long and moderately fine coat. Upon maturity the body weight of the boar varies from 300–400 kg whereas the body weight of the sow varies from 230–320 kg. This is the most widely reared exotic breed of pig in India. This breed is an excellent choice for crossbreeding programmes.

1.10.2 Middle White Yorkshire

Middle White Yorkshire breed was developed by crossing between the Large White Yorkshire and a smaller breed of Yorkshire extraction. This breed is recognized as an outstanding pork pig, attaining slaughter weight first and with a high proportion of lean meat to bone. These animals are hardy in nature, with quick growth, but not as productive as the Large White Yorkshire. This breed of pig is used extensively to upgrade local pigs due to their smaller size. The average body weight of an adult boar varies from 250–340 kg and that of the sow ranges 180–270 kg.

1.10.3 Landrace

The origin of the Landrace breed of pig is from Denmark. This breed is reared scientifically to produce the best quality bacon in the world. The salient characteristics of this breed are large and long body, legs are relatively short, white in colour with white bristles, black skin spots 'freckles' are also common on white body. The breed is also known for their long, deep side; square ham. The carcass is leaner than that of the meat. The matured body weight of the boar varies from 270–360 kg and that of sow ranges 200–320 kg.

1.11 HORTICULTURE

It covers olericulture (vegetable cultivation), floriculture (flowering and ornamental plants), pomology (cultivation of fruit crops), medicinal and aromatic plants. India is second in most of the horticultural production after China (Table 1.3). Vegetables are grown in kitchen gardens, homestead areas and in large fields for market purposes. Market-driven vegetable farming is a profitable venture, in which early varieties of seasonal vegetables are cultivated to meet the market demand. The planting time of vegetables are adjusted such that the early harvest catches the high price of the market due to higher consumer demand and less production/glut during that period. Generally, prices of seasonal vegetables such as cabbage, cauliflower, radish, broccoli, tomato and coriander are higher during the beginning of the winter season; therefore, farmers who harvest during this period fetch a better price for the produce and hence earn higher income. Such vegetable farmers target the nearby city/urban markets and produce the vegetables as per demand. In recent years, off-season vegetables are cultivated in green houses, glass houses under protected cultivation technology, known as vegetable forcing. Protective cultivation technology allows growing of vegetables such as capsicum, cucumber, tomato, all year. Some farmers produce specialized vegetables in large areas for sale in distant markets, where the demand is very high; such vegetable farming is called truck gardening. Vegetable cultivation can be done with the specific objective to produce particular varieties of vegetables suitable for processing, such as dehydration, canning, fermentation, freezing and pickling meant for processing factories.

The area and production of major horticultural crops are given in Tables 1.4–1.6. In India, among the vegetables, potato covers 20.9% area, followed by onion (12.5%), tomato (7.7%) and brinjal (7.1%). Similarly, the production share of potato is also maximum (27.8%) followed by onion (12.6%), tomato (10.7%) and brinjal (6.9%). Among the fruit crops, the coverage of mango is the highest (34.7%), followed by citrus (15.4%) and banana (13.6%), whereas in terms of production, banana contribute 31.6%, followed by mango 22.4% and citrus 12.9%. Cumin, chilli, coriander, garlic and turmeric cover more than 70% area and production under spices and condiments. Coconut is the dominant plantation crop with 56% area and 91% production. India ranks second in the world for the production of potatoes, onions, cauliflowers, brinjal, cabbages, and other vegetables, but it is the leading producer of okra and ginger. In terms of fruit production, the nation leads the world in the production of bananas (25.7%), papayas (43.6%) and mangoes (including guavas and mangosteens) (40.4%). India can take advantage of great export opportunities due to its large production base. India exported fruits and vegetables totalling Rs. 9,182.88 crores, or 1,277.38 USD million, in 2019–2020. The value of the fruits and vegetables included Rs. 4,832.81 crores, or 668.75 USD million, and Rs. 4,350.13 crores, or 608.48 USD million. The majority of fruits exported from the nation are grapes, pomegranates, mangoes, bananas and oranges, whilst the majority of vegetables exported are onions, mixed vegetables, potatoes, tomatoes and green chillies. Major markets for Indian produce include Bangladesh, the United Arab Emirates, the Netherlands, Nepal, Malaysia, the United Kingdom, Sri Lanka, Oman and Qatar. Fruit and vegetable seeds are another product that India exports; in 2019–2020, exports totalled Rs. 723.43 crores, or $101.50 million USD. India's main markets for its fruit and vegetable seeds were the Netherlands, the United States, Bangladesh, Thailand and Pakistan.

Vegetables and fruits cultivation provide income to the farming family as well as meet the daily family requirements of 100 g of fresh fruits and 300 g of vegetables per person. Fruits and vegetables are rich sources of micronutrients (iron, and calcium), vitamins (C, folic acid, B complex, carotenoids) and macronutrients (carbohydrates/starch, fibre). Natural fruit juices provide energy, vitamins (beta-carotenes, vitamin C) and minerals (K and Ca). Fruit juices being rich in K are ideal beverage for those suffering from hypertension.

Fruit crops are grown in specialized areas called an orchard. A plantation refers to a large-scale farm that specializes in cash crops. Orchards are established in a specific layout such as square, rectangular, hexagonal, diagonal/quincunx, triangular and contour systems. Contour planting is

TABLE 1.3
Status of World and Indian Agriculture in 2017

Item	India	World	% Share	India's Rank	Next to
1. Area (Million Hectares)					
Total area	328.73	13486.03	2.44	Seventh	Russian Federation, Canada, USA, China, Brazil, Australia
Land area	297.32	13002.86	2.29	Seventh	Russian Federation, China, USA, Canada, Brazil, Australia
Arable land	156.46	1390.70	11.25	Second	USA
2. Population (Million)					
Total	1338.68	7547.86	17.74	Second	China
Rural	889.22	3410.07	26.08	First	
3. Crop Production (Million Tonnes)					
(A): Total cereals	313.61	2980.17	10.52	Third	China, USA
Wheat	98.51	771.72	12.77	Second	China
Rice (paddy)	168.50	769.66	21.89	Second	China
(B): Pulses	23.24	95.98	24.21	First	
(C): Oilseeds					
Groundnut (with shell)	9.18	47.10	19.49	Second	China
Rapeseed	7.92	76.24	10.38	Third	Canada, China
(D): Commercial Crops					
Sugarcane	306.07	1841.53	16.62	Second	Brazil
Tea	1.33	6.10	21.72	Second	China
Coffee (green)	0.31	9.21	3.39	Eighth	Brazil, Vietnam, Colombia, Indonesia, Honduras, Ethiopia, Peru
Jute	1.97	3.53	55.69	First	
Cotton (Lint)*	6.05	24.77	24.43	Second	China
Tobacco Unmanufactured	0.80	6.50	12.30	Third	China, Brazil
4. Fruits and Vegetables Production (Million Tonnes)					
(A): Vegetables primary	127.14	1094.34	11.62	Second	China
(B): Fruits primary	92.30	865.59	10.66	Second	China
(C): Potatoes	48.61	388.19	12.52	Second	China
(D): Onion (Dry)	22.43	97.86	22.92	Second	China
5. Livestock (Million Heads)					
(A): Cattle	185.10	1491.69	12.41	Second	Brazil
(B): Buffaloes	113.33	200.97	56.39	First	
(C): Camels	0.33	34.83	0.93	Sixteenth	Chad, Somalia, Sudan, Kenya, Niger, Mauritania, Ethiopia, Mali, Pakistan, Saudi Arabia, U.A.E, Yemen, Mongolia, Algeria, Eritrea
(D): Sheep	63.07	1202.43	5.25	Third	China, Australia
(E): Goats	133.35	1034.41	12.89	Second	China
(F): Chicken	783.27	22847.06	3.43	Sixth	China, Indonesia, USA, Brazil, Iran
6. Dairy Products (Million Tonnes)					
(A): Milk Total	176.27	827.88	21.29	First	
(B): Eggs (Primary) Total	4.85	87.01	5.57	Third	China, USA
(C): Meat, Total	7.30	334.23	2.18	Sixth	China, USA, Brazil, Russian Federation, Germany

Source: FAOSTAT (as on 08.01.2020)
* Data relates to 2013

TABLE 1.4
Area and Production of Major Vegetable Crops in India During 2017–2018

Vegetable Crops	Area (000 ha)	Production (000 t)	Share in Area (%)	Share in Production (%)
Potato	2142	51310	20.9	27.8
Brinjal	730	12801	7.1	6.9
Tomato	789	19759	7.7	10.7
Cabbage	399	9037	3.9	4.9
Cauliflower	453	8668	4.4	4.7
Onion	1285	23262	12.5	12.6
Okra	509	6095	5.0	3.3
Peas	540	5422	5.3	2.9
Tapioca	173	4950	1.7	2.7
Radish	209	3061	2.0	1.7
Chillies (green)	309	3592	3.0	1.9
Others	2721	36437	26.5	19.9
Total	10259	184394	100.0	100.0

Source: Government of India (2019).

TABLE 1.5
Area and Production of Major Fruit Crops in India During 2017–2018

Fruit Crops	Area (000 ha)	Production (000 t)	Share in Area (%)	Share in Production (%)
Banana	884	30808	13.6	31.6
Mango	2258	21822	34.7	22.4
Citrus	1003	12546	15.4	12.9
Papaya	138	5989	2.1	6.2
Guava	265	4054	4.1	4.2
Apple	301	2327	4.6	2.4
Pineapple	103	1706	1.6	1.8
Grapes	139	2920	2.1	3.0
Sapota	97	1176	1.5	1.2
Litchi	92	686	1.4	0.7
Jackfruit	185	1830	2.8	1.9
Pomegranate	234	2845	3.6	2.9
Aonla	93	1075	1.4	1.1
Muskmelon	54	1231	0.8	1.3
Watermelon	101	2520	1.6	2.6
Others	559	3823	8.7	3.8
Total	6506	97358	100.0	100.0

Source: Government of India (2019).

followed in hilly and mountain ecosystems. The planting distance/spacing varies with the type of fruit crop, season and layout. The planting distance and population of some of the important fruit crops are given in Table 1.7.

TABLE 1.6
Area and Production of Major Spice and Plantation Crops in India During 2017–2018

Spice/plantation Crops	Area (000 ha)	Production (000 t)	Share in Area (%)	Share in Production (%)
Chillies (dry)	752	2149	19.4	26.4
Turmeric	238	1133	6.1	13.9
Garlic	317	1611	8.2	19.8
Coriander	532	710	13.7	8.7
Ginger	160	1118	4.1	13.8
Cumin	966	689	24.9	8.5
Fenugreek	149	202	3.8	2.5
Pepper	134	66	3.4	0.8
Fennel	66	104	1.7	1.3
Cardamom	84	28	2.2	0.3
Tamarind	48	201	1.2	2.5
Other spices	432	113	11.3	1.5
Total spices	3878	8124	100.0	100.0
Arecanut	497	833	13.3	4.6
Cashewnut	1062	817	28.4	4.5
Cocoa	89	20	2.4	0.1
Coconut	2097	16413	55.9	90.8
Total plantation	3745	18082	100.0	100.0

Source: Government of India (2019).

TABLE 1.7
Plant Spacing and Density in Major Fruit Crops Under Different Orchard Layout

Crop	Planting Distance (m)	Tree Density ha^{-1} in Different Layout		
		Square	Hexagonal	Triangular
Mango	10 × 10	100	115	89
Sapota	8 × 8	156	180	139
Coconut	7.5 × 7.5	178	205	159
Clove	6 × 6	277	320	248
Guava	6 × 6	277	320	248
Acid lime	5 × 5	400	461	357
Papaya	1.8 × 1.8	3086	3549	2747
Banana *Kharif*	1.5 × 1.5	4444	5111	3955
Banana *Rabi*	1.5 × 1.2	5555	6388	4944

The hexagonal system of planting accommodates about 15% more plants than the square system, whereas the triangular system accommodates 11% fewer plants. Adoption of scientific orchard management practices allows closer spacing than recommended spacings for some of the fruit crops. This is called high-density planting (HDP) system. In an HDP system, the productivity is maximized by sacrificing the quality. In banana production, under HDP, plant population of 5555 numbers per ha result in yields of about 50–60 t ha^{-1}.

Intercropping in fruit orchards (Table 1.8) is recommended during the initial years of establishment to secure income from the intercrops, when the main crop has not started fruiting. A

TABLE 1.8
Suitable Crops for Intercropping in Fruit Orchards During the Initial Growth Period

Fruit Crop	Age Up To	Intercrops
Mango	7 years	Papaya, legume vegetables, onion, tomato, carrot, pineapple
Grapes	8 months	Snake gourd, bitter gourd
Apple	5 years	Potato, cabbage
Banana	4 months	Sun hemp, onion, yam, basella, cucurbits, marigold, mushroom
Tapioca	3 months	Onion, beans, lab lab, black gram
Arecanut	10 years	Pineapple, turmeric
Coconut	3 years	Banana, tapioca, vegetables, turmeric, guinea grass
Guava	3 years	Pea, cowpea, gram, beans
Pomegranate	3 years	Spinach, fenugreek, radish, carrot
Date	5 years	Gram, lentil, pea, vegetables
Ber	2 years	Gram, chilli, leafy vegetables, mung, guar
Custard apple	5 years	Groundnut, linseed, pea, gram, marigold
Sapota	4 years	Cowpea, cluster bean, pea, pigeon pea
Aonla	3 years	Turmeric, colocasia, gram, chilli, leafy vegetables, pea, mung, cowpea
Jackfruit	4 years	Lady's finger, brinjal, chilli, tomato, gram

multi-tier cropping system is also possible with crops such as arecanut, banana, coconut, litchi and sapota. Growing of shade tolerant fodder crops creates opportunity for integration of dairy farming. Guinea grass (*Panicum maximum*) is a highly suitable fodder crop for cultivation as an intercrop with coconut. Guinea grass is shade tolerant, nutritious and palatable fodder for dairy cattle. The management of an orchard involves manuring, training, pruning and control of diseases, insects and weeds. Pruning is done to remove branches of a tree, to cut out damaged parts, to invigorate growth or to correct or maintain tree structure. Training is a practice where the growth of a tree, usually when young, is managed to initiate growth to achieve a desired shape, size and form. Bending is a practice done in guava to regulate the flowering. Scientific bending is done after the third year of planting and up to the eighth year of an orchard. The off-season flowering can be induced by summer (April–May) bending, which produces fruits during the autumn to winter season. Autumn (October) bending produces fruits during early summer. This artificial flowering initiation is done by taking into consideration the local market demand.

1.11.1 Medicinal Plants

Cultivation of plants such as neem, bel, cinchona, sandalwood, tamarind, bahera, pudina, tulsi, chirayata, ashwagandha, calamus, etc. for medicinal purpose also fetch a better market price today. The important uses of some of the medicinal plants are given in Table 1.9.

1.11.2 Floriculture

Floriculture deals with growing of plants for production of flowers, ornamental foliage, their value addition and marketing. It includes plants of different growth habits such as herbs, shrubs, trees, cacti, succulents, etc. Cut greens, cut flowers, houseplants, bedding plants and flowering plants are all included.

India is blessed with a number of agroclimatic zones that are ideal for growing delicate and sensitive flowers. According to the National Horticulture Board's National Horticulture Database, India's

TABLE 1.9
Important Medicinal Plants to be Included in Horticultural Component of Farming System

Common Name	Scientific Name	Uses
Calamus	*Acorus calamus*	Dried rhizomes act as carminative, expectorant, promotes flow of bronchial secretions and is useful in asthma
Bel	*Aegle marmelos*	Fruit pulps used for sweet drinks as soothing agent for intestines of patients recovered from bacillary dysentery and prevents constipation
Kalmegh	*Andrographis paniculata*	Above ground parts useful in curing fevers, worms, dysentery etc.
Neem	*Azadirachta indica*	Leaves used to cure skin diseases and boils in human beings. In crops useful to control insect pests and diseases.
Brahmai	*Bacopa monnieri*	Used for preparation of medicines for treatment of nervous disorders, mental diseases and constipation.
Centella	*Centella asiatica*	Used to treat skin problems, wounds and revitalizing the nerve and brain cells. Used as a tonic for brain/brain food.
Cinchona	*Cinchona* spp.	Bark used for the medicine quinine, for effective treatment of malaria.
Mint (Pudina)	*Mentha pipertia*	Peppermint from the dried leaves and flowering tops used in treatment of flatulence, vomiting, diarrhoea and nausea. Peppermint oil (menthol) is used for preparation of medicines for stomach disorders, in ointments for headache, rheumatism, cough drops, inhalation, mouth washes etc.
Tulsi	*Ocimum sanctum*	Leaves/leaf extract useful in bronchitis, catarrh, digestive complaints, cough, cold.
Ishabgul	*Plantago ovata*	Seeds and seed husks soaked in water used to control constipation.
Sarpagandha	*Rauvolfia serpentina*	Dried roots with bark is used to prepare medicines for treatment of hypertension, snake and reptile bites, ferver, constipation etc.
Sandalwood	*Santalum album*	Oil from seed is used on skin diseases, treatment of dysuria, cystitis, gonorrhoea and cough. The wood grounded up with water into a paste is applied on local inflammations, on forehead in fever and on skin diseases.
Chirayata	*Swertia chirayita*	The whole plant at flowering stage are dried to prepare medicines to treat fever, diarrhoea and weakness.
Tamarind	*Tamarindus indica*	Fruit pulp has laxative properties.
Ashvagandha	*Withania somnifera*	Dried roots of the plant is used to prepare medicines for sexual and general weakness, rheumatism, root powder is applied locally on ulcers and inflammation.

floriculture production area in 2015–2016 was 249 thousand hectares, producing 1659 thousand tonnes of loose flowers and 484 thousand tonnes of cut flowers. Many states, including Tamil Nadu (20%), Karnataka (13.5%), West Bengal (12.2%), Madhya Pradesh, Mizoram, Gujarat, Andhra Pradesh, Odisha, Jharkhand, Haryana, Assam and Chhattisgarh, are major producers of flowers.

Flowers including roses, tuberoses, gladioli, anthuriums, carnations, marigolds and others are grown in India. Modern poly and greenhouses are used for cultivation, in addition to open agricultural environments. In 2019–2020, India's total floriculture export revenue was ₹541.61 crores, or USD 75.89 million. The United States, the Netherlands, Germany, the United Kingdom and the UAE were the top importers. In India, there are about 300 units focused on exports. Over half of the floriculture establishments are situated in Tamil Nadu, Karnataka and Andhra Pradesh. The international enterprises' technical cooperation has put the Indian floriculture sector in a position to raise its share of global trade. The infrastructure for the floriculture business has been established by the Government of India (GOI) in key cities such as New Delhi, Mumbai, Kolkata, Chennai, Bengaluru and Pune. In and around Hyderabad, Bengaluru, Hosur, Doddaballapur and Devanahalli

TABLE 1.10
Products Specific Intensive Floriculture Zones in India

Sl. No.	Zone	Flower Crops
1	Area around Delhi, UP and Punjab	Rose, Carnation, Chrysanthemum and Gladiolus
2	Area around Bengaluru	Rose, Carnation, Chrysanthemum, Ornamental foliage plants and seeds
3	Area around Trivandrum	Orchids, Anthurium and foliage plants
4	Area around Pune/Nashik	Rose, Carnation, Gladiolus, Dahlia, Chrysanthemum, Aster and Tuberose
5	North Eastern region including Sikkim	Orchids, Gladiolus, Liliums, Gerbera, Salvia, Anthurium and other foliage plants
6	Area around Kolkata	Lotus, Tuberose, Jasmine, Chrysanthemum and Dahlia
7	Area around Srinagar	Gladiolus, Lilies, Carnation and Roses
8	Area around Solan	Gladiolus, other bulbous plants and seeds
9	Area around Coimbatore including Nilgiris	Jasmine, Tuberose, Chrysanthemum, Rose, Carnation and Orchids

in the south; Gurugram in the north; and Lonavala and Pune in the west, there are 100% export-oriented units (EOUs). The government has granted 100% EOUs the ability to sell 50% of their produce in local markets and a 100% tax exemption on greenhouse implements and raw materials. For the purpose of conducting targeted research and developing floriculture, the GOI has designated product-specific zones (Table 1.10).

1.11.3 Agriculture/Mushroom Cultivation

Mushroom production is a small-scale agri-business venture with low investment and quick return. The edible mushrooms are free from fat, cholesterol and gluten; very low in sodium, low in calories, yet rich in minerals (potassium, iron, copper, zinc and manganese), protein and vitamin. It has the potential to find a place in the dietary requirements of the world population to meet nutritional and medicinal needs. This can be integrated in farming systems near cities/towns; however, training should be acquired and input availability should be ensured for effective production. Apart from mushroom production, one can also engage in the production of mushroom seeds, called spawn, for supply to nearby mushroom growers in remote areas. It requires a small hut, chaff cutter, sprayer, poly bags, a water boiling drum, clean water, sterilization and other minor inputs. Paddy straw is used as one of the substrates in mushroom production. The major commercially cultivated mushrooms species of the world are given in Table 1.11.

Paddy straw mushrooms are suitable for warm (30°C) areas, therefore they can be cultivated during summer and rainy seasons (March to October). Oyster mushrooms are cultivated during winter (November to February) when the air temperature is 20–30°C. Button mushrooms are suitable for low temperature (<20°C) areas. In India, Punjab is the leading state in mushroom production followed by Odisha, Haryana, Maharashtra and Gujarat.

Crop residues can be used for mushroom production; however, cereal straws are the most suitable substrate for commercial mushroom production. Cereal grains such as wheat are the best for spawn production especially for button mushroom, but other cereal grains like maize, paddy and millets, such as jowar, have also been used as substrates for spawn production. Careful selection of the paddy straw, growing season, mushroom species/strain and growing conditions are important for commercial mushroom production. Location of the unit is also important for convenience to access the market. A mushroom production unit of size 20 ft × 15 ft requires an approximate initial

TABLE 1.11
Commercial Mushroom Species and Their Contribution to Global and Indian Production

Sl. No.	Mushroom Species	Scientific Name	High Yielding Strains	Contribution to Production (%) World	India
1	White button	*Agaricus bisporus*	DMR-Button-03, U_3-54, NBS-1, NBS-5	15	73
2	Brown button	*Agaricus bisporus*	DMR-Button-06		–
3	Oyster/Tropical	*Pleurotus sajarcaju/ ostreatus*	–	19	16
4	Paddy straw	*Volvariella volvacea*	DMRO-247, DMRO-484	5	7
5	Milky	*Calocybe indica*	DMR-334	[a]	3
6	Shiitake	*Lentinula edodes*	DMRO-38, DMRO-388s	22	1
7	Enokitake	*Flammulina velutipes*	–	11	–
8	Wood ear	*Auricularia auricula-judae*	–	18	–
9	Macrocybe	*Macrocybe gigantea/titans*	DMR-Macrocybe-01	[a]	–

[a]Other mushroom species contribute about 10% to world mushroom production.
Source: Singh et al. (2017); Sharma et al. (2017).

investment of ₹88,700 (Table 1.12). It takes about two years to get back the investment cost and the profit starts from the third year.

1.11.3.1 Brief Procedure for Production of Paddy Straw Mushroom

The paddy straw mushroom can be produced inside a thatched house or under the shade of trees. For this, fresh, good quality paddy straw is used as the substrate. About 10–15 kg paddy straw (25–30 bundles), bottle spawn (400 g) and 250 g arhar/gram powder are required to prepare one bed. A raised bed of height 1 ft from ground is made with the help of bricks or wooden logs. Paddy straw bundles of about one kg are made and soaked in Bavistin mixed water (8 g Bavistin in 100 litres of water) for 12 to 16 hours. Take out the soaked straw and keep in a shade for 2 to 3 hours to drain out the excess water. The moist paddy straw bundles are kept systematically over a raised platform layer-wise. The thickness of each layer may be kept about 9 inches and the root side of the straw outside. In the second layer the paddy straw bundles are kept in opposite directions. After the second layer, spread a small quantity of pulse flour on the paddy straw and then spread the spawn at the margin (10–12 cm from the side) at a spacing of 10–15 cm throughout the length of the bed in a rectangular fashion. The third layer of paddy straw bundles are arranged in a criss-cross manner and in the fourth layer the bundles in opposite direction. Over this layer again apply the spawn and cover with another layer of paddy straw. The bed so formed is pressed, covered with a polythene sheet and kept undisturbed for few days. Remove the polythene cover after 7–8 days. On day 9 spray water 3–4 times a day. The mushrooms start appearing from all sides from 12 to 14 days as tiny buttons, which can be harvested in the next 4 to 5 days. Harvesting should start just at the button stage since the sporocarp that has opened will be more fibrous. Typically, 10 kg of substrate can yield 1–2 kilograms of mushrooms.

1.11.3.2 Brief Procedure for Production of Oyster Mushroom

The oyster or Dhingri mushroom is produced over a mass of chopped paddy straw made into a bundle by the binding of mycelial growth of the mushroom. The materials required for filling one polythene of size 80 cm length and 40 cm width (4 sq. ft) are fresh paddy straw 3–4 bundles (20 kg),

TABLE 1.12
Estimate for Establishment of a Mushroom Production Unit

Items	Quantity	Rate (₹)	Amount (₹)
A. Non-recurring Expenditures			
1. Polythene sheet	150 m^2	50	7500
2. Sprayer	1 no.	5500	5500
3. Mushroom house made of wooden poles, bamboo mat and thatch roofing	1 unit	–	30000
4. Drum for boiling (200 litres capacity)	3 nos.	1200	3600
5. Wooden frame (0.5 m×0.25 m×0.18 m)	2 nos.	1500	3000
6. Wire cage for straw	2 nos.	2000	4000
7. Punching machine	1 no.	1000	1000
8. Nylon ropes for hanging	–	–	2000
9. Thermometers	–	–	1000
Total Non-recurring	–	–	57,600
B. Recurring Expenditures			
1. Labour cost	30 man-days	300	9000
2. Paddy straw	1 tonne	3000	3000
3. Spawn	80 kg	70	5600
4. Poly bags	10 kg	100	1000
5. Pesticides	–	–	2500
6. Miscellaneous items viz. hand chopper, binding thread (jute/plastic), pressing board, plastic buckets, disinfectants etc.	–	–	10,000
Total recurring	–	–	31,100
Investment – 1st year	–	–	88,700
Production – 1st year	600 kg	100	60,000
Investment – 2nd year	–	–	31,100
Production – 2nd year	600 kg	100	60,000

spawn 200 g (half bottle) and wheat grain 250 g. Other materials required are a sprayer and straw cuttler/sickle. Paddy straw are chopped into small (1–2 cm) pieces, soaked in water for 12 to 16 hours and then pasteurized by hot water (80°C) for 2 hours. Then dried in shade for 3 to 4 hours so that the excess water drains out from the straw. The spawn is made into three parts. Soak the wheat grains in water for 2–3 hours and boil for 25 minutes so that the grains don't split. Then dry the boiled wheat grains in shade and make three parts. The boiled paddy straw is made into four parts. Spawning is done while filling in the bags. The first part of straw is filled in polythene bag and over this one part each of spawn and wheat grains are spread. Similarly, the second and third layers of paddy straw, spawn and wheat grains are added. Over this the 4th layer of paddy straw is filled and the polythene bag is tied. Spawn running (mycelial growth) occurs in dark room (24–28°C) within a period of 18–30 days. Holes (15–20) are made over the polythene bag and are kept in a hanging position. Water is sprayed 3–4 times a day. Pinhead initiation occurs in 2 to 7 days after opening by maintaining 13–30°C, and 80–85% humidity, light and ventilation. One can harvest 1–2 kg of mushroom from one bag and it is possible to handle 10–15 bags in a low-cost small investment proposition.

1.11.3.3 Brief Procedure for Production of Button Mushroom

Compost material is used for button mushroom production from where the mushroom derives the nutrients for growth. The materials required for composting are wheat/paddy straw (300 kg), wheat bran (15 kg), calcium ammonium nitrate (9 kg), urea (4 kg), muriate of potash (3 kg), single super phosphate (3 kg) and gypsum (20 kg). The wheat/paddy straw is kept on the floor for 24–48 hours with water sprayed several times. The all the above ingredients except gypsum is mixed to make a stack of about 5 feet height. This stack is turned on day 6 and day 10 and moisture is maintained by spraying water. On day 13, gypsum is mixed and then turning is done on every third day up to day 28. The moisture content of the compost should be around 70%. The spawn is mixed with the compost @ 0.5–0.75% i.e. for 100 kg compost 0.5–0.75 kg spawn. When the surface of the compost is covered with white mycelial growth of the mushroom fungus, 3–4 cm thick layer of casing soil is covered to maintain optimum moisture content. Mushroom starts appear at 10th to 12th day of seeding and can be harvested after 50–60 days of the initiation of the process.

1.12 BIODYNAMIC FARMING

Biodynamic farms seek to produce soil fertility by the use of special composting practices, combining animals, cover cropping and crop rotation. Through composting, plant matter, soil and animal manures are brought into harmony and transformed into a powerful source of fertility and vigor for the farm operation. The primary distinction between biodynamic farming and organic farming is that biodynamic farming uses different principles that add vitality to the plant, soil and/or livestock.

The principles of biodynamic farming include:

- Crop diversification: Selection of best suitable crops in place of high resource demanding, low remunerative crops.
- Crop rotation: Combining succession crops of a diverse nature to generate soil health.
- Animal husbandry: Integrating livestock in farming to generate income and input resources required for biodynamic farming.
- Biodynamic compost/bio-compost preparations: Rudolf Steiner's philosophy of spiritual science (Anthroposophy) describes nine biodynamic preparations based on cow manure, silica and various herbal plants. The soil is opened to receive cosmic energy flowing around. Dynamization is the process of stirring while preparing specific compost, which is also used as nutrient supplying spray solutions. These nine bio-compost formulations numbered from 500 to 508. Out of these, six composts provide nutrients and three are used for control of diseases and pests.
- Use of forces of earth: Principles of chemistry, physics and biological sciences are applied in agriculture. Phases of moon and cycles which are seasonal to use the positive effect of cosmic forces in agriculture.

1.12.1 Preparation of Biodynamic Compost

The site should be flat and dry with provision of air circulation into the compost heap by placing PVC pipes at the centre of a rectangular plot. Supply of oxygen is ensured through these pipes. The plant biomass (green and dry) is piled with a mixing of lime and rock phosphate at the centre layers. The entire pile is covered with a paste of cow dung and soil. Proper moisture is maintained by regular watering.

BD 500 Cow horn manure: For this cow horn and cow dung (50–150 g/horn) are required. It is used to increase soil fertility of degraded soils. Fill cow horns with cow dung during October/November and bury in burial pits with base downward and spaced at 1 inch and surrounded by compost and soil. Cover with soil for 4 to 6 months. The burial place should be kept moist, free from weeds and earthworms. BD 500 is applied when earth is breathing in (dew falling), i.e., during late afternoon to evening (descending moon). BD 500 is used @ 62.5 g in 37.5 litres of rain water. Stir for 1 hour alternatively clockwise and then anti-clockwise and then spray.

BD 501 Cow horn silica: For this cow horn and powdered silica quartz crystal (200–300 g/horn) are required. Prepared during the summer season (April/May to September) when the earth is breathing out and cosmic energy is most active. Powdered silica quart crystals are made into paste by use of water and filled in cow horn. Bury the horns with the same procedure as BD 500. It is applied to crops during the early morning by using 1 g BD 501 in 15 litres of water. Dissolve in water before sunrise and stir for 1 hour alternatively in clockwise and anticlockwise direction. Spray by the use of low pressure knap-sack sprayers.

BD 502 Yarrow (*Achillea millifolium*): For this, flowers of yarrow and urinary bladder of stag is required. The preparation is made under the planetary influence of Venus. Inside the stag bladder yarrow flowers are packed and moistened with plant extract. Cotton thread is used to stitch the packed bladder.

BD 503 Chamomile (*Matricuria chamomilla*): For this cow intestine and flowers of chamomila is used. The flowers are picked when the petals are horizontal during the morning. The uncleaned cow intestine is filled with these flowers and buried in soil.

BD 504 Himalayan stinging nettle (*Urtica parviflora*): Dried leaves moistened with leaf extracts are packed in mud pots and covered with a lid. The filled pots are kept under the influence of Mars in the month of March and are harvested during the month of September.

BD 505 Himalayan oak (*Quercus glauca*) bark: This requires the skull of an animal and the bark of an oak tree. This compost is made to supply calcium both from bark and from the skull of the animal. The crushed oak bark is filled into the brain cavity of the skull. Then it is covered with a bone piece and buried in a swampy place during the month of September and harvested in March.

BD 506 Dandelion (*Taraxicum officinalis*): This uses the mesentery of bovine species such as cow. Dandelion flowers, which are sensitive to light, are placed in the mesentery of a cow and wrapped with jute fibre. It is placed inside a pot with a mixture of soil and compost during the month of September and harvested in March.

BD 507 Valerian (*Valeriana officinalis*): Flower juice of valerian is used for this preparation. The flowers are made into a paste with the help of a mortar and pestle. This paste is mixed with water in the ratio of 1:4 and stored in a cool place. This preparation is applied along with BD 502-506 after mixing @ 5%.

BD 508 *Equisetum arvense*: In this preparation horsetail herb or Casuarina is used for controlling fungus. Tincture is made by boiling in water for 2 hours. This is diluted @ 50 g in 10 litres of water and sprayed for controlling mild fungal diseases in plants.

Cow pat pit: This is a specialized biodynamic compost prepared by mixing cow manure with crushed egg shell and basalt dust. The mixture is put into a 12 inch deep pit lined with bricks. It is used as a soil inoculant, as foliar feed on crops at all stages except close to harvest. It is applied on fruit trees before any mulching material is used around the trees.

1.13 ECO-FARMING

Eco-farming is farming which employs modern innovations in agriculture while preserving and conserving agro-biodiversity. Such farming evidences respect for ecology, wild life and biodiversity, produces less aerosols and less environment polluting chemicals and prevents greenhouse gas emissions and their effects. Therefore, this farming system is based on the principles of agroecology. The seven principles of eco-farming are (1) Food sovereignty (2) Helping farmers and rural communities (3) Smarter food production and yields – that lower the global production and consumption of amount of animal products (4) Biodiversity (5) Sustainable soil health and cleaner water (6) Ecological pest management and (7) Resilient food production systems.

Eco-farming system involves the creation of appropriate vegetation by combination of trees and shrubs in the farming system. Soil and water erosion is restricted by growing erosion-protecting crops along the contour. The farm may be divided into number of small fields by planting hedges. Regeneration of degraded land is carried out by afforestation, intercropping, organic manuring, integration of livestock, integrated plant protection and selective weed control measures. Soil health is taken care of by application of vermicompost, vermiwash, biofertilizers, etc. Plant health is taken care of by generation of botanical pesticides and rearing beneficial insects such as lady bird beetles, spiders, parasites of *Trichogramme,* etc. and released in the field as bio-control agents.

The success of an eco-farming system depends on community participation in its implementation over large areas. Eco-farming practices of pest control as described above, rodent eradication, developing biofertilizers such as *Rhizobium, Azotobacter, Azospirillum* cultures, Blue Green Algae (BGA) and *Azolla* needs community participation. Large-scale use of eco-friendly inputs creates a symbiotic relation between man and environment, conserve natural resources and respect the rights of animals.

1.14 BEE KEEPING

Bee keeping, also known as apiculture/apiary/honey bee farming, is a profitable eco-friendly, gender-inclusive subsidiary enterprise for production of honey, bee-wax, propolis, royal jelly, bee venom, pollen, bee brood, queen pheromones, etc. This enterprise requires low investment (₹50,000–70,000 for 10 bee colonies) but gives quick and high returns (₹70,000–2,00,000 for 10 bee colonies). Landless farmers in rural areas can also adopt beekeeping for augmenting their income. Several small- and large-scale industries/services can be based on apiculture, such as fabrication of bee hives and other beekeeping equipment, honey packaging, marketing of honey and other bee hive products, custom/advisory services for bee management, bee health and use of honey bees or crop pollination, queen bees breeding for sale of quality queen and export of honey and other bee products.

Bee wax is used for the manufacture of cosmetics, perfumes, polishes and pharmaceutical preparations such as ointments, capsules and pills covering and vitaminized sweets. Propolis is a complex substance containing proteins, vitamins, various minerals and volatile ether. It has antibiotic properties and is used as an ointment for healing wounds. Royal jelly is used to gain weight, treatment of ulcers and prevent ageing. Bee venom is used against several diseases called Apitherapy. Bee venom is used in homoeopathy. Apitherapy is done via a natural bee sting or administered through injection, ointment, inhalation of venom vapours and tablets of extracted bee venom. It cures rheumatism, arthritis, neurosis, polyarteritis, myositis, migraine, etc. Apitoxin, another name for bee venom, is a secretion from the acid and alkaline glands and stored in venom sacs at the base of the sting apparatus of honey bees.

Bee hives with bee colonies has to be integrated with crops such as maize, sorghum, cucumber, pumpkin, melon, mustard, sunflower, niger, safflower, sesame, cashew, coconut, litchi, neem, acacia, sisso, eucalyptus, jamun, papaya, guava and banana (Sarangi et al. 2021). Mutual benefit due to this integration results in crop pollination and nectar collection, increased crop yield by about 60% apart from honey production. There are mainly four types of honey bees – the rock bee (*Apis dorsata*), the Indian bee (*Apis cerana indica*), the little bee (*Apis florea*) and the European or Italian bee (*Apis mellifera*). The rock bees are found in high altitude areas up to 2700 m and produce about 36 kg honey per comb/year. However, these bees are difficult to rear due to their ferociousness. The Indian bees are domesticated species, construct multiple parallel combs and yield about 8 kg honey per colony per year. However, they are also prone to swarming. The little bees build single vertical combs with very low honey production (about 0.5 kg hive^{-1} year^{-1}). These are also not domesticated as they change their place frequently. The European bees are similar in habits as the Indian bees with high production (25–40 kg honey colony^{-1} year^{-1}). They are less prone to swarming.

The honey bee boxes, also known as hive bodies/deeps/brood boxes, are the essential requirement for honey bee keeping. The Langstroth hive, consisting of vertically hanging frames, on which bees build their comb, are suitable for Italian bees. Newton's hives, BIS hives, and Marthandam hives are suitable for rearing the Indian bees. Apart from hives, other equipment such as smoker, protective dress, bee brush, decapping knife, honey extractor are required for bee keeping. There are government institutions and schemes to boost beekeeping enterprises in India; some of these are described below.

1.14.1 National Bee Board (NBB)

NBB is a registered society with around 700 permanent members who are beekeepers, beekeeping groups, enterprises, companies, etc. The primary goal of NBB is to advance scientific beekeeping nationwide in order to boost crop yield through pollination and boost honey output, which will raise the income of the farming community. NBB imparts training on modules such as basics on beekeeping, scientific beekeeping, post-harvest management, processing and production of high value beehive products. As a role model for beekeeping, NBB has commissioned 16 integrated beekeeping development centres (IBDCs), one in each of the following states: Haryana, Delhi, Bihar, Punjab, Madhya Pradesh, Uttar Pradesh, Manipur, Uttarakhand, Tamil Nadu, Karnataka, Himachal Pradesh, West Bengal, Tripura, Andhra Pradesh, Arunachal Pradesh and UT of Jammu and Kashmir. Projects are also sponsored under *Rashtriya Krishi Vikas Yojna* (RKVY) for adoption of scientific beekeeping by farmer as a business and promoting pollination support by beekeepers/ beekeeping entrepreneurs on a custom-hiring basis. Banks finance beekeeping enterprise by providing credit limits/kisan credit cards (KCC).

1.14.2 National Beekeeping and Honey Mission (NBHM)

This is a central sector initiative for advancement and development of scientific beekeeping to achieve the goal of 'Sweet/Golden Revolution.' For continuation of NBHM, the Government of India has approved ₹500 crores for three years (2020–2021 to 2022–2023). NBHM is implemented through NBB.

1.14.3 Mission for Integrated Development of Horticulture (MIDH)

This scheme is executed by the state horticulture/agriculture departments, in which assistance is provided for producing high-quality nucleus stock, producing bee colonies by bee breeders, distributing honey bee colonies, beehives, beekeeping equipment, organizing training/exposure visits, seminars, workshops, etc. Therefore, farmers and young entrepreneurs should take the opportunity to examine different schemes to integrated beekeeping in farming systems and so increase income.

1.15 AGNIHOTRA FARMING

It is also known as *Homa* farming or Vedic farming, based on ancient Vedic principles of ecological cycles by means of performing '*Yajnya*' or '*Homa*' or '*Havan*.' This is done to purify the environment with the power of fire by burning organic substances such as cow's ghee, cow's dung, rice grains, etc. The specific Vedic mantras dealing with the plant kingdom known as '*Banaspati Sukta*' are chanted while performing '*Yajnya*.' The burning of the organic material is done in a specific copper pyramid during the circadian rhythm of nature, i.e., sunrise and sunset on a regular basis. This process creates positive energy around it and neutralizes the negative forces. The ash generated during the process of agnihotra is used for reclamation of degraded soil, growth promoter of plants and as insecticide.

The soil is fertilized with agnihotra ash @ 12.5 kg ha^{-1} alone or @ 25 kg ha^{-1} mixed with powdered cow dung cakes. Seed treatment is done with a mixture of 20 g agnihotra ash, 1 kg seed and cow urine. Agnihotra ash can also be applied to soil along with irrigation water by suspending a cotton bag filled with agnihotra ash @ 12.5 kg ha^{-1} in 1000 litres of irrigation water. The mixture of agnihotra ash and water is sprayed on crops before flowering and during the ripening stages. During the storage of cereals and pulses grains, agnihotra ash is mixed to protect form stored grain insect pests.

1.16 NATURAL FRAMING OR PERMACULTURE

The natural way of farming/green philosophy in farming was devised by Masanobu Fukuoka, known as the father of one-straw revolution. He is a Japanese national from the island of Shikoku in southern Japan. His concept of natural farming involves very minimal weeding, no chemicals, and no machinery. He does not till the soil or use ready-made compost on his farm. Sowing of grain crops is done at a time when it would naturally have fallen on the ground after maturity, instead of ploughing to control weeds, ground cover of white clover and a mulch of barley straw are used. Scattering of the straw across the entire field is done after sowing seeds. Vegetable growing is done by mixing seeds of various vegetables and scattering the mixed seeds everywhere on the farm. Vegetables grow in locations as per their suitability and reseed themselves afterwards. This system of vegetable cultivation gradually reverts to the form of their semi-wild ancestors. Thus, this concept is basically based on continuous grain growing without tillage and cultivating vegetables like wild flora.

1.17 ZERO BUDGET NATURAL FARMING (ZBNF)

ZBNF originated in Karnataka state with a concept developed by Subhash Palekar in treating his way of natural farming. This is a farming method that relies on the significance of co-production of crops and animals so that mutual benefits of different parts of the system can be obtained, effectively utilizing the easily available farm ingredients to produce crop treatments on farm and microorganisms or mycorrhizae to build fertility of the soil (FAO 2019; Khadse and Rosset 2019). The nine principles of ZBNF are: 1) Let crops cover soil for 365 days (with natural root system), 2) Minimize soil disturbance, 3) Use bio stimulants as necessary catalysts, 4) Grow indigenous seed, 5) Diversify crops and trees (15–20 crops), 6) Include animals in the farming, 7) Raise organic residues in the soil, 8) Manage pests and diseases through botanical extracts and 9) Completely avoid synthetic fertilizers, pesticides and herbicides. There are mainly five preparations used in ZBNF.

1.17.1 Jivamrita/Jeevamrutha/Jiwamrita

Microbial inoculum to stimulate microorganism activity to make nutrients available to plants and shield crop against pathogens. It is a fermented microbial culture prepared by a 48 hour fermentation process. The aerobic and anaerobic bacteria found in the cow dung and cow urine multiply as they consume organic ingredients (pulse flour). Small amounts of virgin soil from farm bunds are mixed to provide natural native microbes and organisms to the preparation. First, 200 litres of water are taken in a plastic barrel or drum to which 10 kg of fresh desi cow dung and 5–10 litres of aged cow urine, 2 kg jaggery/gur, 2 kg of pulse flour (besan) and a handful (1 kg) of soil is added. The mixture is stirred and kept in shade for fermentation. Stirring is done two times daily clockwise for two minutes to release harmful gases such as ammonia, carbon monoxide, carbon dioxide, methane, etc. The preparation is used within seven days. The dose of application is 500 litres ha^{-1}, applied fortnightly with irrigation water or to the crops as 10% foliar spray.

1.17.2 Bijamrita/Beejamrutha/Beejamrita

'Beej' means 'seed' and 'amrut' mean 'devine nectar' and it is a preparation to protect nascent roots from fungal, soil-borne and seed-borne diseases with microbial culture. It is used as seed/seedling/planting material treatment. The ingredients are desi (local breed) cow dung (5 kg), which act as a natural fungicide, cow urine (5 litres), which acts as anti-bacterial liquid), lime (250 g), undisturbed soil (1 kg) and water (20 litres). The ingredients are mixed with water and kept for 24 hours, with stirring done two times with a wooden stick. The seed treatment is done by mixing by hand for a uniform coating, then drying in the shade before sowing. The seeds germinate quickly and grow faster with very good root growth.

1.17.3 Ghanjeebamrita

'*Ghan*' means 'concentrated,' therefore, in this preparation more quantities of desi cow dung (100 kg) are used. Other ingredients are jaggery/gur (1 kg), pulse (arhar, gram, green gram, black gram) flour (2 kg), undisturbed soil from bunds (1 kg) and cow urine (5 litres). All the components are added meticulously to prepare a solid paste, which is kept covered with a gunny bag for two days with sprinkling of water over it. Small balls are made out of this and applied to the soil while sowing the seeds.

1.17.4 Acchadana

Acchadana means mulching the soil to avoid it disturbance by protecting the top soil by without deep ploughing/tilling. Soil mulching is done to promote aeration and water retention. Straw mulching is done by covering the soil with dried biomass waste of crops, which on decomposition form humus through the activity of the soil biota, activated by microbial cultures. Live mulch (symbiotic intercrops and mixed crops) is done by intercropping monocots and dicots in the same plot to make available all essential elements to the soil and crops. Legumes supply nitrogen, whereas cereals such as rice and wheat supply P, K and S. There is production of stabilized soil organic matter through *acchadana*.

1.17.5 Whapasa

Whapasa means managing soil moisture by improving soil structure and reducing tillage. In this concept, crop roots need water vapour not water as such. It refers to the state of soil that contains both water and air molecules. This concept encourages reducing the amount of irrigation water application and irrigating crops only at noon in alternate furrows. Other components of ZBNF are practicing contour bunding, using local species of earthworms, etc.

1.18 HOMESTEAD FARMING SYSTEM

A homestead farming system (HFS) is an agricultural production system adjoining the residence of the farming family. Though the area under HFS is relatively small, it includes cultivation of diverse crops (vegetables, fruits, medicinal plants, flowers, etc.), pisciculture in the pond, small-scale animal husbandry and in some cases integration of allied activities such as mushroom production, honey bee rearing, etc. Resource recycling such as use of by-product of one component as input for others makes it a perfect farming system. Approximately 768 million (118 million more than 2019) people worldwide suffered from hunger in 2020 and nearly 2.37 billion people did not have access to adequate food; 2020 saw a rise in global hunger as a result of the COVID-19 epidemic (FAO et al. 2021). As per a FAO report in 2021, more than one-half (418 million) and more than oneothird (282 million) of the 768 million undernourished people worldwide in 2020 resided in Asia and Africa, respectively, with around 8% (60 million) living in Latin America and the Caribbean. During the period of 2015–2030, one of the Sustainable Development Goals (SDG) of the United Nations is to eradicate hunger, ensure food security and improve nutrition and sustainable agriculture. To fulfil this objective, promotion of HFS, particularly in South Asia, is one of the options.

Different HFS models strive to provide family needs in different agro-ecosystems. In coastal region, vegetables such as Indian spinach (*Basella alba*), brinjal (*Solanum melongena*), arum (*Colocasia esculenta*), elephant foot yam (*Amorphophallus paeoniifolius*), chilli (*Capsicum annuum*), bitter gourd (*Momordica charantia*), pumpkin (*Cucurbita maxima*), okra (*Abelmoschus *esculentus*), potato (*Solanum tuberosum*), onion (*Allium cepa*), etc. are cultivated to meet family nutrition. Among other horticultural components, drum stick (*Moringa oleifera*), guava (*Psidium guajava*), mango (*Mangifera indica*) and coconut (*Cocos nucifera*) find a place either in the pond dyke or in the boundary. Pisciculture is the dominant component (occupy 45–50% area) in coastal HFS. Fish species include Rohu (*Labeo rohita*), Catla (*Catla catla*), Mrigal (*Cirrhinus mrigala*), Silver carp (*Hypophthalmichthys molitrix*), Grass carp (*Ctenopharyngodon Idella*), Japani punti (*Puntius japonicus*) and other various small local fish. A portion of the pond area is used for growing water spinach (*Ipomoea aquatica*) to supply green vegetables to the family. Animals such as cows, goats, sheep, poultry birds and ducks are domesticated in the HFS. Coconut-based fodder production systems enable rearing of dairy cattle in the homesteads of Kerala. Guinea grass (*Panicum maximum*) is an excellent fodder crop for intercropping in coconut orchards. Guinea grass is shade tolerant in nature and provides nutritious and palatable fodder for dairy cattle. In the tropical regions of Indonesia, *Pekarangan*, a traditional homestead garden, is an ideal and sustainable form of agroforestry system (Arifin et al. 2014). In Sri Lanka, Kandyan Home Gardens (KHGs) include fruit trees such as jackfruit, mango, cashew, citrus, guava, sweet orange, rambutan and avocado.

1.19 MIXED FARMING SYSTEM

It is the traditional farming system involving enterprise combinations to produce diverse products to meet subsistence needs. Advantages are less external input requirement, diverse outputs, risk aversion, etc.; however, it lacks a focus on surplus production and marketing. The traditional crop + livestock system comprises diverse crop cultivation and rearing of domestic animals by the use of crop by-products and use of animal manure for soil enrichment. Cattle rearing is practiced by most of the rural households; however, the demerit of the use of cow dung as fuel cake creates household ambient air pollution and loss of manure as a soil enricher. The by-products, such as cow urine, are today valued both for making various soil enriching products as well as for pest management options. Crop component consists of a variety of crops, cereals, millets, pulses, oilseeds, vegetables, fruits, flowers, medicinal plants, etc. Reducing the risk of failure in the era of climate change can be achieved by practicing a mixed farming system more than component farming.

Forms of mixed farming depend on external (weather, market, technology, policy, etc.) and internal (soil, land, farm family size, etc.) factors. The most common form of mixed farming is crop and livestock, which is different from a diversified system. In a diversified system, components coexist independently from each other; they reduce risk but resource recycling is limited unlike a mixed farming system. A mixed farming system may be classified into expansion agriculture (EA), low external input agriculture (LEIA), high external input agriculture (HEIA) and new conservation agriculture (NCA).

EA: When there is no constraint of land availability, with depletion of soil fertility new land is acquired or used, e.g., mixed farming practiced in West Africa and European grazing systems. Animals graze in pasture, bunds, open fields and fertilize the crop fields.

LEIA: In this system, most of the inputs are met from in-situ generation and soil fertility is maintained by organic manures as land is a limited resource and hence to be used sustainably. Animal feed is generated from crop by-products by growing fodder crops in field bunds. Sources of energy for farm operations are human and animal power.

HEIA: This mixed farming system uses synthetic sources of fertilizers, feeds and other inputs (pesticides and agro-chemicals) from markets for increasing production. Exchange or purchase of inputs also takes place from other farms; however, due to high level of input use, this affects the environment and there is emission from crop-animal production systems. There is a constraint of waste disposal in this mixed farming system.

NCA: This is followed to reduce the energy crisis, level of pollution and waste disposal constraints. The by-products are used to reduce the cost of inputs and accumulation of wastes. The term 'waste' in agriculture should be a misnomer and it should be recycled to generate wealth.

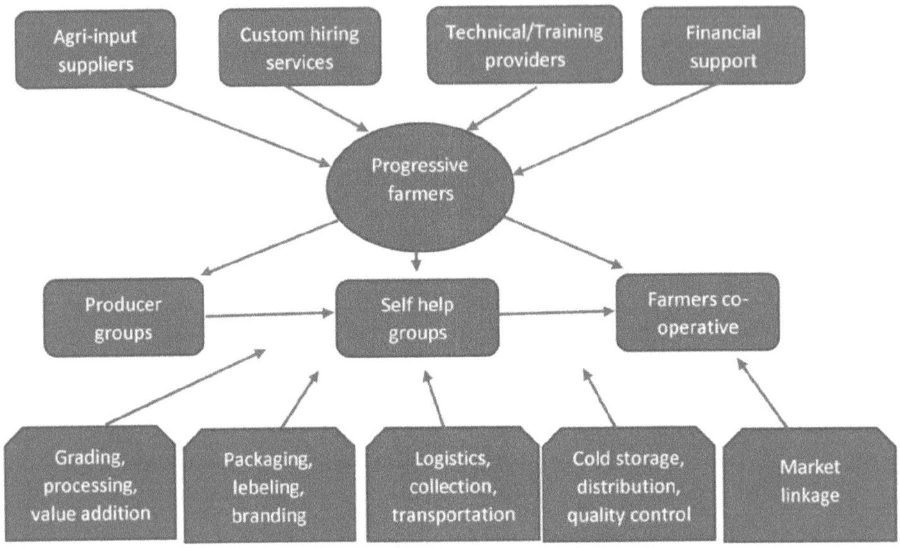

FIGURE 1.4 Backward and forward linkages required for a mixed farming system.

1.19.1 Mixed Farming System: Subsistence to Surplus

India's livestock population during 2019 was 192.9 million cows, 148.9 million goats, 109.9 million buffaloes, 74.3 million sheep, 9.1 million pigs and 729 million poultry birds. These livestock are mostly reared under a mixed farming system practiced by rural households. Cattle are reared by more than 65 million households, buffaloes by 39 million households, goats by 33 million households, sheep by 4.5 million households and pigs by 2.5 million households. Mixed farming system with three cows/buffaloes yielding at least eight litres of milk day^{-1} could generate an annual income of ₹60,000–75,000 household^{-1}. Similarly, a mixed farming system with 8–10 goats could generate an annual income of ₹50,000–60,000 besides meeting the household consumption needs.

TABLE 1.13
Financial and environmental comparison of mixed and specialized farming systems

Parameters	Dairy Farm	Beef Farm	Crop Farm	Mixed Farm
Overall gross margin (€ ha^{-1})	1135±235^{a}	625±359^{b}	477±140^{b}	607±278^{b}
Total cost (€ ha^{-1})	865±580^{a}	237±102^{b}	218±74^{b}	241±194^{b}
Fertilizer cost (€ ha^{-1})	87±32^{a}	43±31^{b}	77±30^{c}	71±54^{c}
Feed cost (€ ha^{-1})	620±371^{A}	56±35^{B}	-	113±150^{c}
N_{in} (kg ha^{-1})	135.4±46.9^{a}	55.4±27.0^{b}	102.8±28.5^{c}	82.5±37.5^{d}
N_{out} (kg ha^{-1})	74.9±27.9^{a}	17.5±10.7^{b}	114.8±24.4^{c}	57.8±32.3^{a}
N_{S} (kg ha^{-1})	60.5±25.3^{a}	37.9±23.3^{b}	-11.9±34.2^{c}	24.6±26.1^{b}

Source: After Ryschawy et al. 2012.

Differences between treatments are indicated as follows: a, b for $P < 0.05$; A, B for $P < 0.01$. NS (nitrogen surplus) is calculated as follows: NS = (N_{in} – N_{out})/Utilized Agricultural Area. N_{in} is the total amount of nitrogen in purchased inputs: feed concentrate, forage (including leys), straw, animals, mineral fertilizers and manure. N_{out} is the total amount of nitrogen outputs of the system: animal products such as milk, animals and animal manure; crops, straw and forage that are sold. N_{in}, N_{out} and NS are expressed in kg N/year.

Animal rearing under this system needs adaptable high-yielding breeds, appropriate health care (vaccination and deworming), nutrition (mineral mixture and silvi-pasture development) and market linkage. Fodder crops such as hybrid napier bajra can be raised on field bunds to meet the green fodder requirement of animals.

The advantages of a mixed farming system include regular income from livestock, adaptation to climate change, efficient nutrient recycling, nutritional supplementation for children and women, reduction of drudgery, scope of growing high value organic vegetables/fruits and medicinal plants such as ginger, turmeric, aloe vera, brahmi, tulsi, aswagandha, pudina, etc. through efficient water resource management. Backward and forward linkage is necessary to make the system profitable (Figure 1.4).

A comparison of mixed farming system with specialized farming in France revealed that mixed crop-livestock systems reduce farmer's financial risks, are relatively environmental-friendly and are appropriate at larger scales (Table 1.13).

REFERENCES

Arifin, H.S., Kaswanto, R.L., Nakagoshi, N. 2014. Low carbon society through pekarangan, traditional agroforestry practices in Java, Indonesia. In: Nakagoshi N., Mabuhay A.J. (eds) *Designing Low Carbon Societies in Landscapes*. Ecological research monographs. Springer, Tokyo. pp 129–143. DOI: https://doi.org/10.1007/978-4-431-54819-5_8.

Economic Survey. 2020–2021. Volume 2, Government of India, Ministry of Finance, Department of Economic Affairs, Economic Division, North Block, New Delhi.

FAO. 2019. *52 Profiles on Agroecology: Zero Budget Natural Farming in India*. http://www.fao.org/3/a-bl990e.pdf.

FAO, IFAD, UNICEF, WFP and WHO. 2021. *The State of Food Security and Nutrition in the World 2021. Transforming Food Systems for Food Security, Improved Nutrition and Affordable Healthy Diets for All*. FAO, Rome. DOI: https://doi.org/10.4060/cb4474en. Accessed on 29 September 2021.

Gangwar, B., Singh, A.K. 2011. *Efficient Alternative Cropping Systems*. Project Directorate for Farming Systems Research, Modipuram, Meerut, India. pp 339.

Government of India. 2019. *Pocket Book of Agricultural Statistics 2019*. Ministry of Agriculture & Farmers Welfare, Department of Agriculture, Cooperation & Farmers Welfare, Directorate of Economics & Statistics, New Delhi.

https://www.greenfacts.org/en/fisheries/l-2/06-fish-consumption.htm. Accessed on 5 July 2021.

Khadse, A., Rosset, P.M. 2019. Zero budget natural farming in India-from inception to institutionalization. *Agroecology and Sustainable Food Systems* 43:848–871. DOI: https://doi.org/10.1080/21683565.2019.1608349.

Lampkin, N. 1990. *Organic Farming*. Farming Press Books, IPSWICH. pp 701.

Ryschawy, J., Choisis, N., Choisis, J.P., Joannon, A., Gibon, A. 2012. Mixed crop-livestock systems: An economic and environmental-friendly way of farming? *Animal* 6(10): 1722–1730. DOI: https://doi.org/10.1017/S1751731112000675.

Salim, S.S. 2016. Fish consumption pattern in India, exports-Overview. *Food and Beverages News*. 8th edition. pp 25–28.

Sarangi, S.K., Islam, M.R. 2019. Advances in agronomic and related management options for Sundarbans. In: Sen H. (eds) *The Sundarbans: A Disaster-Prone Eco-Region*. Coastal Research Library, vol 30. Springer, Cham. pp 225–260. DOI: https://doi.org/10.1007/978-3-030-00680-8_8.

Sarangi, S.K., Sundaray, J.K., Prakash, A. 2021. Journey towards eco-sustainable farming systems in South Asia. *Journal of Applied Zoological Researches* 32(1):71–80.

Sharma, V.P., Annepu, S.K., Gautam, Y., Singh, M., Kamal, S. 2017. Status of mushroom production in India. *Mushroom Research* 26(2):111–120.

Singh, M., Shwet, K., Sharma, V.P. 2017. Status and trends in world mushroom production-I. *Mushroom Research* 26(1):1–20.

2 Sustainable Integrated Farming Systems for Food and Nutrition Security

2.1 INTEGRATED FARMING SYSTEM (IFS)

IFS is a scientific combination of two or more components that makes use of cutting-edge agronomic management technologies (e.g., conservation agriculture, precision farming, artificial intelligence, machine learning, etc.) and the cardinal principles of minimum competition and maximum complementarity. The goals of IFS are to increase family nutrition, farm income, and ecosystem services in a sustainable and environmentally friendly manner. It promotes maximal recycling, agricultural diversification and biodiversity preservation. Perfect integration results in zero wastage and full utilization of byproducts as inputs, resulting in significant reduction in the cost of cultivation. Therefore, there is an increase in net income, employment, and reduction in risk due to market price fluctuations as well as weather/climate-related aberrations.

The modern IFS should follow the principles of Climate Smart Agriculture (CSA). The Food and Agriculture Organisation (FAO) in 2010, at the Global Conference on Agriculture, Food Security and Climate Change held in The Hague, defined CSA as an integrative approach for addressing the interconnected issues of climate change and food security through the three dimensions (economic, social and environmental) of sustainable development. The economic dimension calls for raising agricultural productivity in a sustainable manner in order to promote fair increases in farm incomes, food security and development. Social dimension signifies adapting and strengthening the resilience of agriculture and food security systems to climate change at various levels, and environmental aspects include reducing greenhouse gas (GHG) emissions resulting from crops, livestock and fisheries sectors in agriculture (FAO 2013).

By intensifying and diversifying crops and integrating related enterprises, the IFS offers the chance to raise the economic yield per unit area per unit time (Singh et al. 2007). The integration of crop (0.80 ha) with pigeons (10 pairs on 0.01 ha), goats (5 female + 1 male on 0.05 ha), agroforestry (0.10 ha) and farm ponds (0.04 ha) was shown to be a good alternative farming with recycling of composted goat manure to the crop for small and marginal farmers who are short on cash (Shekinah and Sankaran 2007).

Experience and guidelines on integration of field and fodder crops with fish, livestock, duck, etc. from different countries such as China, Germany, Switzerland, the United States and Australia can be incorporated according to the given location and condition. Several models could be weaved together to create an integration web through longitudinal and transverse nutrient cycling and energy flow in accordance with local conditions in order to improve the efficiency of energy use, lower costs and boost production and income.

It is important to identify and make use of the diversity found on farms in IFS. For example, 30–40 different types of edible herbs can be found in large quantities in the rice fields of coastal India. In eastern Indian local markets and today even in city markets there is a supply and demand for lesser known leafy vegetables such as water spinach (*Iopmoea aquatica*), Indian pennywort (*Centella asiatica*), brahmi (*Bacopa monnieri*), water clover (*Marsilea quadrifolia*), tropical spiderwort (*Commelina benghalensis*) and marsh barbel (*Hygrophila auriculata*), among other herbs

DOI: 10.1201/9781003239963-2

(Chatterjee 2009). Through recycling of the byproducts like organic wastes/residues, the various components of the system must be mutually synergistic, complementary and efficient, making the system more ecologically safe and profitable (Yadav 2006).

Nationally as well as internationally research is still continuing to formulate suitable IFS models. Compared to cropping alone, the integration of improved cropping system with components like fish, poultry, pigeons and goats in Tamil Nadu produced higher output, including better profitability, employment and the potential to recycle nutrients. This model generated better employment opportunity of 576 person-days (Jananthi 2002). In the case of resource poor, small and marginal holders with less investment ability, the IFS model with crops (0.80 ha) integrated with birds such as pigeons (10 pairs on 0.01 ha), small livestock like goat (5 female + 1 male on 0.05 ha), agroforestry (0.10 ha) and farm ponds (0.04 ha) was found to be a good option with efficient recycling of manures and crop residues (Shekinah and Sankaran 2007). In Bhubaneswar, Behera and Mahapatra (1999) it was found that an IFS involving crops, pisciculture, poultry, duckery, mushroom cultivation and apiary on a 1.25-hectare plot of land produced an employment equivalent to 573 person-days and resulted in a net return of ₹58,360.00 from an initial investment of ₹49,286.00. Typically, marginal and small farmers have more than enough family labour to complete the farming operations. According to Behera et al. (2014), they choose enterprises that need less land and capita and generate a recurring income stream to cover daily family needs. There is an opportunity to diversify farming in the coastal region while taking biodiversity, soil, climate and landforms into account.

IFS looks for ways to produce food in a more natural way, leveraging the biodiversity and recycling resources and nutrients while optimizing external inputs for fertility and productivity. IFS can lead to productive, sustainable and climate-resilient agricultural systems compared to high-input, high-intensity farming systems.

2.2 APPROACHES TO INCREASE FOOD SECURITY, NUTRITION SECURITY AND SAFE FOOD PRODUCTION, MARKETING AND CONSUMPTION

Food and nutrition security are key indicators of the well-being and survival of individuals and populations. Ensuring that all individuals have access to enough nutritious food is essential for good health and for promoting sustainable development.

Food and nutritional security refer to the availability, access and adequacy of food to fulfill the dietary requirements and food choices. This includes having enough food for all members of the household, as well as access to a diversity of foods that meet all the necessary nutrients for good health. To improve food and nutrition security, a number of strategies can be used, including accelerating food production and distribution, provisioning access to nutritious food and educating individuals and communities about healthy eating. Additionally, addressing the underlying socioeconomic reasons of food and nutrition insecurity like poverty, lack of education or access to health care, and poor infrastructure, is crucial. However, food security and nutrition security can be affected by a variety of factors, including poverty, conflict, natural disasters and climate change. For example, during times of war or natural disasters, food may become scarce and difficult to access, leading to food insecurity. Similarly, during times of economic hardship, families may not have enough money to buy nutritious food, leading to nutrition insecurity.

Safe food production, marketing and consumption are interrelated and essential components of a food system that ensures that food items are safe for consumption and free from harmful contaminants or pathogens. This includes implementing good agricultural, manufacturing, and hygiene practices throughout the entire food production process, from growing and harvesting to processing, packaging and storage.

Food marketing refers to the promotion and sale of food products to consumers. This can include advertising, labelling and packaging of food products. To ensure that food products are safe for consumption, food producers must adhere to strict regulations and guidelines when marketing their products. This includes ensuring that the information on food labels is accurate and not misleading, and that food products are packaged in a way that preserves their safety and quality.

To ensure safe food production, farmers and food producers must adhere to strict regulations and guidelines, such as those set by the FDA and USDA in the United States. These regulations cover everything from the use of pesticides and fertilizers to the handling and processing of food products. For example, there are guidelines for preventing cross-contamination in food production facilities, such as separating raw and cooked food products and regularly sanitizing equipment. In addition to regulations and guidelines, food producers must also implement good agricultural practices (GAP) and good manufacturing practices (GMP) to ensure safe food production. GAPs focus on growing and harvesting food products in a way that minimizes the risk of contamination, while GMPs focus on processing, packaging and storing food products in a way that preserves their safety and quality. Traceability and record-keeping are also important aspect of safe food production. The ability to trace food products back to their origin is crucial in the event of a food safety incident, as it allows for quick identification and removal of contaminated products from the market. Safe food consumption refers to the safe handling, preparation and storage of food products by consumers. This includes proper cooking temperatures, storage temperatures and refrigeration practices, as well as ensuring that food is not expired or spoiled. Consumers also have a role in identifying and reporting food safety issues, such as food recalls and alerts issued by government authorities.

There are several approaches that can be taken to increase food security, nutrition security, and safe food production, marketing and consumption.

2.2.1 Agroecological Farming

Agroecological farming is an approach that focuses on using ecological principles towards better food and nutrition security, while also safeguarding natural capitals. This can be achieved through practices such as crop rotation, intercropping, agroforestry and the use of organic fertilizers and pesticides. World hunger can be addressed through agroecological approaches (Patel 2021).

2.2.2 Community-Based Food Systems

Food as well as nutritional security can be achieved through community-based food production systems by promoting local food production and consumption. This can be accomplished through initiatives such as community gardens, farmers' markets and food cooperatives.

2.2.3 Food Processing and Value Addition

Food processing and value addition can address and increase food and nutrition security by extending the shelf life of food products, minimizing post-harvest losses and increasing the value of food products. This can be achieved through initiatives such as training and technical assistance for farmers and small-scale processors and the development of food processing facilities and infrastructure.

2.2.4 Food Safety and Quality Assurance

The assurance for food safety and quality are critical for ensuring that food products are healthy for consumption and meet the needs of consumers. This can be achieved through initiatives such as food safety regulations, food safety training and technical assistance and food safety certification schemes.

2.2.5 Partnerships and Collaborations

Partnerships and collaborations between farmers, governments, private sector, civil society and other stakeholders can help to increase food security, nutrition security and safe food production, marketing and consumption. This can be achieved through initiatives such as public-private partnerships, community-based organizations and knowledge sharing and networking among stakeholders.

A combination of these approaches can be used to increase food security, nutrition security and safe food production, marketing and consumption. By focusing on sustainable, community-based and ecologically based practices, as well as partnerships and collaborations, it is possible to create a more resilient and equitable food system. In this era of climate change, the farming systems are facing an unpredictable future. Farming different crops in an interlinked system supporting each other is the natural way for ensuring livelihood, nutritional security and sustainability of the food production system. It enables resource-poor communities to ensure gender equity, equitable distribution and other forms of equity in addition to enabling them to sustainably produce adequate food for themselves and others.

Food security as defined by FAO is the availability of a sufficient quantity of food with appropriate quality supplied through domestic outputs or imports to reach a state of nutritional well-being where all physiological needs are met at all times (FAO 2006. Food Security: Policy Brief). Access to food containing adequate calories and nutrients to meet dietary requirements is not guaranteed by food availability alone. Moreover, premium foods frequently fetch higher prices on the market, which may lead to a rise in global food insecurity, particularly for the poor who cannot afford these goods.

Having regular and fair access to wholesome, secure and reasonably priced foods that are necessary for optimum health and well-being is what is meant by nutritional security. On the other hand, hunger is influenced by food quality just as much as quantity. Today, the nutritive value is highly compromised in food, resulting in various ailments. An increased risk of obesity, diabetes and heart disease, as well as more general effects like greater health-care expenditures and decreased productivity, are all associated with poor nutrition, which is a major cause of sickness. A few of the main obstacles include the depletion of production factors, the impact of climate change and the sustainability of natural resources. The profitability and sustainability of farming are seriously threatened by the trend of falling land holding sizes. Smallholders, or farmers who own less than 2.0 hectares of land, who make up the great majority of farmers in developing nations, are unlikely to be able to make a living from a single agricultural venture, such as the traditional mono-cropping system.

2.3 ENHANCING/DOUBLING FARMERS' INCOME AND PROFIT IN THE CONTEXT OF IFS

In India, doubling farmers' income has drawn lots of attention and debates as well. However, one cannot ignore the potential of IFS in this context. Five approaches can be followed for doubling income and profit of the farming community in the context of IFS in India.

1. ***Crop diversification and value addition:*** By diversifying crops and incorporating value addition, farmers can increase their income by accessing new markets and selling higher value products. For example, farmers can grow high-value crops, such as horticultural crops, and process them into products with extra value, such as jams and jellies. Growing of crops for supply of biomass energy can reduce the production cost in agriculture (Acaroglu and Aksoy 2005).
2. ***Agroforestry:*** Agroforestry can help farmers to increase their income by providing multiple products and services, such as timber, fuelwood, fruits and medicinal plants. Additionally, agroforestry can help to improve soil health and water conservation, which can increase crop yields and improve farm productivity.
3. ***Livestock integration:*** Integrating livestock into IFS can help farmers to increase their income by providing additional products and services, such as milk, meat and manure. Additionally, by using livestock to control pests and weeds, farmers can reduce the need for chemical inputs, which can lower production costs and increase profits.
4. ***Watershed management:*** Watershed management can help farmers to increase their income by improving water availability, which can increase crop yields and improve farm

productivity. Additionally, by improving soil health, farmers can reduce their need for chemical inputs, which can lower production costs and increase profits.

5. ***Farmer Producer Organizations (FPOs) and market linkages:*** FPOs can help farmers to boost their income by providing access to credit, markets and other resources. By forming FPOs, farmers can accrue better prices for their products and access new markets. Additionally, by forming FPOs, farmers can also access technical assistance and training, which can improve their farming skills and productivity.

Overall, by implementing IFS, farmers can increase their income and profits by diversifying their crops, integrating livestock, managing their watersheds and forming FPOs. Additionally, by incorporating value addition, farmers can access new markets and sell higher value products. Furthermore, by adopting sustainable and organic farming practices, farmers can improve their livelihoods and achieve conservation of natural resources. Considering the priority to double farmers' income, it can be recommended that when farmers adopt IFS, their net annual income is significantly higher than what they earn by cultivating paddy as a monocrop (Behera and France 2016). Implementing IFS is one way to reduce risk as well as double farmer income. The terms 'integrated farming system' and 'integrated agriculture' are frequently and widely used to refer to farming practices that are more integrated than monoculture techniques. 'Integrated bio-systems' are sometimes used to describe agricultural systems that incorporate fish, horticulture, apiculture, sericulture, fungiculture, lac culture, livestock and other related activities. This method uses a network of interconnected components or enterprises such that the "waste" from one component can be used as an input for another, lowering costs and increasing output and/or revenue. IFS makes sure that agricultural wastes from one type of farming are used as resources for another. We not only get rid of wastes but also guarantee a general boost in production for the entire agricultural system because it turns wastes into resources.

Seven pillars comprise the main components of Intensive Integrated Farming Systems (IIFS): soil health care, water harvesting and management, energy management, crop and pest management, post-harvest management, crop selection, farm animals and other farming components such as organization, skill, information and management empowerment. In an expanded symbiotic or synergistic system, these systems integrate aquaculture, livestock, agriculture and agro-industry. Wastes from one process can be treated or used as an input for other processes to provide energy, fertilizer and feed for maximum productivity at the lowest possible cost. To ensure the sustainable use of natural resources for the benefit of both current and future generations, IFS is a component of a larger plan (NABARD 2017). Given that the genetic potential of all main crops is plateauing and that the productivity of rain-fed areas is diminishing, the strain on Indian agriculture to supply food and nutritional security for its rapidly growing population is becoming an increasingly difficult task. The severe lack of green fodder brought on by the division of landholdings has a negative impact on dairy production and cattle. The scarcity of drinking water and fodder, in particular, is causing a crisis in rural areas related to food and livelihood. Agrarian hardship is being caused by the shrinking size of landholdings and the lack of alternative revenue augmentation opportunities, which is creating a reduction in agricultural income. To increase their incomes, a significant portion of smallholders are forced to relocate to cities or engage in non-farming activities.

2.4 IFS TO ADDRESS UNDERNUTRITION AND MALNUTRITION

To solve problems through varied agricultural enterprises and reduce risks brought on by environmental limits, the IFS requires effective farmer engagement. It is a dynamic, gender-sensitive, socially responsible, environmentally sustainable, location-specific, farmer-oriented and holistic approach. By putting IFS into practice, risks would be reduced, wastes and residues would be recycled, two or more businesses would be integrated, all resources would be used as efficiently as possible, productivity and profitability would be maximized, the ecological balance would be maintained, jobs would be created, input efficiency would increase and end products from one business would be used as input in another. It has been observed that farmers who use IFS have far greater net incomes and are also more protected from crop failure risks than farmers who do not use IFS.

The contribution of agriculture to food and nutrition security of the human and animal kingdom is beyond doubt. It serves as the foundation for food production, contributes to nutrition clearly and is thought to be a major factor in stunting and micronutrient deficiencies (McDermott et al. 2015). However, undernutrition still exists and poses a serious risk to health in low- and middle-income nations despite the expansion of agriculture. Part of the problem stems from agriculture and food policies' historical emphasis on supply-side production of staple commodities (Tontisirin et al. 2002; McDermott et al. 2015). Furthermore, the great majority of these previous agricultural strategies did not take micronutrient outcomes into account (Tontisirin et al. 2002). The majority of these strategies concentrated on supplying macronutrients, such the energy-giving carbohydrates and the protein found in staple meals, especially grains like rice, wheat and maize. For instance, the Green Revolution accelerated the rise of food production, reduced poverty and raised household energy use. This primarily happened in the 1940s and 1960s with the goal of addressing the issues of food scarcity that many developing nations were dealing with at the time. However, the Green Revolution frequently resulted in a monoculture of rice, wheat and maize because efforts were primarily directed towards boosting the production of staple crops. These had a significant negative impact on nutrition outcomes since they drove out other nutrient-rich crops like millet and decreased agricultural diversification in some developing nations (Pingali 2023). With a focus on women and children in the first 1,000 days of life, from conception to age two, agricultural policies and programs today aim to address malnutrition by using both nutrition-specific interventions and agriculture that is sensitive to nutrition (Ruel et al. 2018). According to Cusick and Georgieff (2016) and Prado and Dewey (2014), the first 1,000 days are crucial for promoting newborns' and early children's optimal growth and development as well as preventing growth retardation, micronutrient deficiencies and childhood illnesses. Through iron or folic acid supplementation to expectant mothers, micronutrient fortification of staple foods and nutrition education, nutrition-specific diverse interventions directly target the immediate causes of undernutrition, such as inadequate dietary intake and ill health (Fiorella et al. 2016; Ruel et al. 2018). Due to their affordability and ease of implementation, supplementation and fortification are approaches often used to alleviate micronutrient deficiencies (Allen 2003).

Nutrition-sensitive agriculture, according to the Food and Agricultural Organization (2018) of the United Nations, is a method that aims to guarantee the production of a range of reasonably priced, healthful, safe and culturally appropriate foods in sufficient quantity and quality to meet population dietary needs in a sustainable way. The phrase 'food-based' approaches refers to nutrition-sensitive interventions that attempt to address the underlying drivers of micronutrient deficiencies, i.e., ensuring adequate consumption and availability of a range of safe and nutritious foods (Tontisirin et al. 2002; Ruel et al. 2018). More recently, as the worldwide agricultural community, among others, works towards global nutrition targets, coordination and collaboration throughout agricultural sectors and across food systems have come to be seen as crucial. To (1) determine how and to what extent integration, as opposed to specialization of agricultural subsectors, could relieve the ecological, sociopolitical and economic pressures that farming households face, and (2) assist in addressing the multiple burdens of malnutrition, research aiming to achieve these goals is highly relevant and becoming increasingly important. It is believed that economically and environmentally sustainable approaches to agriculture development, like integrated aquaculture-agriculture (IAA) systems like fish-rice farming and fish-livestock farming, can simultaneously reduce food insecurity, malnutrition and poverty (Ahmed and Turchini 2021; Ahmed et al. 2014). In particular, these systems have garnered considerable interest in the literature since the late 1990s.

'Integrated agriculture' involves diversifying agricultural production from smallholder plots to the landscape level, including farm management techniques like crop rotation and intercropping.

This refers to the integration of agricultural subsystems, such as aquaculture, fruit and vegetable production and livestock husbandry, which are believed to enhance resource management and space utilization (Hendrickson et al. 2008).

2.5 ADVANTAGES OF IFS VS. COMPONENT FARMING, PEST AND DISEASE MANAGEMENT, CONTINGENCY PLANS UNDER NATURAL CALAMITIES

IFS have several advantages compared to component farming and traditional farming practices. Some of these advantages include increased crop diversity and productivity. IFS involves the integration of different farming components, such as crops, livestock and agroforestry, which can increase crop diversity and productivity. This can lead to improved food and nutrition security, as well as increased income for farmers. Improved soil health and water conservation: IFS practices, such as agroforestry, intercropping and crop rotation, can improve soil health and water conservation. This can help to increase crop yields, reduce the need for chemical inputs and improve the resilience of farms to drought and other climate-related challenges. IFS practices, such as crop rotation, intercropping and agroforestry, can help to reduce the incidence of pests and diseases. Additionally, by integrating livestock into farming systems, farmers can use natural pest control methods, such as using chickens to control insects. IFS practices can help farmers to be more resilient to natural calamities, such as drought, floods and storms. For example, by incorporating water conservation practices, such as rainwater harvesting, farmers can be better prepared for drought, and by following agroforestry, farmers can reduce soil erosion and landslides. In contrast, component farming systems are based on the cultivation of a single crop in large areas, which can lead to monoculture and high dependence on external inputs. Therefore, it is more vulnerable to diseases, pests and natural calamities.

In summary, IFS offers several advantages over traditional and component farming systems, including increased crop diversity and productivity, improved soil health and water conservation, natural pest and disease management and greater resilience to natural calamities. These advantages make IFS an attractive option for farmers who are looking to improve their livelihoods, while also promoting sustainable and organic farming practices.

The IFS has many benefits, including high productivity together with nutritional security, income diversity, protection of natural resources and strengthening the agricultural production system's resilience to the climate and the economy. IFS encourages more efficient use of natural resources and fertilizers, which increases food output per unit of land and inputs and improves nutrient recycling. The primary cause of this is the existence of fish and animal components, which alter the nutrient fluxes at the interface between the soil, plants and atmosphere. The primary centralizing compartment of the various synergistic activities that take place in the agro-ecosystem is the soil. As various agricultural production systems integrate energy and nutrients, animals serve as catalysts by bringing in variability and novel paths for water and nutrient flows. The kind of crops in the cropping system, crop nutrient management, and livestock component will determine how much the processes and nutrient fluxes are impacted. Crop rotation is another crucial component to take into account when diversifying IFS arrangements because it permits the introduction of plants with varying root systems and nutritional requirements. This can enhance soil quality and decrease nutrient losses while also increasing nutrient cycling (Tiecher et al. 2017). Because they offer distinct production potentials, root systems and waste inputs, adding crops like cowpea, moong, baby corn and chilli to the crop cycle after rice is a crucial tactic to reap the benefits of crop rotation in this situation. In the coastal lowland ecosystem, integrating dairy, fisheries and poultry components with diverse cropping systems is crucial for enhancing soil fertility and optimizing nutrient usage (Paramesh et al. 2021). The problem of feeding a growing world population persists even with increased food production brought about by the Green Revolution in agriculture and the matching Blue Revolution in aquaculture. The Blue Revolution's social responsibility, food productivity, and environmental sustainability have all drawn criticism. In a similar vein, the social and environmental effects of the green revolution have drawn criticism. Integrated rice-fish farming, or the 'Blue-Green Revolution,' provides a means of addressing the problems of food production, environmental sustainability and socioeconomic advantages. Through integrated rice-fish production, a type of sustainable intensification that may produce more food from the same area of land and water with minimal or no environmental impact, the Blue Revolution (aquaculture) and the Green Revolution (agricultural) can be combined. Rice-fish co-culture has the potential to boost overall productivity by up to 27% when compared to rice monoculture. Including other crops (such vegetables, livestock and ducks) could boost the production of a wider variety of foods. Nonetheless, one of the possible obstacles to rice-fish production is water management. For rice-fish farming to become more widely used, further study on water management is required, with the cooperation of important stakeholders.

2.6 NUTRITION MANAGEMENT UNDER IFS

Nutrition management is an important aspect of IFS, as it aims to improve food and nutrition security for farmers and their communities. Here are a few examples of how nutrition management can be incorporated into IFS.

Crop Diversification: IFS involves the integration of different farming components, such as crops, livestock and agroforestry, which can increase crop diversity. This can lead to improved food and nutrition security as farmers have access to a wider variety of nutritious foods. Crop rotations have better effect on soil quality than continuous monocropping (Karlen et al. 2006).

Livestock Integration: Integrating livestock into IFS can help farmers to improve their nutrition security by providing additional protein and micronutrients through milk, meat and eggs. Additionally, by using livestock to control pests and weeds, farmers can reduce the need for chemical inputs, which can lower production costs and increase profits.

Home Gardens: Home gardens are an important component of IFS as they provide a source of fresh, nutritious fruits and vegetables for farmers and their families. Home gardens can also be used to grow medicinal plants, which can help to improve health and nutrition outcomes

Agroforestry: Agroforestry can help farmers to improve their nutrition security by providing fruits, nuts and other non-timber forest products, which are rich in micronutrients. Additionally, agroforestry can help to improve soil health and water conservation, which can increase crop yields and improve farm productivity

Nutrition Education and Training: Nutrition education and training are critical for ensuring that farmers and their families have the knowledge and skills to make healthy and nutritious food choices. This can be achieved through initiatives such as nutrition education and training for farmers and the development of nutrition education materials and resources.

The U.N.'s Food and Agriculture Organization has identified ten key principles for communities pursuing agroecology (FAO 2018).

Human and Social Values: Attention to equity, inclusion and justice ensures that no one goes hungry. Women, who in most cultures are responsible for feeding the family, are among the key innovators and leaders of agroecology projects. Whereas industrial systems devalue farm labour, food production becomes an enjoyable, shared task and a vehicle for enhancing ties with nature and the community. Protecting the environment enables future generations to inherit intact natural systems.

Culture and Food Traditions: Industrial agriculture and global trade have generated imbalanced diets worldwide, with cardiovascular diseases and diabetes existing in tandem with starvation and malnutrition. Agroecology allows local food cultures to thrive, restoring a direct relation between food production and consumption.

Recycling: Agroecology recycles natural resources and nutrients. Like nature, it produces no waste. Recurring cycles include the movement of atmospheric nitrogen into and out of the soil, the flow of rainwater from groves to fields and back, the use of farm animal manure as soil fertility builder and the conversion of crop residues like straw into animal feed.

Circular and Solidarity Economy: To directly connect food production to demand, agroecology aims to establish connections between local food producers and customers. These localized economies increase the earnings of farmers, give consumers access to fresher product, strengthen ties within the community and utilize less energy for transportation while also reducing food waste.

Responsible Governance: Systems of governance at global, national and local levels that protect the rights of food-growing communities to land, clean water, intellectual property (such as traditional seeds and know-how) and other resources are essential for agroecological systems to thrive. Existing systems of trade and governance militate against the well-being of small farmers, who are among the hungriest and most indebted people in the world.

Resilience: A diversified crop promotes a quicker recovery from natural disasters and other climatic calamities. Furthermore, the ability to cultivate food using recycled and locally sourced materials protects communities against outside shocks to the economy.

Efficiency: Half the nitrogen provided by synthetic fertilizers ends up polluting the environment, while thirsty hybrid or genetically modified crops deplete water tables. In contrast, agroecological systems use resources and nutrients with optimal efficiency. Locally adapted seeds, water harvesting, nitrogen-fixing microbes and natural predators help produce food with minimal external inputs. Energy efficiency in agriculture can be estimated from an equivalent value of chemical energy (MJ ha^{-1}) of inputs and outputs using published values (Calilung 1985; Mittal et al. 1985; Chamsing et al. 2006; Pimentel and Pimentel 2008; Pishgar-Komleh et al. 2011; Quilty et al. 2014).

Synergies: The various subsystems of agroecology work together synergistically to create something that is far greater than the sum of its parts. Agroecology not only lowers poverty and hunger but also protects and improves nature, culture, soil fertility, water availability and the community for the benefit of current and future generations, providing marginalized people with justice and equity.

Diversity: Diverse species intercropping boosts resilience and productivity since one crop may fail in circumstances that allow another to flourish. Improving biodiversity, for example by creating habitats for birds, frogs and other predators and aids in the management of pests. Moreover, agroecological systems are themselves diverse, each being tailored to the local ecology and culture.

Co-creation and Sharing of Knowledge: Communities work together to identify and share the agroecological practices that fit best with their particular ecological and cultural circumstances. There is no universal, top-down prescription of what should be grown and how; instead, there are guidelines for a process of shared experimentation and discovery that lead to different solutions in different places.

2.7 MECHANIZATION, SELF-RELIANT FARMING SYSTEM

Mechanization and self-reliant farming systems are important aspects of IFS as they can help farmers to increase productivity and efficiency, while also reducing labour costs. Here are a few examples of how mechanization and self-reliance can be incorporated into IFS:

Mechanization: By using appropriate and locally available machinery, farmers can reduce labour costs and increase productivity. For example, farmers can use tractors, power tillers, threshers and other machinery to reduce labour inputs and increase efficiency.

Custom Hiring Centres: Custom hiring centres can provide farmers with access to machinery and equipment at an affordable cost. This can help farmers to reduce the need for large capital investments, while still being able to take advantage of mechanization.

Self-Reliance: By developing a self-reliant farming system, farmers can reduce their dependence on external inputs and increase their resilience to market fluctuations. For example, farmers can use agroforestry to provide their own fuel, fodder, and other inputs, which can reduce their dependence on external sources.

Maintenance and Repair: Self-reliance can also be achieved by developing the capacity of farmers to maintain and repair their own machinery and equipment. This can be achieved through training and technical assistance programs, which can help farmers to acquire the knowledge and skills needed to maintain and repair their own machinery and equipment. Mechanization and self-reliance are also important for farmers who want to achieve profitability and sustainability. Achieving self-reliance in farming can provide farmers with the ability to produce their own inputs, reducing the dependency on external inputs and increasing the resilience of the farm to market fluctuations.

Despite its enormous significance, the agriculture sector has lagged well behind all other sectors in terms of digital transformation and technological integration. The agricultural industry has enormous potential for IoT systems and sensors. The majority of farmers strategically set IoT sensors and cameras in their fields so that they can easily check on their crops with just a click whenever they feel the need to. These photos are sent anywhere in the world for examination, among other uses; however, to extract meaningful information from these images, specialized image recognition technologies must be applied. This allows farmers to always have access to real-time farm data, which is helpful in making better decisions regarding crop nutrition. By enabling farmers to respond swiftly to any outbreak that could harm their crop, technology is helping to reduce the quantity of agricultural products that are wasted annually.

Integration of IoT with Farm Equipment: In a similar manner to the IoT systems and sensors deployed in the fields, farmers can also put IoT sensors on their agricultural machinery.

Drones for Crop Monitoring: Manual observation of crop canopy for health in a small area such as in a homestead garden can be done quickly by the survey of all of the plants and can be determined for management practices such manuring, trimming, watering, etc., but this is not the case with large crop fields. It is difficult for most farmers to monitor the crop all at once or even over an extended period of time because they work on fields that spread over many acres. To know what's happening in such a scenario, there is requirement of some sort of aerial perspective of the crop. With 3D imaging technology, modern drones can also be used for more sophisticated tasks, such as assisting farmers in forecasting soil quality and seeding/planting patterns. Drones are also used in the agriculture sector to spray pesticides and other crop production inputs that are essential for plant growth and development.

Use of Robotics in Farming: Even though artificial intelligence and robots are arguably the most terrifying technical developments now in existence, they have been unparalleled in their ability to assist humanity in overcoming obstacles. Startups are now developing robots led by lasers and cameras that can detect particular kinds of weeds in a crop and eradicate them without the need for human assistance. This entire concept has the potential to be revolutionary since it might completely eliminate the need for the chemicals that were previously used to eradicate weeds. This would lower operating costs while also improving the health and organic content of our crops. Robots are also used in horticulture, particularly in vegetable and fruit farms, where labour-intensive tasks like harvesting fruit from plants require a lot of time and labourers. In an effort to streamline, reduce costs and expedite this process, an increasing number of companies are developing different robotic solutions. Certain robots that are undergoing testing are capable of tasks as delicate as gathering nuts in orchards.

Radio Frequency Identification (RFID) Tracking in Farming: RFID is a technology that tracks and identifies objects or people wirelessly using radio waves. It is made up of two primary parts. (1) RFID tag: a tiny electronic gadget with additional data and a unique identity. (2) RFID reader: this is a gadget that sends out radio waves in order to pick up signals from nearby RFID tags. Like many other cutting-edge technologies, RFID has countless uses in the agricultural industry.

Each has its own advantages; it might involve anything from RFID marking farm products till they are sold to tracking the machinery in the enormous agricultural areas. A work environment as large as a farm increases the likelihood of equipment misplacing more than any other place. RFID tagging and real-time equipment tracking can guarantee that no equipment is ever lost, save thousands of dollars that would have otherwise been spent on new equipment and speed up processes because all the equipment is readily available. Product RFID tagging may also be very helpful in tracing the product's origins.

Machine Learning (ML) and Big Data Applications in Agriculture: These days, every industry is undergoing a digital revolution, and a key component of that change is ML and data analytics. Speaking of industry 4.0 and digital transformation would be incomplete without discussing the applications and advantages of big data analytics and ML. This technique is both incredibly basic and extremely sophisticated; fresh, more optimized procedures are designed using historical manufacturing and market data.

2.8 AVAILABLE AND FUTURE MODELS OF IFS

2.8.1 Traditional Farming System of Hill Agro-Ecosystem

Women are essential to the conservation of biodiversity and the preservation of seeds. The primary crop of the *jhum* system is rice, but it also includes other pulses like rice beans (*Vigna umbellata*), oilseeds like sesame and vegetables like cucumbers, bell peppers and bhindi. The entire land is sown with rice mixed with maize crop. Crops such as millet are planted along the *jhum* land's edge. The harvest of rice and maize is followed by the sowing of root crops like colocasia and local yam. First-year *jhum* land yields are often higher than second-year yields. Nonetheless, the wetland rice farming is unique in the hill agro-ecosystem.

In the valleys, wetland rice culture (WRC) is consistently more prevalent than *jhum* production. For second year *jhum*, first year *jhum*, and WRC, the corresponding output:input ratios are 1.25, 1.50 and 1.75, respectively. Due to their extensive farming experience, farmers in this agro-ecosystem have developed a thorough understanding of the nutritional value and therapeutic qualities of numerous plants. For these tribal people of Northeast India, pseudocereals, small millet, native pulses, oilseeds and several other forest plants are an essential part of their diet. Daily use of the nearby forest products as nutritional supplements continues, and some are also frequently sold in nearby markets. From ancient times, these species have been utilized as both medicines and food to support life. Different native fruit species grown at different periods of the year ensure a year-round source of nourishment in addition to diversifying the food base. The Adi tribes of Arunachal Pradesh's West Siang district follow their own custom of traditional farming, raising a diverse range of crops for the community's use as food, fuel, fodder, fibre, nutrition and medicine. During times of drought, famine and other natural disasters, these underutilized crops and vegetables play a crucial role as a staple sustenance for people. Farmers in hilly areas have a unique method for storing food by applying their traditional knowledge (Sarangi et al. 2009).

2.8.2 Farming System of Coastal Saline Region

This region has climatic hazards, high humidity, saline soil and water, shallow saline water table, tidal water intrusion, deep waterlogging in *Kharif* with poor drainage infrastructure and, inadequate good quality irrigation water resulting in mono-cropping. Rice is the predominant crop grown under rainfed conditions during *Kharif* season and some farmers grow fish in ponds. Catching of fish from open water resources such as rivers, canals and common water bodies are practiced to supplement the household needs. Vegetables are cultivated in small areas both during *Kharif* and during *Rabi* seasons. Summer/*Boro* rice area is increasing in some of the coastal states of India with provision of irrigation facility, either with the introduction of canal irrigation or by withdrawing the ground water. Fruit crops are grown in patches along boundaries of the land or in the homestead area. Small and marginal farmers practice a homestead farming system to meet the family needs with very small or negligible amounts of marketable surplus.

2.8.3 Farming System of Indo-Gangetic Plains

Rice-wheat is the predominant cropping system in the northern Indian conditions and in the Indo-Gangetic region. The major states include Uttar Pradesh, Punjab, Haryana, Madhya Pradesh and parts of West Bengal. Availability of an irrigation facility, fertile soil and congenial weather conditions favor this cropping system. The landholding size is comparatively larger than the rest of India. In recent years due to heavy use of inputs, viz., fertilizers, pesticides and irrigation water, there are environmental issues, poor input use efficiency as well as declining productivity, organic carbon status of soils, lowering of ground water table and residues of chemicals in agricultural produce. Intensive tillage operations such as puddled transplanting of rice and thorough ploughing for wheat sowing involve energy, time and soil-related stresses. Issues include overexploitation of natural resources (land, water and forest), loss of biodiversity, emission of greenhouse gases (GHGs) and global warming. A rice-wheat cropping system consumes about 11,650 m^3 ha^{-1} water out of which 7650 m^3 ha^{-1} is by rice, thus, the water table in some pockets is declining at alarming rates (Bhatt et al. 2016). Large-scale *in-situ* burning of the rice crop residues also results in huge emissions of GHGs, resulting in a decline in soil organic carbon (SOC) sequestration in the R–W system of the region (Singh et al. 2020). Continuous following of this system has also resulted in weed menace, resistance development of weeds to herbicides, formation of hard pan and declining soil fertility status with emergence of multi-nutrient deficiencies. This chemical intensive agriculture is moving away from nature and there is stagnation/unsustainable crop yields due to heavy dependence on chemicals. Therefore, to tackle these issues, a future farming system models involving smart technologies such as crop rotations, residue recycling and conservation tillage practices are essential with integration of compatible enterprises.

2.9 FUTURE FARMING SYSTEM MODELS

The future farming systems have to address several challenges for meeting nutritionally rich sufficient food for ever increasing population as well as meeting environmental aspirations. To meet these needs, agriculture needs sophisticated technological options such as drones, chips, smart phones, satellites, IOTs, GIS, GPS, robotics, sensors, gene-edited crops/varieties, etc. The key drivers of future farming system models are given below (Table 2.1).

Based on risk factors, climatic, market, social and other considerations, such as food habit changes, profitability, land use changes, etc. aspects of future farming system models for major agro-ecological regions are discussed below.

2.9.1 Coastal Integrated Farming System Models

Rice based system: Rice is cultivated as rainfed/irrigated in the low-lying lands concentrated in Odisha, West Bengal, Andhra Pradesh, Telangana, Tamil Nadu, Karnataka and Kerala in India. Coastal zone of Bangladesh includes Jessore, Narrail, Gopalganj, Shariatpur, Chandpur, Satkhira, Khulna, Bagerhat, Pirozpur, Jhalakati, Barguna, Barisal, Patuakhali, Bhola, Lakshmipur, Noakhali, Feni, Chittagong and Cox's Bazar. Southern Sri Lanka covers Colombo, Gampaha, Kalutara, Galle, Matara and Puttalam. In this system, cattle and small ruminants such as goats are integrated. This system provides rice, vegetables and legumes as livelihoods. The major constraints are excess water during the rainy season and water scarcity during the dry season, soil salinity and fertility issues and limited fodder production. These problems can be managed by practicing paddy-cum-freshwater fish system in the low land with provision of rainwater harvesting, raising

TABLE 2.1
Drivers of Future Farming System Models

Drivers	Cause/Effect
Increased food demand	Population and income growth
Diverse food needs	Change in food habits, health consciousness, demand for millet (nutri-cereals), fruits, vegetables, dairy products, fish and meat.
Change in farm size	Fragmentation, consolidation, industrialization, degradation
Improved technological options	Farm mechanization, reduction in labour availability, efficiency in input use
Altered crops/varietal requirement	Climate change, biotic and abiotic stresses
Water scarcity	Water quality degradation, change in rainfall pattern, competitive use of water for industry/demand for other uses
Soil health	Salinization, chemical use, soil erosion, deforestation
Innovations	Anticipated challenges
Credits, insurances, contracts	Government interventions/policy changes, weather aberrations, floods, draught, tsunami
Post-harvest processing, value additions	Diverse food needs, production losses, gluts, export, import
Market changes	FPOs/SHGs/Start ups
Climate changes, off farm production	Protected cultivation, green/glass house cultivation
Competition for farming	Involvement of private sectors for innovative products, better seeds, fertilizers, plant protection chemicals, customized farm machinery, feed for animals
Reduction in land/cultivated area	Soilless cultivation, e.g., hydroponics, aeroponics, aquaponics, strawponics, rooftop farming, sea water farming, cage culture, 3D farming, Biofloc etc.
Enhancement of food quality, safety, efficiency in input use	Nano-technology, precision farming, fertilizer/input responsive crop varieties, dual culture, e.g., Rice+fish
Digital retailing	Time factor, effectiveness, consumer satisfaction, transit
UN's SDGs: Zero hunger, nutrition, climate action	Increasing hungry population, pandemics, malnutrition
Climate change	Reduction in crop duration, changes in photosynthesis, crop respiration, pest dynamics, diseases, soil microbes, SOM, soil erosion.

the dikes for cultivation of fruits and vegetables and integration of dairy and vermicomposting (Figure 2.1).

Coastal artisanal fishing: Fishery is one of the major avocations in the coastal region of India, Bangladesh and Maldives. Fishery is supplemented by rice, coconut, livestock and vegetables. This system provides livelihoods in the form of fish, coconut, rice, legumes and livestock. Constraints include storms, cyclones, waterlogging, soil salinity, poor fertility, sea water intrusion and sea level rise.

Jute-based system: This is predominant in the lower Gangetic plains/delta with rich loam and riverine silt deposits on soil. The best composition of soil and mineral structure make it suitable for growing jute. India and Bangladesh are the major jute-producing countries in the world. Jute in this region is mostly planted during March–May (pre-*Kharif*) depending upon the jute species. After jute, during *Kharif* rice is grown followed by Rabi crops such as potato, mustard, wheat, pulses and vegetables. Jute-wheat, jute-rice-sunflower, jute-mustard, jute-rapeseed-rice systems are also practiced in some of the areas. Apart from a jute-based system, other double/triple cropping systems of this region are rice-wheat, rice-potato, rice-mustard, sesamum-groundnut-sunflower, pulses-wheat, rice-vegetables, vegetables-vegetables-vegetables, rice-wheat-summer rice, etc. Thus, intensive triple cropping is possible in this agro-ecological region. However, there are constraints

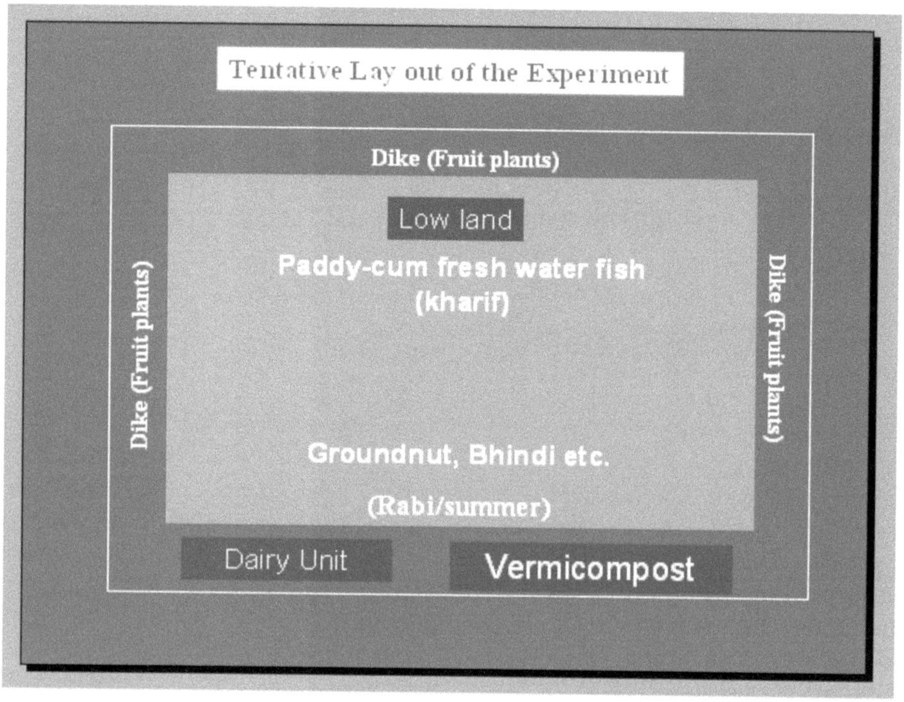

FIGURE 2.1 Farm layout for waterlogged coastal land.

of water availability for cultivation and retting of jute, the adverse impact of climate change and market-related issues.

2.9.2 Resilient Fishery-Based Farming System Models for the Coastal Region

The coastal region is climatically vulnerable due to nearness to the sea, low-lying topography and high population density. Under the influence of climate change there will be heavy intensity rain, resulting in flooding of cultivated land. Under this situation, well planned fishery-based component should be an integral part of the IFS. The capacity of the water harvesting structure should be designed to capture the anticipated high-intensity rain. There should be provision of net fencing to enclose the fish resources within the desired farm area. Floating cage culture fishery may be planned in the water body along with dwelling house. Floating houses with provision of fishery below may be planned on a scientific basis.

The floating houses are made up of locally available materials such as water-resistant wood, bamboo, polyvinyl chloride (PVC) pipes, polythene sheets, etc. Below the floating house, nets are fenced (Figure 2.2) for fish culture in enclosure, with proper water circulation. Bamboo/PVC pipes and tyres keep the house afloat. Inside the house, large trap doors provide access from the raft's platform floors to the cages below to feed the fish and harvest the fish. Fish species such as *Puntius, Oxyeleotris, Pangasius,* etc. may be cultivated under this system. These fish species make their own fish feed from agricultural waste and small fish.

2.9.3 Float Farming and IFS Model for Coastal Saline Region

Floating agriculture gardens are an integration of aquaculture and crop production in the waterlogged/swampy areas. Floating organic beds prepared over the water by the use of plant products, such as banana pseudostem, bamboo, water hyacinth, etc. over this soil and organic matter are

FIGURE 2.2 Floating house with cage culture fishery – a climate resilient system.

laid to grow vegetables. This an adaptation strategy to climate change in flood vulnerable coastal regions. With an increase in the water level, the floating beds are not affected and very short duration vegetables, such as multicut greens, amaranthus, coriander, etc., are grown depending upon the season. Upon drying of the water body, the beds sink and touch the muddy soil, which provides an opportunity to grow new vegetables to cover the ground.

The IFS model for a coastal region should be location specific with improved salt and submergence tolerant crop variety, modern irrigation methods, management practices that encourage better utilization of natural resources, multiple use of water, complementary use of land, recycling of byproducts, reduction in use of chemicals, reduction in salinity build up, reduction of migration, employment opportunity in farm, higher income and reduction of risk through diversification of crops/enterprises. Livestock, pisciculture and duckery are the most important components of an IFS model for coastal saline region (Figure 2.3).

2.9.4 Integrated Agri-horticulture (Orchard) with Short and Sturdy Varieties

Winds, storms and hail significantly affect the crop and defoliate and damage the vegetation, particularly tall plants/trees. However, with the availability of short statured varieties with strong stems, lodging tolerant species should be considered for this ecologically fragile region. Coconut is an important horticultural crop providing significant income to farm families in the coastal region. In the future IFS system, farmers should go for dwarf varieties to avoid lodging. Coconut trees upon lodging also cause damage to nearby infrastructure; however, the recently released dwarf varieties, viz., Malaysian Dwarf, Chavakadan Dwarf, Ganga Bondam, D×T and Sunnaki Dwarf not only reduce the chance of lodging but also pose less risk to the nearby infrastructure.

2.9.5 Local and Dwarf Animal Breeds to Tolerate Weather Extremes

Local breeds have greater tolerance to weather fluctuations and extremities, as a result they survive better. Conservation of native cattle breeds is essential to meet climate resilience. Indigenous dairy breeds have disease tolerance, better quality of milk and larger numbers of calving. The cost of management of hybrid cows are high; therefore, it involves higher risk under changing climate. Goat and sheep may be preferred over swine and cattle in hot climates.

2.10 FUTURE IFS MODEL FOR DRY AND ARIS REGION

The constraints of scarcity of soil moisture, wind erosion, erratic rainfall resulting in drought and extreme heat, necessitates a well-planned IFS with soil and water conservation, stress tolerant crop

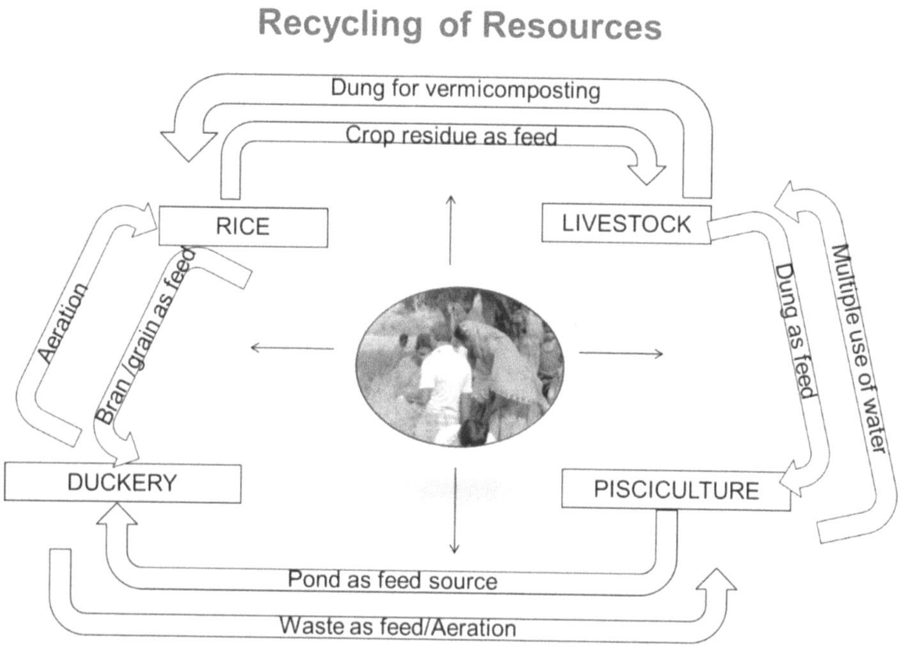

FIGURE 2.3 Recycling of resources in an IFS model in coastal saline region.

varieties, solar energy use and checking evaporative loss of soil water as well as agronomical, mechanical and engineering measures to restrict soil erosion. Livestock are an integral component of arid region farming with agroforestry. The limited amount of rainfall of this region should be captured *in-situ* and harvested for further use. Rooftop rain harvesting helps in reducing the runoff as well as restrict soil and nutrient erosion. Land slopes may be managed by terracing to reduce runoff velocity. Improve irrigation water application through a drip/sprinkler system for crop production to increase water productivity. Inclusion of commercial crops such as henna (*Lawsonia inermis*), cumin (*Cuminum cyminum*), psyllium (*Plantago ovata*), etc. improves income. For the agri-silvicultural system, trees such as *Prosopis cineraria*, *Hardwickia binata*, *Zizyphus* sp., *Albezzia lebbek*, *Cassia siamia*, *Emblica officinalis* and *Moringa oleifera* may be included. Strip cropping under an agri-pastural system reduces soil erosion. Horticulture is an important component of arid region farming. Ber, karonda, kair, cactus, bael, aonla, khejari, neem, datepalm, pomegranate, guava, khair and jojoba are potential crops to be integrated in an IFS of this region. Cactus has a special type of photosynthetic mechanism called Crassulacean Acid Metabolism (CAM), so that the stomata remain closed during the daytime to avoid water loss. These plants capture carbon dioxide at night. The cultivated species *Opuntia ficus indica* (var. Gigante, Redonda and Clone IPA-20) and *Nopalea cochenillifera* (var. Doce) have high forage biomass yield (220 t ha^{-1} green biomass in two years) and high-water productivity of 100–150 kg of water per kg dry matter (Edvan et al. 2020).

Khejri (*Prosopis cineraria* L.) is a multipurpose (food, fodder, timber, fuel, etc.) tree for the arid region. It is an N-fixing plant belonging to Fabaceae and has deep roots adapting to dry climates. The tender pods (sangari) are used as vegetables. The agri-silvi-horti system of Khejri+guava+cluster bean-barley could result in a net return of ₹76,650 ha^{-1} significantly higher than traditional cropping of cluster bean-barley with a net return of ₹15,953 ha^{-1} (Kaushik et al. 2017).

Jojoba (*Simmondsia chinensis*) is a perennial heliophytic shrub of the family Simmondsiaceae. This plant has a special quality for drought and heat tolerance. It grows as a bush and may attain a maximum 6 m height. The oil has a great commercial importance. It is propagated both by seeds as well as by cuttings. This crop can be grown in the dry regions of Rajasthan, Haryana, Punjab, Uttar

Pradesh, Gujarat, Odisha and Andhra Pradesh. The oil is used in cosmetics and skin care products, as lubricants and for medicinal uses, etc. During the initial sapling growing period, intercropping of field crops such as barley, sorghum, alfalfa and safflower may be taken up to get continuous income.

Khair (*Acacia catechu*) is a multipurpose tree, which provides fodder, fuel and katha, used in Ayurvedic preparations and Khair is edible along with betel leaf and nut. Timber is used for preparation of furniture, household articles and agricultural implements. The leaves/lopping are very good fodder for camels and goats. Dried small branches are used as fuel. Lac insects can be grown on this plant for production of lac. Katha and cutch have great medicinal value are products of this plant. It is a hardy tree having drought tolerance. It is propagated from seeds and vegetatively from cuttings (coppicing). It is a thorny tree and can attain a maximum height of about 15 m. Khair tree-based farming system involving field crops (pigeon pea, linseed, sesamum) and cattle may be adopted in this region (Kabir et al. 2016).

Mesquite/Kikar (*Prosopis juliflora*) is a thorny shrub grows to a height of 12 m belong to the Fabaceae family. It is a phreatophytic, shade tolerant, fuel wood-, timber- and forage-yielding plant. It can be grown for phytoremediation of contaminated soils as it is tolerant to drought, salinity, alkalinity and waterlogging. However, it has been reported as an invasive species, making livestock rearing extremely difficult (Mehari 2015).

2.11 IFS MODEL FOR EASTERN REGION OF INDIA

The eastern part of India covers about 22% of the geographical area of India with more than 30% of human and bovine population. Natural calamities like cyclones, floods and droughts occur frequently. Farmers are marginal with 83% rural households. Holding size is small, scattered with monocropping the predominant practice. Soils are poor and problematic. Therefore, most of the people have low income (Table 2.2) and there is concern for food and nutritional security. There is low seed replacement particularly in pulses and millet. Farm mechanization is less, climate change impacts agriculture significantly. Very high numbers of non-descript livestock population density (230 nos. km^{-2}) with low productivity (0.8–1.9 kg day^{-1} milk). Out of 150 economically most backward districts, 69 districts are confined in the eastern region of India. Poverty is a major issue with landless and marginal rural folk and with migration of male farmers, land degradation and poor extension services.

The IFS models for the eastern region of India are crop and fish integration with various combinations of horticulture (banana), cattle, buffalo, goat, duck, poultry, pig, etc. In the tribal population

TABLE 2.2
Population Density (numbers km^{-2}), Per Capita Income, Number and Size of Operational Holdings in Eastern India

State	Population Density (nos. km^{-2})	Per Capita Income (₹)	Number of Operational Holdings (million)	Size of Operational Holdings (ha)
Assam	398	60952	2.72	1.10
Bihar	1106	34168	16.19	0.39
Chhattisgarh	189	84767	3.75	1.36
Jharkhand	414	62816	2071	1.17
Odisha	270	68293	4.67	1.04
Eastern UP	931	48520	10.03	0.60
West Bengal	1028	78903	7.12	0.77
Eastern Region	619	–	47.19	0.72
India	325	94130	138.35	1.15

Source: Agricultural Census, 2010–11.

dominated areas, IFS such as rice+vegetables+fruit trees+duck+vermicompost+ hedge row species, rice+vegetables+mushroom+vermicompost+hedge row species, ricefruit trees+vegetables+hedge row species+apiculture, rice+fruit trees+backyard poultry+apiculture+hedge row species, fingermillet+vegetables+goat+mushroom, rice+fruit trees+vegetables+piggery, rice+finger millet+vegetables+piggery+hedge row species are recommended. In some of the areas, makhana (*Euryale ferox*) based IFS is recommended to promote makhana as well as water chestnut (*Trapa bispinosa*) cultivation in lowlying fields/shallow ponds. In makhana, the dried seeds called fox nut are edible and have market value. Water chestnut has medicinal properties. Integration of fish+makhana+water chestnut is possible in a low-lying agroecosystem, where water chestnut is taken as a tertiary crop. For landless farmers avenues such as backyard poultry, goat, milch cattle/buffalo are to be taken up in a scientific basis.

2.12 IFS FOR HILL AGRO-ECOSYSTEM

The estimated area and population of the hill states and the designated hill areas in India is 6.91 lakh square km and 62.16 million, respectively, covering 35% of the total geographical area of the country and the majority of these areas have slopes above 15% (Barah 2010). In this region, soil erosion due to slopy land and high rainfall is the major constraint. Special IFS models are required to address this constraint and meet the livelihood requirement and business models for location specific high-value crops.

Hill IFS models are based on a micro-watershed farming system models with soil and water conservation measures and maintaining the natural forest at the top portion of the hills. The various models are systems based on dairy, mixed forest, silvi-pastoral, agro-pastoral, agri-horti-silvi-pastoral, silvi-horticultural, natural forest and timber. In a dairy-based system, planned land use is followed in terraces and terrace risers to provide fodder to the animals. In terraces, annual fodder crops (maize, cowpea, rice bean, oat, etc.) are grown and in raisers, guinea (*Panicum maximum*) and broom (*Thysanolaena latifolia*) grasses are grown. Broom grass not only provides fodder but also the inflorescence has economical value as it is used for the manufacture of broom. In this system, a vermicompost unit is made at an appropriate place, preferably near the dairy unit. Fodder production is planned for year-round supply to dairy cattle. During the kharif season (June–November), annual fodder crops and grasses; during December–March, broom grass during April–May, stover of crops, guinea grass may meet the requirements. Holstein Friesian cows may be integrated for milk production by feeding crop residues, straw, fodder of cowpea, hybrid napier (*Pennisetum purpureum* X *Pennisetum typhoides*) grass. The dung and refuge of cow are used as compost/vermicompost for crop nutrition. Dairy animal provides daily income to the farmer by the sale of milk.

For the agro-pastoral system, creation of terraces in the mid and bottom of the hills provides the cultivated area both in the terrace plains and in the raisers. The plain terraced area is used for crop cultivation and risers are used for cultivating perennial fodder crops such as guinea grass, elephant grass (*Pennisetum purpureum*), broom grass, etc. These grasses provide fodder as well as prevent soil erosion due to the soil binding nature of the roots. Field crops and spices are grown year-round from which stover or byproduct also supplements the fodder requirement of dairy cattle, besides the pasture.

Agri-horti-silvi-pastoral system provides multiple products, including timber from forest trees such as *Alnus, Schima*; *Ficus* and *Symingtonia* produce green fodder. Broom and guinea grass in the interspaces of trees provide green fodder. Horticultural crops such as pineapple, guava, chilli, papaya, brinjal, tomato, bhindi, broccoli, cauliflower, capsicum, pea and cabbage may be grown on the terraces. Legume crops (French bean, cow pea) should be included in the system. Animals such as pigs, cows and goats are suitable. In the top natural forest area, black pepper may be integrated. Spice crop-based cropping systems include turmeric, ginger, cardamom. Intercrops such as cucumber, bottle gourd, sponge gourd and squash may be taken up in turmeric crop. Vertical cropping involving turmeric+bottlegourd/squash/cucumber/sponge gourd with turmeric at the bottom, and other crops as climbers over turmeric may be taken up for effective utilization of space.

Ray et al (2019) reported that IFS models in NEH region could provide a net income of ₹2,46,160 per year with highest contribution of income from piggery (31.25%), followed by poultry (25.34%), vegetables (12.47%), mushroom (9.87%), spices (8.1%), field crops (6.2%), fishery (4.05%) and fruits (2.72%).

2.13 AGRO-ECOTOURISM-BASED (AET) IFS

Agro-ecotourism is an important aspect in devising the hill farming systems. The landscape of hill farming is quite attractive; however, it has great potential to integrate and plan in the scheme of the farming systems. A portion of the land is to be developed for accommodation and recreation of tourists in an eco-friendly manner. Here the organic components of the hill region not only offer a point of visit but also provide exposure for the potential buyers and creation of awareness. Organic turmeric and ginger of Meghalaya, cardamom of Sikkim, pineapple of Tripura, tea gardens of Assam and Darjeeling area, bamboo, black turmeric of Arunachal Pradesh need to be popularized through the IFS models.

The AET-based IFS may have components such as an environmental area (arboretum, vegetable garden, Indigofera garden, herbal garden), recreation and sports area (bicycle tracks, outbound facilities, sports field, water recreation), livestock area with animals and birds, processing area, manure production and waste processing area, etc. AET-based IFS provides employment and independent communities could increase their land resources and farmers would prosper in a sustainable manner (Indrawati et al. 2020).

2.14 EXPORT-ORIENTED HIGH-VALUE CROPS AND OFF-SEASON VEGETABLES-BASED IFS

The hill and mountain ecosystem has the special potential in terms of producing high-value crops such as passion fruits, kiwi, pachauli, lemon grass, citronella, saffron, squash, capsicum, broccoli, etc. For cultivation of off-season vegetables specially designed poly houses, compatible crops and varieties, good quality seeds/planting materials and technology are required. An organic way of production will attract additional benefits. Special agroforestry trees such as the Rudraksha (*Elaeocarpus ganitrus* Roxb.) known as blue-marble tree or the Utracham bead tree has commercial value also it requires conservation measures. In Indonesia, people of all religions wear Rudraksha for health purposes (Sharma et al. 2015).

Strawberry (*Fragaria vesca*) is another high value crop suitable for growing in the hilly region. Availability of well-drained soil and plenty of direct sunshine make this ecosystem suitable for strawberry growing. Varieties such as Chandler, Tioga, Torrey, Selva, Belrubi, Fern, Pajaro, Premier, Red Cost, Local Jeolikot, Dilpasand, Bangalore, Florida 90, Katrain Sweet, Pusa Early Dwarf, Blakemore, Etna, Sweet Charlie, Douglas, Confictura, Dana, Gorella, Addie, Majestic, Phenomenal and Brighton are suitable. Mulching either with crop residues (paddy straw) or with polyethylene suppresses weeds and conserves soil moisture. The net profit from one bigha (one-fifth of an acre) area may result in ₹4.185 lakhs per annum in hills (Table 2.3) with the cost of cultivation of ₹2.72 lakhs per bigha (Pramanick et al. 2013).

Millet are the neglected crops, but have high nutritional value, particularly rich in dietary fibre, minerals and vitamins. These crops (Table 2.4) should find significant place in the future hills farming system models (Sarangi 2009). These are the nutri-cereals having adaptation capacity to climate change, tolerance to drought, low water requirements and are photo-insensitive crops. These crops have relevance in addressing lifestyle diseases of human beings such as diabetes, obesity, heart diseases, cancer, anaemia and hypertension (Rao et al. 2017).

TABLE 2.3
Economics of Strawberry Cultivation in the Hills for 1 bigha (800 m²) Area

Capital Costs	**Amount in INR (₹)**
Cost of drip irrigation system	10,000.00
Cost of 300 L refrigerator	30,000.00
Cost of plastics and infrastructure for tunnels	25,000.00
Total capital cost	65,000.00
Recurring costs (over 6-year period)	
Cost of strawberry plants in 1st and 4th year @ ₹2 plant^{-1}	20,000.00
Field preparation (twice)	5,000.00
Cost of chemicals, manures, fertilizers etc.	10,500.00
Labour charges @ ₹2000 p.m.	1,50,000.00
Marketing cost	21,000.00
Total recurring cost	2,06,500.00
Total (Capital + Recurring)	2,71,500.00
Projected Returns	
Description	**Amount in INR (₹)**
Strawberry fruit yield	
1st year @ 200 g plant^{-1}, ₹60 kg^{-1}	60,000.00
2nd year @ 300 g plant^{-1}, ₹60 kg^{-1}	90,000.00
3rd year @ 200 g plant^{-1}, ₹60 kg^{-1}	60,000.00
4th year @ 200 g plant^{-1}, ₹60 kg^{-1}	60,000.00
5th year @ 300 g plant^{-1}, ₹60 kg^{-1}	90,000.00
6th year @ 200 g plant^{-1}, ₹60 kg^{-1}	60,000.00
Sale of strawberry plants	2,70,000.00
Total receipts	6,90,000.00
Net profit @ 6th year in the hills per 800 m² area	4,18,500.00

Source: Pramanick et al. 2013

Large cardamom (*Amomum subulatum* Roxburgh) is a cash crop cultivated in Arunachal Pradesh, Sikkim, Nagaland, Uttarakahand and the Darjeeling hills of West Bengal. Sikkim is the leading state for large cardamom cultivation in India. Tradtional cultivars, e.g., Seremna, Dzongu-golsa and Bharlangey are disease-tolerant and high yielding (Sharma et al. 2016). The major adaptive measures to cope with the impacts of climate change and revive cardamom cultivation are applying traditional knowledge, improved or disease tolerant cultivars, replanting on new farm land, bringing cardamom into home gardens, using manure to maintain soil fertility, provision of irrigation during dry seasons, and managing diseases and pests by uprooting, drying and burning infected plants (Sharma et al. 2016). Harvesting of rainwater, its conservation, *in-situ* soil moisture conservation will help in overcoming the issue of dry spells and erratic rainfall.

2.15 RAISED AND SUNKEN BED-BASED IFS MODEL TO ADDRESS WATERLOGGING AND UNSEASONAL HIGH RAINFALL IN VARIOUS AGRO-ECOSYSTEMS

To adapt to the future problems of high intensity rain, seasonal/perennial waterlogging issues, water scarcity and soil degradation, raised and sunken bed technology holds promise in diverse agro-ecosystems. This land modification-based technology addresses the issues of waterlogging,

TABLE 2.4
List of nutri-cereals suitable for hill agroecosystems

Sl. No.	Common Name	Scientific Name
1	Sorghum	*Sorghum bicolor* (L.) Moench
2	Bajra/Pearl millet	*Pennisetum glaucum* (L.) R. Br.
3	Finger millet	*Eleusine coracana* (L.) Gaertn
4	Proso millet	*Panicum miliaceum* (L.)
5	Kodo millet	*Paspalum scrobiculatum* (L.)
6	Foxtail millet	*Setaria italica* (L.) P. Beauvois
7	Barnyard millet	*Echinochloa crusgalli* (L.) P. Beauvois
8	Little millet	*Panicum sumatrense*

salinity/sodicity, declining watertables, etc. through provision of drainage and water conservation. Soil waterlogging occurs when excess water saturates the soil pores with either no layer of water or a very fine one on the soil surface (Arduini et al. 2019). In the global assessment of soil degradation (GLASOD) estimate, waterlogging affects 4.6 million ha area, largely in the irrigated areas of India and Pakistan; 4.2 million ha affected by salinization, nearly all in the dry zone (FAO 1994). It has been estimated that out of 14.06 million ha culturable command area of 42 canal irrigation projects of India, about 1.6 million ha (about 11%) suffer suffering from waterlogging (Verma et al. 2016). Estimates of salt affected area in India have been reported as high as 6.73 million ha (Mondal et al. 2010). In tubewell irrigated areas, due to overexploitation of water, in excess of the rates of recharge, the ground water table has been progressively lowered (FAO 1994). To address all the above issues, a raised and sunken bed technology has been recommended (Table 2.5).

2.16 MULTI-ENTERPRISE MODEL FOR PARTIALLY RECLAIMED SODIC SOILS OF NORTH INDIA

Multi-enterprise agriculture model in 1 ha land, involving fisheries, dairy, horticulture, poultry, duckery and mushroom cultivation resulted in a total gross income of ₹400–700 day^{-1} and net income of ₹250–500 day^{-1} (Singh et al. 2009). The average net income from crop and subsidiary components together was ₹34,85,595 out of which ₹72,020 came from cereal crops (including fodder), ₹35,880 from vegetables and fruits, and ₹1,95,650 from subsidiary components, from an area of 2 ha, which was substantially higher than a conventional rice-wheat cropping system (RWS) of ₹3,02,250 (Sharma 2021). In this enterprise mix diversification system, the BCR of rice-wheat-moong, maize-wheat-moong, dairy enterprise and the model as a whole was 3.54, 4.20, 1.78 and 1.48, respectively (CSSRI Annual Report 2019). The multi-enterprise model involving grain, vegetables, horticulture, fodder, fisheries, livestock and poultry enterprises produced about 20% higher (13.3 t ha^{-1}) rice equivalent yield (REY) than RWS (11.1 t ha^{-1}), generated 15.3% higher average (of eight years 2007–2015) net income ($ 5363) than RWS ($ 4650) and 875 man-days/annum employment and the system used about 50% lesser water than RWS and also improved soil health (Yadav et al. 2021).

2.17 BANANA-BASED IFS MODEL

The banana plant has multiple uses; all its parts are useful for the farming family as well as for the farm, making it a best suitable crop for an important component of IFS. India is the pioneer producer of banana with contribution of about 29% of global banana production. It contributes about 37% to total fruit production in India. This is an important crop in Maharashtra in terms of production and popular in the south Indian states of Tamil Nadu, Karnataka, Andhra Pradesh, Telangana, Kerala,

TABLE 2.5
Raised and Sunken Bed Technology for Diverse Agro-Ecological Regions of India

Sl. No.	Dimensions	Agro-Ecological Region and Benefits	Reference
1	Alternate sunken and raised beds (1:1) each of 30 m length and 5 m width	Coastal Odisha. Additional income per ha per year ₹70,000	Indian Institute of Water Management 2014
2	Alternate ridges (1.5 m top width × 1.0 m height × 3.0 m bottom width) and furrows (3.0 m top width × 1.5 m bottom width × 1.0 m depth)	Coastal West Bengal. Furrow for fish+rain water harvesting. Ridges for vegetable cultivation	Bandyopadhyay et al. 2009
3	Beds of 4–5 m width and 1 m height and furrow of 5–6 m width and 1 m deep with a provision of (2 m × 4 m × 1 m) fish shelter at the end of the furrow	*Tsunami* affected areas of Andaman & Nicobar islands. Raised beds for vegetable cultivation round the year. Furrows for fish cultivation. *In-situ* rain water harvesting of about 3800 m^3 ha^{-1}.	CSSRI, NAIP 2014
4	Raised beds of 6 m width and 0.30 to 0.35 m in height. Sunken beds of 6 m width running in parallel to the raised beds.	Vertisols of Central India. Soybean in raised beds and paddy in sunken beds in rainy season. In post rainy season, chickpea, linseed, safflower in raised beds and wheat in sunken beds.	Tomar et al. 1996
5	Top width of raised beds 2.0 m and bottom width 4.0 m, length 60.0 m, height 1–2 m. Width of sunken beds 7.0 m, length 60.0 m. Depth of sunken beds for water chestnut cultivation 0.50 m, for fish fingerling raising 0.75 m to 1.0 m and for fish stocking 1.0–1.5 m. Side slopes of raised and boundary beds 1:1.	Waterlogged sodic soils of North India. Raised beds for cultivation of vegetables. Sunken area for water harvesting and fish rearing. Fruits (guava, lemon, mango) and trees (eucalyptus, acacia, neem, teak) on peripheral bunds of raised and sunken bed system.	Verma et al. 2016
6	Raised and sunken beds of 1 m width each (1:1). Height of raised beds 40–60 cm, made by cutting soil from sunken bed area, 50% runoff water from raised beds drain into sunken bed for interplot water harvesting. Beds are permanent for 4–5 years.	North Eastern Hill region. Vegetables in raised beds round the year. Rice followed by lentil/pea in sunken beds. Net return from the system US $ 1741 ha^{-1} compared to US $ 383 ha^{-1} under farmer's practice of rice monocropping.	Das et al. 2015
7	Raised beds of 60 cm wide at the base, 30 cm wide at top, with mid furrows 60 cm apart and furrow depth of 20 cm.	Coastal low-lying salt affected land. Maize grown on raised beds during *Rabi* season after *Kharif* rice. Raised bed sowing of maize resulted system benefit ratio of 1.83 compared to 1.48 and 1.63 under zero and conventional tillage respectively	Sarangi et al. 2020

the northeastern states of Assam, Arunachal Pradesh, Tripura and the western state of Gujarat and eastern state of Odisha. Being a wide-spaced (1.8–2.0 m) crop, it provides ample scope for intercropping as well as integration of other enterprises such as mushroom production *in-situ*. Diverse varieties such as Poovan, Monthan, Saba, Rasthali, Karpooravalli, Kalki, Uthayam, Nendran and Cavendish are available for different farm situations and uses. Dwarf Cavendish (G 9) variety with height of plant about 6 feet is suitable for areas prone to natural calamities.

Banana varieties in India are also have Geographical Indication (GI) tags. Mysuru and Chamarajnagar area of Karnataka obtained registration in 2005 for GI tag 35 for the Nanjanagudu Rasabali banana and Kalaburgi area of Karnataka for the Kamalapur Red banana (GI tat 133), while the Dindigul district of Tamil Nadu has a GI tag 124 for Virupakshi hill banana. Apart from its valuable fruits, a banana pseudostem can be used as a raw material for vermicompost production by the use of cow dung from dairy cattle enterprises and used for soil fertility buildup. Banana leaves are utilized as fish feed for herbivorous fish such as grass crap, as a cattle feed and for household uses like serving plates and wrapping material for snacks. The banana flower is a nutritious vegetable rich in phosphorus, iron and calcium. Fruits apart from consumption as raw or ripe and have a great value as processed food products such as chips, squash, pickles, jam, sauce, etc.

Banana crop can be intercropped with coconut and arecanut plants in coastal areas. In the banana based multi-tier cropping, vegetable crops such as beans, tomato, chilli, okra, gourds in the first tier, drum stick plants in the second tier followed by arecanut/coconut in the third tier intercropped with banana crop. In the boundary of the third tier fodder crops such as guinea grass, cowpea, glyricidia are planted with provision of vermicomposting unit. In Kerala, the banana crop is integrated with dairy and fishery units. The fodder crops are integrated to the dairy cattle, and manure is used for growing the crops. In Assam, North Eastern Hill (NEH) region and the Andaman and Nicobar Islands, other fruit crops such as lemon, orange, pineapple and arecanut are also intercropped with banana and integrated with fishery, apiary and sericulture. Agri-horti banana-based IFS model of 0.7 ha area involving cropping system in 0.38 ha (rice-wheat+mustard/chickpea-dhaincha, okra-cauliflower-baby corn+cowpea, sesbansia seed production-chickpea-green gram), horti-pasture in 0.18 ha (kinnow+fodder crops), agri-horti in 0.04 ha (banana+field crops), integrated with dairy (one buffalo-Murrah and two cows-Gir) with boundary plantation of Karonda (Pink strain) and guava (Allahabad Safeda) and vermicomposting (50 m^2) provided a gross return of ₹3,44,785 and net return of ₹1,81,324 in which agri-horti system has 22% of net income of the model (Kashyap et al. 2021).

2.18 ROOFTOP FARMING

In urban areas the rooftops of buildings, terraces and window sheds can be efficiently utilized for agriculture with the use of a growing medium, manure mixed soil or by hydroponics. It has advantages of harvesting and utilizing the rainwater, reducing the heating effects on the houses and providing fresh vegetables/flowers to the family. Being grown on harvested rainwater and with self-monitored biological and herbal pest management measures, the products are less contaminated. Greenhouses can be built to increase yield, water productivity and profitability. The hydroponics method of cultivation may be followed as a medium of production.

Ecofriendly management measures to reduce the effect of insects or diseases on the crops (mostly vegetables and ornamental flowers) grown on rooftops are available. Organic sources of nutrients such as oilcakes, vermicompost and manures are better used than synthetic fertilizers.

2.19 CONCLUSION

Sustainable integrated farming systems offer a promising approach to improving food and nutrition security, while also reducing environmental impact and promoting social and economic well-being. However, implementing SIFS can be challenging and requires a multi-stakeholder approach, including partnerships between farmers, researchers and policymakers. Investing in research and development and promoting policies and incentives that support the adoption of SIFS will be critical to their success. Implementing SIFS can be challenging, particularly in the face of competing demands for land and resources. However, there are a number of opportunities to overcome these challenges, including building partnerships between farmers, researchers and policymakers to promote the adoption of SIFS; investing in research and development to improve the understanding of SIFS and develop appropriate technologies; promoting policies and incentives that support the adoption of SIFS, such as agroforestry and sustainable land management; and developing value chains that allow farmers to access new markets and increase their income.

REFERENCES

Acaroglu, M., Aksoy, A.S. 2005. The cultivation and energy balance of *Miscanthus giganteus* production in Turkey. *Biomass and Bioenergy* 29(1):42–48. DOI: https://doi.org/10.1016/j.biombioe.2005.01.002.

Ahmed, N., Turchini, G.M. 2021. The evolution of the blue-green revolution of rice-fish cultivation for sustainable food production. *Sustainability Science* 16:1375–1390. DOI: https://doi.org/10.1007/s11625-021-00924-z.

Ahmed, N., Ward, J.D., Saint, C.P. 2014. Can integrated aquaculture-agriculture (IAA) produce "more crop per drop"? *Food Security* 6:767–779. DOI: https://doi.org/10.1007/s12571-014-0394-9.

Allen, L.H. 2003. Interventions for micronutrient deficiency control in developing countries: Past, present and future. *Journal of Nutrition* 133(11 Suppl 2):3875S–3878S. DOI: https://doi.org/10.1093/jn/133.11.3875S. PMID: 14672284.

Arduini, I., Kokubun, M., Licausi, F. 2019. Crop response to waterlogging. *Frontiers in Plant Science* 10:1578. DOI: https://doi.org/10.3389/fpls.2019.01578.

Bandyopadhyay, B.K., Burman, D., Sarangi, S.K., Mandal, S., Bal, A.R. 2009. Land shaping techniques to alleviate salinity and waterlogging problems of mono-cropped coastal land for multi-crop cultivation. *Journal of the Indian Society of Coastal Agricultural Research* 27(1):13–17.

Barah, B.C. 2010. Hill agriculture: Problems and prospects for mountain agriculture. *Indian Journal of Agricultural Economics* 65(3):584–601.

Behera, U.K., France, J. 2016. Integrated farming systems and the livelihood security of small and marginal farmers in India and other developing countries. *Advances in Agronomy* 138:235–282.

Behera, U.K., Mahapatra, I.C. 1999. Income and employment generation for small and marginal farmers through integrated farming systems. *Indian Journal of Agronomy* 44(3):431–439.

Behera, U.K., Kaechele, H., France, J. 2014. Integrated animal and cropping systems in single and multi-objective frameworks for enhancing the livelihood security of farmers and agricultural sustainability in Northern India. *Animal Production Science* 55(10):1338–1346. DOI: http://dx.doi.org/10.1071/AN14526.

Bhatt, R., Kukal, S.S., Busari, M.A., Arora, S., Yadav, M. 2016. Sustainability issues on rice–wheat cropping system. *International Soil and Water Conservation Research* 4(1):64–74. DOI: https://doi.org/10.1016/j.iswcr.2015.12.001.

Calilung, E.J. 1985. *Comparison of Land Preparation Equipment Used on Small Rice Farms in the Philippines*. Thesis No. AE-85-2. Asian Institute of Technology, Bangkok, Thailand.

Chamsing, A., Salokhe, V.M., Singh, G. 2006. Energy consumption analysis for selected crops in different regions of Thailand. *Agricultural Engineering International: The CIGR Ejournal* VIII:1–18. Available online: https://ecommons.cornell.edu/bitstream/handle/1813/10526/EE%2006%20013%20Salokhe%20final%206Nov2006.pdf?sequence=1. Accessed on 27 August 2023.

Chatterjee, A.S. 2009. Diversified farming systems – learning from past to move into future. *LEISA INDIA* 11(1):8–10.

CSSRI Annual Report. 2019. ICAR- Central Soil Salinity Research Institute, Karnal, Haryana, India. pp 34. Available online at https://krishi.icar.gov.in. Accessed on 16 January 2022.

CSSRI, NAIP. 2014. *Final Report of NAIP Sub-Project on Strategies for Sustainable Management of Degraded Coastal Land and Water for Enhancing Livelihood Security of Farming Communities (Component 3, GEF funded)*. Burman D., Mandal S., Mahanta, K.K. (eds) Central Soil Salinity Research Institute, Regional Research Station (CSSRI, RRS), West Bengal. p 104.

Cusick, S.E., Georgieff, M.K. 2016. The role of nutrition in brain development: The golden opportunity of the "First 1000 Days". *Journal of Pediatics* 175:16–21. DOI: https://doi.org/10.1016/j.jpeds.2016.05.013.

Das, A., Layek, J., Ramkrushna, G.I., Patel, D.P., Choudhury, B.U., Chowdhury, S., Ngachan, S.V. 2015. Raised and sunken bed land configuration for crop diversification and crop and water productivity enhancement in rice paddies of the north eastern region of India. *Paddy and Water Environment* 13:571–580. DOI: https://doi.org/10.1007/s10333-014-0472-9.

Edvan, R.L., Mota, R.R.M., Dias-Silva, T.P., de Nascimento, R.R., de Sousa, S.V., da Silva, A.L., de Araujo, M.J., Araujo, J.S. 2020. Resilience of cactus pear genotype in a tropical semi-arid region subject to climatic cultivation restriction. *Scientific Reports* 10:10040. DOI: https://doi.org/10.1038/s41598-020-66972-0.

FAO. 1994. *World Soil Resources Reports*. Available online at fao.org/3/v4360e/V4360E06.htm#Chapter. Accessed on 14 January 2022. FAO, Rome, Italy.

FAO. 2006. Policy brief: food security. Issue 2. Published by FAO's Agriculture and Development Economics Division (ESA) with support from the FAO Netherlands Partnership Programme (FNPP) and the EC-FAO Food security programme. https://www.fao.org/fileadmin/templates/faoitaly/documents/pdf/pdf_Food_Security_Cocept_Note.pdf

FAO. 2013. *Climate Smart Agriculture Source Book*. FAO, Rome. http://www.fao.org/docrep/018/i3325e/i3325e.pdf. Accessed on 30 September 2021.

Fiorella, K.J., Chen, R.L., Milner, E.M., Fernald, L.C.H. 2016. Agricultural interventions for improved nutrition: A review of livelihood and environmental dimensions. *Global Food Security* 8: 39–47. DOI: https://doi.org/10.1016/j.gfs.2016.03.003.

Food and Agricultural Organization (FAO). 2018. *The 10 Elements of Agroecology: Guiding the Transition to Sustainable Food and Agricultural Systems*.FAO, Rome.

Hendrickson, J., Hanson, J., Tanaka, D., Sassenrath, G. 2008. Principles of integrated agricultural systems: Introduction to processes and definition. *Renewable Agriculture and Food Systems* 23(4):265–271. DOI: doi:10.1017/S1742170507001718.

Indian Institute of Water Management. 2014. *Technology Brochure*. Compiled and Edited by Das, M., Verma, O.P. and Kumar, A. pp 6–7. Directorate of Water Management, Bhubaneswar, Odisha, India. p 12.

Indrawati, E., Ruchiat, R., Indrawati, D., Febriyani, S. 2020. Integrated agriculture based agrotourism model with eco-friendly environmentalism on Carik Injeman land in Cibodas village. *Journal of Environmental Science and Sustainable Development* 3(1):177–194. DOI: https://doi.org/10.7454/jessd.v3i1.1031.

Jananthi, C. 2002. *Project Report Submitted to ICAR on Sustainable Farming System Models for Low Land Farms of Tamil Nadu*, India.

Kabir, M.A., Billah, K.M.M., Parvez, M.M. 2016. *Acacia catechu* trees in rice fields: A traditional agroforestry system of Northern Bangladesh. *International Journal of Agriculture System* 4(2):107–120.

Karlen, D.L., Hurley, E.G., Andrews, S.S., Cambardella, C.A., Meek, D.W., Duffy, M.D., Mallarino, A.P. 2006. Crop rotation effects on soil quality at three northern corn/soybean belt locations. *Agronomy Journal* 98(3):484–495.

Kashyap, P., Ravisankar, N., Panwar, A.S., Ghasal, P.C., Nath, A., Malik, S., Prusty, A.K., Chandrabhanu, Shamim, M., Dutta, D. 2021. Sustainable integration of banana in farming systems for enhanced income and nutritional security. *Indian Horticulture* 66(6):18–22.

Kaushik, N., Tikkoo, A., Yadav, P.K., Deswal, R.P.S., Singh, S. 2017. Agri-silvi-horti systemsfor semi-arid regions of North-West India. *Agricultural Research* 6:150–158. DOI: https://doi.org/10.1007/s40003-017-0247-9.

McDermott, J., Johnson, N., Kadiyala, S., Kennedy, G., Wyatt, A.J. 2015. Agricultural research for nutrition outcomes – rethinking the agenda. *Food Security* 7:593–607. DOI: https://doi.org/10.1007/s12571-015-0462-9.

Mehari, Z.H. 2015. The invasion of *Prosopis juliflora* and a far pastoral livelihoods in the Middle Awash area of Ethiopia. *Ecological Processes* 4:13. DOI: https://doi.org/10.1186/s13717-015-0039-8.

Mittal, V.K., Mittal, J.P., Dhawan, K.C. 1985. *Research Digest on Energy Requirement in Agriculture Sector*. Co-ordinating cell, AICRP on energy requirements in agricultural sector, Punjab Agricultural University, Ludhiana.

Mondal, A.K., Sharma, R.C., Singh, G., Dagar, J.C. 2010. *Computerized Data Base on Salt Affected Soil in India*. Technical Bulletin: CSSRI/Karnal/2/2010. Central Soil Salinity Research Institute, Karnal, Haryana, India. pp 27.

NABARD, 2017. Sustainability report 2017–18. National Bank for Agriculture and Rural Development, Mumbai, India. 92p. https://www.nabard.org/auth/writereaddata/tender/2004222638sustainability-report2017-18.pdf

Paramesh, V., Chakurkar, E.B., Bhagat, T., Sreekanth, G.B., Kumar, H.B.C., Rajkumar, S., Gokuldas, P.P., Mahajan, G.R., Manohara, K.K., Ravisankar, N. 2021. Impact of integrated farming system on residue recycling, nutrient budgeting and soil health. *Indian Journal of Agricultural Sciences* 91(1):44–48.

Patel, R. 2021. Agroecology is the solution to world hunger. *Scientific American,* November (Volume 325 Issue 5).

Pimentel, D., Pimentel, M. 2008. *Food, Energy and Society*, third Ed. CRC Press, Boca Raton, FL.

Pingali, P. 2023. Are the lessons from the green revolution relevant for agricultural growth and food security in the twenty-first century? In: Estudillo J.P., Kijima Y., Sonobe T. (eds) *Agricultural Development in Asia and Africa*. Emerging-Economy State and International Policy Studies. Springer, Singapore. pp 21–32. DOI: https://doi.org/10.1007/978-981-19-5542-6_2.

Pishgar-Komleh, S.H., Sefeedpari, P., Rafiee, S. 2011. Energy and economic analysis of rice production under different farm levels in Guilan Province of Iran. *Energy* 36:5824–5831.

Prado, E.L., Dewey, K.G. 2014. Nutrition and brain development in early life. *Nutrition Reviews* 72(4):267–284. DOI: https://doi.org/10.1111/nure.12102.

Pramanick, K.K., Kishore, D.K., Sharma, S.K., Das, B.K., Murthy, B.N.S. 2013. Straw berry cultivation under diverse agro-climatic conditions of India. *International Journal of Fruit Science* 13(1–2):36–51. DOI: https://doi.org/10.1080/15538362.2012.696969.

Quilty, J.R., McKinley, J., Pede, V.O., Buresh, R.J., Correa Jr., T.Q., Sadro, J.M. 2014. Energy efficiency of rice production in farmers' fields and intensively cropped research fields in the Philippines. *Field Crops Research* 168:8–18. DOI: http://dx.doi.org/10.1016/j.fcr.2014.08.001.

Rao, B.D., Bhasharachary, K., Christina, G.D.A., Devi, G.S., Tonapi, V.A. 2017. *Nutritional and Health Benefits of Millets*. ICAR Indian Institute of Millets Research, Rajendranagar, Hyderabad, India. p 112.

Ray, S.K., Chatterjee, D., Rajkhowa, D.J., Baishya, S.K., Hazarika, S., Paul, S. 2019. Effects of integrated farming system and rain water harvesting on livelihood improvement in North-Eastern region of India compared to traditional shifting cultivation: Evidence from action research. *Agroforestry Systems* 94:451–464. DOI: https://doi.org/10.1007/s10457-019-00406-3.

Ruel, M.T., Quisumbing, A.R., Balagamwala, M. 2018. Nutrition-sensitive agriculture: What have we learned so far?. *Global Food Security* 17:128–153. DOI: https://doi.org/10.1016/j.gfs.2018.01.002.

Sarangi, S.K. 2009. Indigenous farming system of Adi tribes: Source of food, nutrition and medicine. *LEISA INDIA* 11(1):22–23.

Sarangi, S.K., Singh, R., Singh, K.A. 2009. Indigenous method of rat proof grain storage by Adi tribes of Arunachal Pradesh. *Indian Journal of Traditional Knowledge* 8(2):230–233.

Sarangi, S.K., Singh, S., Kumar, V., Srivastava, A.K., Sharma, P.C., Johnson, D.E. 2020. Tillage and crop establishment options for enhancing the productivity, profitability, and resource use efficiency of rice-rabi systems of the salt-affected coastal lowlands of eastern India. *Field Crops Research* 247:1–15. DOI: http://doi.org/10.1016/j.fcr.2019.03.016.

Sharma, D.K. 2021. Integrated farming system: An approach for sustainable management of natural resources. *Journal of Natural Resource Conservation and Management* 2(1):1–5. DOI: https://doi.org/10.51396/ANRCM.2.1.2021.1-5.

Sharma, G., Partap, U., Dahal, D.R., Sharma, D.P., Sharma, E. 2016. Declining large cardamom production systems in the Sikkim Himalayas – Climate change impacts, agroeconomic potential and revival strategies. *Mountain Research and Development* 36(3):286–298. DOI: https://doi.org/10.1659/MRD-JOURNAL-D-14-00122.1.

Sharma, T.V.R.S., Jaisankar, I., Kumar, J., Chander, M.P. 2015. Gardening of *Elaeocarpus* (Rudraksha) for conservation and tourist interest in Andaman and Nicobar Islands. *Journal of the Andaman Science Association* 20(2):172–177.

Shekinah, D.E., Sankaran, N. 2007. Productivity, profitability and employment generation in integrated farming systems for rainfed vertisols of western zone of Tamil Nadu. *Indian Journal of Agronomy* 52(4):275–278.

Singh, G., Jat, H.S., Kumar, A., Sekhawat, K., Yaduvanshi, N.P.S., Tripathi, R.S., Batra, L., Pandey, R.S., Sharma, P.C., Chaudhari, S.K., Singh, S.K., Kundu, S.S., Sirohi, N.S., Mohan, S., Singh, D., Kailash. 2009. *Eco-Friendly Integrated Multi-Enterprise Model for Livelihood Security in Small Farm Holdings*. Central Soil Salinity Research Institute, Haryana, India. Technical Bulletin: CSSRI/Karnal/2009/05. p 28.

Singh, K., Bohra, J.S., Singh, T.K., Singh, J.P., Singh, Y., Singh, C.S. 2007. Productivity and economics of integrated farming system in irrigated agro-ecosystem of eastern Uttar Pradesh. *Indian Journal of Agronomy* 52(1):11–15.

Singh, P., Singh, G., Sodhi, G.P.S. 2020. Energy and carbon footprints of wheat establishment following different rice residue management strategies vis-à-vis conventional tillage coupled with rice residue burning in north-western India. *Energy* 200:117554. DOI: https://doi.org/10.1016/j.energy.2020.117554.

Tiecher, T., Calegari, A., Caner, L., Rheinheimer, D.S. 2017. Soil fertility and nutrient budget after 23-years of different soil tillage systems and winter cover crops in a subtropical Oxisol. *Geoderma* 308:78–85.

Tomar, S.S., Tembe, G.P., Sharma, S.K., Tomar, V.S. 1996. Studies on some land management practices for increasing agricultural production in Vertisols of central India. *Agricultural Water Management* 30(1):91–106. DOI: https://doi.org/10.1016/0378-3774(95)01195-1.

Tontisirin, K., Nantel, G., Bhattacharjee, L. 2002. Food-based strategies to meet the challenges of micronutrient malnutrition in the developing world. *Proceedings of the Nutrition Society* 61(2):243–250. DOI: https://doi.org/10.1079/PNS2002155.

Verma, C.L., Singh, Y.P., Damodaran, T., Singh, A.K., Mishra, V.K., Sharma, D.K. 2016. Developing design guidelines for calculation of width and height of raised bed and depth of sunken bed system in water-logged sodic soil. *Journal of Soil Salinity and Water Quality* 8(1):59–66.

Yadav, G., Jat, H.S., Raju, R., Yadav, R.K., Singh, S.K., Chaudhari, S.K., Sharma, D.K., Singh, G., Sharma, P.C. 2021. Enterprise mix diversification: An option for ecologically sustainable food and nutritional security of small holders in Indo-Gangetic plains. *International Journal of Agriculutral Sustainability* 20(3):1–11. DOI: https://doi.org/10.1080/14735903.2021.1912978.

Yadav, J.S.P. 2006. Improved farming and ecological security in coastal region. *Journal of Indian Society of Coastal Agricultural Research* 24(2):229–240.

3 Integration of Enterprises and Efficient Resource Use in Diverse Farming Systems

3.1 CONCEPT OF INTEGRATION IN INTEGRATED FARMING SYSTEM (IFS)

A sound integrated farming system (IFS) enhances farm productivity, reduces environmental degradation, improves farmer's quality of life and ensures sustainability of the farm. Within a farm unit, integration occurs when outputs (usually byproducts) of one subsystem are used as inputs for another (Lightfoot and Minnick 1991). The integrated farming system is defined as an aspect of farming that minimizes risk, increases productivity and maximizes profits while improving organic waste and crop residue utilization (Radhammani et al. 2003).

A farming system requires a holistic approach in order to be sustained. It is a mix of farm enterprises (crop, livestock, aquaculture, agro-forestry, agri-horticulture and sericulture) in which farm families allocate resources so that the existing enterprises can be used more efficiently by enhancing productivity and profitability. While crop and other enterprises coexist in such diversified farming, the main goal is to reduce risk, whereas in IFS, a strategic combination of one or more enterprises along with cropping creates a complementary effect through efficient recycling of wastes and crop residues, which includes an additional source of income for the farmer (Paramesh et al. 2021). The IFS activity focuses on a few selected interdependent, interrelated and interlinking production systems involving crops, animals and related professions. The IFS approach has the potential for high productivity and resource recycling, as well as ecological soundness leading to sustainable agriculture.

Integrated crop and livestock production systems are highly efficient; crop residues can potentially be used as livestock feed; waste products (e.g., faeces and urine) are fed into biogas digesters and the effluent is used to fertilize ponds for plankton/algae production, with fish farming serving as the final activity. These systems are worth investigating as a means of supplying nutrients/fuel for the family, lowering fossil fuel burning and methane production and so reducing environmental pollution (Preston 1990). One of the best examples of mixed farming is integrated fish aquaculture. This sort of farming is one of the key ecologically sustainable technologies, and it is practiced in many forms mostly in East and South East Asian nations. The technology involves a combination of fish polyculture integrated with crop/livestock production. The integration of aquaculture with livestock or crop farming aids in the provision of high-quality protein food, resource use, agricultural waste recycling, job creation and economic development (Gupta and Acosta 2004; Shoko et al. 2011; Megerssa Endebu et al. 2016). Integrated aquaculture systems are used in various parts of the world (Neori et al. 2004) and have a long history of improving fish production, lowering operational costs, protecting the environment from pollution and waste management and increasing income by producing fish feed from waste materials (Prein 2002). Global fishery and aquaculture related production statistics, information on management, impact on food security and other issues are available (FAO 2021).

Several IFS models such as rice (*Oryza sativa* L.) – fish – on-dyke horticulture-vegetable (Mohanty et al. 2004, 2009, 2017b), integrated system of rice intensification (ISRI) with fish and horticulture crops (Thakur et al. 2015), pond-based IFS (Srivastava et al. 2004), self-reliant agri-based IFS

DOI: 10.1201/9781003239963-3

(Rautaray et al. 2020) are the most prevalent, lucrative and replicable systems in the Indian context. Aquaculture-based IFS integrating aquaculture (carp polyculture), agriculture (*Kharif* rice followed by green gram and vegetables), on-dyke horticulture (banana and papaya) and poultry is one of the most water-saving, sustainable, productive and replicable models in terms of water management. These systems, as a whole, make a net profit of ₹1,10,000–70,000 ha^{-1} with an output value: cost of cultivation ratio of 2.7–2.8 and net consumptive water productivity of ₹9.8–16.2 m^{-3}. Further, when fish and prawn are included as a key component, this system also helps to improve water production and water use efficiency as well as minimize water footprint (Mohanty et al. 2009, 2010). Integrated aqua-farming not only increases productivity but also creates employment opportunities, raises income and provides nutritional security to farming communities with limited resources. It also distributes risk (both biological and economic).

3.1.1 Basic Principle

IFS is a set of interconnected, often interlocking production systems based on a few crops, animals and related subsidiary enterprises that maximize the utilization of nutrients in each system while minimizing the negative environmental impact of these enterprises. The interconnected, interdependent-interlocking aspect of IFS entails using primary and secondary produce from one system as fundamental input for the other system, so integrating them as one single unit (Parajuli et al. 2018). In this, an interconnected network of production systems is employed to transfer 'waste' from one component to another. This decreases costs while increasing productivity and/or income. Farmers not only eliminate waste by using trash as a resource, but they also ensure an overall rise in the productivity of the farming system. Improved farming techniques such as conservation technology, integrated nutrient management, site-specific nutrient management, zero tillage, use of biofertilizers, crop rotation and the use of agricultural systems are examples of new integrated methods (Behera and France 2016; Goverdhan et al. 2020).

The fundamental idea behind an IFS is to increase ecological diversity by (1) choosing the right crop method with crop rotation, crop mixing and intercropping so that there is less competition for water, nutrition and space, and (2) using a multi-story layout so that the entire available area can be used effectively and has a high level of interaction between biological and abiotic components, (3) a combination of these methods and (4) tree species that produce pods or leaves throughout the year. Further, during the wet season, surplus fodder leaves, crop leftovers, etc., should be kept as silage or grass for the lean season.

3.1.2 Aim and Goals of IFS

It is a synthesis of various agricultural activities on a unit area of land with the following goals: (1) Maximum return from unit area of all component enterprises to provide steady and stable income, (2) Maintenance of soil fertility, i.e., rejuvenation/improvement of system's productivity and attainment of agro-ecological equilibrium, (3) Establishing supplemental and complementary business partnerships such that byproducts from one farming system component can be used as inputs for the other, (4) Manage natural cropping systems to prevent the buildup of insect pests, diseases and weed populations and keep them at a low degree of intensity, (5) Supply society with healthy produce and an environment free of chemicals by lowering environmental contamination.

3.1.3 Components in IFS

Agriculture, fish farming, fodder production, goat and sheep rearing, horticulture, duck and pigeon rearing, forestry, mushroom cultivation, sericulture, azolla farming, poultry, kitchen gardening,

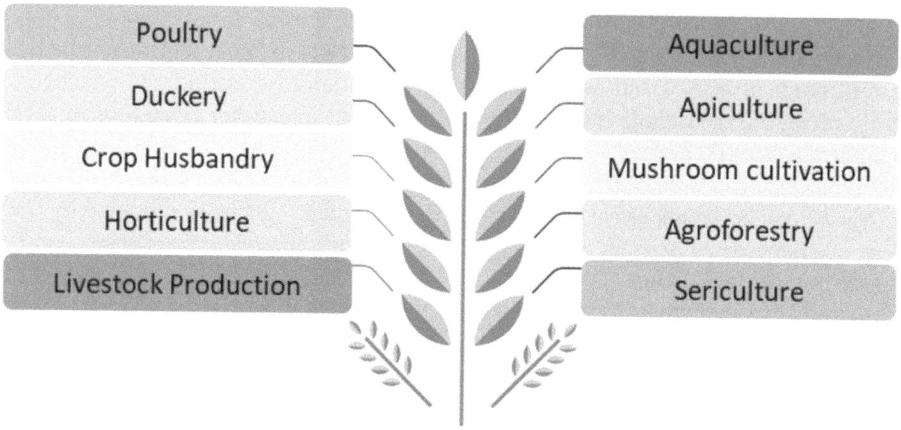

FIGURE 3.1 Components of integrated farming system.

seed production, piggery, vermiculture, rabbitry and value addition are just a few of the various IFS components (Figure 3.1). Four major categories can be used to group these elements.

- **A**: Crop – Planting Crops, Cereals, Oilseeds, Pulses, Fruits, Vegetables, Flowers, Spices, Sugarcane, Fiber Crops, Fodder/Forage Crops, Agricultural Forests.
- **B**: Livestock and Poultry – Cows, Buffaloes, Goat, Sheep, Pigs, Poultry, Duck.
- **C**: Fishery – Composite Fish Culture, Fingerling Production, Paddy cum Fish Culture.
- **D**: Secondary Agriculture – Mushroom Cultivation, Bee Cultivation, Food Processing, Azolla Cultivation, Vermicomposting, Silk Cultivation, Biogas Production.

It is feasible to create an integrated farming system by combining the four categories of components listed above (A + B, A + C, B + C, A + D, B + D, C + D, A + B + C, A + B + D, B + C + D, A + C + D, A + B + C + D).

3.1.4 Elements of IFS

- Farm ponds
- Bio-fertilizers
- Green manuring
- Biological pesticides
- Biogas
- Solar energy
- Vermicomposting
- Rainwater harvesting
- Watershed management

3.1.5 Factors Determining the Implementation of IFS

- Soil and climate feature of particular areas.
- Availability of resources and labour.
- The current level of resource use.
- Economics of the planned integrated farming.
- Farmer's management skills.

3.2 ADVANTAGES OF IFS

IFS encourages diversification of enterprises, increasing cash income, improving the quality and quantity of productive food and assisting in the exploitation of underutilized resources, nutrient recycling, nature conservation, environmental safety, land and water management (Figure 3.2). Major advantages of IFS are:

- This approach introduces changes to farming techniques to increase productivity (increasing economic yield per unit area by intensifying crops and related businesses) and maximize resource utilization.
- Improved cost-effectiveness achieved mainly by way of reduced costs due to recycling of wastes of one enterprise as energy inputs for other systems.
- Greater sustainability in agricultural output as a result of the integration of many businesses with varying economic value. Recycling of wastes helps to lessen reliance on external high-energy inputs, saving natural and limited resources.
- Integration of diverse production systems and production of variety of food products provides an opportunity to reduce malnutrition problem.
- Recycling of animal waste such as dairy, pig, poultry, etc. for production helps to avoid waste stacking, subsequent contamination and can effectively minimize environmental pollution.

FIGURE 3.2 Integrated farming system and farm management approach.

- The farming system provides a steady flow of money to farmers throughout the year through the sale of eggs, milk, edible mushrooms, honey, silkworm cocoons and so on. Assist resource-poor farmers in escaping from the clutches of moneylenders.
- Because of the linkage of dairy/mushrooms/sericulture fruit crops/vegetable crops/flower cultivation, etc., cash available all year could incentivize small and marginal farmers to adopt modern technologies.
- Recycling field waste in a production-oriented system. Dairy farming, poultry farming, pig farming, fisheries, silk farming, etc. can bring prosperity to agricultural activities when combined with given agro-climatic conditions and socioeconomic status.
- Requirement of chemical fertilizer is reduced through recycling of organic wastes. Further, biogas production can fulfil daily household energy requirement. Therefore, IFS goes a long way in solving energy crises.
- Energy saving – The IFS system efficiently reduces the further dependence on fossil fuels as a source of energy by providing substitute fuel sources as byproducts such as biogas.
- Meeting the fodder crisis – Perennial legume fodder trees can be grown inside the farm boundaries. These trees not only fix nitrogen *in-situ* but also provide quality fodder for the animals.
- Solving the fuel and timber crisis – Incorporation of Silvi component provides fuel and timber wood, reduces deforestation and helps in protecting the natural ecosystem.
- Employment generation – Integration of diverse components in IFS generates and provide year-round employment.
- Integrated farming system improves the literacy level as the entrepreneur tries to know more things in the process.
- Agro-industries – Production of agro-products in IFS contributes greatly to the growth of agro-industries and agribusiness.
- Increased input efficiency – As reliance on external inputs such as fertilizers, food, agro-chemicals and energy has decreased, input efficiency in this agricultural system has increased dramatically.
- The overall benefit of IFS is an improved farmer's standard of living due to products such as edible mushrooms, fish, fruits, eggs, milk, honey, vegetables and so on.
- Annual revenue – Because of the range of enterprises in IFS, the farmer generates income all year. It improves farmers' lifestyle components such as food, shelter, health and education.
- One of the most significant advantages of integrated farming is the better production system. Increased productivity indicates that economic yield per unit area per unit grows over time as agricultural and associated farming enterprises intensify.
- As productivity rises, so will profit margins. This is due to the fact that we are utilizing a waste material or byproduct from one farming enterprise as an input to another farming enterprise.

3.3 POSSIBLE OUTPUTS OF IFS

IFS is an interconnected complex matrix of soil, water, plant, animal and environment interactions; therefore, the system is more viable and profitable than an arable farming system. IFS contributes to food security, nutritional security, nutrient recycling, biodiversity protection, improved employment, higher profitability and climate resilience (Figure 3.3). It leads to the production of high-quality food and the strengthening of the food chain. It is necessary to eradicate nutritional disorders caused by a lack of mineral nutrients and vitamins in the food consumed. Horticultural and vegetable crops may produce two to three times more energy than cereal crops on the same area of land, and their incorporation in the existing system will ensure nutritional security. Similarly,

FIGURE 3.3 Ecosystem services provided by integrated farming system.

because of space constraints, inclusion of beekeeping, fisheries, sericulture and mushroom cultivation provides additional high-energy food without harming food grain output.

The efficient recycling of resources, especially agricultural wastes, is essential in farming systems (Sarangi et al. 2016) as 80–90% of the micronutrients still exist in the biomass. About 16 million tonnes of paddy straw are produced in the Indo-Gangetic plains, where it is not recycled effectively, and even in Punjab, where rice farming is practiced on 2.6 million ha, it is destroyed by burning. The use of second-generation machinery for effective crop residue management to preserve moisture, promote soil microorganism activities, regulate soil temperature, check soil erosion, suppress weed development and decomposition improves soil fertility is necessary to reduce such precious input loss. The crop residue can also be utilized as dry feed, compost, floor thatch for cattle sheds, and for cultivating mushrooms. Multiple water uses for fisheries, farming and the production of crops, fruits and vegetables may also increase water productivity. Adoption of ideas such as social forestry, water harvesting and recycling fisheries and stall feeding for goatry/piggery will thereby increase the profit margin along with several other indirect benefits, such as increased employment and enhanced local ecosystem. Such forms of enterprise integration produce extra income that ranges between ₹20,000–40,000 for rainfed ecosystems and ₹50,000–75,000 for irrigated ecosystems per hectare. Income increase of 25–50% owing to integration of processing and on-farm value addition, yield increase of 0.5–1.0 t ha^{-1} due to enhanced soil health, cost reduction of ₹2500-4,000 ha^{-1}, and employment generation of 50–75 man-days per household have also been reported.

3.4 LIMITATIONS OF INTEGRATED FARMING SYSTEM

- The fragmentation of landholdings makes implementation challenging.
- Individual farmers' resources and needs differ greatly.
- A lack of understanding of the sustainable farming method.
- Lack of technical know-how of various farming system models.
- Requires an initial investment and a lack of simple and acceptable lending facilities.
- The absence of some marketing facilities, particularly for perishable goods.
- Inadequate storage facilities and insufficient input availability.
- A lack of education, understanding and encouragement in farming areas, particularly among rural youngsters.
- Pathogenic bacteria may be found in livestock manure.

3.5 PRODUCTIVITY OF INTEGRATED FISH FARMING

Planktons, which are the natural food source for fish, are typically used to assess the biological productivity of any aquatic body. Through a variety of pathways, animal wastes boost the biological productivity of ponds, which, in turn, increases fish production (Oribhabor and Ansa 2006). Productive capacity measures the 'net community production' of the system by referring to the biomass generated in kg ha^{-1} (Dalsgaard et al. 1995). This capacity is determined by the type of manure used, the quality of the pond water and the quality of the underlying soil. The type of manure has an impact on the organization of the plankton population. According to Odum (1989), the amount of biomass maintained per unit of energy flow in a given ecosystem indicates how mature the system is and how well it can utilize the energy that is available. Excreta and waste food from animals, such as cattle, pigs or chickens, increase water productivity, generate natural feed and help in the growth of fish, as well as provide other symbiotic benefits (Figure 3.4) for animals. Furthermore, Edwards (2000) and FAO (2003) have emphasized the need of incorporating aquaculture-based IFS into future rural development efforts. Aquaculture-based IFS is more profitable than monoculture,

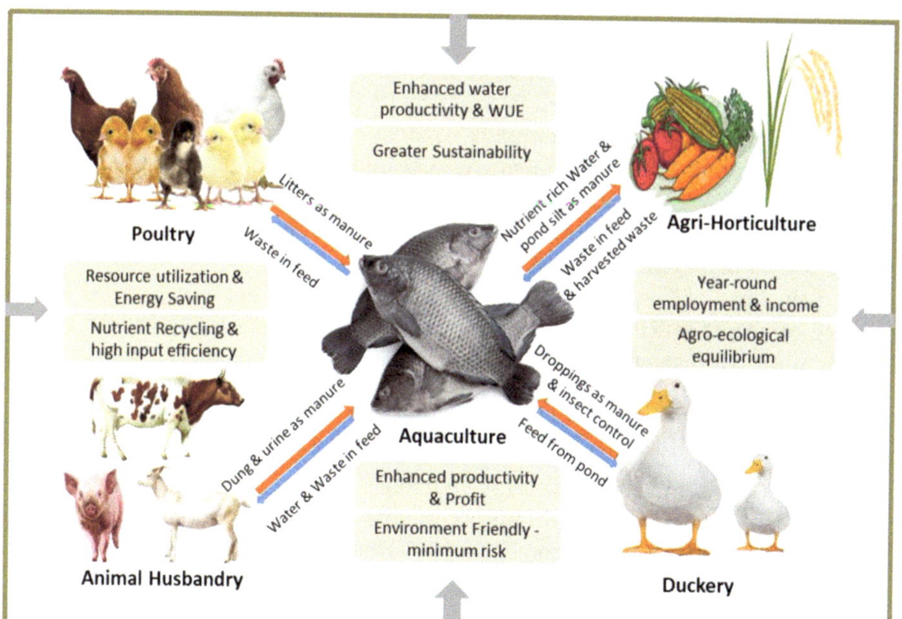

FIGURE 3.4 Schematic representation of major components, pathways of material flow and benefits in integrated farming system.

which contributes to the economic empowerment of farm families by allowing year-round maximization of production (Nnaji et al. 2003; Tipraqsa et al. 2007; Gabriel et al. 2007).

3.6 RICE-BASED CROPPING AND FARMING SYSTEMS

Rice is a crop of global importance with more than 164 million ha area worldwide. The eight Asian countries (India, China, Bangladesh, Indonesia, Thailand, Vietnam, Myanmar and Philippines) together cover more than 75% of world rice acerage (FAOSTAT 2022). India is the leading country in rice acreage with 45.00 mha area, followed by China (30.08 mha), Bangladesh (11.42 mha) and Indonesia (10.66 mha) during 2020 (Figure 3.5a). However, with respect to production, China is the leading country with 211.86 mt, followed by India (178.31 mt), Bangladesh (54.91 mt) and Indonesia (54.65 mt) during 2020 (Figure 3.5b). The average rice productivity is highest for China, Japan and Brazil with 6.6–7.0 tha^{-1}, followed by Vietnam (5.9 tha^{-1}), Indonesia (5.1 tha^{-1}), Bangladesh (4.8 tha^{-1}) and the Philippines (4.1 tha^{-1}). Though, India has the highest rice area, with respect to productivity, it lags behind many countries as well as in world average productivity (Figure 3.5c).

Rice-wheat is the predominant cropping system covering about 10.5 mha area followed by rice-rice system (5.9 mha) in India. Relay/*paira*/*utera* cropping is a traditional system in rice, where one of the subsequent crops such as horse gram, green gram, black gram, grass pea (Lathyrus/*Khesari*), gram, pea, lentil and linseed are sown to the standing crop of rice to utilize the residual soil moisture and to escape from drought/salinity hazards, which occur during the latter part of the *Rabi* season. Mixed cropping of one variety of rice with another is also seen in some of the states of India (Assam, West Bengal, Tamil Nadu and Kerala). Intercropping of crops such as green gram, black gram, pigeon pea, sesame, maize and finger millet is also practiced in India. Intensive triple cropping systems of rice-rice-rice, rice-rice-cereal, rice-rice-pulses, rice-wheat-pulses, rice-toria-wheat, etc., are practiced under some irrigated conditions.

3.6.1 Sustainability of Rice-Based Cropping System

Rice-wheat is the backbone of food security in the Indian subcontinent, particularly in the Indo-Gangetic plains (IGP). This cropping system covers an area of 26 million hectares spread over the IGP in South Asia and China (Chauhan et al. 2012). Imbalanced and indiscriminate chemical fertilizer use in this system has resulted in major (N, P, K and S) and micro (Zn, B and Fe) nutrient disorders (Ullah et al. 2021). A substantial amount of crop residues of this system in Punjab, Haryana and Uttar Pradesh is burned *in-situ*, as a result both the number of fires and $PM_{2.5}$ emissions during the rice harvest roughly double than those from wheat residue burning (Montes et al. 2022). Establishment methods in a rice-based cropping system as well as nutrient management plays a significant role in management of greenhouse gas emissions. At par grain yield (6.62 tha^{-1}) and reduced methane flux (1.57 mg m^{-2} h^{-1}) in rice was observed with application of NPK + Sulphur + direct sowing compared to NPK + Sulphur + puddled transplanting (grain yield 6.5 tha^{-1}, methane flux 2.25 mg m^{-2} h^{-1}) (Singh et al. 2009).

3.6.2 Conservation Agriculture for Rice-Wheat Cropping System

Conventional rice-wheat system has several limitations with respect to input productivity, yield, economics and adverse environmental impacts. A conservation system has been found in recent years as an alternative sustainable technology for this mega cropping system. Zero tillage (ZT) sowing is possible by the use of different machines like happy seeder, super seeder and rotavator. Happy seeder performs ZT sowing at the same time by the straw management unit it shreds and clears the residues in front of the tyne openers and then deposits the residue around the seeded row as mulch (Dhruwe and Victor 2021). Happy seeder (weight 550 kg) requires a tractor of 45–55 hp, which covers 0.25–0.35 ha area in one hour. It sows seeds at a row spacing of 22.5 cm and covers 9–12 rows.

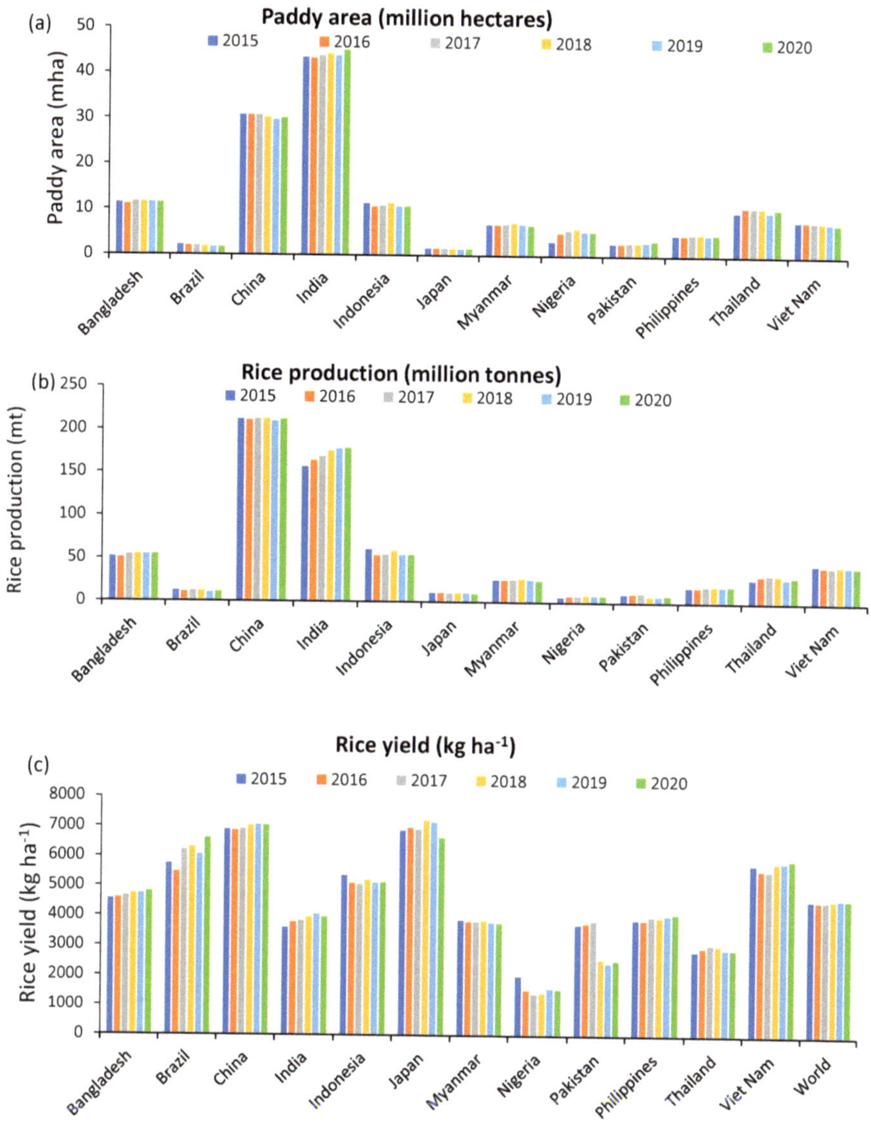

FIGURE 3.5 Country-wise (a) rice acreage (b) production and (c) yield during 2015–2020 (Source: https://www.fao.org/faostat/en/#data/QCL).

The cost of operation of happy seeder and zero till drill (straw collected manually) are ₹2098.65 and ₹3106.38 per hectare respectively (Dhruwe and Victor 2021). Super seeder is an advanced machinery to plough anchored rice stubbles into soil and sow wheat seeds simultaneously in a single operation. It has less seed requirement, tackles the heavy load of rice residues to incorporate in soil. However, super seeder requires tractor of 65 hp to pull the machine. In a happy seeder, stubbles are not ploughed back into the soil, therefore, there are problems in wheat seed germination. a rotavator is used both for primary and secondary tillage operations, such as for good seedbed preparation and pulverization in a single operation. It has a tractor-mounted disc harrow with two gangs of discs for thorough cutting of the soil and incorporating the straw. It can be operated under dry as well as wet soil conditions. Super seeders and happy seeders are better in terms of getting high wheat yield compared to zero till drill and rotavators/disc harrows (Kirandeep et al. 2020).

Six years (20092015) of study of rice-wheat and rice-maize systems at Patna, Bihar, revealed that ZT crop establishment of both *Kharif* and *Rabi* crops along with crop residue retention (~33%) had a positive impact on soil organic C-pools, macro-aggregate formation, and aggregate carbon content over conventional agricultural practices (Nandan et al. 2019). In the Eastern Gangetic plains (EGP) inclusion of a mung bean (*Vigna radiata*) as a third crop to rice-wheat system resulted in 24.5–29.6% increase in rice equivalent yield, 2.8–8.0% increase in irrigation water input and 21.8–29.8% increase in irrigation water productivity (Chaki et al. 2021).

3.6.3 Conservation Agriculture for Rice-Based Cropping System

Three years (2013–2016) of study of rice-based cropping systems under coastal saline soils (ECe 1.9-14.3 dS m^{-1}) reported positive effect of conservation tillage (DSR/non-puddled transplanting) over conventional tillage (puddling) in rice on subsequent *Rabi* crops, 25–44% higher yield recorded in rapeseed and 8–13% higher yield in maize (Sarangi et al. 2020a). The advantages of DSR and crop residue retention (~40%) in terms of reducing cost of cultivation, increasing system yield, reducing irrigation water and nitrogen requirement, improving irrigation water productivity and soil quality (BD, OC, microbial biomass carbon) was observed in DSR-raised bed planted maize (RBPM) with crop residues retention over puddled transplanted rice-RBPM with removal of all crop residues (Sarangi et al. 2020b). Conservation agriculture involving ZT planting and crop residue mulching in coastal saline soils enables cropping system intensification. Cultivation of potato after harvest of *Kharif* rice under ZT planting and paddy straw mulching (~ 12 t ha^{-1}) resulted in 37% higher tuber yield, 27% reduction in cost of cultivation, reduction of soil salinity from 5 to 3 dS m^{-1}, conserved 4-8% soil moisture, reduced soil BD from 1.49 to 1.44 Mg m^{-3} and increased soil organic carbon from 0.39 to 0.44% (Sarangi et al. 2021a). Profitability of cropping system intensification options also enhanced through conservation agriculture as evidenced in rice-garlic system with increase in net return from ₹14,415 ha^{-1} under conventional agriculture to ₹16,99,488 ha^{-1} under conservation agricultural practices with corresponding increase in BCR from 1.19 to 3.36 (Sarangi et al. 2021b). In tropical coastal regions, to get higher water productivity on a net return basis, cropping systems such as rice-vegetables (tomato, potato, onion, eggplant, chilli), rice-maize and rice-sunflower are recommended (Sarangi et al. 2017). In coastal regions, a rice-based system under conventional tillage functioned as GHG emission source, whereas under reduced tillage with residue retention functioned as a GHG sink (Mandal et al. 2021).

3.6.4 Integrated Rice-Based Farming System

Rice-fish is the preferred farming system in the low-lying cultivated areas particularly in the coastal region with high annual rainfall of more than 1800 mm. In coastal West Bengal at Canning Town, the 30 years (1992–2021) annual average rainfall observed was 1887 mm (Figure 3.6), which is congenial for rice-fish farming system. This system is integrated with various other enterprises such as poultry, duckery, horticultural crops, etc. for effective utilization of land, maximization of productivity and minimization of risk. Rice-fish diversified farming system in one hectare of area can produce 16–18 tonnes of food crops, 0.6 tonnes of fish and prawn, 0.55 tonnes of meat and 8000–12000 numbers of eggs in addition, flowers, fuelwood, crop residues (paddy straw and other crops residues), fodder and feed for farm animals providing additional employment of about 300 man-days over sole rice cultivation (Pandey et al. 2010). Therefore, this system is an efficient technology to sustainably increase the food production in the rice-based systems by achieving ecological agriculture. However, conversion of rice land to shrimp farming has adverse effects on soil such as an increase in soil salinity, acidity, depletion of soil organic carbon content as well as calcium,

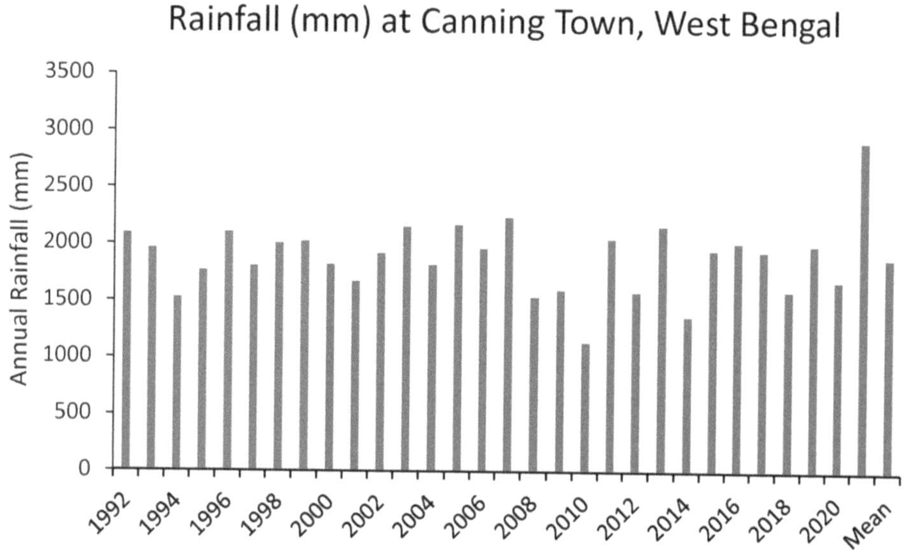

FIGURE 3.6 Annual rainfall 1992–2021 recorded at Canning Town (Lat. 22°15′ N; Long. 88°40′ E), West Bengal India.

potassium and magnesium as well as results in soil degradation (Ali 2006). Introduction of fish in to the rice field increases rice yield and improves soil fertility (Lin and Wu 2020).

3.6.4.1 Rice-Livestock Integrated Farming System (RLIFS)

RLIFS consists of rice as the main crop integrated with animals such as cow and buffalo in an environment friendly manner with zero waste, due to efficient recycling process and minimization of chemical inputs. Farm family utilizes the rice grains and the paddy straw/bran are fed to livestock. The manure, urine and farmyard residues of livestock are valuable inputs through manure, composts, vermicompost for making a healthy soil for rice-based crop production. However, an appropriate policy environment, particularly the regulation of prices of inputs (seeds, fertilizer, medicines, vitamins, manures etc.) and outputs (rice, milk, milk products), has an impact on the sustainability of the system (Lindawati et al. 2018).

3.6.4.2 Rice+Fish+Poultry IFS

This system is different from pond-based fishery with poultry farming. In this case, the entire paddy area is integrated with a fish and poultry system. During the rainy season when the entire area is covered with water, fish occupy the whole paddy and a poultry shed is built in the paddy field. In this system the paddy area serves as the pond area for the fish. The risk element in rice production in lowlands can be eliminated by this diversified farming system, and this system is a better bargain than single rainfed lowland rice cultivation under the risk of vagaries of climate change (Sarangi 2004). In this system, trenches of 1–2 m wide and deep are dugout along the boundary of the plot and the excavated soil is used to make dykes and borders of about 1 m high for raising vegetable/fruit crops (Sarangi 2004). Rice varieties having high tillering capacity may be selected for better weed suppression (Tuong et al. 2000). Rice varieties such as Jaladhi-1, Jaladhi-2, Jalmagna, Utkal Prabha, Manika, Mahalaxmi, Panidhan etc. may be selected. Generally Indian major carps i.e. Rohu (*Labeo rohita*), Catla (*Catla catla*) and Mrigal (*Cirrhinus mrigala*), common carp (*Cyprinus carpio*) and grass carp (*Ctenopharyngodon Idella*) are recommended for this system. When the rice seedlings attain a greenish colour, the fish fingerlings of 8–10 cm long are released to the paddy field with stocking density of 5000 to 6000 fingerlings per hectare. Poultry cages of dimension 1.8 m × 1.2 m × 0.9 m accommodating 20 broiler birds in each cage @ 1 bird per 0.3 m^2 of cage floor space and 10 m^2 of rice field area installed; this system increased the annual household income by 97% over the monocropping of rice, with significant nutritional advantage to the household due to increase in intake of vegetables, fish and meat from 15, 2.17, 1.5 kg $month^{-1}$ (baseline) to 31.54, 3.52 and 4.11 kg $month^{-1}$ after adoption of this IFS model (Ramanathan et al. 2020).

3.6.4.3 Rice + Fish + Duck IFS

Rice and fish/duck has a symbiotic relationship, fish/duck loose the soil around the rice plant, creating aeration in the rhizosphere of rice plants, eating away mosquito larvae and pupae. A paddy field provides an aquatic environment and shelter for fish/ducks. In this system, a duck coop may be built over a small part of the rice field adjacent to the bund. In this model, a duck component is introduced to the rice+fish system. Ducks are found to be environment friendly in the lowland rice ecosystem and have a synergistic effect. The ducks graze the rice field during the daytime, eat away small weeds, snails, eggs/larvae/pupae/adults of rice crop insect pests and, in turn, improve soil and water aeration, add soil fertility, and duck droppings are used as feed for fish (Sarangi et al. 2016). The rice crop gets the benefit of better aeration, less weeds and higher yield due to soil health improvement. The total system output is increased from higher rice and fish yields with additional income from duck eggs and meat. Foraging of duck in the paddy field reduced the feed consumption from 15 kg day^{-1} without foraging to 10 kg day^{-1} with foraging and the corresponding reduction in cost was from ₹365 to 250 day^{-1} for 100 ducks (Thomas et al. 2022). The net return from rice monoculture, fish, rice-fish and rice-fish+duck from 0.4 ha area was ₹40,700, 114,590, 16,10,888, 2,50,527, respectively, with corresponding cost of cultivation of ₹50,872, 92,410, 1,43,828 and 1,78,948 (Thomas et al. 2022).

This model creates a beautiful landscape and is environment friendly and productive; however, application of harmful chemicals such as insecticides, fungicides and herbicides should be restricted. The pest control measures should rely on predators, parasites, botanicals, plant extracts, animal excreta and manual measures. The beneficial effects of a rice-duck farming system have been reported from several parts of the world (Table 3.1).

3.6.4.4 Diversified Integrated Rice-Based Farming System

In the aquatic environment of the paddy field other organisms such as frogs, crabs, prawns/shrimps can also be grown as co-culture with rice. In a sub-tropical paddy soil, the mean grain yield of rice over nine years (2008–2016) in rice-frog-fish (RFF), rice-fish (RF) and rice monoculture (RM) was 5402.54, 5058.31 and 4422.31 kg ha-1, respectively with increasing trend in RFF and RF, and decreasing trend of rice yield in RM; the increasing yield trend was due to better soil phosphorus availability (Lin and Wu 2020). Co-culture of rice with other aquatic animals such as crab/prawn/shrimp in the coastal lowland rice ecosystem is also feasible. Increase in grain yield of rice by 22% was observed in rice-crab co-culture with 6000 crabs ha^{-1} with application of additional feed for crabs; this system controlled drained out water nutrient concentration, which helped in water pollution control (Bashir et al. 2021). The freshwater prawns like *Macrobrachium* species and brackish water shrimps like *Penaeus monodon*, crabs such as *Oziotelphusa senex senex* or *Parathelphusa hydrodromus* (naturally occurring freshwater rice field crabs), *Eriocheir sinensis* (Chinese mitten crab) and *Scylla serrata* (mud crab) can be used for a rice-aquatic animal integrated farming system (Reddy and Kishori 2018). Cultivation of both tiger prawn and crab in the paddy field produced a prawn yield of 300 kg ha^{-1} (price of prawn ₹600 kg^{-1}) and crab yield of 250 kg ha^{-1} (price of crab ₹1000 kg^{-1}) with total gross return of ₹4,30,000 and investment of ₹2,05,000 ha^{-1} (Sreelatha et al. 2022).

The *Palayamanan* model of diversified integrated rice-based farming system consists of field crop production, ornamental garden, nursery area for growing seedlings, passion fruit, multi-animal production (2 cows, 10 goats, 5 pigs, 50 chicken and 50 ducks) and small farm reservoir (Corales et al. 2004). Small farm reservoir is used for Tilapia culture, with fruit trees around it and water in the pond is used as supplemental irrigation during the dry season. In coastal ecologies of eastern India,

TABLE 3.1
Comparison of Rice-Duck Farming System (RDFS) vs Rice Monoculture (RM) in Terms of Productivity and Environmental Benefits

Parameters	Advantage of Rice-Duck Farming System (RDFS) Over Rice Monoculture (RM)	References
Soil enzyme (urease, phosphatase, sucrase, catalase) activity	Soil enzyme activity in RDFS is 1.13–1.40 times more compared to RM	Teng et al. 2015
Available N	107.33 mg kg^{-1} in RDFS vs 84.00 mg kg^{-1} in RM	
Available P	23.35 mg kg^{-1} in RDFS vs 15.70 mg kg^{-1} in RM	
CH_4 emission	8.80-16.68% less in RDFS compared to RM	Xu et al. 2017
Rice yield	6456.25 kg ha^{-1} in RDFS vs 3403.81 kg ha^{-1} in RM	Teng et al. 2015
	Increased by 0.76-2.43% in RDFS over RM	Xu et al. 2017
	Increased up to 9000 kg ha^{-1} in RDFS compared to 4200 kg ha^{-1} in RM	https://www.equatorinitiative.org
	20% higher rice yield in RDFS over RM	https://agroinsight.com
Cost of production	Reduced by 30% in RDFS compared to RM	https://www.equatorinitiative.org
	Increased by 29% in RDFS over RM	https://agriculture.com.ph
Net income	Increased by 80% in RDFS over RM	https://agroinsight.com
	Increased by 5.2 times in RDFS than RM	https://agriculture.com.ph

integration of fishery component contributes 63% to net return followed by *Kharif* rice (19%), *Rabi* rice (10%), fruit crop (8%), however, *Kharif* rice generated highest employment (288 man-days) followed by *Rabi* rice (154 man-days), horticulture (40 man-days) and fishery (25 man-days) per year (Behera et al. 2017).

3.6.5 Land Modification for Rice-Based Farming System

In coastal West Bengal, the annual excess rainfall of about 420 mm over reference crop evapotranspiration is lost as runoff (Sarangi et al. 2014). After the cessation of monsoonal rain, the soil salinity builds up due to the capillary movement of salts from the underground brackish water and subsequent deposition on soil surface. The soil cracking upon drying is also an important issue affecting crop growth and input use efficiency during the dry season. These issues can be addressed by scientific modification of land, resulting in the creation of water harvesting structures, high and medium lands free from waterlogging. There are basically three types of land modifications: farm pond (FP), deep furrow and high ridge (DF) and shallow furrow and medium ridge (SF).

3.6.5.1 Farm Pond-Based Technology

In this technology, a pond is excavated in the lowland area by the use of about 15–20% of the total area (Bandyopadhyay et al. 2009; Ambast et al. 1998). The pond should have strong, wide (about 3.0 m) and high (about 1.5–2.0 m) bond to provide scope for growing vegetable crops on the side of the bonds. The design dimensions of the farm pond should be developed based on the amount and pattern of rainfall, soil type and diversified cropping system (Table 3.2).

The freshwater pond is generally non-saline, with relatively quiet waters, abundant phytoplankton, zooplankton, large plants and animals. In this ecosystem, the littoral zone (shallow water where light is plentiful) is relatively large and the limnetic (open water zone up to the deposit where effective sunlight penetrates) and profundal (deep-water area which is beyond the depth of effective light

TABLE 3.2
Farm Pond Design Dimensions for Water Harvesting and Rice-Based Diversified Farming in the High Rainfall Coastal Region

Parameters of the Pond	Dimensions/Details
Pond area	40 m × 40 m
Depth	3.0 m
Dead storage	0.5 m
Depth of water for agriculture	2.1 m
Water storage for agriculture	2700 m^3
Total design storage	4700 m^3
Bond width	3.0 m
Bond height	1.5–2.0 m
Gap from cut portion	1.0 m
Slope of dyke	1:1

penetration) zones are small or absent. Based on the water depths, ponds are of two types, temporary and permanent ponds. In temporary ponds, the water dries up for a part of the year, whereas permanent ponds contain water throughout the year. Periodical desilting/renovation of ponds are essential to restore the storage capacity, efficient production of aquatic output and using the pond silts for soil fertility improvement.

The cross-sectional view of the pond is trapezoidal in shape, wider at the top and narrower at the bottom. The length-width ratio of on-farm reservoir (OFR) is taken 1:1 as it is having minimum perimeter and therefore, attains maximum storage (Ambast 2019). It is recommended to convert 17% or 20% of the farm area into OFR, without or with bund, respectively, and the length/width of OFR (20% area) should be 44.7 m (Ambast 2019).

3.6.5.2 Deep Furrow and High Ridge System

Alternate ridge and furrow are created in the low-lying land to create water harvesting structure (furrow) and land free from waterlogging (ridge). In this land modification system, about 50% of the farm area may be converted into ridges (1.5 m top width × 3.0 m bottom width × 1.0 m height) and furrows (3.0 m top width × 1.5 m bottom width × 1.0 m depth), in which the ridges are made by using the soil excavated for creation of furrows (Bandyopadhyay et al. 2009). The ridges are used for year-round cultivation of vegetables/fruits. During the rainy season, the runoff water is stored in the furrows for pisciculture and during the dry season, this fresh water is used for irrigating the crops grown on the ridges. In this system fresh water of about 2250 m^3 ha^{-1} is harvested in the furrows. Three types of land based on elevation are created viz. furrows, ridges and original lowland. During the *Kharif* season, the furrows become a part of paddy-cum-fish cultivation. The original lowland is used for rice cultivation under paddy-cum-fish during the *Kharif* and during the *Rabi* season, lowland is used for growing low-water requiring crops such as potato, mustard, sunflower, green gram.

3.6.5.3 Shallow Furrow and Medium Ridge System

In this land modification system, the dimensions of ridges and furrows are smaller than those under a deep furrow and high ridge system. The purpose is to allow rice cultivation in shallow furrows during the *Rabi*/Summer season with lesser use of supplemental irrigation water. The medium ridges have 1 m top width, 2 m bottom width and 0.75 m height and shallow furrows are 2 m top width, 1 m bottom width and 0.75 m depth (Bandyopadhyay et al. 2010). During the *Kharif* season, the furrows are used for rainwater harvesting and pisciculture under a paddy-cum-fish system. Fruits/vegetables are cultivated on the ridges throughout the year. The original low land is used for rice cultivation with fish during *Kharif* and low water requiring field/vegetable crops during *Rabi*/Summer season. This system may be brought to the extent of 40–75% of farm land to harvest about 1200 m^3 ha^{-1} of rainwater.

3.6.6 Modern Innovations in Rice-Based Farming System

During the last two decades several technologies advances have been made in rice cultivation practices which address challenges in the rice-based farming system. Water-saving practices, high yielding genotypes, stress tolerant varieties and climate smart cultivation practices are some innovations to transform this system.

3.6.6.1 Drip/Micro-Irrigation for Rice

Rice is an irrigation intensive crop as conventional puddled transplanted rice requires about 1300–1500 mm water per season (https://www.knowledgebank.irri.org//step-by-step-production/growth/water-management). Farmers in irrigation commands apply even much more (~2000 mm) water and a major portion of it goes to waste. Irrigated rice consumes about 34–43% of the world's total irrigation water, which is about 24–30% of the entire world's developed freshwater resources (https://www.knowledgebank.irri.org//step-by-step-production/growth/water-management). Inefficient use of water in agriculture has depleted aquifers and reduced river flows; therefore, efficient use of irrigation water in the agriculture sector is an absolute necessity to increase land and water productivity of crops (Sharma et al. 2018).

The global water footprint of rice production is 784 km^3 $year^{-1}$ (784 billion m^3 water $year^{-1}$) with average water use of 1325 m^3 t^{-1} (0.755 kg m^{-3}), which is 48% green (rainwater), 44% blue (irrigation) and 8% grey (wastewater) water (Chapagain and Hoekstra 2011). The national water footprint of rice production is the highest for India (255.5 billion m^3 water $year^{-1}$) followed by China (172.5 billion m^3 water $year^{-1}$), with average water use of 2020 m^3 t^{-1} for India and 971 m^3 t^{-1} for China, whereas the highest water footprint (2874 m^3 t^{-1}) per unit of paddy rice produced is for Pakistan (Chapagain and Hoekstra 2011). It has been reported that among the three major irrigation methods, highest field application efficiency (90%) is achieved in drip irrigation, followed by sprinkler irrigation (75%) and lowest efficiency of 60% in surface irrigation (FAO.org). It has been demonstrated that through the use of drip irrigation 20–66% water could be saved in field, fruit and vegetable crops (Table 3.3).

The technology of drip irrigation and fertigation with mulching (paddy husk/polythene) in rice has resulted in saving of 50.0–66.3% irrigation water, increased mean water productivity from 0.12 kg m^{-3} under flood irrigation to 0.58 kg m^{-3} under drip irrigation and rice yield increased by 7.34–29.9% under drip irrigation studied in locations in Andhra Pradesh, Telangana, Chhattisgarh, Rajasthan, Punjab, Tamil Nadu and Maharashtra states of India (Padmanabhan 2019). Studies at Uttarakhand revealed that the water requirement of drip irrigated transplanted rice ranged between 938 and 1836 litres per kg as compared to over 4000 (4250–5508) litres per kg under surface check basin irrigation (Singh et al. 2019).

3.6.6.2 Stress Tolerant and Bio-fortified Rice Varieties

Development, release and popularization of abiotic and biotic stress tolerant and bio-fortified rice varieties have opened up new vistas in rice farming. Rice varieties have been developed to withstand drought as well as having weed competitiveness for successful cultivation in the drought-prone ecology (7.5 mha). Soil salinity and sodicity is increasingly affecting crop production globally in about 1130 mha area, out of which, about 52 mha area in South Asia and more than 6.7 mha in India. Submergence and flooding adversely affects rice cultivation in coastal, rain-fed low-lying areas of more than 12 mha in India. Apart from abiotic stresses, abiotic stresses such as insect pests, diseases and weeds compete with rice crop and adversely affect the rice yield and economics of cultivation.

TABLE 3.3
Evidences of Irrigation Water Saving Through Adoption of Drip Irrigation in Various Crops

Crop	Water Saving (%) in Drip Irrigation Over Conventional Flood Irrigation	Reference
Rice	50–60	Padmanabhan (2019)
Sugarcane	20–60	Kaushal et al. (2012)
Banana	38–39	Sharma and Kispotta (2016)
Papaya	45–66	Deshmukh and Hardaha (2014)
Citrus	34	Panigrahi and Srivastava (2017)
Cotton	20	Amrutrao et al. (2018)
Wheat	26	Umair et al. (2019)
Maize, barley	36–40	Fonteyne et al. (2021)
Vegetables	40–60	Mahanta et al. (2019)

TABLE 3.4
Stress Tolerant and Bio-Fortified Rice Varieties

Stress/Trait	Tolerant Varieties	Reference
Drought	Vandana, Sahabhagi dhan, Naveen, Anjali, Kalinga III, Virendra, Annada, Pyari, Satyabhama, NDR 97, Heera, Dhalaheera	Swain et al. (2014)
Coastal salinity	Canning 7, Sumati, Utpala, Bhutnath, Amal-Mana, Panvel 1, Panvel 2, Vytilla 2, Vytilla 3, Vytilla 4, Lunisree, Bidhan 2, WGL 20471, Luna Suvarna, Luna Sampad, Luna Barial, Luna Shankhi	Sarangi and Maji (2017)
Inland salinity and sodicity	CSR 10, CSR 13, CSR 23, CSR 27, CSR 30, CSR 36, CSR 46	Sharma et al. (2016)
Submergence	BR 11-Sub1, Ciherang-Sub1, CR 1009-Sub1, IR 64-Sub1, Samba-Sub1, Swarna-Sub1	Sarangi et al. (2020)
Insect pests	Ambika, Triveni, CSR 10, CSR 30, Indira, Mahalaxmi, Mahamaya, Punshi, PR-115, Tara, Tripti, VLdhan-221, Kshira, Seema, Saktiman, Saket-4 and Mandira	Subudhi and Padhi (2008)
Zn rich	DRR Dhan 45 (Zn 22.6 ppm)	Yadava et al. (2017)
Protein rich	CR Dhan 310 (10.2% protein), CR Dhan 311 (10.1% protein)	Chattopadhyay et al. (2019)
Climate-smart rice	CR Dhan 801, CR Dhan 802	Pradhan et al. (2019)

However, recently, varieties with tolerance to abiotic and biotic stresses have been developed, tested in farmers' fields and released for large-scale cultivation (Table 3.4).

3.6.6.3 Innovation for Rice + Vegetable Cultivation in Paddy Field

Rice is a semi-aquatic crop, therefore it is cultivated in a waterlogged situation, where other aerobic crops cannot survive. Therefore, in the vast area of coastal lowland, rice is the only crop choice. However, recent innovations with modification of rhizospheric environment by the introduction of soil- and manure-filled bags enabled intercropping of vegetables with rice in this ecosystem. In this system, climbing types of vegetables such as ridge gourd, bitter gourd, cucumber and bottle gourd are cultivated along with rice (Figure 3.7). Locally available used bags of about 100 kg capacity are filled with about 90 kg soil and 10 kg farm yard manure, mixed thoroughly and placed in a

FIGURE 3.7 Cultivation of cucumber crop in waterlogged paddy field. (a) vegetative stage (b) reproductive stage.

waterlogged paddy field at a spacing of 3 × 3 m in such a way that the top of the bags remains about 0.3 m above the paddy field water level. This technology has been found profitable in the Sundarbans region of West Bengal, India, with BCR of 2.54–2.61 compared to 1.201.–81 under sole rice cultivation in paddy fields.

The climbing types of vegetable crops such as cucumber grow on a wire net placed well above the rice crop canopy by the use of locally available wooden poles (Figure 3.7a). The effect of the intercrop (cucumber) on the main crop (rice) is negligible as the bags and the wire net are placed in such a way that there is no shading for the rice crop (Figure 3.7b).

3.6.6.4 C4 Rice

The C4 pathway is characterized by a complex combination of biochemical and anatomical specialization that ensures high CO_2 partial pressure at RuBisCo sites in the bundle sheath cells (Ermakova et al. 2020). Rice has the C3 photosynthetic mechanism, as its leaves does not have Kranz anatomy. C3 plants are less efficient in fixation of CO_2 as photorespiration is not suppressed, whereas C4 plants (e.g., maize, sugarcane, sorghum) have high photosynthesis rate as photorespiration is reduced. The C4 pathway can raise photosynthetic efficiency by 50%, improve nitrogen use efficiency and treble water use efficiency in rice. Because their rate of photosynthesis significantly increases with increasing light intensity (measured as photosynthetic photon flux density) and does not appear to saturate (Rizal et al. 2012), C4 plants can utilize the maximum amount of sunlight. By 2050, each hectare of rice-growing land will need to feed at least 43 people (up from 27 now), necessitating a 50% increase in rice yield to prevent mass malnutrition among the 700 million Asians who currently rely on rice for more than 60% of their daily calorific intake (Hibberd et al. 2008). Success of the development of C4 rice will be a great step towards meeting the food security of the poor people in the underdeveloped and developing countries.

3.6.6.5 Climate Smart Practices for Sustainability and Resilience

3.6.6.5.1 Climate Smart Agronomy

Due to global warming, there is a change in rainfall patterns; diurnal and seasonal variations of temperature; increases in CO_2 concentration; occurrences of cyclones, storms, droughts, floods as well as increase in abiotic (salinity, water scarcity) and biotic (diseases, insect pests, weeds) stresses. Therefore, commensurate adaptation strategies are required to make modern rice cultivation sustainable to feed the increasing population. Under flooding/submergence events in the coastal areas, a combination of tolerant rice varieties and matching agronomic management practices could reduce the risk. Flood tolerant varieties such as Amal-Mana and Sabita performed better under stagnant flooding, whereas, under flash flood situations, cultivars such as CR 1009-Sub1, BR11-Sub1 and Swarna-Sub1 have yielded advantages (Sarangi et al. 2016; Sarangi et al. 2020).

Decision support systems (DSS) will be increasingly helpful in agriculture for the rational use of inputs as well as to deal with abiotic and biotic stresses. Rice doctor is an interactive tool for researchers, extension workers, students and others who want to learn about, diagnose and find solutions to pest, disease and other problems in rice (https://www.irri.org/rice-doctor). For nutrient management in rice, Rice Crop Manager, launched in 2013 by IRRI, gives an yield advantage of 0.5 t ha^{-1} $crop^{-1}$ (≈USD 150 ha^{-1} cropping $season^{-1}$) in India and 0.4 t ha^{-1} $crop^{-1}$ (≈USD 100 ha^{-1} cropping $season^{-1}$) in the Philippines (https://www.irri.org/crop-manager). To manage insect pests, diseases, nematodes, weeds, nutrient deficiencies and toxicities in rice, a mobile App 'riceXpert' has been developed by the ICAR-National Rice Research Institute, Cuttack, Odisha, India (Mohapatra et al. 2019). This App is developed for Android mobiles, which can be downloaded from Google Play Store or from www.nrri.in or from www.crri.nic.in for use as a diagnostic tool for a rice field by sending text, photo and recorded voice, and the solution is received through SMS (https://icar-nrri.in/). However, DSSs, used in agriculture, are focused on increasing the supply of individual soil functions such as nutrient cycling, while neglecting other important soil functions such as water purification and regulation, climate regulation, carbon sequestration, soil biodiversity and habitat provision (Debeljak et al. 2019).

Sustainability of a system is defined through the triple bottom line of environmental, social and economic aspects, whereas resilience is viewed as the ability of a system to prepare for threats, absorb impacts, recover and adapt following persistent stress or a disruptive event (Marchese et al. 2018). Diversification of a rice-wheat cropping system (RWCS) may be achieved through one or other options such as replacing rice, replacing wheat, replacing whole cropping system, addition of a third crop to make RWCS sustainable, introduction of break crop/vegetable crop/medicinal and aromatic plants through farming system approach (Shahane and Shivay 2019). In coastal regions of India, diversification of dry season rice crop (*Boro*) to crops such as maize, mustard and potato saves irrigation water, increases income and improves soil health (Sarangi et al. 2020a; Sarangi et al. 2021a). Diversification to legumes reduces the use of external inputs, fixes atmospheric nitrogen, releases in the soil high-quality organic matter and facilitates soil nutrients circulation, water retention, reducing GHG emission in addition to serving as high-quality food and feed (Stagnari et al. 2017). Diversification of farming to include enterprises such as livestock, fishery, tree species and orchards adds to resilience as well as maintaining ecosystem services.

3.6.6.5.2 Agromet Advisory, Market Intelligence and Crop Insurance

Agromet advisory and use of ICT play significant roles in providing information about the abiotic and biotic stresses such as floods, drought, dry spells, hail storms, high winds and pest outbreaks. Information on market-related aspects such as availability of inputs, prices and scope for sale of outputs also empowers farmers to make decisions and adopt precautionary measures to reduce the shock of these events. The accurate and timely availability of this information helps farmers in effectively implementing the farm operations. The agro-advisory services play a big role in making the rice-based farming system climate smart. Use of ICTs such as mobile base SMS, WhatsApp, mobile calls, web-based materials in local languages, community radios, e-bulletins, etc. will help farmers in the vulnerable regions respond effectively to weather- and climate-related risks/uncertainties. Contingency measures such as switching to alternate climate resilient crops, scheduling the time of application of inputs such as fertilizers, pest control products and soil amendments could be well planned provided the information related to weather reaches the farming community in a timely way.

Use of market intelligence in Punjab by the basmati growers during 2012–2013, earned additional income of ₹381 crores (avg. ₹25,400 per hectare) by reducing the area under basmati rice (Sidhu et al. 2014). In India crop insurance schemes such as *Pradhan Mantri Fasal Bima Yojana* (PMFBY) is implemented from 2016 to extend financial support to the affected farmers. Critical barriers to broader access and availability of crop insurance still need to be addressed, including increasing efforts to sensitize the farmers to its advantages (Kumar et al. 2021). A website (https://pmfby.gov

.in) has been created to facilitate implementation of PMFBY and it is available in 12 languages (English, Hindi, Assamese, Bengali, Gujarati, Kannada, Malayalam, Marathi, Odia, Punjabi, Tamil and Telugu). The website gives the detailed operational guidelines for weather-based crop insurance scheme (WBCIS), unified package insurance scheme (UPIS), national agricultural insurance scheme (NAIS), national crop insurance programme (NCIP), modified national agricultural insurance scheme (MNAIS) and coconut palm insurance scheme (CPIS).

3.6.6.5.3 *Climate-Smart Storage System*

Rice seeds should be stored at 12% moisture content, whereas rice to be used for grain purposes may be stored at 14% moisture content. Modern hermetic storage methods help in safe storage of rice seeds without pests and with extended germination life. The cocoon is a commercially available hermetic storage container that consists of two plastic halves that are joined together with an airtight zipper after the cocoon is loaded with sacks filled with rice grains (https://www.knowledgebank.irri.org/step-by-step-production/postharvest/storage/grain-storage-systems/hermetic-storage-systems/cocoon). IRRI super bags are farmer-friendly storage bags that extend the germination life of seed for planting from 6 to 12 months, control insect grain pests (without using chemicals) and improve the head rice recovery by 10% (https://www.knowledgebank.irri.org/step-by-step-production/postharvest/storage/grain-storage-systems/hermetic-storage-systems/irri-super-bag).

3.6.6.5.4 *Ecosystem Services of Rice-Based Farming Systems*

Rice grows in a unique habitat with ponded water, surrounded by a network of bunds, landscapes with shrubs and trees and other enterprises such as livestock, birds, fishes and the household. This makes the exclusive ecosystem with a wide array of plants, animals and biodiversity to function in several ecosystem services. However, in modern chemical-based farming, due recognition and respect is underestimated as per the lives in and around the rice-based farming system. Issues include indiscriminate use of harmful pesticides and other agro-chemicals, absence of awareness of safe handling and use as well as non-adoption of eco-friendly methods of pest management, selective use of high-yielding varieties of crops has made the ecosystem prone to pest attacks as well as resulting in the loss of aesthetics of this beautiful landscape called rice farming. In the pursuit of increasing the cropped area, other landscapes (highlands, hillocks, forests) are cleared, which not only reduces the vegetation diversity but also results in loss of habitat of other creatures such as birds, reptiles, foxes, mongooses, etc.

In this backdrop, rice-based farming system needs an ecological engineering with integration of non-chemical and natural pest management strategies, trap cropping, and restoration of soil fertility through organic approaches such as green manuring, crop residue recycling which adds to the ecosystem value of this farming system. Productive utilization of the rice field bunds for growing legumes such as moong bean, urd and pigeon pea as well as for growing *Dhaincha* (*Sesbania* spp.) enhances ecosystem services. Growing such plants in the field bunds serves as a resting place for beneficial birds (Figure 3.8), which predate on the harmful insects/caterpillars of rice plants giving ecological benefits to the farmers. In the Mekong Delta region of Vietnam, farmers who practice rice-fish co-culture resort to fewer insecticide applications and secure higher income apart from getting fish as food and other services (Horgan et al. 2022).

3.7 FRESHWATER AQUACULTURE-BASED IFS MODELS

The low-cost aquaculture-based IFS (Figure 3.9) results in productive agricultural waste utilization and ensures sustainable use of available resources and personnel. These IFS models combine agriculture, livestock farming and fish culture, such as fish-cum-pig, fish-cum-duck and fish-cum-poultry farming with vegetable/fruit crops. The introduction of animal manure to ponds encourages the growth of phytoplankton and zooplankton, which serve as a source of nutrients for fish. Fish ingest indigestible feed components found in dung directly. Through fish culture, leftover feed from

FIGURE 3.8 Cultivation of *Dhaincha* (*Sesbania* spp.) on rice field bunds provides resting place for predatory birds.

FIGURE 3.9 An aquaculture-based integrated farming system model (Aquaculture, On-dyke horticulture, Agriculture, Duckery, Poultry, Agroforestry) developed by ICAR-IIWM, Odisha, India.

feeding can also be recycled. The pond bottom soil layer, which is rich in biomass, is occasionally removed and transferred to a field for crop cultivation. Fish pond silt is greatly sought after by farmers as it makes excellent manure for crops.

Farm leftovers like leaves, stalks, or other waste materials can be given to fish. Additionally, livestock waste that accumulates around fish farms can be used to generate plankton, which is a significant source of food for fish. For the sake of sustainability, integrated fish farming combines several agreeable agricultural enterprises into a workable or unified complete farming system. The by-products of one business are recycled into another as input, creating a zero waste, low-cost, and low-energy production system (Ayinla 2003). The production of fish as well as meat (chicken, duck, cattle, hog, etc.), milk, vegetables, fruits, eggs, grains, fodder and mushrooms has the potential to improve the socioeconomic condition of underprivileged groups.

3.7.1 Fish-Duck Integrated Farming

Ducks are quite compatible with farmed fish, therefore raising them in fish ponds works well with the fish culture method. Farmers can gain from this arrangement in a variety of ways. The ducks fertilize the pond with their waste when given free range at the pond's surface. Ducks are known as manuring machines because of their effective way of producing pond manure, which completely eliminates the need for pond fertilizer and supplemental fish feed, which accounts for about 60% of the whole expense of conventional fish culture. Ducks control aquatic plants, loosen the bottom of ponds with their dabbling and aid in nutrient release from the soil, all of which boost pond productivity. Ducks are biological aerators because they aerate the water as they swim. On the pond's dyke, duck houses are constructed; therefore, additional land is not required. Ponds provide the majority of the food that ducks need in the form of aquatic grasses, insects, larvae, earthworms and other such things. They require extremely little food, which is typically provided by farmers in the form of kitchen trash, molasses and rice bran.

In this system, duck density usually varies between 100 and 3500 ha^{-1} (Sin 1980). Woyanarovich (1980) reported that 300–500 ducks ha^{-1} gave fish yield of 500–800 kg ha^{-1} in 150 days in Hungary without supplemental feed where carps were cultured. In Vietnam, rearing 1,000–2,000 ducks ha^{-1} on ponds enhanced the regular fish yield to as much as 5.0 t ha^{-1} yr^{-1}, compared to 1.0 t ha^{-1} yr^{-1} without ducks (Delmendo 1980).

The ducks are allowed to roam freely over the pond's surface from 9 am to 5 pm, when they naturally manure the pond by dispersing their waste throughout it. Between 125 and 150 g of droppings are excreted daily by each duck. Every morning, the excretions are removed from the duck home and added to the pond. The rearing density of 200–300 ducks ha^{-1} in one hectare pond gives 10,000–15,000 kg of droppings, which is recycled every year. On a dry matter basis, these droppings contain 81% moisture, 0.91% nitrogen and 0.38% phosphate. Duck raising in fish ponds reduces the protein demand by 2–3% in duck feeds. The daily waste of duck feed (around 20–30 g $duck^{-1}$) aids as fish feed in ponds or as manure, ensuing in higher fish yield. Ducks' body weight and feed efficiency both rise, and fish may be able to use the spilled meals. Ducks raised in fish ponds have a 3.5% higher chance of surviving due to a clean environment.

In this system, the pond should be water-retentive, holding water of 1.0–2.0 m depth for eight to nine months. The pH of the soil should fall between 6.5 and 7.5. If the soil pH is not at the desired level, it can be adjusted by adding lime in 3 to 4 split doses (about 400 kg ha^{-1} for pH 6.5 to 7.0, 200 kg ha^{-1} for pH 7.0 to 7.5). Lime helps in maintaining pH and kills harmful parasites. The basal dose of lime and cow dung application in one hectare meter of water is 400 kg and 5000 kg, respectively. There should be regular cleaning of ponds to ensure sunlight penetration, oxygen circulation and removal of predators from the pond.

Fish fingerlings can be stocked preferably during the June and July months depending upon the climate and the availability of optimum water level in the pond. In contrast, fingerlings may be stocked in April and harvested in November/December when water bodies remain functional for eight to nine months. Stocking density should range between 5000 and 6000 fingerlings ha^{-1}. Fish species with a macrophytic feeding habit, effective phyto- and zooplankton utilization and human consumption make excellent candidates for integrated livestock fish farming. Fishes are categorized into three groups based on how they feed: bottom feeders, column feeders and surface feeders. Both Indigenous and exotic fish species are advised for an integrated fish farming system. Indigenous species like *Catla catla,* which is a zooplankton feeder, and exotic species *Hypophthalmichthys molitrix,* which is a phytoplankton feeder, are fine example of a surface feeder, whereas *Labio rohita* an indigenous omnivorous species is a column feeder. The Indigenous species *Cirrhinus mrigala, Labeo calbasu* are detritivorous and exotic common carp species (*Cyprinus carpio*), which is detritivorous/omnivorous in nature are bottom feeders. Exotic species like *Ctenopharyngodon idella,* which is herbivorous in nature, cover surface, column and marginal areas of feeding zones. Depending upon the availability of fingerlings, composite fish culture with 3, 4 or 6 species combination can be adopted. In integrated livestock cum fish farming considering the surface, column and bottom feeder the ratio of fishes, viz., Catla, Rohu and Mrigal – 4: 3: 3 (3 species), Catla, Rohu, Mrigal and common carp – 3 : 3 : 2 : 2 (4 species) and Catla, Rohu, Mrigal, Silver carp, Grass carp and common carp – 1.5 : 2.0 : 1.5 : 1.5 : 1.5 : 2.0 (6 species), respectively.

3.7.2 Fish-Chicken Integrated Farming

This system uses chicken droppings for fish cultivation to minimize expenses and increase returns. Fish culture can be combined with the raising of chickens for meat (broilers) or eggs (layers). The chicken waste can be repurposed to fertilize the fishponds by raising chickens over or close to the ponds. Raising chickens over ponds makes the most efficient use of available space and reduces labor-intensive manure transportation. There are two methods for integrating fish and fowl. Installing the coops above the pond's surface is one approach to ensure that the droppings fall into the pond. The alternative is to construct deep litter chicken coops and recycle the waste daily at a rate of 50 kg ha^{-1} into fish ponds. Since fully builtup littre has nearly three times the manurial value of chicken droppings, the deep litter method is typically selected. Without the need for the addition of supplemental feed, poultry dung is a complete fertilizer that possesses the qualities of both organic and inorganic fertilizers (FAO 2003). Poultry waste contains less moisture, fiber and compound like tannins that discolours pond water when used as fertilizers. Fresh chicken dung contains 1.6% nitrogen, 1.5% phosphorus and 0.9% potassium (Woynarovich 1979). Every day, each bird produces about 40 g of dung. Like other organic waste, poultry manure has the ability to transform low-quality inedible materials into high-quality fish feed (Schroeder 1980). By releasing the nutrients and minerals that were once bound, it can promote microbial activity in the water column and at the pond bottom and can act as substrates for the photosynthetic production of essential fish food. For a 1.0 hectare pond, 500–600 birds are typically required to supply the necessary fertilizer. Two batches of broilers are made from chicks that are about two weeks old, and laying hens are made from chicks that are 4.5 to 5.5 months old. Laying hens are raised for 12–18 months in order to produce eggs. Without the need of any additional fish feed or ponds, this system produces 4.5 to 5.0 tons of fish, about 70,000 eggs, and 1.0 tons (live weight) of chicken meat from a 1.0 hectare water area per year. At low to moderate stocking densities, it has been claimed that ponds fertilized with poultry manure produce more fish (2.5–4.5 t ha^{-1}) than those treated with pig or cow manure (Fagberno and Sydenhan 1988). Important practices for profitable chicken rearing:

- Poultry with dual purposes makes a suitable choice for integrated poultry-fish production.
- A separate shed should be used for brooding.
- Using a deep litter management system is preferable instead of building a floor with slats over the pond.
- Comply with the deworming and immunization regimen.
- Rotate your litter at least once a day to prevent the production of cakes.

- Remove your shoes before entering the shed.
- Make a good nest for laying eggs so they are clean.
- Dissuade people from visiting the property.
- Keep the layer's 14-hour light and 10-hour darkness pattern in place.
- Offer sufficient calcium to layers.
- Keep accurate records of the production of eggs and other farm data.

3.7.3 Fish-cum-Cattle Integrated Farming

Fish-cum-cattle integration can cut down/save more fertilizers, fish feeds and increase the income from milk, fish and beef. In this system cow dung is used as manure for fish culture. Farmer can apply 5,000 to 10,000 kg of raw cow dung per ha to a fish pond over time. The growth of plankton, which serves as fish food, is encouraged by this cow dung. The wastewater from cleaning cow sheds can be diverted to a pond that has excellent manurial value, along with the cow dung, urine and leftover feed (Athithan et al. 2001). One cow or buffalo will typically excrete 12000 kilograms of dung and 8000 litres of urine annually. The filter-feeding and omnivorous fish benefit from the urine and feces of cattle. Fish feed can be made from the leftover cattle feed. Building cattle sheds on the fish pond's broad embankment will allow waste and washing to run right into the fish pond. In addition to 9000 l of milk and 3000–4000 kg of fish annually, roughly five to six cows can supply enough manure for a 1 hectare pond. Building a cow house near a fish pond will make handling of cow manure easier. The system gives a net profit of Rs. 3-4 lakhs per year from one-hectare land.

3.7.4 Fish-cum-Rabbit Farming

Rabbits are less susceptible to disease than chickens and ducks. The rabbits can be raised both on the ground and above the pond. According to studies, 400–500 rabbits are sufficient for a one-hectare fish pond, depending on the pond's productivity. Every year, a rabbit excretes 60–70 kg (140–150 g per day), which is turned into acceptable manure for fish farming. The feces and excreta produced by rabbits are completely utilized in fish rearing as powerful organic fish food and manure.

Two species of rabbit are preferred for the baby rabbit production (1) *European wild rabbit*: It is an exotic type rabbit and delivers 80 babies annually and (2) *American cotton tail rabbit*: It delivers 85 to 90 babies per year. One female rabbit produces around 8 offspring in 45 days; 400 female rabbits produce 3200 babies in 45 days; 3200 bunnies are ready after three to five months (0.5 to 5 kg), resulting in 9600 kilogram rabbits. The rabbit, like other animals, consumes cabbage, carrots, dry leaves and hay. A mature rabbit (male or female) consumes only 140 to 200 grams of hay or vegetables per day. In order to ensure that every rabbit is replaced every two years and that there are always rabbits available for farming, only 40 to 50% of the entire rabbit population should be culled annually. Rabbits give birth to young between the ages of three and six years. To prevent the rabbit from contracting any diseases, vaccinations should be done periodically, and the environment should be clean. Heat stress accounts for 73% of rabbit mortality, but under a cooling regime, mortality is reduced to 6%, after accounting for losses owing to conception rate, litter weight and pre-weaning death. To relieve heat stress in rabbits, cool drinking water should be provided. Temperature, humidity and ventilation are the most critical factors influencing a rabbit's physiological stress. In this system, fish yields of 3.5–4.0 tons by stocking 15000 fingerlings ha^{-1} and rabbit production of 9.0–9.5 tons annually can be achieved with gross income of ₹ 6.0 lakh per year.

3.7.5 Fish-cum-Pig Farming

The most common pig breeds for integrating with fish are White Yorkshire, Hampshire and Landrace. Pig dung is useful to condition the soil and supply the nutrients needed to fertilize the pond water. Pigs are viewed as 'costless fertilizer factories' in China, where fish-cum-pig culture is widely practiced. According to reports, 100 pigs weighing 30 kg each were thought to be sufficient in Hong Kong and the Philippines to fertilize one hectare fish pond and maximize total revenue (Bardach 1972). In India, a pig husbandry and fish culture production system that is both economically and scientifically sound has been developed. Pig dung is either dumped directly into the pond or retrieved from the animal house and dumped in the pond. Pig stalls are constructed on or near the back of the pond. In most case, fish consume pig excrement directly, which is a 70% digestible source of food for fish. There is no further need of manuring the pond or feeding the fish with this arrangement, accounting for 60% of the input costs in fish culture. Fish production of about 5.5-6.0 tons ha^{-1} year^{-1} can easily be obtained through this system with rearing density of 20,000 fish fingerlings and 60–80 pigs ha^{-1} pond (Francis et al. 2004). In India, it is found that about 35–40 pigs are sufficient to fertilize one hectare of fish pond, while fish production in pig-fish farming operations ranges between 2.0–5.0 tons ha^{-1} in six months. In some instances, a fish-pig integrated farming system with a rearing density of 80 pigs and 20,000 fish fingerlings per ha of water area was able to produce 6.8 tons ha^{-1} year^{-1} of fish (Hopkins 1982), whereas in Thailand, 150 pigs per pond were able to produce five tons of tilapia annually (Edward 1985). Pig cum fish farmers may experience an economic advantage of up to 28% to 30% above conventional pig farmers (Ayinla 2003).

The system has obvious advantages:

- Pig excrement serves as a great pond fertilizer, increasing the pond's biological productivity and, as a result, the amount of fish produced.
- Some fishes consume pig excrement directly because it provides 70% digestible fish food.
- The pond dikes allow room for the construction of animal housing units.
- No supplemental feed is required for the fish culture, which typically accounts for 60% of the total input cost in conventional fish culture.
- Pond water is used to wash the pigs and clean their stalls.
- This system has special significance in the North Eastern Region of India as it can improve the socioeconomic status of weaker communities, especially the tribals who traditionally raise pigs in their backyards and are fond of eating fish.

3.7.6 Fish-cum-Sheep/Goat Farming

The goat is regarded as a poor man's cow, and goat excreta are regarded as an excellent organic fertilizer. The goat excreta contain 60% organic carbon, 2.7% N, 1.78% P, 2.88% K, and its urine is also equally rich in both N and P. At least 50–60 goats are essential to fertilize one ha pond. The goats should have a dry, safe and comfortable home that is protected from extreme heat. For milk, appropriate goat breeds include Jamanapari, Beetal and Barbari, while Bengal, Sirihi and Deccani are utilized for meat. Goats are choosy eaters, eating Berseem, Napier grass, Cowpea, Soybean, Mulberry, and other plants. This integration can offer 3.5–4.0 tons fish ha^{-1} year^{-1} without the need for additional feeding or fertilizer.

3.7.7 Fish-cum-Sericulture System

Mulberry is the producer in this integration, silkworm is the initial consumer, and fish is the secondary consumer, swallowing silkworm feces directly. Inorganic nutrient in the silkworm feces is utilized by phytoplankton, and filter-feeding fish, in turn, consumes heterotrophic bacteria. The seri-fish system connects the mulberry and pond sub-systems. Mulberry leaves are harvested and given to silkworms, and waste material from silkworm rearing enters the fish pond as a mixture of mulberry leaves and silkworm excrement.

3.7.8 Fish-cum-Horticulture Farming

A significant portion of an aquaculture farm is available in the form of dykes, which can be better used for the production of horticultural produce. An integrated horti-agri-aquaculture farming technique results in improved resource management and higher yields. In general, a land space of about 200 m^2 is required to provide the daily demands of fruits for a family of four to five individuals. A fish pond's embankments typically cover an area of more than 200 m^2. Pond bunds cover 25 to 33% of farmland. A farmer's family can not only meet their daily fruit and vegetable requirements but also make extra money by growing fruits and veggies on pond bunds. Flowers such as marigold, rose and jasmine can also be grown on pond embankments. When bananas or coconuts are grown in rows in wetlands, the ditches formed between the rows serve as supply or drainage channels. These canals serve as fish culture systems owing to their round-the-clock supply of water and rich insect populations. Some fish species such as *Channa marulius, C. strialus*, common carp and tilapia are suitable species for culture in this system. Banana, papaya, coconut, lemon and guava are more suitable among fruit crops for growing on pond embankments.

3.7.9 Fish-cum-Vegetable Farming

In this system vegetables can be grown throughout the year. Several varieties of winter vegetables (cabbage, cauliflower, tomato, brinjal, coriander, turnip, radish, beans, spinach, fenugreek, bottle gourd, potato and onion) and summer vegetables (amaranth, papaya, okra, bitter gourd, sponge gourd, sweet gourd, ridge gourd, chilly, ginger and turmeric) can be cultivated depending upon the size, shape and condition of the dykes. The various crops provide strength and stability to the pond embankment, prevent soil erosion and keep the pond water from becoming excessively murky. Further, the pond helps the system by providing humus, which is pulled out of the pond after each harvest. Humus is an excellent source of fertilizer for crops.

Intensive vegetable growing can be carried out on broad dykes (4 m and higher) adopting the practice of multiple cropping with single or mixed crops year-round can yield 65 to 75 t ha^{-1} $year^{-1}$. Semi-intensive farming is viable on pond dykes (2 to 4 m wide) where regular ploughing, irrigation and weeding are not practical. Crops with a longer growing season, such as beans, ridge gourd, okra, papaya, tomato, brinjal, mustard and chilli, have been found to be ideal for such dykes. Pond dykes (up to 2 m wide) can be used for cultivation of sponge gourd, sweet gourd, bottle gourd, citrus and papaya after initial cleaning, weeding and digging small pits along the length of the dykes. Extensive cultivation of ginger and turmeric is suitable for shaded dykes.

During harvest, leaves of cabbage, cauliflower, turnip and radish can be used profitably as supplemental feed for grass carp. Grass carp can be fed turnip, cabbage and cauliflower leaves in the winter, and amaranth and water-bind weed by fortnightly cutting in the summer as supplementary feed. At a stocking density of 1000 fish ha^{-1}, a monoculture of grass carp fed solely on vegetable leaves has an average annual productivity of roughly 2 t ha^{-1}. While a mixed culture of grass carp, rohu, catla and mrigal (50:15:20:15) at a density of 5000 fish ha^{-1} produces an average of 3 t ha^{-1} $year^{-1}$.

3.7.10 Integrated Fish-Water Chestnut Farming

Water chestnut (*Trapa bispinosa*) is an annual aquatic angiosperm plant native to the tropics and subtropics that can coexist with carp and other air-breathing fish species. It has a large potential in terms of nutritional characteristics for humans. In India it is extensively grown in Bihar, Madhya Pradesh, Uttar Pradesh, West Bengal, Odisha, Jharkhand, Karnataka and Jammu and Kashmir. The plant is commercially cultivated for its nutritious nuts. The dried fruit contains approximately 70%

carbohydrate, 13% protein, 2 to 7% sugar, 0.8% fat, 3% mineral elements (iron, calcium, magnesium, phosphorus) and plenty of iodine, as well as significant amounts of vitamins such as thiamine, riboflavin, nicotinic acid, vitamin C, vitamin A and phosphorylases, whereas the raw fruit contains approximately 70% water, 23.3% starch (carbohydrate), 4.7% protein, 0.3% fat, 0.6% fiber and ash. The fruit can be eaten fresh, boiled, roasted or dried and ground into flour. It is also an important plant of the Indian Ayurvedic system of medicine which is used for problems in the stomach, genito-urinary system, liver, kidney and spleen. It is useful in diarrhoea, dysentery, ophthalmopathy, ulcers and wounds. Due to its nutritive, strengthening and cooling properties it increases fertility and chances of conception for women, stabilizes pregnancy and prevents abortion.

Water chestnut is transplanted at a rate of 10,000–15,000 plants ha^{-1} in this dual production system. The 10% space in the centre of the pond/waterbody is kept empty as an open space/refuge region for appropriate oxygenation and fish growth. Catfish (*Clarious batrachus*) at 15,000–20,000 fingerlings ha^{-1}, murrels (*Channa* spp) at 10,000–15,000 fingerlings ha^{-1}, and carp (Rohu, Catla, Common carp, and Mrigal) at 5000–6,000 advanced fingerlings ha^{-1} can be integrated in the ratio of 40:20:20:20 between March and April, and fish harvesting is completed between December and January. Water chestnut is harvested from October to November. Growth of fish is found appreciably higher in combination with water chestnut cultivation and 25 to 30% of the supplemental fish feed could be replaced with natural feed from the growth environment. The fish-water chestnut system can yield water chestnuts at 15–18 t ha^{-1}, carp at 3.5 t ha^{-1}, and air-breathing fish at 1.7–1.8 t ha^{-1} each crop. This co-production system can easily generate a net profit of ₹1.2–1.5 lakh ha^{-1}. Fish-water chestnut co-production has the potential to increase overall output, water productivity, reciprocal advantage and efficient usage of shallow waterlogged areas for improved productivity. This system can reduce by more than 65% pest attacks by the water chestnut beetle (*Galerucella birmanica*), as *magur* acts as predator to *G. birmanica*.

3.8 INTEGRATED RICE-FISH FARMING

The rice-fish/rice-animal co-culture system (Figure 3.10) is a unique agro-landscape found throughout the world, particularly in tropical and subtropical Asia. Considering rice-fish farming's outstanding contribution to food and livelihood security, its importance in terms of biological diversity and genetic resources, landscape diversity, aesthetic beauty, cultural values, other ecosystem goods and services and Indigenous knowledge of land and water management, it has been designated as a Globally Important Agricultural Heritage System (GIAHS) by FAO and UNESCO. Rice-fish farming has been used in Asian countries for over 2000 years (Anita et al. 2014; Islam et al. 2015; Lu and Li 2006) and is widely acknowledged as a promising alternative to rice mono-cropping (Suloma and Ogata 2006). This IFS can improve productivity (Mohanty et al. 2009; Xie et al. 2011; Saikia et al. 2015), soil fertility by increasing nitrogen and phosphorus availability (Giap et al. 2005), nutrient use efficiency and reducing nutrient loss to the environment (Feng et al. 2016; Li et al. 2008; Oehme et al. 2007).

Rice cultivation, which consumes up to 90% of total irrigation water in Asia and feeds nearly 50% of the world's population, is a vital component of food and environmental security (Liu et al. 2018). Rice production system is a major source of GHG emissions and waste of available water resources (FAO 2013). Rice and aquatic animal co-culture in paddy rice systems has been proposed as a strategy to improve the utilization of land and water resources to provide both grains and meat to humans, while reducing the risks of environmental pollution associated with rice production (Hu et al. 2016; Ahmed and Garnett 2011). The animals used in the co-culture system may vary from site to site, depending on the tradition and farmers' interest (Zhang et al. 2017; Halwart and Gupta 2004). This system may improve food safety by requiring less/no pesticide treatment because cultured animals can help to reduce weeds and devour numerous insects and pests as a feed source (Frei et al. 2007; Hu et al. 2016; Rahman 2016). Furthermore, the nitrogen (N) and phosphorus (P) released from aquatic creature feeding in a rice-animal system are available for plant uptake. As a

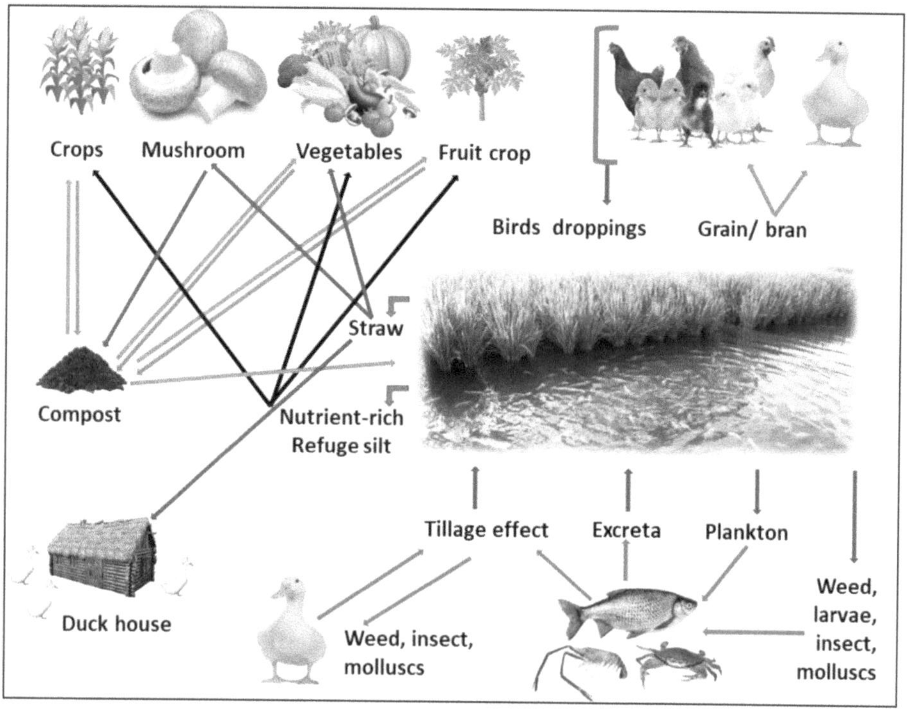

FIGURE 3.10 Schematic representation of major components, pathways of material flow and benefits in integrated rice-animal co-culture system.

result, the amount of N fertilizer given to the rice crop can be reduced (Feng et al. 2016; Zhang et al. 2016).

There are four fundamental rice agro-ecosystems, each with its own set of edaphic conditions: irrigated ecosystem, upland and lowland rainfed environment and flood-prone or deep water ecosystem (Figure 3.11). In this environment, fish culture is carried out concurrently or rotationally with rice at four intensities: traditional (catch), low-intensity culture (without feed and fertilizer), medium-intensity (just fertilization) and high-density culture (with feed and fertilizer). Fish incorporation into rice growing provides important protein, particularly for subsistence farmers who manage rain-fed agricultural systems. Rice fields provide shade and organic matter for fish, which oxygenate soil and water, consume rice pests and promote nutrient recycling (Mohanty et al. 2009; Brahmanand et al. 2006).

In most countries, rice-fish/animal co-culture farming is characterized by intensive rice-culture and extensive fish-culture, where fish yields are usually very low, ranging between 0.3–2.5 t ha^{-1} (Nhan et al. 1998; Mohanty et al. 2008), although there are examples of higher production levels up to 7.03 t ha^{-1} (Chen et al. 1995). In these systems, rice yield usually ranges between 2.60–3.25 t ha^{-1} (Mohanty et al. 2009), reaching as high as 4.84 t ha^{-1} (Feng et al. 2016), which is even higher than the mean yield of the world's rice production of 4.49 t ha^{-1} (FAOSTAT 2014). Usually, farmers adopt different types of field design for integrated rice-fish culture in low and medium land ecosystems, with or without integration of other components like duckery, poultry, on-dyke horticulture, etc. In lowland situations, in-field refuge with peripheral trench is generally practiced, where fish move all around the field and when water recedes, they aggregate in the refuge (Mohanty et al. 2004). In a medium land ecosystem, out-field refuge at the downstream of the field is preferred, where excess water from the rice field is collected in the refuge for fish culture separately. However, yield performance seems to be higher in a medium land ecosystem due to less water submergence in a rice field,

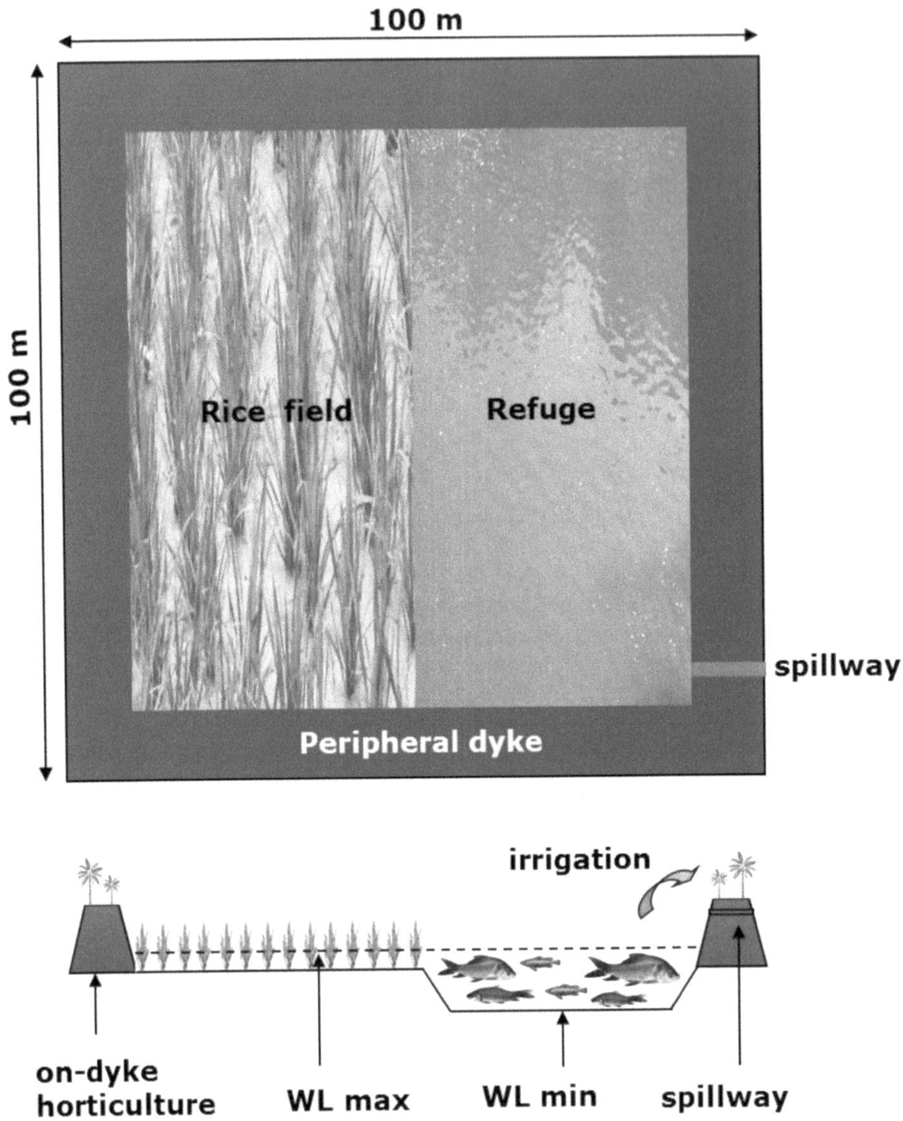

FIGURE 3.11 Layout of deepwater rice-fish system (1 ha unit).

more water available to fish in the refuge and better operational management (Mohanty et al. 2004; Brahmanand et al. 2006; Mohanty et al. 2009).

Recently, a system of rice intensification (SRI) approach (Thakur et al. 2015) and triggering compensatory growth (CG) response in fish through cyclic feed restriction and refeeding (Turano et al. 2008; Rubio et al. 2010; Stumpf et al. 2010; Mohanty 2015) have proved beneficial in terms of significantly higher yield. In this system, rice yields of 6.16 t ha^{-1}, fish yields of 2.6 t ha^{-1} and net water productivity of ₹18.9 m^{-3} were achieved (Thakur et al. 2015). The strategy of adopting SRI, i.e., low planting density, wider spacing, a small population and less fertilizer application helps in better root zone development, increasing the number of tillers and improving ecological conditions (ventilation and illumination) that prevent lodging and help produce heavier grains, high and stable yields. Similarly, CG has the potential to improve the growth rate of fish by carefully selecting a feeding plan that alternates between periods of feed deprivation and periods of satiation feeding

(Nikki et al. 2004; Wang et al. 2005; Montserrat et al. 2007; Mattila et al. 2009). Many fish have the ability to actively regulate their growth as a defense mechanism against negative environmental changes, and they have the capability to elicit above-normal growth rates during stress recovery periods (Mohanty 2010a). If CG can entirely compensate for lost growth during hunger, there may be a way to save money on fish feed by starving the fish and making up for lost growth when feeding starts. These developments serve as an impetus for promoting rice-fish farming.

3.8.1 Mutualism of Rice and Fish

Compatible ecological requirements of rice and fish provide the foundation for their synchronized growth within the same habitat. Water is required for fish farming as well as for rice growth and development. Introduction of fish, particularly herbivorous and omnivorous fish, into rice fields adds a new link to the existing food chain. They feed on primary producers and therefore reduce energy losses, improve the use of photosynthetic products and promote transformations in the rice field ecosystem that increase the carrying capacity of the rice field. Rice stabilizes water temperature and quality in rice-fish fields, creating an environment that is suitable to the growth of natural fish food species. These main producers turn solar energy into food energy, which fish require for existence. Fish in rice fields consume available plankton (which competes with rice for fertilizer), weed (which competes with rice for nutrients), insect and bacteria (which hurt rice plants) and mosquito larvae (which harm humans).

Only 3–4% of the feeds are assimilated by fish, with the remainder being discharged into rice fields as manure. Fish play a crucial role in recovering lost energy and modifying energy flow because they consume phytoplankton, zooplankton and weeds that compete with rice. When they swim through water, they emit carbon dioxide, increasing the quantity of carbon accessible to plants. They also break the soil surface and oxidize soil layers, which increases the supply of oxygen, boosts root growth and increases the rice plant's tillering capability. It is reported that under this system, paddy yield has been increased by 8–47% (Brahmanand et al. 2006; Mohanty et al. 2009). The increase in the yield of paddy probably results due to better aeration of water and greater tillering effect caused by the presence of fish. The excreta of fish, left over supplemental feed and additional fertilizer used also help in increasing the soil fertility and productivity. A comparative presentation of various ecological components and pathways of material flow in conventional and integrated rice-fish co-culture system is presented in Figure 3.12.

3.8.2 Production Intensification Approach

The rice-fish system's rice production approach is low planting density (20–30% lower than the density utilized in conventional fields), and minimal fertilizer application. This promotes better root zone development, increases the number of tillers and improves ecological conditions (ventilation and illumination) to minimize lodging and produce bigger grains with higher and more consistent yields. The ability of fish species to filter and feed on plankton (bacteria, phytoplankton and zooplankton) and to tolerate low levels of dissolved oxygen should be used to select fish species for stocking into rice-fish farming systems. The proper stocking density of fish/prawn species is crucial for achieving large cumulative fish yields and meeting the system's upper carrying capacity. Classification of a productive rice-fish system based on water and soil characteristics is presented in Table 3.5.

Management options such as high-density rearing (stocking with a larger initial fish biomass) followed by phased/selective harvesting, when the growth curve of stocked fish/prawn begins to slow down, are used to increase fish production from integrated rice-fish farming systems. With increasing integration of duck and nitrogen-fixing aquatic fern *Azolla*, the productivity of current rice-fish farming systems has a huge potential for improvement. *Azolla* can be used not only as an

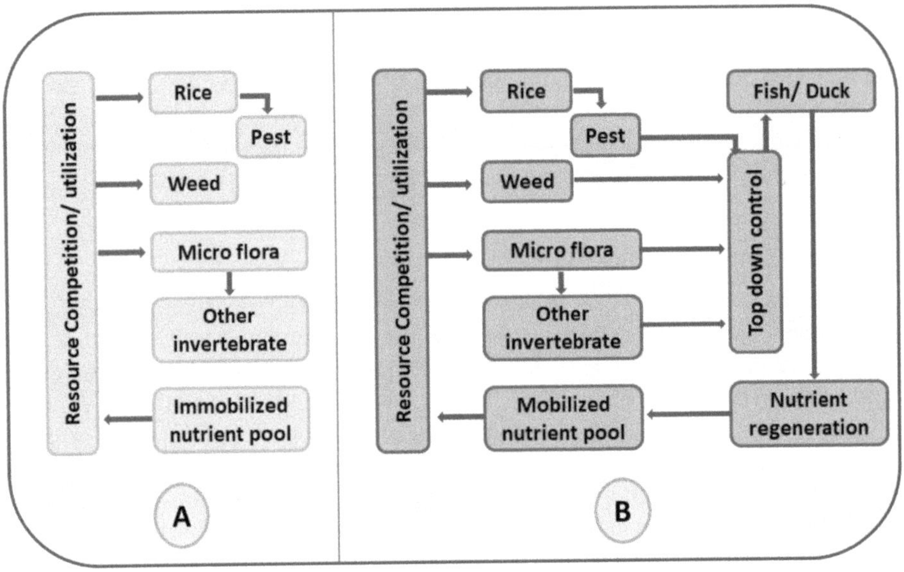

FIGURE 3.12 Schematic representation of ecological components and pathways of material flow in conventional (A), and integrated rice-fish co-culture system (B).

TABLE 3.5
Classification of Productive Rice-Fish System Based on Water and Soil Characteristics

Parameter	Unproductive	Low Productive	Highly Productive
1. Water			
pH	6.0–7.0	7.0–7.5	7.5–8.5
Total alkalinity (ppm)	10–30	30–90	90–170
Inorganic nitrogen (ppm)	<0.1	0.1–0.3	0.3–0.7
Orthophosphate (ppm)	<0.05	0.05–0.2	0.2–0.6
Dissolved oxygen (ppm)	2.0–4.0	4.0–6.0	6.0–9.0
Plankton (number l^{-1})	5–35	35–70	>70
2. Soil			
pH	6.0–7.0	7.0–7.5	7.5–8.5
Available Nitrogen (mg 100 g^{-1})	5–15	15–35	35–90
Avl. Phosphorus (mg 100 g^{-1})	3–6	6–20	20–50
Exchangeable calcium (mg 100 g^{-1})	20–50	50–150	150–300
Bottom fauna (g m^{-2})	1–15	15–20	>20

organic fertilizer for crops but also as animal and fish feed. Fish, *Azolla* and ducks integrated with a rice farming system can result in nutrient enhancement, pest (weed, insect, golden apple snail) control, feed supplementation and biological control. Nutrient recycling in an integrated rice-fish-*Azolla*-duck farming system is better and more efficient compared to a rice-duck or rice-fish farming system resulting in higher productivity. Fish and duck cultivated alongside rice have their own economic worth, resulting in greater farm output. Rice-duck farming combined with fish and *Azolla* boosts the productivity, bringing sustainability and addressing ecological challenges related to rice field ecology and aquatic biodiversity protection.

3.8.3 Species Suitability for Integrated Rice-Fish Culture

Fish species should be adaptable, compatible, resistant to environmental changes, high-yielding and be able to tolerate heavy doses of fertilizer. Since the water column in the refuge, perimeter canal and the paddy field in the renovated system is suitable for rearing of carp, three Indian major carp, i.e., *Catla catla, Labeo rohita and Cirrhinus mrigala,* Tilapia, small Indigenous fish species (SIFS), prawn (*Macrobrachium rosenbergii* and *M. malcolmsonii*) and exotic carp like silver carp and common carp may be stocked for culture in the integrated system. An improved high-yielding, tall, long-duration, submergence and pest resistant variety of rice with in-built characteristics of photoperiod-sensitivity, strong seedling vigor should be integrated. Some of the promising medium deep to deep water varieties of rice in different states of India are listed as follows:

Odisha	Durga, Tulasi, Gayatri, Sarala, Panidhan
West Bengal	Mahananda, Sabita, Purnendu, Sasi, Mandira
Assam	Prafulla, Joymati, Rupsundari
Andhra	Badava Mahsuri
Bihar:	Santosh, Janaki, Sudha, Vaidehi
Tamil Nadu	TNR1, TNR2
Uttar Pradesh	Jal-Lahari jalapriya, Madhukar, Chakia-59, Jalamagna
Chhatisgarh	Jaldubhi

3.8.4 Rice Cultivation under Integrated Rice-Fish Culture

Land preparation for a rice crop is done at the beginning of the season with minimum tillage and leveling for uniform crop establishment. Sowing of rice is usually preferred over transplanting, as sowing helps in the establishment of a good crop stand of rice by the time the field is waterlogged during the monsoon. Sowing also helps in avoiding puddling (for transplanting), which would otherwise lead to a siltation of refuge, besides creating turbidity problem for the fish. Normally line sowing and skip-row method of sowing are followed to provide fish passage and to encourage movement of fish inside the rice field when it is flooded with water. Interestingly, the skip-row method, though it reduces the plant population; it does not affect the overall rice yield significantly. When sowing of rice is not feasible due to unprecedented rainfall or some other reason, transplanting is done with 25–30 days old rice seedlings, after proper land preparation, preferably when the water level is around 5–10 cm in rice field.

Sowing can be done in dry conditions well before the monsoon, with a 20 cm gap between rows. Transplanting can be done early if necessary by using aged and robust seedlings. Fertilizer @ 40:20:20 kg NPK ha^{-1} should be administered during seeding (add 50% N at seeding and the remaining 50% after weeding). Organic manure (Table 3.6), FYM @ 5 t ha^{-1}, is also required during soil preparation. To monitor and manage the yellow stem borer, use of sex pheromone trap and a neem extract-based composition such as Nethrin or Nimbecidine @ 1% is advisable. To ensure better productivity, higher planting density must be avoided to provide greater ventilation and illumination, which helps in enhancing both plant and plankton growth that ultimately helps fish growth and yield.

Although fish in rice fields can consume some pests and help with biological management, they cannot completely replace pesticides; therefore, integrated management is required. To avoid fish/prawn mortality, chemical plant protection should be avoided. In fact, pesticide barely works in a wet condition because standing water dilutes the concentration. In an emergency, however, broad spectrum compounds with minimal toxicity, low residue and great effectiveness can be used. Chemicals

TABLE 3.6
Nutrient Composition of Different Organic Manures Used in IFS and Rice-Fish Culture System

	Nutrient Composition (%)				Nutrient Composition (%)		
Manure	N	P_2O_5	K_2O	**Manure**	N	P_2O_5	K_2O
Animal Origin				Plant Origin			
Cattle dung	0.5	0.4	0.2	Groundnut cake	6.5	1.0	1.0
Pig dung	0.6	0.6	0.4	Mustard cake	4.5	1.5	1.6
Poultry dropping	1.7	0.9	0.7	Mahua cake	2.5	0.8	1.9
Duck dropping	0.9	0.5	0.6	Neem cake	5.2	1.1	1.5
Farm yard manure	0.6	0.5	0.1				

in powder form should be used early in the morning while there is still dew on the ground, and sprays should be administered after the due dissipates.

3.8.5 Fish Culture under Integrated Rice-Fish Culture

Following proper refuge preparation, liming at 0.50–0.75 t ha^{-1} and manuring with raw cattle dung at 5.0–8.0 t ha^{-1} as a basal dose should be performed at the start of the monsoon in June. Weeds and predatory fishes are cleared from the rice field refuge before releasing fingerlings. Manual removal of floating and emergent weeds is preferable to chemical/herbicidal removal. Mahua oil cake @ 250 ppm is used to eradicate predatory/unwanted species and as a manuring substance at the onset of the monsoon when rainwater starts accumulating in the perimeter canal.

During the months of July and August, when precipitation begins to accumulate in the refuge and paddy field, fingerlings of Catla, Rohu, Mrigal, Silver carp, Common carp and prawn juveniles with a 30:25:45 (surface feeder: column feeder: bottom feeder) composition may be stocked. In this method, *Labeo bata* can also serve as a stocking component. Due to the short-culture duration, fingerlings (>15 g size) should be stocked at a higher density of 15,000–20,000 ha^{-1} for continuous rearing for four to five months or advanced fry (1.5 g size) at a higher density of 75,000–1,00,000 ha^{-1} for advanced fingerling production, using the phased harvesting principle. In order to increase fish production from an integrated rice-fish farming system, management measures such as high-density stocking (stocking with a larger starting fish biomass) followed by phased harvesting are used when the growth curve of the stocked fish/prawn begins to decelerate. This will help in reducing the population pressure and enhance the growth of the remaining stock. To augment growth, supplementary feed may be given to fish daily @ 3% in the initial two months and then 2.5% rate of mean body weight of stocked fish/prawn. In this culture system the fish/prawn will grow for a period of three to four months in the entire area and then two to three months in the confined area of infield refuge/perimeter canal. Under this system, production ranges between 1.4–1.8 t ha^{-1} (if stocking size <1.5 g) or 2.8–3.2 t ha^{-1} (if stocking size >15 g) per crop with a survival rate of about 65–90%, respectively.

3.8.6 Prawn Culture under Integrated Rice-Fish Culture

Among the several intrinsic and extrinsic factors connected with scampi culture, heterogeneous individual growth (HIG) poses a substantial challenge to farmers, resulting in significant economic loss. Mohanty (2010b) describes *M. rosenbergii* as having a sophisticated social organizational hierarchy with morphologically distinct dominant, subdominant and subordinate groups. The male morphotype's preponderance of a distinct social hierarchy promotes the unequal growth pattern within these prawns. Males have a high degree of HIG, whereas females have a more homogenous size distribution. Typically, three male morphotypes (SM-stunted male, OCM-orange clawed male, and BCM-blue clawed male) indicate three developmental stages of male maturity and are known to transition from SM >> OC >> BC. SMs are subordinate, not territorial and are at the beginning of the developmental route. OCMs are subdominant, indicating rapid somatic development. BCMs represent the terminal stage in the morphotypic transformation pathway. When a group of prawns achieves the terminal morphotype stage, it prevents other morphotypes from progressing to the next stage. This results in a large range of diversity in the growth pattern (Mohanty 2010a). In ponds/refuges, 50% of the male population consists of SM with a weight range of 5–20 g, while OCM (40%) and BCM (10%) have a weight range of 30–180 g and 180–250 g, respectively. To overcome the relative proportion of SM, OCM and BCM morphotypes in the harvested population, the following suggestions may be considered.

- Adopt two-stage (nursery and grow-out) rearing management.
- Stock post larvae ($PL_{15\text{-}20}$) in nursery for 50–60 days till they attain 4–5 g size.
- Sex segregation should be done at nursery site and only males should be stocked in rice-fish system (all male culture).
- Cull harvesting should be carried out from second month onwards (BCMs should be targeted first followed by OCMs) in grow-out phase.

3.8.7 Water Management in Rice-Fish Culture

Water quality is a dynamic property of a system affected by chemical, biological and physical factors, which ultimately influences the aquatic environment and production of rice-based farming systems. Management of water quality is a daily monitoring process, which includes the behavior of fish, nature of water and inputs of diligence from farmers. Since scientific water quality management and maintenance of tolerance limit (Table 3.7) of hydrological parameters cannot be expected at the farmers end, the minimum adoptable techniques for water quality monitoring at the farmers end should be encouraged. This includes periodic liming @ 0.10–0.15 t ha^{-1}, phased manuring

TABLE 3.7
Suggested water quality criteria for fish rearing

Chemical Parameters	Tolerance Ranges
Ammonia (NH_3)	<0.125 ppm (un-ionized form)
Calcium	10.0–160.00 ppm
Carbon dioxide	0.00–10 ppm
Chlorine	0.03 ppm
Hydrogen sulphide	0.002 ppm
Iron (total)	0.0–0.15 ppm
Nitrate (NO_3)	0.0–3.0 ppm
Nitrite (NO_2)	0.1 ppm in soft water, 0.2 ppm in hard water
Dissolved oxygen	5.0–9.0 ppm
pH	7.5–8.5
Phosphorous	0.01–3.0 ppm
Total suspended solids	80.0 ppm or less
Total Alkalinity	90–170 ppm
Water colour	Green to brown
Temperature	26-32°C

with raw cattle dung @ 2.0–3.0 t ha^{-1} to maintain plankton bloom, which, in turn, regulates dissolved oxygen and pH of the eco-system. As these two parameters are too critical for fish growth and survival, utmost care must be taken for bloom management. As the optimum ratio of phyto and zooplankton is 10:1, periodic estimation of plankton is essential. In the case of plankton crash, re-inoculation should be carried out followed by fertilization with Urea + SSP (1:1) @ 3–4 ppm or systematic water replenishment in the case of excessive bloom buildup. To maintain a clear aquatic environment, meal to meal feed management is essential in reducing the organic load in the system that affects water quality.

3.8.8 Integrated Pest Management (IPM) in Rice-Fish Culture

Pest management options are divided into four categories: mechanical, chemical, cultural and biological. Weeding is likely the best example, although cultural approaches such as water level control are also important. Insecticide applications in rice have proven to be a serious issue since these destabilize the environment and stimulate pest resurgence, creating an even more critical scenario than would have occurred without their use. Biological pest control has a wide range of applications, from promoting specific organisms that are predators of specific pests to using disease-resistant rice types. Natural management is the most profitable choice for farmers, especially when pesticide-related health effects are considered. An integrated approach using various management options is the preferred choice for plant protection in rice.

Stocking fish in rice fields minimizes pest infestation, reducing or eliminating the need for herbicides, insecticides and molluscicides. The efficiency of fish as a bio-control agent is determined on their distribution throughout a rice field. If fish spend most of their time in the pond refuge, they will be ineffective at controlling rice pests. In such a case, the use of pesticides and other control strategies should be weighed against the potential costs and losses in rice productivity and fish harvest. The important characteristics to be considered in the selection of any pesticide to be applied in a rice-fish farm are (1) relative fish safety and effectiveness against the target bug species, (2) should not accumulate in rice; instead, it should be converted into non-toxic molecules; and (3) should either volatilize, bio-degrade or chemically degrade immediately after application.

3.8.9 Weed Management in Rice-Fish Culture

Land preparation, water depth fluctuation, mechanical weeding, pesticide use and herbivorous fish stocking are all viable techniques for weed management in rice fields. Mechanical weeding is probably the most common method of weed control. It is, however, an extremely labor-intensive method of weed control, and as such, it frequently has a large opportunity cost. Herbicides are widely utilized, however they are not regarded as a severe issue in rice-fish farming. Herbicides are typically used soon after transplantation. Fish are stocked 10 to 14 days after being applied. Introducing fish to the rice field can help to eliminate weeds in a variety of ways. Weeds are an important part of the diet of herbivorous fish. Weeds are just in the way of bottom-feeding creatures. The muddy bottom of a rice field is tilled in the process of seeking for nourishment, leaving little possibility for submerged weeds to embed their roots in the soil, impeding their growth and proliferation. Fish have been proven to be more successful in weed management than either manual weeding or chemical application. After five weeks, the introduction of fish reduced the amount of weeds in one rice field from 101 kg to just 20 kg, whereas in an adjacent rice field with no stocked fish, weed biomass grew from 44 kg to 273 kg (Wu 1995).

3.8.10 On-Dyke Horticulture

To make the system more economically viable, the dykes that will be built to prevent fish escape from the integrated system may be utilized to produce vegetables and other fruit trees such as papaya and banana. Vegetables such as gourd, radish, brinjal, leafy vegetables during pre and *Kharif* season and vegetables such as tomato, french beans, radish, bitter gourd, cucumber, cauliflower, cabbage, brinjal, pumpkin and leafy vegetables (coriander, amaranthus and Indian spinach) can be grown during winter. Vegetables such as snakegourd, bittergourd, ridgegourd, bottlegourd or ashgourd can be grown throughout the year on raised platforms. Banana-based farming systems involving cultivation of compatible intercrops along with the integration of enterprises like poultry, dairy, mushroom, fishery, crops, etc. offer tremendous opportunities to efficiently increase the yield and income of farmers in a shorter time (Kashyap et al. 2021).

3.8.11 Advantages of Rice-Fish Integrated System

- Fish/prawn culture in paddy fields can convert material and energy into fish/prawn production, accelerate rice growth and boost solar energy fixation, hence increasing paddy field productivity (Figure 3.13).
- Fish performs as a bio-controller of insects, lowering the cost of chemical input.
- Weeds compete with rice because photosynthesis requires CO_2, H_2O and other nutrients. Fish aids in the biological management of aquatic weeds, lowering herbicide input costs.

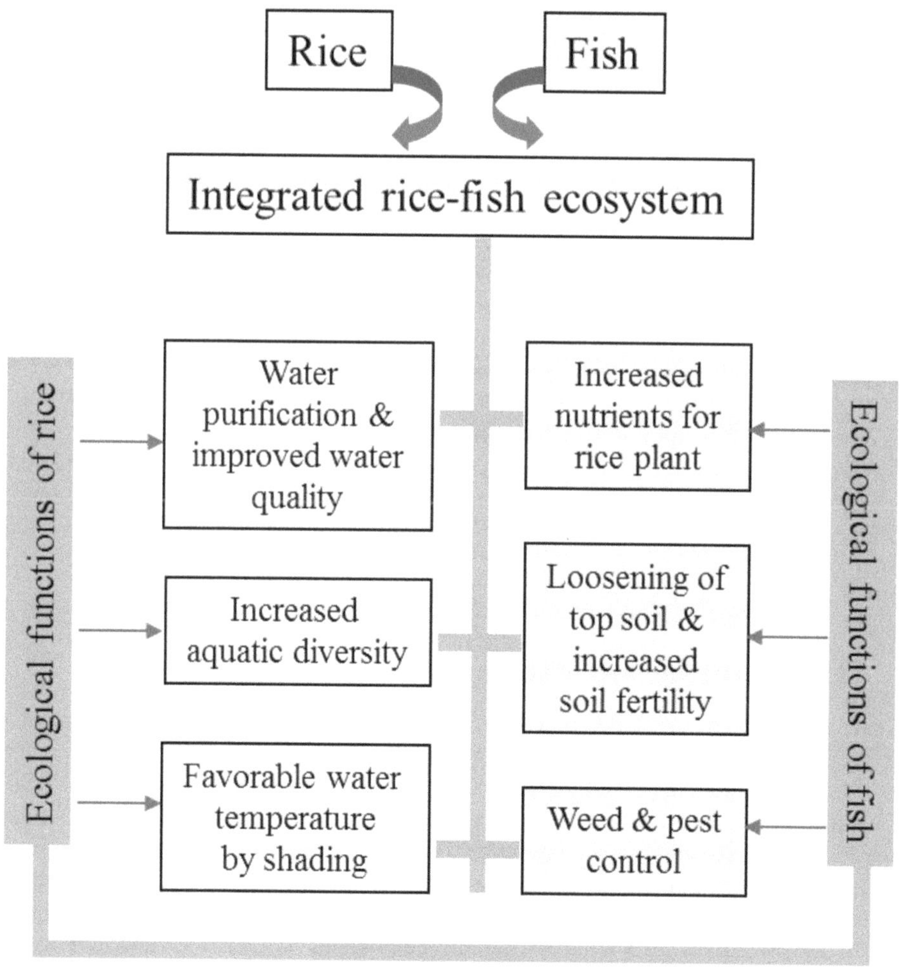

FIGURE 3.13 Ecological functions and inter-component interaction in rice-fish system.

- Fish in a rice field can convert insoluble nitrogen in the soil to soluble nitrogen, increasing soil fertility. Fish farming in rice fields increases the quantity of nitrogen in the soil and the amount of nitrogen absorbed by the rice plant, resulting in more nitrogen being transferred to the rice field.
- The stirring movements of fish aerate the soil and improve its topsoil structure and porosity. This increases the oxygen content in the soil and elevates its oxidation and reduction potential during the period of rice growth.
- Rice plants protect fish and prawns from avian predation, and fish obtain enough oxygen from rice and phytoplankton to survive and flourish.
- Fish infections are uncommon in rice-fish farming because to the clear aquatic environment, high oxygen content and abundant natural food, which produces strong and disease-resistant fish/prawn.
- Following rice harvest, the roots and residual straw (straw contains 9–13% cellulite, 1.5-–% potassium and 30–40% cellulose) offer organic matter and promote the growth of microorganisms, the ultimate natural food of fish/prawn.
- The feces of poultry/duck/animals could be recycled as fish feed, increasing the biological productivity of water. The addition of animal/bird droppings to water reduces the cost of fish feed up to 40% and fertilizer by 60–65%.

3.9 INTEGRATED SYSTEM OF RICE INTENSIFICATION (ISRI)

A system of rice intensification (SRI) procedures provides new opportunities for food security and poverty reduction under the rainfed production, as evidenced by their constant outperformance of conventional practices. A modified rice cultivation method suggests keeping rice fields moist but un-flooded during the crop's vegetative stage, typically with alternate-wetting-and-drying (AWD), and then maintaining shallow flooding during the crop's post-vegetative stage. SRI rice farming (single plant, wider spacing 25 x 25 cm) can be used to maximize land and water productivity in rice-fish culture. Harvesting excess rainwater and storing it inside the rice field in a built pond or refuge (10% of rice area) allows farmers to integrate aquaculture with rice farming as well as produce horticulture crops on the refuge bunds. Overall, this modified method of rice crop management produced 58% higher grain yield with 16% less water use. In this system, paddy yields of 6.2-6.5 t ha^{-1} (42–44% more than conventional rice) and fish yields (Indian Major Carps: *Catla, Rohu, Mrigal*) of 2.6–2.8 t ha^{-1} can be achieved with B:C ratio of 2.9–3.1 and net water productivity of ₹18-20 per m^3 of water. The Integrated SRI system has the potential to generate a net profit of ₹2,70,000 ha^{-1} in two crops (Thakur et al. 2015). In this increasingly water-scarce world, ISRI offers opportunities to reduce rice farmers' need for water while enhancing their grain production. This significantly improves water productivity and, therefore, offers smallholders an option to enhance their income while improving their food security and livelihoods.

3.10 IFS USING ALTERNATE RAISED AND SUNKEN BED SYSTEM

The raised and sunken bed (RSB) is a land layout created for low-lying areas using the cutting and filling procedure. Such beds are created by removing the surface soil layer from one area and placing it on the neighboring area to the desired height. Adoptions of a permanent RSB system of cultivation with proper cropping sequence in rice paddies in lowlands or canal commands will not only increase the production and productivity of vegetables and cereal crops but also enhance cropping intensity, employment and income substantially. This system helps in crop diversification, increases cropping intensity and provides year-round employment opportunity and profitability. This replicable system can be well adopted in irrigated canal command and medium and low land rice ecosystems.

It is possible to diversify the cropping system with rice–pea or rice–lentil in sunken beds and vegetables such as tomato–brinjal–broccoli, tomato–okra–broccoli, etc., on raised beds with effective land utilization. Further, growing rice + fish in sunken beds of alternate raised and sunken bed system (30 m length and 5 m width) instead of rice alone can give more yield (rice: 5.2 t ha^{-1}; fish: 3.7 t ha^{-1} 150 d^{-1}) and return. A combination of rice-fish in sunken beds with cabbage (28.5 t ha^{-1}), pointed gourd (4.7 t ha^{-1}) and snake gourd (20.2 t ha^{-1}) crops in an adjacent raised bed can give the highest rice equivalent yield (48.8 t ha^{-1}), water use efficiency (174.54 kg ha-cm^{-1}), net water productivity (Rs.13.3 m^{-3}) and benefit-cost ratio (4.78). In this system fish fingerlings @ 5000 ha^{-1} is recommended. Fish culture in sunken bed helps in improving soil nutrient status while rice cultivation with fish improves water quality. This system has the potential to enhance the production by sixfold to sevenfold and water saving of up to 20%. Highest B:C ratio of 4.78 is recorded for rice-fish + snake gourd + pointed gourd + cabbage combination. When compared to rice monocropping, RSB increases WUE and water productivity by around three to four times.

3.11 COASTAL AQUACULTURE-BASED IFS

The coastal belt is vulnerable to climate change and must adapt to sustain and improve productivity and stability. Integrated coastal aquaculture-based farming activities have the potential not only to increase productivity but also to generate employment opportunities, increase income and provide nutritional security to resource-poor farming communities, as well as to distribute risk (both biological and economic). Coastal IFS are characterized by the temporal and spatial mixing of crops, livestock, fishery, and allied activities in a single farm and are aimed at meeting multiple demands (impart farm resilience, farmer livelihoods, food security, ecosystem services and making farms adaptive and resilient, etc.). It is anticipated that these complex farms are more productive at the system level, less subject to volatility and generate less negative externalities than simpler farms, and they cater to the needs of small and marginal farmers. Further, multiple uses of water in aquaculture-integrated agriculture systems (AIAS) are inevitable to produce more food per drop of water to address water shortage, food insecurity and climate change.

Shrimp monoculture/polyculture with brackishwater fish/crab-fish polyculture, has been quite profitable and promising in the coastal belts of India and other South Asian countries. Shrimp (*Penaeus monodon* and *Litopenaeus vannamei*) polyculture with brackishwater fish (*Mugil cephalus, Liza tade, Liza persia, Chanos chanos, Mystus gulio*, etc.) or its monoculture (Mohanty et al. 2017a, 2018), and fish-crab (*Scylla olivacea* and *S. serata*) polyculture (Mondal et al. 2020) in the coastal belt of India (West Bengal, Odisha, Andhra Pradesh, Tamil Nadu, Kerala, Karnatak, Mahrastra, Goa, Gujrat) is found to be very lucrative and promising. Farmers with few resources, unemployed youths and migrant laborers in India might participate in these activities by taking advantage of existing government programs. The IFS approach in the coastal area should have as its goal improving farm income, family nutrition and livelihood in a sustainable manner for small and marginal farmers.

3.11.1 Coastal Agro-ecosystem

The coastal system is economically and environmentally significant because it includes a variety of physiographic features (mountains, valleys and coastal plains), riverine systems, climatic conditions, soil and water bodies and flora ranging from lush tropical rainforests to coastal mangroves. Agriculture, horticulture, agroforestry and silviculture are all significant activities in this ecosystem. The coastal zones' agro-climatic characteristics are ideal for growing a variety of field and horticultural crops. The region has a remarkable export record for agricultural produce, spices and marine products. Low-lying coastal agriculture consists mostly on rice, legumes in rice fallows and sugarcane as rotational crops. Plantation crops such as coconut, arecanut, cocoa, spices (black pepper, ginger, turmeric, clove, nutmeg, etc.), and fruit crops (mango, banana, pineapple, guava, etc.) dominate horticulture. Tapioca, sweet potato and elephant foot yam are among the tuber crops farmed. Appropriate crop management practices and cropping systems aimed at making the best use of land and water resources could help boost production and improve the rural economy. Plantation crops have enormous potential for product diversification, value addition and export earnings, as well as for providing raw materials for a variety of agro-based enterprises. Though there are challenges in resolving problems of excessive salinity of soil and groundwater, coastal aquaculture is growing as a key producing activity due to the abundance of brackish water. However, output from these sources is only 45 to 50% of its capacity. All this calls for an integrated coastal systems research management strategy, based on technologies to increase the overall productivity of the systems, thereby enhancing the quality of life of the people.

3.11.2 Integrated Mangrove Fishery Farming System (IMFFS)

In India, the majority of prawn farms are rectangular-shaped clay ponds measuring around one hectare in size, and a cluster of farms typically has between 5 and 20 of these ponds. They are pump-fed, and a sort of semi-intensive type of culture system is used. To preserve the quality of the water, brackish water is periodically pumped into and out of the ponds. Hatchery produced post-larvae of prawn are transferred to ponds where they are fed with artificial feed until they reach marketable size, which takes about four to six months. As a result of this practice, inputs cost is always high and artificial feed increase organic load leading to pollution. Earthen ponds in the IMFFS are constructed to allow for the growth of saline-tolerant, economically significant vegetation like mangroves and halophytes (Figure 3.14). Bunds can be built inside the pond in a zigzag pattern or as little mounds to create space for plants. The soil from the pond's bottom is dug up to make these mounds and bunds. The pond deepens as a result, and the bottom level drops below the tide level. As a result, tidewater gravitationally enters the pond at high tide and exits at low tide. The pond can be made deeper to allow 3 to 4 feet of water to remain in the pond as standing water for fish/prawn/crab culture. The tidal water inlet and outlet are established at opposite ends of the farm with net barrier to prevent escape of fish/prawn/crab or a single structure can be used both as inlet and as outlet.

The ponds are designed in such a way that nearly 30 to 35% of the space is left for planting mangroves (mainly *Rhizophora* and *Avicennia marina* saplings) and halophytes (*Sesuvium portulucastrum* etc.) whereas remaining 70 to 65% is left for holding seawater for crab/shrimp/fish culture (*Lates calcarifer*, *Mugil cephalus*, *Liza tade*, *Liza persia*, *Chanos chanos*, *Mystus gulio* etc.). *Rhizophora* saplings can be planted in two rows along the bottom borders of the bund, while *Avicennia* saplings can be planted around 2 m above the *Rhizophora* plantation.

Both *Rhizophora* and *Avicennia* should be planted at 5 m intervals in a row. Succulent halophyte can be planted on the top and sides of the bunds. Stem cuttings of about 15 cm in length are suitable

FIGURE 3.14 A view of Integrated Mangrove Fishery Farming System (IMFFS) in India.

for planting at an interval of 1 m. This improved extended culture technique requires no electricity (for pumping and aeration), artificial feed or pesticides, and coastal farmers can earn a net profit of ₹30,000 to 120,000 ha^{-1}. The mixed mangrove-shrimp/fish systems have several advantages over other systems, including minimal capital requirements, livelihood diversification through polyculture, consistent revenue and recognition as as an organic farming method.

3.11.3 Fish-Crab/Fish-Shrimp: On-Dyke Horticulture System

Mud crab farming can be highly profitable when combined with commercially valuable shrimp and finfish (Shelley and Lovatelli 2012), as it improves resource utilization, productivity and water quality (Feng et al. 2014), as opposed to monoculture of crabs, which loses significant nutrients due to mud crab feeding habits (Aubin et al. 2015). Finfish such as *Lates calcarifer, Mugil cephalus, Liza tade, Chanos chanos, Oreochromis niloticus*, and *Mystus gulio* are considered suitable species for brackish water polyculture with crabs due to their lowest trophic level feeding habit and ability to accept artificial supplementary diet, in addition to high consumer preference for flesh quality, flavour (Biswas et al. 2012), and market price. Co-culture of mud crab (*S. olivacea*) with finfish can be taken up as a profitable venture for sustainable coastal aquaculture. Different studies' on water quality, growth, production performance and economic factors (Mondal et al. 2020) show the technical and economic viability of crab-catfish, crab-tade mullet and crab-grey mullet co-culture. However, because both species are readily available locally, the crab-catfish combination has been proven to be the most profitable and recommended for coastal farmers. Polyculture of shrimp (*P. monodon*) with finfish is another economical and sustainable kind of diversified coastal aquaculture. Keeping in view the water quality, growth, production performance and technical and economic viability, shrimp-fish polyculture combinations of (1) *P. monodon* + *M. cephalus* + *L. parsia,* (2) *P. monodon* + *M. cephalus* + *C. chanos* + *L. parsia* and (3) *P. monodon* + *M. cephalus* + *C. chanos* culture can be taken up. However, *P. monodon* + *M. cephalus* + *L. parsia* combination is found to be the most compatible, profitable and recommended for the coastal farmers. Polyculture of crab/shrimp with economically important fish can be very profitable and sustainable since it maximizes the usage of underutilized trophic resources and results in higher productivity than monoculture. Because the growth rate of finfish at an early stage is sluggish, stocking advanced fingerlings could boost survival, output and profitability even more. More research is needed, however, to optimize the stocking density of each species in these polyculture systems. On pond embankments, farmer can plant a variety of winter vegetables, summer vegetables and flowers such as marigold, rose and jasmine. Adjacent pond areas can also be utilized for salt tolerant paddy in *Kharif.* This system not only provides year-round employment opportunity but also gives a higher net return of ₹1.0 2.0 lakh per crop.

3.11.4 Shrimp-on-Dyke Horticulture System

In the Indian environment, coastal shrimp aquaculture of *Penaeus monodon, P. indicus*, or *Litopenaeus vannamei* is favored (Mohanty et al. 2017a, 2018). Semi-intensive ponds (1–5 ha) are routinely stocked at a rate of 5 to 20 PL m^{-2} using hatchery-produced seeds. Water exchange is carried out on a regular basis by tide and is reinforced by pumping. The shrimp eat natural foods supplemented by artificial diets, which are increased by pond fertilization. The annual yield ranges from 0.5 to 4.0 t ha^{-1}. Intensive farms are typically located in non-tidal areas where ponds can be entirely drained and dried prior to stocking. Ponds are usually small (0.1 to 1.0 ha) with a square or rectangular shape. Stocking density ranges from 20 to 60 PL m^{-2}. For internal water circulation and oxygen provision for animals and phytoplankton, heavy aeration, whether supplied by diesel engines or electric motors, is essential. Four to five times a day, artificial diets are fed, and then the feed tray is checked. Final FCR typically ranges from 1.2:1 to 1.8:1. Due to their lower risk of transmitting viral illnesses through intake water, reduced water exchange and closed systems have been popular since the advent of white spot disease. However, feed and phytoplankton blooms need to be carefully monitored and managed to avoid deterioration of the pond bottom and water quality. Efficient feed management is the major criteria for a successful crop, since feed represents over 50–60% of the production costs in intensive systems. Regular measurements should be made of water quality indicators such pH, salinity, dissolved oxygen, alkalinity, Secchi disc, H_2S and unionized ammonia. If a closed system culture is used, stocking should be kept to a minimum because doing so would allow for early harvesting of the pond (in 3.5 months as opposed to four to five months) and smaller shrimp (20 g as opposed to 30–35 g, as obtained in semi-intensive and intensive systems with water exchange). Depending on the management conditions, production of 4 to 15 t ha^{-1} $year^{-1}$ is possible.

Several winter vegetables (tomato, brinjal, coriander, cabbage, cauliflower, turnip, radish, fenugreek, bottle gourd, potato, beans, spinach and onion) and summer vegetables (amaranth, okra, bitter gourd, sponge gourd, water bind weed, papaya, sweet gourd, ridge gourd, chilly ginger and turmeric) can be cultivated depending upon the size, shape and condition of the dykes. Pond embankment can also be utilized to raise pulse and oil seed crops. Flowers such as marigold, rose and jasmine can also be grown on pond embankments. Adjacent pond areas can also be utilized for salt tolerant paddy in *Kharif.* This system not only provides year-round employment opportunities but also gives a higher net return of ₹1.5–3.5 lakh per crop.

3.12 INTEGRATED MULTI-TROPHIC AQUACULTURE (IMTA)

In IMTA, inorganic (macroalgae/aquatic weed) and organic extractive (bivalves) species, also known as un-fed species, are integrated with cultivated or fed species in the same system (Chopin 2006). Multiple marine or freshwater organisms from various trophic levels or nutritional levels can be farmed simultaneously in the same environment using IMTA (Barrington et al. 2009; Angel and Freeman 2009). IMTA is the farming of aquaculture species with complementary ecosystem functions, in a way that allows one species' uneaten feed and wastes, nutrients and byproducts to be recaptured and converted into fertilizer, feed and energy for the other crops, and to take advantage of synergistic interactions between species. IMTA is a method that combines the farming of fed aquaculture species (such as finfish and shrimp) with organic extractive aquaculture species (such as shellfish and herbivorous fish) and inorganic extractive aquaculture species (such as seaweed) to create systems that are balanced for economic stability (product diversification and risk reduction), social acceptability (better management practices) and environmental sustainability (bio-mitigation). The IMTA is a sort of polyculture system that effectively utilizes each trophic level and converts the waste produced at one trophic level into food for the next. It is an advanced system and sustainably better with greater production and more desirable water quality than the age-old polyculture practice (Biswas et al. 2019). Usually, the higher trophic levels are occupied by carnivorous fish or shrimp. They excrete particulate organic carbons (fish fecal matter), orthophosphate (P), and soluble inorganic ammonia. These inorganic nutrients (inorganic N and P) can be readily absorbed from the environment by seaweed and other species that are comparable. Shellfish and bottom feeders consume the organic nutrients (fish feces and uneaten feeds, POM), which are then consumed by seaweed along with their excretory inorganic byproducts. Thus, the IMTA provides a new farming approach to increase the food production by efficient utilization of trophic levels in a sustainable way which ensures better biodiversity, biosecurity and ecosystem (Figure 3.15).

Integrated Multi-Trophic Aquaculture = Fed Aquaculture (Finfish) + Extractive Aquaculture, i.e., Organic (Shellfish) Inorganic (Seaweed)

Theoretically, in an IMTA farm, low-trophic extractive species like bivalves, macroalgae and detritivores might utilize metabolic waste and uneaten feed from fed species, reducing any potential environmental consequences. Although the configuration of an IMTA farm can vary greatly (Chopin et al. 2012), three major types of extractive species have been considered in the literature: seaweed, bivalves and bottom feeders. According to Troell et al. (2009), seaweeds can lessen the amount of dissolved nutrients in the water column. Bivalve farms' ability to filter water could help

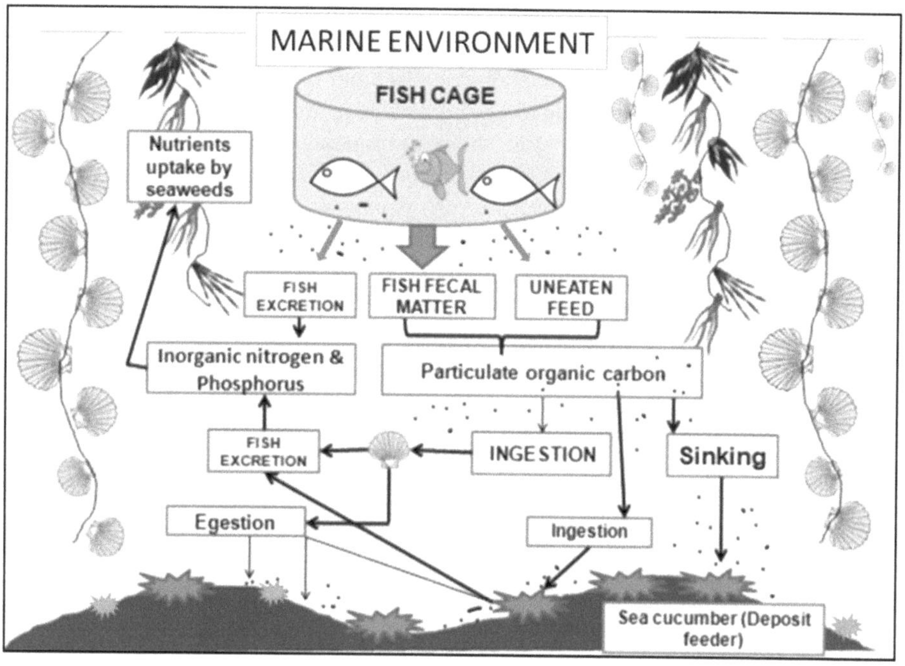

FIGURE 3.15 Flow diagrammatic representation of Integrated Multi-Trophic Aquaculture System (IMTA).

to counteract the negative impacts of particulate organic matter (Troell et al. 2009). Bottom feeders like sea cucumbers could be included into the IMTA system to reduce particle organic loading (Zhang and Kitazawa 2016). IMTA has been promoted as a farming method to diversify products, reduce operational risks and achieve mitigation potential (Carras et al. 2020).

3.12.1 Concept of IMTA

In IMTA, fed species are grown alongside extractive species that benefit from the inorganic and organic wastes produced during aquaculture. This farming technique differs from 'polyculture' of finfish, where the fishes share the same biological and chemical processes that may cause an ecosystem to change. Multi-trophic describes a system that includes animals from several trophic levels. IMTA is more intensive cultivation of many species close to one another, linked by fertilizer and energy transfer through water and the multi-trophic sub-systems are integrated. The choice of species and environmental sustainability are key factors in IMTA. Feed, which consists of pellets or garbage fish, nourishes fed species like carnivorous fish and shrimp. Organisms that remove nutrients from the environment are called extractive organisms. Bivalves and seaweed are the two economically significant cultural subgroups that fit within this category. Co-cultured species combinations will need to be carefully chosen based on a number of requirements and factors: (1) *Complementary roles with other species in the system*: Use species that complement each other at various trophic levels. For example, species must be able to feed on the waste of other species in order for newly integrated species to improve water quality and flourish efficiently. Not all species can be farmed efficiently together. (2) *Adaptability in relation to the habitat*: It is possible to use indigenous species that are well within their typical geographic range and for which technology is accessible. This will reduce the possibility that invasive species would damage the local environment and perhaps other economic enterprises. Additionally, native species have developed to become well adapted to the environment. (3) *Culture technologies and site environmental conditions*: Particulate organic matter and dissolved inorganic nutrients should be considered, as well as the size range of particles, when selecting a farm site. (4) *Ability to provide both efficient and continuous bio-mitigation*: Use species that can produce a substantial amount of biomass. This characteristic is crucial if the organisms are to function as a bio-filter that removes most of the surplus nutrients from the water. A species with a very high value is the other option, in which case smaller volumes can be grown. But with the latter, the bio-mitigating function is diminished. (5) *Market demand for the species and pricing as raw material or for their derived products*: Use species that have an established or perceived market value. (6) *Commercialization potential*: Use species, for which regulators and policymakers will facilitate the assessment of new markets, not enforce new regulatory impediments to commercialization. (7) *Contribution to improved environmental performance,* (8) *Compatibility with a variety of social and political issues.*

3.12.2 Inorganic Extractive Sub-system in IMTA

Since aquatic plants' biofiltration involves assimilation, it increases the environment's potential to absorb nutrients. Plants photosynthesize new biomass using solar energy and extra nutrients, primarily C, N and P. The process creates a miniature ecosystem in the culture system where, in a well-balanced system, fish and microbial heterotrophy are balanced out by plant autotrophy in terms of nutrients as well as oxygen, pH and CO_2. Thus, plant bio-filters can maintain the culture environment and significantly lessen the total environmental impact of fish culture. Furthermore, farming of species that are low in the food chain and that extract their nourishment from the water involves relatively low input. Seaweed is the best choice for bio-filtration because it can be inexpensively grown and has high production. This are widely used for human nutrition, as phycocolloids, feed additives, agrichemicals, nutraceuticals and medicines. In marine cage culture zones, seaweed farming has long been advocated in China as a means of bio-extracting nutrients from the seawater. According to FAO aquaculture statistics, there are 37 different kinds of seaweed. *Eucheuma* seaweed, which are grown in tropical and subtropical waters and account for 8.44 million tonnes, are dominant, followed by Japanese kelp (5.94 million tonnes).

Only a handful of seaweed types have been thoroughly investigated for their aquaculture and/or bioremediation potential. Genus *Ulva*, the flat sheet morphotypes have correspondingly high growth rates as well as high N contents, making them very good candidates for remediation. In mid- to large-scale animal mariculture systems, *Ulva* have been successfully integrated because their life cycle and its regulations are generally well understood. While *Laminaria, Porphyra* and *Gracilaria* cultivations are also regarded as having the ability to produce a sustainable seaweed mariculture and integrated aquaculture. Along the Indian coast, *Kappaphycus alvarezii* is employed in integrated open-sea aquaculture (IMTA) and has proven to be a promising species.

3.12.3 Organic Extractive Species Sub-system in IMTA

Filter-feeding bivalves are cultured adjacent to meshed fish cages in a conceptual open-water integrated culture system, reducing nutrient loadings by filtering and assimilating particulate wastes (fish feed and fecal matters) as well as any phytoplankton production stimulated by introduced dissolved nutrient wastes. Waste nutrients are collected from the farmed bivalves rather than being lost to the local environment as in typical monoculture. A native bivalve species must be considered to suit the local ecology, potential markets, and the need to engineer IMTA systems to accommodate them. According to the literature, 95% of particles emitted by aquaculture systems, fish farms, and closed recirculation systems have an approximate diameter of 20 microns (5–200 microns range) and will settle. There is evidence that filter-feeders collect particles from the water column selectively, rejecting the remainder. As a result, it is critical to understand the particle size of wastes from an IMTA system and to select from a wide range of bivalves that will select the desired particle size and kind. The marketability of these secondary products is a consideration, but it does not have to be the most important. One could add a fish range alongside the primary fed fish to act as a first-stage bio-filter and use a marine bivalve as the second stage. Studies have shown that bivalves are skillful of utilizing fish farm wastes as an additional food supply. In India, the bivalve mussel, *Perna viridis,* and the oyster *Crassostrea madrasensis* that are commercially produced along the Indian coast can economically mitigate eutrophication in IMTA.

3.12.4 Fed-Aquaculture Species Sub-system in IMTA

The only fed component in the majority of IMTA systems are finfish, releasing significant amount of nutrients to the system. Fish contribute dissolved and particulate nutrients, oxidation reduction potential reducing chemicals and revenue to the industry in an IMTA system. Species, size and feed composition are only a few of the variables that affect the amount and form of these nutrients. Feed formulation provides perhaps the most obvious route for fish effluent modification for the extractive components, for an IMTA system. The IMTA systems that are semi-closed (recirculation aquaculture systems) and open to the environment (cage based) are distinct from one another. Most open systems provide environments that are both essential and adequate for raising extractive species, but semi-closed systems demand considerably tighter coupling of the various trophic levels that are being cultivated.

3.12.5 Benefits and Advantages of IMTA

- Effluent bio-mitigation: Mitigation of effluents through the use of bio-filters solves a number of the environmental challenges posed by monoculture aquaculture.
- Increasing profitability through diversification: Sale of commercial byproducts/additional goods increases the overall economic value.
- Improving local economy: Economic growth through employment (both direct and indirect) and product processing and distribution.
- Ensure 'natural' crop insurance: Product diversification offers financial protection and decreases economic risks when price fluctuations occur, or if one of the crops is lost to disease or inclement weather.
- Disease control: Due to their antibacterial activity against fish pathogenic bacteria, various seaweeds help prevent or reduce disease in farmed fish.
- Premium prices and increased profits: Potential for differentiation of the IMTA products through eco-labelling or organic certification programmes.
- Offers tremendous potential for food and nutritional security.
- Eco-friendly: Promotes environmental sustainability by converting byproducts, wastes and uneaten feeds, efficient utilization of all the trophic levels, reducing eutrophication.

3.13 HYDROPONICS

Hydroponics is a soil-free agricultural system (Figure 3.16) where plants are grown in nutrient-rich water on a substrate or medium. Vermiculite, coconut coir, perlite, brick fragments, polystyrene packing peanuts and wood fibre are examples of the aggregate substrate, or growing media. Plant support and moisture retention are provided by these mediums. This system, which uses less water as compared to soil faming, has successfully been used for cultivation of different vegetables (cucumbers, lettuce, spinach, tomatoes) as well as ornamental crops as these respond well to hydroponics due to low nutrient demands and a short growth period (Cifuentes-Torres et al. 2020). It also has the potential to increase yields while occupying less space and promotes food security by growing food vertically (Joshitha et al. 2021). Large-scale hydroponics facilities currently run under climate, lighting and irrigation controls provided by a variety of sensors, web platforms, softwar, and mobile applications.

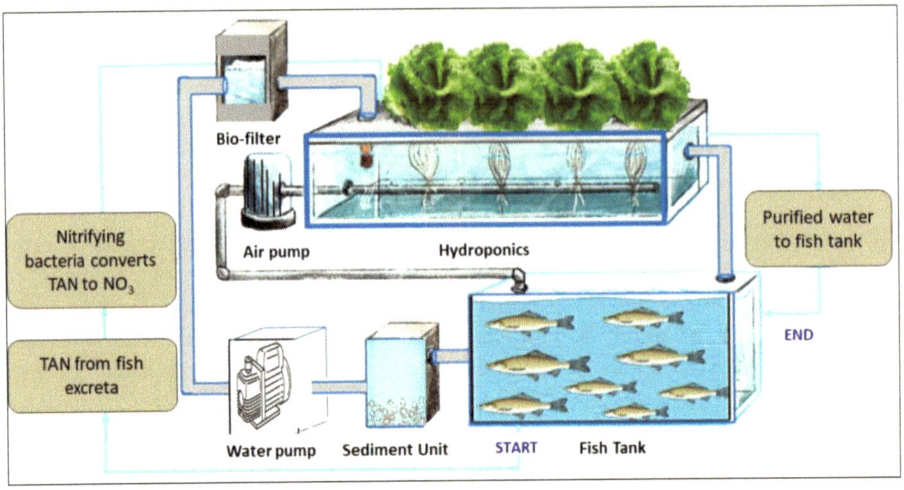

FIGURE 3.16 Flow diagrammatic representation of hydroponics system.

A nutrient solution is introduced to the root zones of the plants in this system by integrating irrigation systems into the substrate or growing media. This mixture offers all the nutrients required for plant growth. Plants grow with bare roots in an aqueous media or on a substrate. By preventing plant contact with soil and allowing soil-less media to be cleaned up and used again between harvests, hydroponics can actually manage soil-borne illnesses and pests. The specific requirements of intense production are met by this reuse of substrates. Some substrates perform significantly better than soil, especially in terms of the ability to retain water and the availability of oxygen in the root zone. Nutrient availability at plant roots is better manipulated, monitored and real-time controlled, leading to higher quantitative and qualitative productions.

Additionally, because the nutrient solution is recycled, the majority of soil-less culture techniques use a fraction of the water required for conventional soil-based production. There has been a rise in demand for out-of-season and high-value crops. To enhance production capacity and extend the supply of crops throughout the year, this growth in demand has prompted the expansion of numerous types of protected agriculture systems. To meet the evolving needs of agriculture, there has been a transition from soil production to soil-less culture. In conclusion, there are four main factors that make soil-less culture a growing agricultural practice: reduced presence of soil-borne diseases and pathogens due to sterile conditions; improved growing conditions that can be adjusted to meet optimal plant requirements leading to increased yields; increased water- and fertilizer-use efficiency; and the ability to establish agriculture in areas where suitable land is not readily available. Additionally, there has been a great deal of investigation into organic and soil-less techniques due to the rise in demand for produce free of chemicals and pesticides as well as more sustainable agricultural practices.

3.13.1 Nutrient Solution for Hydroponics and Its Managements

All needed macro and micro nutrients are delivered to hydroponic plants in the form of a nutrient solution, which is made up of dissolved fertilizer salts in water. The success or failure of hydroponic systems is mostly determined by a precise nutrient management scheme. An effective hydroponic practice will involve carefully managing the nutrient solution pH level, temperature and electrical conductivity (EC) as well as changing the solution as needed. The larger the deviation of a nutritional solution's pH from the prescribed pH range, the lower the chances of success. The ideal EC range for hydroponics is between 1.5–2.5 dS m^{-1}. Higher EC will prevent nutrient absorption due to osmotic pressure and lower EC severely affect plant health and yield. Though hydroponic farmers can create their own fertilizer combinations to generate nutrient solutions utilizing entirely water soluble nutrient salts, the market too offers a variety of formulations. Some fertilizer salts react with one another, resulting in insoluble precipitations. In the tank, for example, ammonium sulphate and potassium chloride combine to generate less soluble potassium sulphate. In the presence of high calcium and magnesium concentrations, phosphate fertilizers cause precipitation of low soluble phosphates. As a result, choose fertilizers that are compatible with one another.

3.13.2 Different Types of Hydroponics

Hydroponic farming includes nutrient film technique (NFT), wick system, drip system, ebb and flow and deep-water culture (DWC). The wick system is the simplest hydroponic method which uses wicks to draw nutrients from the reservoir without use of pumps or timer. NFT hydroponics is a method where shallow channels are used to supply the nutrient solution to the bare plant roots through a recirculation process (Nisha et al. 2019). In DWC plant roots are suspended directly into the nutrient rich water solution, while the 'drip system' uses micro emitters to drip the nutrient and water directly to the plant roots with the help of a pump (Verdoliva et al. 2021). Ebb and flow involves flooding the plant tray with the nutrient solution using a pump that is connected to the solution tank at given time intervals with the use of a timer. The solution is later drained back to the nutrient tank.

3.13.3 Advantages of Hydroponics

- It can be employed in situations where in-ground agriculture or gardening are not possible (such as dry desert areas or cold climatic regions).
- Greater control over fertilizer content, pH and the growing environment.
- Lower water and nutrient expenses related with recycling water and nutrients.
- Faster growth due to increased oxygen availability in the root region.
- The eradication or decrease of soil-borne insects, fungus and bacteria.
- Significantly increased crop yields.
- There is no need for weeding or cultivation.
- Some crops, such as lettuce and strawberries, can be lifted from ground level to a much better height for planting, cultivation and harvesting. This provides much better working conditions and hence lowers labor costs.
- Crop rotation/fallowing is not necessary.
- Reduced transplant shock.

3.14 AQUAPONICS SYSTEMS

Aquaponics is a name derived from the words 'aquaculture' (the cultivation of fish in a closed environment) and 'hydroponics' (the cultivation of plants in a soil-free environment). Aquaponics combines aquaculture and hydroponics symbiotically in a balanced recirculatory environment (Figure 3.17). It is a type of integrated agri-aquaculture systems, or IAAS (Gooley and Gavine 2003). The rationale of IAAS is to take advantage of the resources shared between aquaculture and plant production, such as water and nutrients, to develop and achieve economically viable and environmentally more sustainable primary production practices. Aquaponics has been said to have evolved from relatively ancient agriculture practices associated with integrating fish culture with plant production (Komives and Junge 2015). Although the practice of modern aquaponics was first developed in the USA in the 1970s, it is believed that James Rakocy and his team's work, which began in the early 1980s, is responsible for the development of practically all current systems.

Ecological, water and nutrient efficiency drivers, as well as optimization, are now being added to the concept of aquaponics, along with economic drivers (Goddek and Körner 2019; Reyes Lastiri et

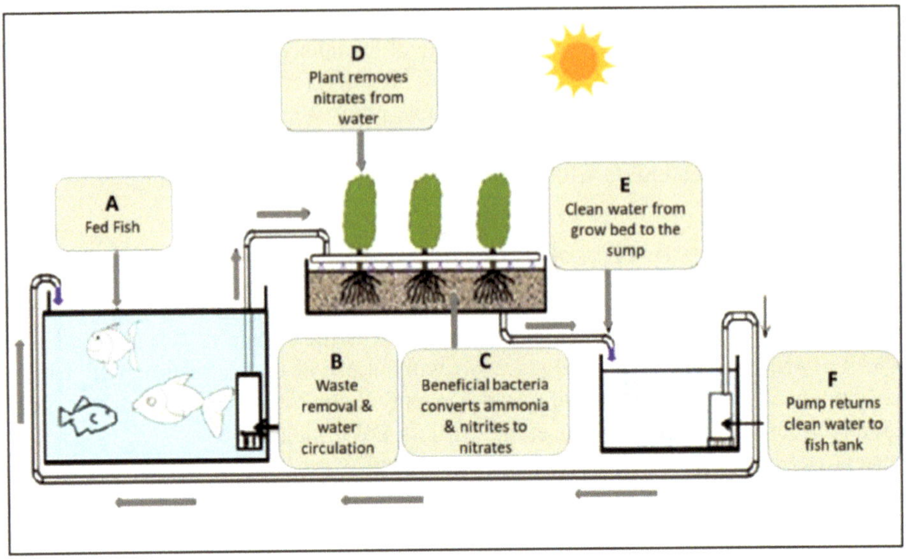

FIGURE 3.17 Flow diagrammatic representation of aquaponics system.

al. 2016; Yogev et al. 2016). According to numerous studies (Buzby and Lian-shin 2014; Wongkiew et al. 2017; Suhl et al. 2016), aquaponics has benefits in terms of water and nutrient use efficiency, sustainability, its capacity to produce two crops from a single input source (fish feed) and its reduced environmental impact.

In aquaponics system, nutrient-rich water from fish tanks is used as liquid fertilizer to fertilize hydroponic production beds. These nutrients in the water are produced from fish manure, algae and decomposing fish feed which otherwise increases the toxic levels in the fish tanks affecting the fish growth. As a biofilter, the hydroponic beds remove ammonia, nitrates, nitrites, and phosphate from the water so that it can be recirculated back into the fish tanks. The nitrifying bacteria that live in the gravel and in close proximity to plant roots are essential for the cycle of nutrients. These nitrifying microorganisms change ammonia into nitrate, which the plants use as a source of nitrogen. As a result, the fish can tolerate the nitrogen level when the water is returned to the fish tanks. Contrary to traditional farming, the aquaponics system maintains a constant flow of water and supplies nutrients to the plants continuously. This process allows the fish, plants and bacteria to thrive symbiotically and to work together to create a healthy growing environment for each other, provided that the system is properly balanced.

3.14.1 Different Methods of Aquaponics

- **Deep water culture** (DWC): DWC or *raft-based* growing uses a foam raft that is floating in a channel filled with fish effluent water that has been filtered to remove solid wastes. Plants are placed in holes in the raft and the roots dangle freely in the water. This method is most appropriate for growing salad greens and other fast-growing, relatively low-nutrient plants.
- **Media-based aquaponics**: This involves growing plants in inert planting media such as expanded clay pellets or shale. The media provides both the biological filtration (conversion of ammonia to nitrates) and the mechanical filtration (removal of solid wastes) in the same system. Media-based systems are great for home and hobby-scale systems to grow a wide variety of crops.
- **Nutrient film technique** (NFT): NFT systems work by flowing nutrient-rich water through a narrow trough, such as a PVC pipe. Plants are placed in holes drilled in this pipe, and

the roots dangle freely in this stream of water. This method of growing works very well for plants that need little support, such as strawberries and other herbs. NFT is also a great way to utilize unused space because plants can be hung from ceilings above other growing areas.

- **Vertical aquaponics**: Plants are stacked on top of each other in tower systems such as the AquaVertica. Water flows in through the top of the tower and flows through a wicking material from which the plants roots absorb water and nutrients. The water then falls into a trough or directly into the fish tank. This form of agriculture makes the most of each square foot of space and works very well with leafy greens, strawberries and other crops that do not require support to grow.

3.14.2 How Does Aquaponics Work?

In aquaponics, fish are kept in a fish tank while plants are grown in a grow bed. Millions of naturally occurring helpful bacteria in the grow bed convert the ammonia in the nutrient-rich water from the fish tank that contains fish waste into nitrites and ultimately into nitrates. These nitrates and other nutrients are absorbed by plants to aid in their growth. Before the water is pumped back into the fish tank where the fish will live, the plant's roots clean and filter it. The water is recirculated back to the fish tank, where the cycle will start over with new, clean and oxygenated water.

3.14.3 What Type of Plants Can Be Grown in Aquaponics?

Based on the aquaponics system crops such as vegetables, leafy greens, small root vegetables, microgreens, fruiting plants, herbs and flowers all thrive well in aquaponics systems. Selection of plants corresponds to an aquaponics technique.

- **Media-based aquaponics system**: Lettuce, tomato, ginger, eggplant, cucumber and any plants that will fit your grow bed.
- **Raft system**: Lettuce, basil, kale, cabbage, swiss chard, bok choy, mint, watercress and other small rooting plants.
- **Nutrient film technique (NFT)**: Lettuce, strawberry, spinach, parsley, dill and other small rooting plants.

3.14.4 Components of an Aquaponics System

Fish: Fish play an essential role in aquaponics because their waste acts as a natural fertilizer for the plants. To achieve a maximum growth output from fish, select the best fish for the system. The most common fish raised in aquaponics systems are: (1) Edible fish: tilapia, carp, catfish, trout, large-mouth bass, salmon, yellow perch, bluegill, jade perch, barramundi; (2) Ornamental fish: goldfish, koi, tetras, guppies; (3) Other aquatic animals such as shrimp.

Plants: Easy-to-grow plants are leafy greens, lettuce and herbs. When the fish growth in the system is larger, start adding tomatoes, peppers and other fruit-bearing plants. Aquaponics involves growing plants in pipes, floating rafts or grow beds. Make sure the grow bed container is sturdy and the grow media is deep enough to accommodate the plants. Using net pots is recommended because it allows the plant roots to absorb ample nutrients and securely hold the plant. If planting is done in floating foam, it should be lightweight, buoyant and be able to hold the plants upright.

Bacteria: The bacteria in an aquaponics system can be present in the biofilter, grow beds, and fish tanks. Bacteria convert fish wastes into nutrients absorbed by the plants. So it is essential to maintain a healthy bacterial colony in your system.

3.14.5 Factors That Determine a Successful Aquaponics System

The first and most important factor in having a successful aquaponics garden is ensuring that the fish are kept healthy. A robust aeration system is crucial since all fish require oxygen to survive and grow. For fish to survive, the water's temperature is also very important. The fish may experience issues with any unusual temperature increase or decrease. A pH of 7 to 7.5 is optimum and it is crucial to address the problem as soon as the pH changes, either rising or falling. This could be due to overfeeding, in which case uneaten fish food disintegrates and gathers at the aquarium's bottom. By regularly testing the water quality, you can ensure any problems that arise can be resolved long before they have the chance to cause actual damage.

3.14.6 Advantages and Benefits

- Aquaponics is a sustainable and intensive production system (Azad and Salam 2016).
- It offers incomes from fish and plant production.
- It does not require soil or large space of land.
- Easy to setup, significant increase in farm productivity and offers profitability without any net increase in water consumption (water efficient).
- Reuse of otherwise wasted on-farm resources (e.g., capture and reuse of nutrients and water), no need to use artificial fertilizer.
- Provides an artificial filtration system of fish culture.
- Sustainable and environment friendly.
- Naturally organic and free from chemicals.
- Minimum damage from pests and disease.
- No weeding and thus a cleaner form of gardening.
- Plants grow faster, allows year-round production by regulating the temperatures and using greenhouses.
- Eco-friendly, reduction of net environmental impacts of semi-intensive and intensive farming practices.

3.15 CRITERIA/PARAMETERS TO JUDGE THE EFFICIENCY AND PROFITABILITY OF IFS MODELS

The performance of IFS models are judged based on productivity, economics, employment generation, input use efficiency, energy use efficiency, nutritive value of products generated, environmental foot prints, climatic parameters, etc. Some of the major criteria are described below.

1. *System productivity*: In a cropping/farming system, there are different crops/components with different economic yields and value. To evaluate such systems, there is a need to convert the yields in to a particular crop yield called crop equivalent yield (CEY). In rice-based cropping systems, generally the yields of other crops are converted to rice equivalent yield (REY). The formula to convert to CEY is given below.

   CEY = [Production of the component (kg or litres or any other unit) × Price (₹/unit)]/ Price of equivalent crop ((₹/unit)

2. *Economic analysis:* This analysis takes into account the costs and returns involved in the system. The costs are of two types: fixed costs and variable costs, together called total cost. Fixed costs include rental value of land, expenditure incurred on implements/tools, machinery, farm buildings, animals, etc. Variable costs are expenditure on seeds, fertilizers, manures, pesticides, irrigation, labour, interest paid on working capital, etc. Gross return is calculated from the amount of produce of different components and the market

price of these produce. Gross margin is the gross return minus variable costs and net income is the gross return minus total costs. Benefit-cost ratio (BCR), is the ratio of gross retun to total cost. In long-term analysis of farming systems, economic parameters like net present value (NPV) or net worth, internal rate of return (IRR) and BCR are used. NPV is the difference between present value of cash outflows and present value of cash inflows over a period of time. NPV of an IFS model to be implemented for a number of years could be estimated from fixed cost, variable cost, total returns and discount rate data for each year. Net return is calculated as total return minus total cost. The discount rate is the interest rate used to determine the present value of future cash flows. The discount factor is calculated for each year by the following formula.

Discount factor = $1/(1+\text{Discount rate})^n$, where n is the number of compounding periods. By the use of discount factor, discounted total cost (DTC) and total returns (DTR) are calculated for each year. Then the sum of DTC and DTR are calculated for the entire period under consideration. Now NPV is calculated by deducting the sum of DTC from sum of DTR.

$$\text{NPV} = \sum \text{DTR} - \sum \text{DTC}$$

IRR is the rate of return that would result in an NPV of zero, when used as the discount rate. IRR is estimated by trial-and-error method, by putting a different discount rate in the calculation of discount factor to achieve NPV of zero, i.e., the discount rate for which the sum of DTR and sum of DTC are equal. IRR gives the rate of return from the investment, if IRR is greater than the discount rate, then the proposed IFS model will be profitable. For the long-term assessment of BCR of IFS model, the ratio of the sum of DTR and DTC is taken into account.

$$\text{BCR} = \sum \text{DTR} / \sum \text{DTC}$$

3. *Employment generation*: It indicates the labour required for accomplishing various activities of a system. A man working for a period of eight hours in a day is considered as one man-day and an woman working for the same period is treated as two-thirds man-day. In IFS, the man-days created per unit area per annuum is calculated from different components and compared with other systems. Involvement of family members is included to calculate the total man-days created.
4. *Water requirement*: Water smartness could be measured by various parameters such as water use efficiency (WUE) and water productivity (WP). WUE is an efficiency term; therefore, it has no unit, expressed in percentage. WP is the amount of output (e.g., grain yield) per unit amount of water used. If only irrigation water is considered, then it is called irritation WP (IWP) and if both irrigation and other sources such as rainfall is considered it is called total WP (TWP). IWP and TWP are expressed as kg m^{-3}. When WP is expressed in economic terms, i.e., output is conveted into economic terms by multiplying with its market price and then divided by the amount of water used, it is called economic WP (EWP), e.g., expressed as ₹ m^{-3}.
5. *Energy efficiency*: Energy input and output are worked out for individual components of IFS and then computed for the whole system.

   Energy input from different sources in the form of seed, chemical and organic fertilizer, pesticides, human power, fuel consumed, machinery used, etc. for various operations in the crop production, and output in the form of yields is converted into equivalent value of chemical energy (MJ ha^{-1}) by using the standard values (Table 3.8) given by various researchers [Mittal et al. (1985); Calilung (1985); Pimentel and Pimentel (2008); Pishgar-Komleh et al. (2011); Quilty et al. (2014)]. The energy input, output, output-input ratio as well as energy productivity are calculated based on the following formulae given by Chamsing et al. (2006).

TABLE 3.8
Energy Conversion Factors for Agronomic Inputs and for Yields of Crops Used in the Energy Calculation

Agronomic Input/Output	MJ Unit^{-1}	Reference
Diesel (L^{-1})	47.7	Pimentel and Pimentel (2008)
Power tiller (h^{-1})	123.0	Calilung (1985)
Tractor (h^{-1})	542.0	Quilty et al. (2014)
Adult man (h^{-1})	1.96	Mittal et al. (1985)
Adult woman (h^{-1})	1.57	Mittal et al. (1985)
Electricity (KW h^{-1})	11.93	Mittal et al. (1985)
Nitrogen (kg^{-1})	63.5	Pimentel and Pimentel (2008)
Phosphorus (kg^{-1})	13.95	Pimentel and Pimentel (2008)
Potassium (kg^{-1})	6.69	Pimentel and Pimentel (2008)
Farmyard manure (kg^{-1})	0.30	Mittal et al. (1985)
Water for irrigation (m^3)	1.02	Acaroglu and Aksoy (2005)
Herbicide (kg^{-1})	303.8	Pimentel and Pimentel (2008)
Insecticide (kg^{-1})	418.4	Pimentel and Pimentel (2008)
Fungicide (kg^{-1})	115.0	Pishgar-Komleh et al. (2011)
Seed (kg^{-1})	15.5	Pimentel and Pimentel (2008)
Rice grain yield (kg^{-1})	15.2	Pimentel and Pimentel (2008)
Rice straw yield (kg^{-1})	12.5	Mittal et al. (1985)
Maize grain yield (kg^{-1})	14.7	Mittal et al. (1985)
Maize stover yield (kg^{-1})	18.0	Mittal et al. (1985)
Rapeseed seed yield (kg^{-1})	25.0	Mittal et al. (1985)
Rapeseed stover yield (kg^{-1})	18.0	Mittal et al. (1985)

$$\text{Energy input (GJ ha}^{-1}) = E_f + E_s$$

where

E_f (GJ ha^{-1}) = energy input in farm operations (physical + chemical + biological energy inputs in farm operations)

E_s (GJ ha^{-1}) = energy sequestered in machinery (energy sequestered in manufacturing of machinery in GJ h^{-1} × machine working hour in h ha^{-1}

$$\text{Energy output (GJ ha}^{-1}) = (\text{Yield} \times E_{eqm}) + (\text{By-product} \times E_{eqb})$$

where,

E_{eqm} = energy equivalent value of main product

E_{eqb} = energy equivalent value of by-product

$$\text{Energy productivity (kgGJ}^{-1}) = \frac{\text{Crop yield (kg ha}^{-1})}{\text{Energy inputs to crop production, GJ ha}^{-1}}$$

$$\text{Energy output-input ratio} = \frac{\text{Energy output (GJ ha}^{-1})}{\text{Energy inputs to crop production (GJ ha}^{-1})}$$

6. *Nutritional value*: The IFS which yield better in terms of higher nutritional value such as carbohydrate, protein, fat, vitamin, minerals and natural fibre of the produce will be more profitable. Adoption of new crop varieties with high zinc, protein, iron, minerals, etc., there

will be higher market value of the produce. Protein-rich rice variety (CR Dhan 310), Zn rich rice variety (DRR Dhan 45), Zn- and Fe-rich wheat variety (WB 02), vitamin-rich maize variety (Pusa Vivek QPM9) are some of the examples. Zinc is one the most limiting factor in rice eating population, therefore, Zn fortified rice varieties will solve the Zn malnutrition.

7. *Carbon emission*: Carbon footprint of the IFS model should be either neutral or negative. Carbon balance can be estimated by calculating how much carbon is emitted and how much carbon is absorbed by the system and expressed on yearly basis. IFS should be a sink rather than source of carbon emission.
8. *Climate/ecosystem services*: This can be achieved by good agricultural practices as well as maintaining adequate vegetation by efficient utilization of fallow land, boundary plantations and agroforesty systems. However, crops (rice, wheat, maize, etc.) prone to bird/rodent damage should be planted away from trees/bushes/orchards. If bird damage to crops is severe, the boundary area may be planted with papayas or bananas instead of trees to avoid bird damage. The fish pond should have lined embankments with some aesthetic planning. Housing for livestock/birds should be as per recommendations/scientific guidelines.

3.16 ENERGY SOURCES AND FLOW IN IFS

Energy flows in IFS is an important process (Figure 3.18), for which proper planning is needed to meet the energy requirement of different components. Field crops produce grains and straw/stover. Grains are economic products either consumed by the family or sold in market or used as feed for animals. Straw/stover are fodder for animals, which are converted into dung as a source of organic manure. Animals produce milk, meat, excreta and urine. The excreta are used for decomposition

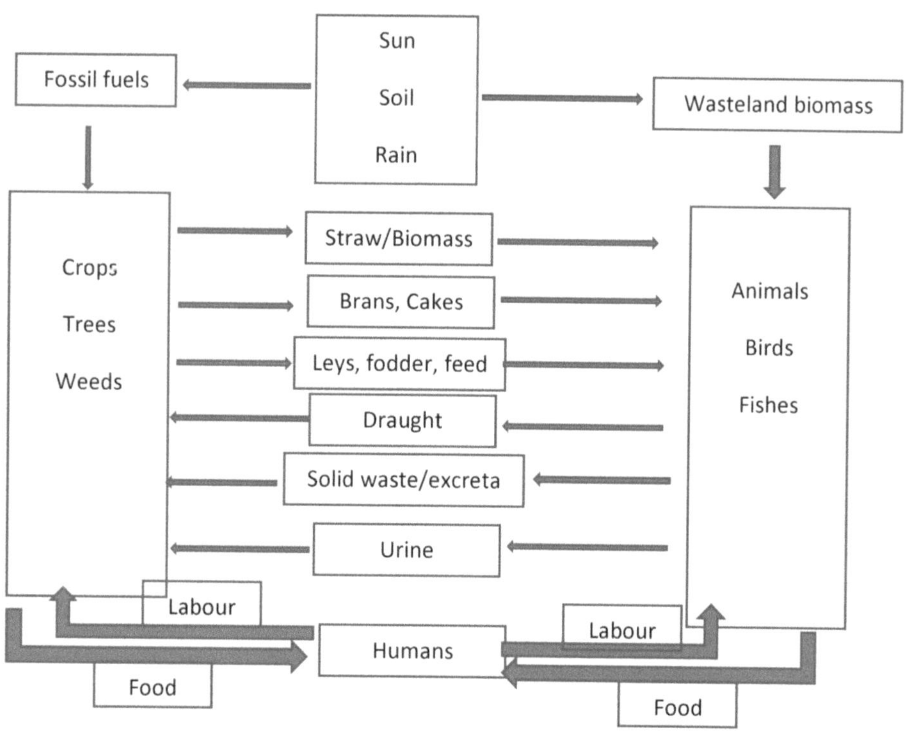

FIGURE 3.18 Energy/resource flow in mixed crop-livestock IFS model.

of biomass in the compost pit and finally used in the crop fields. Cow dung is used for biogas production, as well as slurry. Slurry is used for preparation of compost/vermicompost. Biogas is used by the household, compost/vermicompost is used as a source of manure for crops. Poultry excreta/animal urine/floor/shed washings are properly channelized to fish pond to supplement fish feeds. Irrigation from a fish pond to the cultivated fields provides water as well as nutrients. Tank silt (fish pond mud) is used for strengthening and fertilizing the bunds for growing diverse crops by creating bamboo structures, which partly cover the pond.

The poultry and duckery house should be constructed on or near the fish pond so that the excreta flow down and become a source of feed for the fish. Azolla produced in ponds may be fed to fish/poultry birds/ducks or released in paddy fields to provide organic manure. Vegetable leaves, stems, fruit covers and peels may be fed to animals. Vermiwash is prepared from vermicompost and used as foliar nutrient supplement to crops. Paddy straw may be used for growing mushrooms and after completion of mushroom production, it is used as a manure for crops.

REFERENCES

Acaroglu, M., Aksoy, A.S. 2005. The cultivation and energy balance of *Miscanthus giganteus* production in Turkey. *Biomass and Bioenergy* 29(1):42–48. DOI: https://doi.org/10.1016/j.biombioe.2005.01.002.

Ahmed, N., Garnett, S.T. 2011. Integrated rice-fish farming in Bangladesh: Meeting the challenges of food security. *Food Security* 3:81–92.

Ali, A.M.S. 2006. Rice to shrimp: Land use/land cover changes and soil degradation in Southwestern Bangladesh. *Land Use Policy* 23:421–435. DOI: https://doi.org/10.1016/j.landusepol.2005.02.001.

Ambast, S.K. 2019. Managing land and water resources in Sundarbans India for enhancing agricultural productivity. In: Sen H.S. (ed) *The Sundarbans: A Disaster-prone Eco-Region: Increasing Livelihood Security.* Springer Nature, Switzerland AG 2019. pp 199–224. DOI: https://doi.org/10.1007/978-3-030-00680_8.

Ambast, S.K., Sen, H.S., Tyagi, N.K. 1998. *Rainwater Management for Multiple Cropping in Sundarbans Delta* (W.B.). Bulletin no. 2/98, Regional Research Station, Central Soil Salinity Research Institute, Canning Town (W.B.), India. p. 69.

Amrutrao, M.J., Choudhary, H., Maurya, A., Singh, V. 2018. Drip irrigation system in cotton: One step enroute intended for sustainable agriculture. *Indian Journal of Economics and Development* 14(1a):52–58. DOI: https://doi.org/101.5958/2322-0430.2018.00035-5.

Angel, D., Freeman, S. 2009. Integrated aquaculture (INTAQ) as a tool for an ecosystem approach in the Mediterranean Sea. In: Soto D. (ed) *Integrated Mariculture: A Global Review.* FAO Fisheries and Aquaculture Technical Paper 529:133–183.

Anita, B., Manjeet, K., Yadav, A.S. 2014. Fish culture in intensively cultivated rice fields: Growth performance of *Cyprinus carpio. International Journal of Agricultural Innovations and Research* 3:80–85.

Athithan, S., Francis, T., Ramanthan, N., Ramadhas, V. 2001. Cattle urine as a benefitting source of nutrients for carp production. *Indian Journal of Fisheries* 48(2):231–235.

Aubin, J., Baruthio, A., Mungkung, R., Lazard, J. 2015. Environmental performance of brackish water polyculture system from a life cycle perspective: A Filipino case study. *Aquaculture* 435:217–227.

Ayinla, O.A. 2003. Integrated fish farming: A veritable tool for poverty alleviation/hunger eradication in the Niger Delta Region. In: Eyo A.A., Atanda J.O. (eds) *Conference Proceedings of Fisheries Society of Nigeria.* Owerri, Nigeria. pp 40–41.

Azad, K.N., Salam, M.A. 2016. Aquaponics in Bangladesh: Current status and future prospects. *Journal of Bioscience and Agriculture Research* 7(2):669–677.

Bandyopadhyay, B.K., Burman, D., Sarangi, S.K., Mandal, S., Bal, A.R. 2009. Land shaping techniques to alleviate salinity and waterlogging problems of mono-cropped coastal land for multi-crop cultivation. *Journal of the Indian Society of Coastal Agricultural Research* 27(1):13–17.

Bandyopadhyay, B.K., Mandal, S., Burman, D., Sarangi, S.K. 2010. Soil and water management options for enhancing agricultural productivity of Coastal area of West Bengal. *Journal of the Indian Society of Coastal Agricultural Research* 28(1):1–7.

Bardach, J.E. 1972. *Aquaculture: The Farming and Husbandry of Fresh Water and Marine Organisms.* John Wiley and Sons, New York.

Barrington, K., Chopin, T., Robinson, S. 2009. Integrated multi-trophic aquaculture (IMTA) in marine temperature waters. In: Soto, D. (ed.) Integrated mariculture: a global review. FAO Fisheries and Aquaculture Technical Paper, 529, FAO press, Rome, Italy. pp 7–46.

Bashir, M.A. Wang, H., Sun, W., Zhai, L., Zhang, X., Wang, N., Rehim, A., Raza, Q.U.A., Li, H. 2021. The implementation of rice-crab co-culture system to ensure cleaner rice and farm production. *Journal of Cleaner Production* 316:128284. DOI: https://doi.org/10.1016/j.jclepro.2021.128284.

Behera, U.K., France, J. 2016. Integrated farming systems and the livelihood security of small and marginal farmers in India and other developing countries. In: Donald L.S. (ed) *Advances in Agronomy*. Academic Press, vol. 138, pp 235–282. DOI: https://doi.org/10.1016/bs.agron.2016.04.001.

Behera, U.K., Giri, A., Mishra, A.K., Bhargavi, B., Kar, A., Pal, G. 2017. Rice-based farming systems and its diversification: An economic analysis in coastal ecologies of eastern India. *Indian Journal of Hill Farming* 30(2):295–306.

Biswas, G., Ananda Raja, R., De, D., Sundaray, J.K., Ghoshal, T.K., Anand, S., Kumar, S., Panigrahi, A., Thirunavukkarasu, A.R., Ponniah, A.G. 2012. Evaluation of productions and economic returns from two brackish water polyculture systems in tide fed ponds. *Journal of Applied Ichthyology* 28:116–122.

Biswas, G., Kumar, P., Kailasam, M., Ghoshal, T.K., Bera, A., Vijayan, K.K. 2019. Application of integrated multi-trophic aquaculture (IMTA) concept in brackish water ecosystem: the first exploratory trial in the Sundarbans, India. *Journal of Coastal Research* 86: 49–55.

Brahmanand, P.S., Mohanty, R.K., Kumar, A. 2006. *Integrated Rice-Fish Farming. Water Technology Centre for Eastern Region.* Indian Council Agriculture Research, Bhubaneswar, Orissa. p 78.

Buzby, K.M., Lian-shin, L. 2014. Scaling aquaponic systems: Balancing plant uptake with fish output. *Aquaculture Engineering* 63:39–44.

Calilung, E.J. 1985. *Comparison of Land Preparation Equipment Used on Small Rice Farms in the Philippines.* Thesis No. AE-85-2. Asian Institute of Technology, Bangkok, Thailand.

Carras, M.A., Knowler, D., Pearce, C.M., Hamer, A., Chopin, T., Weaire, T. 2020. A discounted cash-flow analysis of salmon monoculture and integrated multi-trophic aquaculture in eastern Canada. *Aquaculture Economics and Management* 24(1):43–63. DOI: https://doi.org/10.1080/13657305.2019.1641572.

Chaki, A., Gaydon, D.S., Dalal, R.C., Bellotti, W.D., Gathala, M.K., Hossain, A., Rahman, M.A., Menzies, N.W. 2021. Conservation agriculture enhances the rice-wheat system of the Eastern Gangetic Plains in some environments, but not in others. *Field Crops Research* 265:108109. DOI: https://doi.org/10.1016/j.fcr.2021.108109.

Chapagain, A.K., Hoekstra, A.Y. 2011. The blue, green and grey water footprint of rice from production and consumption perspectives. *Ecological Economics* 70(4):749–758. DOI: https://doi.org/10.1016/j.ecolecon.2010.11.012.

Chattopadhyay, K., Sharma, S.G., Bagchci, T.B., Mohapatra, T. 2019. *CR Dhan 310 and CR Dhan 311 – High Protein, Nutrient rich Rice Varieties of NRRI.* ICAR-National Rice Research Institute, Cuttack, India. https://icar-nrri.in/wp-content/uploads/2019/06/2.-leaflet_highprotein_final.pdf. Available online. Accessed on 15 March 2022.

Chamsing, A., Salokhe, V.M., Singh, G. 2006. Energy consumption analysis for selected crops in different regions of Thailand. *Agricultural Engineering International: The CIGR Ejournal* VIII:1–18. Available online: https://ecommons.cornell.edu/bitstream/handle/1813/10526/EE%2006%20013%20Salokhe%20final%206Nov2006.pdf?sequence=1. Accessed on 27 August 2023.

Chauhan, B.S., Mahajan, G., Sardana, V., Timsina, J., Jat, M.L. 2012. Productivity and sustainability of the rice-wheat cropping system in the Indo-Gangetic plains of the Indian subcontinent: Problems, opportunities and strategies. *Advances in Agronomy* 117:315–369. DOI: http://dx.doi.org/10.1016/B978-0-12-394278-4.00007-6.

Chen, D., Ying, H., Shui, M. 1995. Rice-azolla-fish in rice fields. In: MacKay K.T. (ed) *Rice-Fish Culture in China.* International Development Research Centre (IDRC) Ottawa, Canada. pp 169–176; 276.

Chopin, T. 2006. Integrated multi-trophic aquaculture. What it is, and why you should care… and don't confuse it with polyculture. *Northern Aquaculture* 12(4):4.

Chopin, T., Cooper, J.A., Reid, G., Cross, S., Moore, C. 2012. Open-water integrated multi-trophic aquaculture: Environmental biomitigation and economic diversification of fed aquaculture by extractive aquaculture. *Reviews in Aquaculture* 4(4):209–220. DOI: https://doi.org/10.1111/j.1753-5131.2012.01074.x.

Cifuentes-Torres, L., Mendoza-Espinosa, L.G., Correa-Reyes, G., Daesslé, L.W. 2020. Hy- droponics with wastewater: A review of trends and opportunities. *Water and Environment Journal* 35:166–180. DOI: https://doi.org/10.1111/wej.12617.

Corales, R.G., Juliano, L.M., Capistrano, A.O.V., Tobias, H.S., Dasalla, N.V., Canete, S.D., Casimero, M.C., Sebastian, L.S. 2004. Palayamanan: A rice-based farming systems model for small-scale farmers. *Philippine Journal of Crop Science* 29(1):21–27.

Dalsgaard, J.P.T., Lightfoot, C., Christensen, V. 1995. Towards quantification of ecological sustainability in farming systems analysis. *Ecological Engineering* 4(3):181–189. DOI: https://doi.org/10.1016/0925-8574(94)00057-C.

Debeljak, M., Trajanov, A., Kuzmanovski, V., Schröder, J., Sandén, T., Spiegel, H., Wall, D.P., Van de Broek, M., Rutgers, M., Bampa, F., Creamer, R.E., Henriksen, C.B. 2019. A field-scale decision support system for assessment and management of soil functions. *Frontiers in Environmental Science* 7:115. DOI: https://doi.org/10.3389/fenvs.2019.00115.

Delmendo, M.N. 1980. A review of integrated livestock-fowl-fish farming systems. In: Pullin R.S.V., Shehadeh Z.H. (eds) *Integrated Agriculture-Aquaculture Farming Systems*. ICLARM Conference Proceedings 4: pp 59–71.

Deshmukh, G., Hardaha, M.K. 2014. Effect of irrigation and fertigation scheduling under drip irrigation in papaya. *Journal of AgriSearch* 1(4):216–220.

Dhruwe, N.K., Victor, V.M. 2021. Performance evaluation of happy seeder for wheat sowing in combine harvested paddy field. *International Journal of Current Microbiology and Applied Sciences* 10(01):2542–2547. DOI: https://doi.org/10.20546/ijcmas.2021.1001.294.

Edwards, P. 2000. Aquaculture and poverty: Past, present and future prospects of impacts. A discussion paper prepared for the fifth fisheries development donor consultation, Rome, Italy, pp 22–24, February, 1999.

Edward, P. 1985. Pigs over fish ponds. *Pig International* 15(9):10.

Endebu, M., Tugie, D., Negisho, T. 2016. Fish growth performance in ponds integrated with poultry farm and fertilized with goat manure: A case in Ethiopian Rift Valley. *International Journal of Fishery Science and Aquaculture* 3:040–045.

Ermakova, M., Danila, F.R., Furbank, R.T., Caemmerer, S.V. 2020. On the road to C4 rice: Advances and perspectives. *The Plant Journal* 101:940–950. DOI: https://onlinelibrary.wiley.com/doi/10.1111/tpj.14562.

Fagberno, O.A., Sydenham, D.H. 1988. Evaluation of *Clarias isheriensis* (Sydenham) under semi-intensive management in ponds. *Aquaculture* 74:287–291.

FAO. 2021. Fishery and aquaculture statistics. Global aquaculture production 1950–2021 FishstatJ. In: *FAO Fisheries and Aquaculture Department* [online]. Rome. Updated 2021. .https://www.fao.org/fishery/en/fishstat.

FAO. 2003. *The Role of Aquaculture in Improving Food Security and Nutrition*. Committee on World food Security, Food and Agricultural Organization of the United Nations, Romes, Italy.

FAO. 2013. *The State of Food Insecurity in the World*. Available online: http://www.fao.org/3/i3434e/i3434e00.htm.

FAO.org. https://www.fao.org/3/t7202e/t7202e08.htm. Available online. Accessed on 05 March 2022.

FAOSTAT. 2014. http://faostat.fao.org.

FAOSTAT. 2022. https://www.fao.org/faostat/en/#data/QCL.

Feng, J., Li, F., Zhou, X., Xu, C., Fang, F. 2016. Nutrient removal ability and economical benefit of a rice-fish co-culture system in aquaculture pond. *Ecological Engineering* 94:315–319.

Feng, J.X., Gao, Q.F., Dong, S.L., Sun, Z.L., Zhang, K. 2014. Trophic relationships in a polyculture pond based on carbon and nitrogen stable isotope analyses: A case study in Jinghai Bay. *China Aquaculture* 428–429:258–264.

Fonteyne, S., Garcia, A.F., Verhulst, N. 2021. Reduced water use in barley and maize production through conservation agriculture and drip irrigation. *Frontiers in Sustainable Food Systems* 5. DOI: https://doi.org/10.3389/fsufs.2021.734681. Accessed on 05 March 2022.

Francis, T., Padmavathy, P., Ramanathan, N. 2004. Integrated fish farming – A review. *World Aquaculture* 35(3):24–30.

Frei, M., Khan, M.A.M., Razzak, M.A., Hossain, M.M., Dewan, S., Becker, K. 2007. Effects of a mixed culture of common carp, *Cyprinus carpio* L., and Nile tilapia, *Oreochromis niloticus* (L.), on terrestrial arthropod population, benthic fauna, and weed biomass in rice fields in Bangladesh. *Biological Control* 41:207–213.

Gabriel, U.U., Akinrotimi, O.A., Bekibele, D.O., Anyanwu, P.E., Onunkwo, D.N. 2007. Economic benefit and ecological efficiency of integrated fish farming in Nigeria. *Scientific Research and Essay* 2:302–308.

Giap, D.H., Yi, Y., Lin, C.K. 2005. Effects of different fertilization and feeding regimes on the production of integrated farming of rice and prawn *Macrobrachium rosenbergii* (De Man). *Aquaculture Research* 36(3):292–299. DOI: 10.1111/j.1365-2109.2005.01244.x.

Goddek, S., Körner, O. 2019. A fully integrated simulation model of multi-loop aquaponics: A case study for system sizing in different environments. *Agricultural Systems* 171:143–154. DOI: https://doi.org/10.1016/j.agsy.2019.01.010.

Gooley, G.J., Gavine, F.M. 2003. *Integrated Agri-Aquaculture Systems: A Resource Handbook for Australian Industry Development.* RIRDC Publication No. 03/012. Rural Industries Research and Development Corporation, Australia.

Goverdhan, M., Kumari, C.P., Reddy, G.K., Sridevi, S., Alibaba, M.D., Chiranjeevi, K., Kumar, M.S. 2020. Evaluation of integrated farming system model for resource recycling and livelihood security of small and marginal farmers of Telangana State, India. *Current Journal of Applied Science and Technology* 39(34):17–26.

Gupta, M.V., Acosta, B.O. 2004. A review of global tilapia farming practices. *Aquaculture Asia* 9(1):7–12. http://library.enaca.org/AquacultureAsia/Articles/Jan-March-2004/6global-review-tilapia.pdf.

Halwart, M., Gupta, M.V. 2004. *Culture of Fish in Rice Fields.* Food and Agriculture Organization and The World Fish Center, Rome.

Hibberd, J.M., Sheehy, J.E., Langdale, J.A. 2008. Using C4 photosynthesis to increase the yield of rice-rationale and feasibility. *Current Opinion in Plant Biology* 11(2):228–231. DOI: https://doi.org/10.1016/j.pbi.2007.11.002.

Hopkins, K.D. 1982. Outstanding yields and profits from livestock tilapia integrated farming. *ICLARM newsletter* 5(3):13. https://hdl.handle.net/20.500.12348/3623.

Horgan, F.G., Vu, Q., Mundaca, E.A., Martinez, E.C. 2022. Restoration of rice ecosystem services: Ecological engineering for pest management, incentives and practices in the Mekong Delta region of Vietnam. *Agronomy* 12:1014. DOI: https://doi.org/10.3390/agronomy12051042.

https://agriculture.com.ph/2018/10/22/rice-duck-farming-is-profitable-and-good-for-the-environment. Available online. Accessed on 20 February 2022.

https://agroinsight.com/downloads/innovations-in-rural-extension/Innovations-in-Rural-Extension-Chapter12.pdf. Available online. Accessed on 20 February 2022.

https://icar-nrri.in/icar-nrri-developed-mobile-app-ricexpert/#:~:text=Different%20features%20of%20'ricexpert'%20app,workers%20working%20on%20rice%20crop. Available online. Accessed on 07 September 2022.

https://www.equatorinitiative.org/2017/08/08/integrated-rice-duck-farming-and-value-chain. Available online. Accessed on 20 February 2022.

https://www.irri.org/crop-manager. Available online. Accessed on 07 September 2022.

https://www.irri.org/news-and-events/news/rice-doctor-innovative-technological-approach-enhance-rice-production. Available online. Accessed on 07 September 2022.

https://www.knowledgebank.irri.org/step-by-step-production/growth/water-management. Available online. Accessed on 04 March 2022.

https://www.knowledgebank.irri.org/step-by-step-production/postharvest/storage/grain-storage-systems/hermetic-storage-systems/cocoon. Available online. Accessed on 28 April 2022.

https://www.knowledgebank.irri.org/step-by-step-production/postharvest/storage/grain-storage-systems/hermetic-storage-systems/irri-super-bag. Available online. Accessed on 28 April 2022.

Hu, L., Zhang, J., Ren, W., Guo, L., Cheng, Y., Li, J., Li, K., Zhu, Z., Zhang, J., Luo, S., Cheng, L. 2016. Can the co-cultivation of rice and fish help sustain rice production? *Scientific Reports* 6:28728. DOI: https://doi.org/10.1038/srep28728.

Islam, A.H.M.S., Barman, B.K., Murshed-e-Jahan, K. 2015. Adoption and impact of integrated rice-fish farming system in Bangladesh. *Aquaculture* 447:76–85.

Joshitha, C., Kanakaraja, P., Kumar, K.S., Akanksha, P., Satish, G. 2021. An eye on hy- droponics: The IoT initiative. In: *Proceedings of the 7th International Conference on Electrical Energy Systems (ICEES).* pp 553–557.

Kashyap, P., Ravisankar, N., Panwar, A.S., Ghasal, P.C., Nath, A., Malik, S., Prusty, A.K., Chandrabhanu, Shamim, M., Dutta, D. 2021. Sustainable integration of banana in farming systems for enhanced income and nutritional security. *Indian Horticulture* 66(6):18–22.

Kaushal, A., Patole, R., Singh, K.G. 2012. Drip irrigation in sugarcane: A review. *Agricultural Reviews* 33:211–219.

Kirandeep, Sharma, M., Singh, R. 2020. Effect of different sowing techniques and varieties on yield of wheat (*Triticum aestivum* L.). *Journal of Krishi Vigyan* 9(1):92–98. DOI: 10.5958/2349-4433.2020.00194.4.

Komives, T., Junge, R. 2015. Editorial: On the "aquaponic corner" section of the journal. *Ecocycles* 1(2):1–2.

Kumar, A., Saroj, S., Mishra, A.K. 2021. *Crop Insurance and Crop Productivity Evidence from Rice Farmers in Eastern India*. IFPRI Discussion Paper 01996. International Food Policy Research Institute, South Asia Regional Office, New Delhi, India. DOI: https://doi.org/10.2499/p15738coll2.134239.

Li, C., Cao, C., Wang, J., Zhan, M., Yuan, W., Ahmad, S. 2008. Nitrogen losses from integrated rice–duck and rice–fish ecosystems in southern China. *Plant and Soil* 307(1):207–217. DOI: 10.1007/s11104-008-9597-1.

Lightfoot, C., Minnick, D.R. 1991. Farmer-first qualitative methods: Farmers diagrams for improving methods of experimental design in integrated farming systems. *Journal for Farming Systems Research and Extension* 2:11–34.

Lin, K., Wu, J. 2020. Effect of introducing frogs and fish on soil phosphorus availability dynamics and their relationship with rice yield in paddy fields. *Scientific Reports* 10:21. DOI: https://doi.org/10.1038/s41598-019-56644-z.

Lindawati, L., Kusnadi, N., Kuntjoro, S.U., Swastika, D.K.S. 2018. The impact of input and output prices on the household economic behaviour of rice-livestock integrated farming system (Rlifs) and non Rlifs farmers. IOP Conference Series: *Earth and Environmental Science* 122:012020. DOI: https://doi.org/10.1088/1755-1315/122/1/012020.

Liu, J., Liu, H.B., Liu, R., Mostofa, A., Zhai, L.M., Lu, H.M., Wang, H.Y., Zhang, B., Zhang, Y., Zhao, Y., Ding, X. 2018. Water quality in irrigated paddy systems. In: Ondrasek (ed) *Irrigation in Agroecosystems*. Intech Open, London. pp 105–122.

Lu, J., Li, X. 2006. Review of rice-fish-farming systems in China – One of the globally important ingenious agricultural heritage systems (GIAHS). *Aquaculture* 260:106–113.

Mahanta, K.K., Burman, D., Sarangi, S.K., Mandal, U.K., Maji, B., Mandal, S., Digar, S., Mainuddin, M. 2019. Drip irrigation for reducing soil salinity and increasing cropping intensity: Case studies in Indian Sundarbans. *Journal of the Indian Society of Coastal Agricultural Research* 37(2):64–71.

Mandal, U.K., Bhardwaj, A.K., Lama, T.D., Nayak, D.B., Samui, A., Burman, D., Mahanta, K.K., Sarangi, S.K., Mandal, S., Raut, S. 2021. Net ecosystem exchange of carbon, greenhouse gases and energy budget in coastal lowland double cropped rice ecology. *Soil & Tillage Research* 212(2021):105076. DOI: https://doi.org/10.1016/j.still.2021.105076.

Marchese, D., Reynolds, E., Bates, M.E., Morgan, H., Clark, S.S., Linkov, I. 2018. Resilience and sustainability: Similarities and differences in environmental management applications. *Science of the Total Environment* 613–614:1275–1283. DOI: https://doi.org/10.1016/j.scitotenv.2017.09.086.

Mattila, J., Koskela, J., Pirhonen, J. 2009. The effect of the length of repeated feed deprivation between single meals on compensatory growth of pikeperch Sander lucioperca. *Aquaculture* 296:65–70.

Mittal, V.K., Mittal, J.P., Dhawan, K.C. 1985. *Research Digest on Energy Requirement in Agriculture Sector*. Co-ordinating cell, AICRP on energy requirements in agricultural sector, Punjab Agricultural University, Ludhiana.

Mohanty, R.K. 2010a. Dynamics of the compensatory growth response in fish. *Fishing Chimes* 30:78–80.

Mohanty, R.K. 2010b. Impact of phased harvesting on population structure, feed intake pattern and growth performance of *Macrobrachium rosenbergii* DeMan (giant freshwater prawn) in polyculture with carps in concurrent rice–fish culture. *Aquaculture International* 18:523–537.

Mohanty, R.K. 2015. Effects of feed restriction on compensatory growth performance of Indian major carps in a carp–prawn polyculture system: A response to growth depression. *Aquaculture Nutrition* 21:464–473.

Mohanty, R.K., Ambast, S.K., Panda, D.K., Thakur, A.K., Mohanty, S. 2017b. Density-dependent water use in carp polyculture: Impacts on production performance and water productivity. *Aquaculture* 470:32–39.

Mohanty, R.K., Ambast, S.K., Panigrahi, P., Thakur, A.K., Mandal, K.G. 2018. Enhancing water use efficiency in monoculture of *Litopenaeus vannamei*: Impacts on pond water quality, waste production, water footprint and production performance. *Aquacultural Engineering* 82:46–55.

Mohanty, R.K., Jena, S.K., Kumar, A., Sahoo, N., Roy Chowdhury, S. 2008. *Rice-Fish Culture: An Ingenious Agricultural Heritage System*. ICAR Water Technology Centre for Eastern Region. Research Bulletin No.42, p. 54.

Mohanty, R.K., Jena, S.K., Thakur, A.K., Patil, D.U. 2009. Impact of high-density stocking and selective harvesting on yield and water productivity of deepwater rice-fish systems. *Agricultural Water Management* 96(12):1844–1850.

Mohanty, R.K., Mishra, A., Ambast, S.K., Mandal, K.G., Panda, D.K., Panigrahi, P. 2017a. Effects of various shrimp (*Penaeus monodon*) densities on their growth, water and sediment quality, and water budget. *Aquaculture International* 25:2161–2176.

Mohanty, R.K., Thakur, A.K., Ghosh, S., Mohanty, S., Patil, D.U. 2010. Performance evaluation of concurrent rice-fish-prawn culture with and without cull harvesting. *Aquaculture Research* 41:1402–1412.

Mohanty, R.K., Verma, H.N., Brahmanand, P.S. 2004. Performance evaluation of rice-fish integration system in rainfed medium land ecosystem. *Aquaculture* 230:125–135.

Mohapatra, S.D., Tripathi, R., Kumar, A., Kar, S., Mohapatra, M., Shahid, M., Raghu, S., Gowda, B.G., Nayak, A.K., Pathak, H. 2019. Eco-smart pest management in rice farming: Prospects and challenges. *Oryza* 56:143–155. DOI: https://doi.org/10.35709/ory.2019.56.spl.7.

Mondal, A., Bhattacharya, S., Mitra, A., Sundaray, J.K., Mohanty, R.K. 2020. Performance evaluation of mud crab *Scylla olivacea* (Herbst, 1896) coculture with different fish species in confined brackish water ponds. *Aquaculture* 522:735125. DOI: https://doi.org/10.1016/j.aquaculture.2020.735125.

Montes, C., Sapkota, T., Singh, B. 2022. Seasonal patterns in rice and wheat residue burning and surface PM2.5 concentration in northern India. *Atmospheric Environment* X(13):100154. DOI: https://doi.org/10.1016/j.aeaoa.2022.100154.

Montserrat, N., Gabillard, J.C., Capilla, E., Navarro, M.I., Gutierrez, J. 2007. Role of insulin, insulin-like growth factors, and muscle regulatory factors in the compensatory growth of trout (*Oncorhynchus mykiss*). *General and Comparative Endocrinology* 150(3):462–472. DOI: https://doi.org/10.1016/j.ygcen.2006.11.009.

Nandan, R., Singh, V., Singh, S.S., Kumar, V., Hazra, K.K., Nath, C.P., Poonia, S., Malik, R.K., Bhattacharyya, R., McDonald, A. 2019. Impact of conservation tillage in rice-based cropping systems on soil aggregation, carbon pools and nutrients. *Geoderma* 340:104–114. DOI: https://doi.org/10.1016/j.geoderma.2019.0.001.

Neori, A, Chopin, T., Troell, M., Buschmann, A.H., Kraemer, G.P., Halling, C., Shpigel, M., Yarish, C. 2004. Integrated aquaculture: Rationale, evolution and state of the art emphasizing seaweed biofiltration in modern mariculture. *Aquaculture* 231:361–391.

Nhan, D.K., Duong, L.T., Rothuis, A. 1998. Rice-fish farming system research in the Vietnamese Mekong Delta: Identification of constraints. *NAGA* 20:107–111. https://hdl.handle.net/20.500.12348/2615.

Nikki, J., Pirhonen, J., Jobling, M., Karjalainen, J. 2004. Compensatory growth in juvenile rainbow trout, *Oncorhynchus mykiss* (Walbaum), held individually. *Aquaculture* 235:285–296.

Nisha, S., Acharya, S., Kumar, K., Singh, N., Chaurasia, O.P. 2019. Hydroponics as an advanced technique for vegetable production: An overview. *Journal of Soil and Water Conservation* 17:64–371.

Nnaji, C.J., Okoye, F.C., Ogunseye, J.O. 2003. Integrated fish farming practices with special reference to combination rates production figures and economic evaluation. In: Eyo A.A., Ayanda J.O. (eds) *Conference Proceedings of Fisheries Society of Nigerian (FISON)*, Owerri 8th–12th December, 2003. pp 173–178.

Odum, H.T. 1989. Self-organization transformation and information. *Science* 242(4882):1132–1139. https://www.science.org/doi/10.1126/science.242.4882.1132.

Oehme, M., Frei, M., Razzak, M., Dewan, S., Becker, K. 2007. Studies on nitrogen cycling under different nitrogen inputs in integrated rice-fish culture in Bangladesh. *Nutrient Cycling in Agroecosystems* 79:181–191.

Oribhabor, B.J., Ansa, E.J. 2006. Organic waste reclamation, recycling and re-use in integrated fish farming in the Niger Delta. *Journal of Applied Sciences and Environmental Management* 10(3):47–53. DOI: https://doi.org/10.4314/jasem.v10i3.17319.

Padmanabhan, S. 2019. *Drip Irrigation Technology for Rice Cultivation for Enhancing Rice Productivity and Reducing Water Consumption*. Third World Irrigation Forum (WIF3), 1–7 September 2019, Bali, Indonesia.

Pandey, M.P., Verulkar, S.B., Sharma, D. 2010. Rice research: Past achievements, present scenario and future thrust. *Indian Journal of Agricultural Sciences* 80(6):447–469.

Panigrahi, P., Srivastava, A.K. 2017. Water and nutrient management effects on water use and yield of drip irrigated citrus in vertisol under a sub-humid region. *Journal of Integrative Agriculture* 16(5):1184–1194. DOI: https://doi.org/10.1016/S2095-3119(16)61500-9.

Parajuli, R., Dalgaard, T., Birkved, M. 2018. Can farmers mitigate environmental impacts through combined production of food, fuel and feed? A consequential life cycle assessment of integrated mixed crop-livestock system with a green biorefinery. *Science of the Total Environment* 619–620:127–143. DOI: https://doi.org/10.1016/j.scitotenv.2017.11.082.

Paramesh, V., Chakurkar, E.B., Bhagat, T., Sreekanth, G.B., Kumar, H.B.C., Rajkumar, S., Gokuldas, P.P., Mahajan, G.R., Manohara, K.K., Ravisankar, N. 2021. Impact of integrated farming system on residue recycling, nutrient budgeting and soil health. *Indian Journal of Agricultural Sciences* 91(1):44–48.

Pimentel, D., Pimentel, M. 2008. *Food, Energy and Society*, third Ed. CRC Press, Boca Raton, FL.

Pishgar-Komleh, S.H., Sefeedpari, P., Rafiee, S. 2011. Energy and economic analysis of rice production under different farm levels in Guilan Province of Iran. *Energy* 36:5824–5831.

Pradhan, S.K., Pandit, E., Bose, L.K., Reddy, J.N., Pattanaik, S.S.C., Meher, J., Behera, L. 2019. *CR Dhan 801 and CR Dhan 802 Climate-smart Rice Varieties of NRRI*. ICAR-National Rice Research Institute, Cuttack, India. https://icar-nrri.in/wp-content/uploads/2019/06/1.-leaflet_climatesmart_final.pdf. Available online. Accessed on 28 April 2022.

Prein, M. 2002. Integration of aquaculture into crop-animal systems in Asia. *Agricultural Systems* 71:127–146.

Preston, T.R. 1990. Future strategies for livestock production in tropical third world countries. *Ambio Science of Sustainable Development* 19(8):390–393. https://www.jstor.org/stable/4313747.

Quilty, J.R., McKinley, J., Pede, V.O., Buresh, R.J., Correa Jr., T.Q., Sadro, J.M. 2014. Energy efficiency of rice production in farmers' fields and intensively cropped research fields in the Philippines. *Field Crops Research* 168:8–18. DOI: http://dx.doi.org/10.1016/j.fcr.2014.08.001.

Radhammani, S., Balasubramanian, A., Ramamoorthy, K., Geethalakshmi, V. 2003. Sustainable integrated farming systems for dry lands: A review. *Agricultural Reviews* 24:204–210.

Rahman, M.A. 2016. Integration of aquaculture with rice farming: A way to increase farm productivity, food security, livelihood improvement and better environment. In: *6th International Conference on Agriculture, Environment and Biological Sciences (ICAEBS'16)*, December 21–22, 2016 Kuala Lumpur (Malaysia). pp 149–151.

Ramanathan, K., Sangeeviraman, V., Chandrahasan, P., Chaudhary, B.N., Ramachandra, S.S. 2020. Integration of fish culture and poultry rearing in transplanted rice for nutritional security in smallholder farms. *Scientific Reports* 10:10566. DOI: https://doi.org/10.1038/s41598.020.67657-4.

Rautaray, S.K., Dubey, R., Raychaudhuri, S., Pradhan, S., Mohanty, S., Mohanty, R.K., Ambast, S.K. 2020. Sustainable agriwaste management at farm level through self-reliant farming system. *Waste Management & Research* 38(7):753–761. DOI: https://doi.org/10.1177/0734242X20920350.

Reddy, P.R., Kishori, B. 2018. Integrated rice and aquaculture farming. In: Diarte-Plata G., Escamilla-Montes R. (eds) *Aquaculture – Plants and Invertebrates*. IntechOpen. pp 11–32. DOI: https://doi.org/10.5772/intechopen.78062.

Reyes Lastiri, D., Slinkert, T., Cappon, H.J., Baganz, D., Staaks, G., Keesman, K.J. 2016. Model of an aquaponic system for minimised water, energy and nitrogen requirements. *Water Science & Technology* 74(1):30–37. DOI: https://doi.org/10.2166/wst.2016.127.

Rizal, G., Karki, S., Thakur, V., Chatterjee, J., Coe, R.A., Wanchana, S., Quick, W.P. 2012. Towards a C4 rice. *Asian Journal of Cell Biology* 7:13–31. DOI: https://doi.org/10.3923/ajcb.2012.13.31.

Rubio, V.C., Sanchez, E., Cerda-Reverter, J.M. 2010. Compensatory feeding in the sea bass after fasting and physical stress. *Aquaculture* 298:332–337.

Saikia, R., Das, T., Das, D.N. 2015. Physio-chemical qualities of water in high altitude rice fish farming system of Ziro valley Arunachal Pradesh. *Journal of Environmental Biology* 36:1153–1159.

Sarangi, S.K. 2004. Rice-fish farming system – a review. *Agricultural Reviews* 25(1):49–56.

Sarangi, S.K., Maji, B., Singh, S., Sharma, D.K., Burman, D., Mandal, S., Ismail, A.M., Haefele, S.M. 2014. Crop establishment and nutrient management for dry season (boro) rice in coastal areas. *Agronomy Journal* 106(6):2013–2023. doi: 10.2134/agronj14.0182.

Sarangi, S.K., Maji, B. 2017. Sustainable rice cultivation in coastal saline soils: A case study. In: Sasaki T. (ed) *Achieving Sustainable Cultivation of Rice Volume 2: Cultivation, Pest and Disease Management*. Burleigh Dodds Science Publishing Limited, Cambridge. pp 69–103; 406. (ISBN: 978 1 78676 028 9; www.bdspublishing.com). DOI: http://dx.doi.org/10.19103/AS.2016.0003.07.

Sarangi, S.K., Maji, B., Mahanta, K.K., Burman, D., Mandal, S., Mandal, U.K., Raut, S., Lama, T.D., Sharma, D.K. 2016. Integrated farming system: A holistic approach for farm income optimization and risk minimization in coastal region of India. *Journal of the Indian Society of Coastal Agricultural Research* 34(2):15–21.

Sarangi, S.K., Maji, B., Sharma, P.C., Digar, S., Mahanta, K.K., Burman, D., Mandal, U.K., Mandal, S., Mainuddin, M. 2021a. Potato (*Solanum tuberosum* L.) cultivation by zero tillage and paddy straw mulching in the saline soils of the Ganges Delta. *Potato Research* 64(2):277–305. https://link.springer.com/article/10.1007%2Fs11540-020-09478-6.

Sarangi, S.K., Maji, B., Singh, S., Mahanta, K.K., Sharma, D.K. 2017. Water productive and profitable dry season crops for tropical coastal region of India. *E-planet* 15(2):96–101.

Sarangi, S.K., Maji, B., Singh, S., Sharma, D.K., Burman, D., Mandal, S., Singh, U.S., Ismail, A.M., Haefele, S.M. 2016. Using improved variety and management enhances rice productivity in stagnant flood-affected coastal zones. *Field Crops Research* 190(2016):70–81. DOI: http://dx.doi.org/10.1016/j.fcr.2015.10.024.

Sarangi, S.K., Raut, S., Mahanta, K.K., Mandal, U.K. 2021b. Strawponics for successful garlic (*Allium sativum* L.) cultivation in coastal saline soils. *Indian Farming* 71(8):33–37.

Sarangi, S.K., Singh, S., Kumar, V., Srivastava, A.K., Sharma, P.C., Johnson, D.E. 2020a. Tillage and crop establishment options for enhancing the productivity, profitability and resource use efficiency of rice-rabi systems of the salt-affected coastal lowlands of eastern India. *Field Crops Research* 247:107494. DOI: https://doi.org/10.1016/j.fcr.2019.03.016.

Sarangi, S.K., Singh, S., Maji, B., Sharma, P.C., Srivastava, A.K., Burman, D., Mandal, S., Krishnamurthy, S.L., Singh, U.S., Ismail, A.M. 2020. SUB1 varieties increased rice (Oryza sativa) yield in flood-prone rainfed lowlands of coastal regions. *Indian Journal of Agricultural Sciences* 90(11):2064–2070.

Sarangi, S.K., Singh, S., Srivastava, A.K., Choudhary, M., Mandal, U.K., Lama, T.D., Mahanta, K.K., Kumar, V., Sharma, P.C., Ismail, A.M. 2020b. Crop and residue management improves productivity and profitability of rice–maize system in salt-affected rainfed lowlands of East India. *Agronomy* 10(12):2019. DOI: https://doi.org/10.3390/agronomy10122019.

Schroeder, G.L. 1980. Fish farming in manure loaded ponds. *Proceedings of ICLARM Conference in Integrated Agriculture and Aquaculture Farming Systems* 4:73–86.

Shahane, A.A., Shivay, Y.S. 2019. Viable options for diversification of rice in non-conventional rice-conventional wheat cropping system in Indo-Gangetic plains. *International Journal of Bio-resource and Stress Management* 10(4):352–363. DOI: https://doi.org/10.233910/IJBSM/2019.10.4.1979.

Sharma, B.R., Gulati, A., Mohan, G., Manchanda, S., Ray, I., Amarasinghe, U. 2018. *Water Productivity Mapping of Major Indian Crops.* National Bank for Agriculture and Rural Development (NABARD), Mumbai, India; Indian Council for Research on International Economic Relations (ICRIER), New Delhi, India. p 182.

Sharma, D.K., Singh, A., Sharma, P.C., Dagar, J.C., Chaudhari, S.K. 2016. Sustainable management of sodic soils for crop production: Opportunities and challenges. *Journal of Soil Salinity and Water Quality* 8(2):109–130.

Sharma, R., Kispotta, W. 2016. Study on drip irrigation in banana – district of Kaushambi (UP). *IOSR Journal of Business and Management* 18(1):10–12.

Shelley, C., Lovatelli, A. 2012. Mud crab aquaculture: A practical manual. In: Shelley C., Lovatelli A. (eds) *Food and Agriculture Organization of the United Nations (FAO).* CABI publishing, Rome. p 96.

Shoko, A.P., Getabu, A., Mwayuli, G., Mgaya, Y.D. 2011. Growth performance, yields and economic benefits of Nile tilapia, *Oreochromis niloticus* and Kales *Brassica oleracea* cultured under vegetable-fish culture integration. *Tanzania Journal of Science* 37:37–48.

Sidhu, J.S., Singh, J., Kumar, R. 2014. Role of market intelligence in agriculture. A success story of basmati cultivation in Punjab. *Indian Journal of Economics and Development* 10(1a):26. DOI: https://doi.org/10.5958/j.2322-0430.10.1a.017.

Sin, A.W. 1980. Integrated fish husbandry systems in Hong Kong, with case studies on duck-fish and goose-fish systems. In: Pullin R.S.V., Shehadeh Z.H. (eds) *Integrated Agriculture-Aquaculture Farming Systems.* ICLARM Conference Proceedings 4: pp 113–123.

Singh, P.K., Srivastava, P.C., Sangavi, R., Gunjan, P., Sharma, V. 2019. Rice water management under drip irrigation: An effective option for high water productivity and efficient zinc applicability. *Pantnagar Journal of Research* 17(1):19–26.

Singh, S.K., Bharadwaj, V., Thakur, T.C., Pachauri, S.P., Singh, P.P., Mishra, A.K. 2009. Influence of crop establishment methods on methane emission from rice fields. *Current Science* 97(1):84–89.

Sreelatha, A.K., Diya, P.V., Paul, N. 2022. Multilevel integrated farming model in *Pokkali* lands of Kerala. In: Lama T.D., Burman D., Mandal U.K., Sarangi S.K., Sen H.S. (eds) *Transforming Coastal Zone for Sustainable Food and Income Security.* Springer Nature Switzerland AG 2022. pp 301–311. DOI: https://doi.org/10.1007/978-3-030-95618-9_25.

Srivastava, R.C., Singandhupe, R.B., Mohanty, R.K. 2004. Integrated farming approach for runoff recycling system in humid plateau areas of eastern India. *Agricultural Water Management* 64:197–212.

Stagnari, F., Maggio, A., Galieni, A., Pisante, M. 2017. Multiple benefits of legumes for agriculture sustainability: An overview. *Chemical and Biological Technologies in Agriculture* 4(2). DOI: https://doi.org/10.1186/s40538-016-0085-1.

Stumpf, L., Calvo, N.S., Pietrokovsky, S., Lopez Greco, L.S. 2010. Nutritional vulnerability and compensatory growth in early juveniles of the "red claw" crayfish *Cherax quadricarinatus. Aquaculture* 304:34–41.

Subudhi, H.N., Padhi, G. 2008. Field evaluation of rice cultivars against the yellow stem borer *Scirpophaga incertulas* (Wlk.). *Oryza* 45(3):222–225.

Suhl, J., Dannehl, D., Kloas, W., Baganz, D., Jobs, S., Schiebe, G., Schmidt, U. 2016. Advanced Aquaponics: Evaluation of intensive tomato production in aquaponics vs conventional hydroponics. *Agricultural Water Management* 178:335–344.

Suloma, A., Ogata, H.Y. 2006. Future of rice-fish culture, desert aquaculture and feed development in Africa: The case of Egypt as the leading country in Africa. *The Japan Agricultural Research Quarterly* 40(4):351–360. http://www.jircas.affrc.go.jp.

Swain, P., Mohanty, S., Bhattacharyya, P., Baig, M.S., Chattopadhyay, K., Singh, O.N., Sarkar, R.K. 2014. *Rice Cultivation in Rainfed Upland and Drought-Prone Areas under Changed Climate.* CRRI Technology Bulletin – 101. ICAR-National Rice Research Instittute, Cuttack, Odisha, India. p 4. Available online at http://www.nicra-icar.in/nicrarevised/images/publications/Tbu_Crp_Rice%20Cultivation%20in%20Rainfed%20Upland.pdf. Accessed on 05 September 2022.

Teng, Q., Hu, X.F., Cheng, C., Luo, Z.Q., Luo, F. 2015. Establishment of a rice-duck integrated farming system and its effect on soil fertility and rice disease control. *Geophysical Research Abstracts* 17:EGU 2015-892-2.

Thakur, A.K., Mohanty, R.K., Singh, R., Patil, D.U. 2015. Enhancing water and cropping productivity through Integrated System of Rice Intensification (ISRI) with aquacultureand horticulture under rainfed conditions. *Agricultural Water Management* 161:65–76.

Thomas, D., Prabashlal, P., Vighneswaran, V., Sreelatha, A.K. 2022. Sustainable nutritional and income security through integrated farming in the coastal saline *Pokkali* ecosystem of Kerala. In: Lama T.D., Burman D., Mandal U.K., Sarangi S.K., Sen H.S. (eds) *Transforming Coastal Zone for Sustainable Food and Income Security.* Springer Nature, Switzerland AG 2022. pp 301–311. DOI: https://doi.org/10.1007/978-3-030-95618-9_26.

Tipraqsa, P., Craswell, E.T., Noble, A.T., Schmid V.D. 2007. Resource integration for multiple benefits multifunctionality of Integrated farming system in Northeast Thailand. *Agricultural Systems* 94:694–703.

Troell, M., Joyce, A., Chopin, T., Neori, A., Buschmann, A.H., Fang, J.G. 2009. Ecological engineering in aquaculture – potential for integrated multi-trophic aquaculture (IMTA) in marine offshore systems. *Aquaculture* 297(1–4):1–9. DOI: https://doi.org/10.1016/j.aquaculture. 1 December 2009.

Tuong, T.P., Pablico, P.P., Yamauchi, M., Confesor, R., Moody, K. 2000. Increasing water productivity and weed suppression of wet seeded rice: Effect of water management and rice genotypes. *Experimental Agriculture* 36:71–89.

Turano, M.J., Borski, R.J., Daniels, H.V. 2008. Effects of cyclic feeding on compensatory growth of hybrid striped bass (*Morone chrysops* X *M. saxitilis*) food fish and water quality in production ponds. *Aquaculture Research* 39:1514–1523.

Ullah, A., Nawaz, A., Farooq, M., Siddique, K.H.M. 2021. Agricultural innovation and sustainable development: A case study of rice-wheat cropping system in south Asia. *Sustainability* 13:1965. https://doi.org/10.3390/su13041965.

Umair, M., Hussain, T., Jiang, H., Ahmad, A., Yao, J., Qi, Y., Zhang, Y., Min, L., Shen, Y. 2019. Water-saving potential of subsurface drip irrigation for winter wheat. *Sustainability* 11:2978. DOI: https://doi.org/10.3390/su11102978.

Verdoliva, S.G., Gwyn-Jones, D., Detheridge, A.P., Robnison, P. 2021. Controlled comparisons between soil and hydroponic systems reveal increased water use efficiency and higher lycopene and β-carotene contents in hydroponically grown tomatoes. *Scientia Horticulturae* 279:109896. DOI: https://doi.org/10.1016/j.scienta.2021.109896.

Wang, Y., Cui, Y., Yang, Y. 2005. Partial compensatory growth in hybrid tilapia (*Oreochromis mossambicus* X *O. niloticus*) following food deprivation. *Journal of Applied Ichthyology* 21(5):389–393. DOI: https://doi.org/10.1111/j.1439-0426.2005.00648.x.

Wongkiew, S., Zhen, H., Chandran, K., Woo Lee, J., Khanal, S.K. 2017. Nitrogen transformations in aquaponic systems: A review. *Aquaculture Engineering* 76:9–19.

Woynarovich, E. 1979. The feasibility of combining animal husbandry with fish farming, with special reference to duck and pig production. In: Pillay T.V., Wm Dill A. (eds) *Advances in Aquaculture.* Fishing News (Books) Ltd., Farnham, Surrey. pp 203–208.

Woyanarovich, E. 1980. Raising ducks on fish ponds. In: Pullin R.S.V., Shehadeh Z.H. (eds) *Integrated Agriculture-Aquaculture Farming Systems.* ICLARM Conference Proceedings 4: pp 129–134.

Wu, L. 1995. Methods of rice-fish culture and their ecological efficiency. pp. 91–96. In: MacKay, K.T. (ed.) *Rice-Fish Culture in China.* International Development Research Centre (IDRC). Ottawa, Canada. p 276.

Xie, J., Hu, L., Tang, J., Wu, X., Li, N., Yuan, Y., Yang, H., Zhang, J., Luo, S., Chen, X. 2011. Ecological mechanisms underlying the sustainability of the agricultural heritage rice-fish co-culture system. *Proceedings of the National Academy of Sciences U. S. A.* 108:E1381–E1387. DOI: https://doi.org/10.1073/pnas.1111043108.

Xu, G., Liu, X., Wang, Q., Yu, X., Hang, Y. 2017. Integrated rice-duck farming mitigated the global warming potential in rice season. *Science of the Total Environment* 575:58–66. DOI: https://doi.org/10.1016/j.scitotenv.2016.09.233.

Yadava, D.K., Choudhury, P.R., Hossain, F., Kumar, D. 2017. *Biofortified Varieties: Sustainable Way to Alleviate Malnutrition.* Indian Council of Agricultural Research, New Delhi, India. https://icar.org.in/files/BiofortifiedEnglish_.pdf. Available online. Accessed on 15 March 2022.

Yogev, U., Barnes, A., Gross, A. 2016. Nutrients and energy balance analysis for a conceptual model of a three loops off grid, aquaponics. *Water* 8:589. DOI: https://doi.org/10.3390/W8120589.

Zhang, J., Guoming, Q., Benliang, Z., Kaiming, L., Zhong, Q. 2017. *Rice-Duck Co-Culture in China and its Ecological Relationships and Functions,* vol 111. Agroecology in China: Science, Practice, and Sustainable Management.

Zhang, J., Hu, L., Ren, W., Guo, L., Tang, J., Shu, M., Chen, X. 2016. Rice-soft shell turtle coculture effects on yield and its environment. *Agriculture, Ecosystems and Environment* 224:116–122. DOI: http://dx.doi.org/10.1016/j.agee.2016.03.045.

Zhang, J., Kitazawa, D. 2016. Assessing the bio-mitigation effect of integrated multi-trophic aquaculture on marine environment by a numerical approach. *Marine Pollution Bulletin* 110(1):484–492. DOI: https://doi.org/10.1016/j.marpolbul.2016.06.005.

4 Application of Information Technology for Advanced IFS

R&D and Integration of Modern Tools, Including Artificial Intelligence

4.1 INTRODUCTION

Artificial Intelligence (AI), Information Technology (IT) and Internet of Things (IoT) are becoming more and more integrated into daily life, changing the way we work, communicate and engage with the outside world. These are increasingly being used in communication, streaming entertainment services, health care, public transportation, security and home automation, commerce and supply chains, finance, environmental management and, of course, in the field of agriculture and allied fields. These technologies continue to evolve, promising even greater integration into everyday life, with potential benefits and challenges related to privacy, security and ethics. Decision Support Systems (DSS) tools and softwares that assist in decision making by providing data, models and analysis tools are helping in making complex business process decisions. Information Technology (IT) has had a significant impact on agriculture, converting conventional agricultural methods into cutting-edge, data-driven agricultural systems. The use of IT in agriculture, sometimes known as 'AgTech,' or 'precision agriculture,' has several advantages and is now necessary for the efficient and sustainable production of food. Information and communicaiton technologies (ICT) combined with climate smart agriculture (CSA) practices, improve the economic, social and environmental indicators of integrated farming systems (Kakamoukas et al. 2021).

By increasing productivity, profitability and sustainability, the integration of information technology into integrated farming systems holds the potential to completely transform the agricultural industry. Recent developments in drone technology, robotics and geographical information system (GIS) are very useful for the effective adoption of IFS. Farmers are able to maximize resource utilization and make data-driven decisions and adapt to changing environmental conditions, ultimately leading to more resilient and productive farming systems.

The broad areas covered in this chapter include decision-making tools, multi-criteria decision support system, farm management/operation/design, satellite technology, GIS- and RS-based site suitability study, use of drones and artificial intelligence in agriculture, technology adoption, demonstration, optimization of resource combinations, post-harvest operation, transportation, communication, biosafety and health management, inclusion of improved/high yielding variety

4.2 DECISION SUPPORT SYSTEMS

Decision Support Systems (DSS) in agriculture are now available to aid decision making in various operations under IFS. A range of inputs are gathered and processed by DSS, an interactive visualization system (raw data from farm-deployed sensors, agronomic data, crop data, soil data, microorganism data, etc). To present to the farm manager a solution to various aspects of farming. DSS can be used for selection of crops, variety, nutrient dose, insect pests and disease management,

DOI: 10.1201/9781003239963-4

salinity management, irrigation and drainage decisions. Extensive agricultural data, encompassing factors like climate data, soil conditions, market demands and land usage, can be gathered, analyzed and managed using sophisticated information systems. Internet technologies are adopted to assist farmers in making appropriate decisions and obtaining higher profits.

A human-computer system that uses data from multiple sources to give farmers a list of recommendations to help them make decisions in a variety of situations is known as an agricultural decision support system (ADSS) (Zhai et al. 2020).

4.3 PRECISION AGRICULTURE/FARMING

Precision agriculture is a cutting-edge farming method that optimizes many areas of agricultural production by using technology and data-driven strategies. Sensors are used to detect the special variability in rhizosphere, soil and crops and make precise delivery of inputs. The goal of precision farming is to facilitate farm efficiency, reduce resource use and increase overall profitability while reducing adverse effects on the environment. The integration of technology like GPS, remote sensing, automated machinery and IT plays a key role in precision agriculture. Precision farming can increase yields and decrease resource waste by helping farmers optimize planting, irrigation, fertilization, insect pests and disease management. To optimise resource utilization, machine learning algorithms examine crop and soil data to make exact recommendations for fertilizers, irrigation and planting rates for various sections of the field. AI-powered autonomous tractors and machinery can precisely carry out duties like planting, harvesting and weed control. Precision farming has embraced Internet of Things (IoT) technology because of its scalability and environmental friendliness, especially in agricultural mechanization (Popovic et al. 2017).

Precision farming utilizes the latest methodologies in the integrated farming system for effective use of resources. For example, for the best water management scientific irrigation application methods such as a subsurface drip irrigation (SSDI) system is adopted. The SSDI provides solutions for a variety of issues, including poor water use efficiency, declining water quality, lower risk of aquifer pollution due to reduced deep percolation of fertilizers and other chemical compounds and improved yields. This system has the ability to supply water and nutrients to the root zone's most actives sections, while maintaining a dry soil surface for better weed control and crop health. Precision water management based on advanced irrigation technologies saves precious irrigation water and, in turn, brings more crop area under irrigation. Precision water management requires precise measurement of surface and root zone soil moisture deficits to facilitate automation of irrigation systems. Advanced tools such as sensors, models, satellites, machine learning, artificial intelligence and deep learning techniques may be used for accurate measurement of soil moisture.

Precision farming often involves the use of automated machinery and robotics for planting, harvesting and other tasks. These machines can operate with high precision and reduce the need for manual labour. Farmers can remotely monitor field conditions, irrigation systems and equipment performance using smartphones or computers, allowing for real-time adjustments and troubleshooting. Advanced analytics and AI-powered algorithms analyze the data collected from various sources to provide insights into crop health, yield predictions and optimal farming practices. With a data-driven decision-support system, farmers can better plan, plant, harvest and oversee their farms when they have access to up-to-date data and analytical tools.

4.4 GPS AND GIS TECHNOLOGY

Global Positioning System (GPS) and Geographic Information System (GIS) technologies are used to map fields, track equipment and optimize planting and harvesting processes. Data collection and sensors are used to collect data from various sources, including satellite imagery, drones, ground-based sensors and weather stations, to provide information on soil conditions, crop health, weather patterns and other factors which guide decision making and resource allocation. GPS technology is used to precisely map fields, track the movement of farm equipment and enable precise planting, harvesting and other field operations.

GIS tools enable farmers to analyze data spatially, visualizing trends and patterns across their farms. This facilitates the allocation and management of resources with knowledge. GIS and remote sensing data can be used to develop predictive models for crop yields, disease outbreaks and other critical factors, aiding in long-term planning. Incorporating drones, GIS and remote sensing technology into integrated farming systems empowers farmers with a wealth of actionable data. These tools enhance resource efficiency, reduce environmental impact and contribute to more sustainable and profitable agricultural practices.

GIS and remote sensing can assess the impact of farming practices on local ecosystems. This information informs conservation efforts and promotes sustainable land use.

Variable Rate Technology (VRT): These systems enable farmers to customize the application of inputs such as fertilizers, pesticides and water according to the distinct requirements of various sections within a field. This optimization reduces waste and maximizes yields. Tractors fitted with GPS and smart gadgets with high-end machine learning algorithms for differentiating crop and weeds selectively remove weeds either mechanically or by the application of steam, laser beam or micro-doses of herbicides (Yaduraju 2021). Precision nutrient management must involve applying the right amount, at the right time, in the right location, utilizing the right source and employing the right method. For example, for efficient nitrogen management, recently developed leaf colour charts, SPAD meter, Green-seeker, Dualex meter, nutrient expert model, crop simulation model, etc. should be employed.

Smart Sensors and IoT (Internet of Things): Real-time data on crop health, livestock conditions, temperature, humidity and soil moisture is gathered by IoT devices and smart sensors. Using this remotely accessible data, decision-makers can make well-informed choices about animal husbandry, illness control and irrigation. Micro-needles made up of polymers are inserted into plants which transmit electrical currents and reveal the health of the plants depending upon the water availability, nutrients and light (Bukhamsin et al. 2021).

4.5 AI WITH DATA ANALYTICS

Data Analytics and Big Data: Farmers can now analyze vast amounts of data from a variety of sources, such as weather stations, satellites and sensors, thanks to advanced data analytics tools and big data approaches. Predictive modelling, yield forecasting and resource optimization can all be done with this data. Through the analysis of past data, machine learning and artificial intelligence systems forecast crop yields, disease outbreaks and insect infestations, enabling preventative measures. AI-powered optimization models can best distribute labour, water and land among farms in order to optimize output overall and reduce waste and environmental damage.

Environmental Monitoring Weather and Climate Prediction:: Farmers may make informed decisions about irrigation, pest control and disease prevention by using real-time data from IoT sensors that measure temperature, humidity, soil moisture and other variables. AI models can provide accurate short- and long-term weather forecasts, helping farmers make informed decisions about planting, irrigation and pest control. GIS can integrate real-time weather data with farm management systems to provide accurate weather forecasts. Drones can also collect atmospheric data, improving localized predictions. Remote sensing technology contributes to climate modelling by providing historical data on temperature, precipitation and climate trends, helping farmers adapt to changing conditions. Farmers may make well-informed decisions about when to plant and harvest crops, how much water to use and which crops to plant in response to climate change with the help of accurate weather forecasting and climate modelling through the use of IT tools. Accurate weather forecasts obtained from meteorological stations and satellites help farmers schedule planting, harvesting and other activities more efficiently. AI systems that forecast agricultural yields using past data, current weather and real-time monitoring can assist farmers in scheduling harvest and logistics.

4.6 MONITORING AND MANAGEMENT IN DIFFERENT ENTERPRISES

4.6.1 Aerial Surveillance with Drones

Drones are unmanned aerial vehicles (UAV) or unmanned aerial systems (UAS), controlled by remotes, to fly at desired altitude fitted with camera, sensors and a geographical positioning system to collected georeferenced data and imageries or carryout agricultural operations. For aerial spraying in agriculture, drones are employed for several crops. The plant protection measures are effectively implemented with reduced production cost, less labour and timely management of pests and diseases. Large areas of paddy, pigeon pea, groundnut, cotton, plantation crops, orchards and forest areas can be effectively managed by the use of drones. Drones equipped with cameras and sensors can capture high-resolution images and multispectral data of crops. GIS and remote sensing technologies process this data to monitor crop health, identify stress factors and assess overall growth. Combined with GIS tools, detailed maps of fields can be created, indicating variations in soil properties, moisture levels and terrain. Farmers can use these maps to optimize planting, irrigation and fertilization strategies for different areas of the field. Farmers can use this information to apply resources (water, fertilizers, pesticides) precisely where and when needed, minimizing waste and environmental impact. Advantages of use of drones in the farming system include collection of some of the information not observed by the naked eyes in the visible spectrum such as soil moisture content, soil nutrient status, crop health/disease conditions, stresses, etc. The crop health can be monitored by various parameters such as Normalized Difference Vegetation Index (NDVI) values, Leaf Area Index (LAI), biomass, chlorophyll content, etc., which are collected by the sensors fitted to the drones. Drones can also be employed to estimate crop health before and after a natural calamity to estimate the yield losses. Drones are also used for various enterprises in the IFS. Drones are fitted with high-resolution infrared cameras to assess health of animals in a herd. Different imaging sensors such as hyperspectral, multispectral, lidar and RGB can be mounted on drones to acquire sophisticated data on various enterprises in a non-destructive manner. These valuable but huge data sources are further analyzed by artificial intelligence, computer vision and machine learning techniques for practical use. Deep learning techniques are also used for image data analysis.

Artificial intelligence systems can detect early indicators of crop stress, illness or nutritional deficits by analyzing imagery from drones or satellites. This makes prompt interventions possible. By using machine learning models to differentiate between crops and weeds, herbicide applications can be made more precisely and effectively. AI can recommend the most suitable improved or high-yielding crop varieties based on soil quality, climate and market demand. Breeding and genetic improvement can be made possible by AI, which can accelerate the breeding process by analyzing genetic data to identify desirable traits and speed up the development of improved varieties. In India, the National Agriculture Sciences Fund (NASF) has been created for use of advanced technologies for precision management of natural resources, livestock, fisheries, microbes, etc.

4.6.2 Remote Monitoring and Automation

Using IT technologies, farmers can keep an eye on and manage a range of farm operations from a distance. This includes using smartphone apps to regulate irrigation systems, feeding and watering livestock automatically and remotely monitoring security systems. Tasks like planting, harvesting and sorting are carried out by automated machines and robotics that are managed by IT systems.

4.6.3 Farm Management Apps and Software

Farmers can instantly obtain information about crop and animal management with mobile apps. These apps can provide advice on when to plant, how to identify diseases, and available treatments. Apps help farmers access to real-time data, market pricing, weather forecasts, and professional guidance, mobile applications help them make better decisions. Planting and harvesting schedule optimisation, inventory management, and crop growth tracking are all possible using software systems. Farmers may plan, monitor, and manage many parts of their integrated farming operations with the aid of IT programmes such as farm management software. Crop management, livestock management, resource allocation, and financial tracking are all supported by these software programmes. Machine learning can optimize labour allocation based on the specific needs of the farm, improving efficiency and reducing costs.

4.6.4 Animal Health Monitoring and Livestock Behaviour Analysis

IoT sensors and AI can monitor animal health metrics like body temperature, heart rate and feeding patterns, alerting farmers to potential issues. Machine learning can analyze the behaviour of animals to detect signs of distress or illness, helping farmers take timely action. Machine learning can analyze historical data and environmental factors to predict disease outbreaks in crops or livestock, enabling proactive measures. Drones and GIS can be used to track the movement and behaviour of livestock, ensuring their well-being and security. Wearable technology and sensors can observe the well-being and behaviour of livestock, enabling the early detection of illnesses and enhancing animal care. GPS-equipped collars on animals can transmit real-time location data to GIS systems. Remote sensing data can help identify suitable grazing areas, detect vegetation changes and assess livestock's impact on natural habitats.

4.6.5 Soil and Nutrient Management

AI models can assess soil quality and composition, recommending appropriate amendments and nutrient management strategies. Sensors and machine learning can continuously monitor nutrient levels in the soil and suggest real-time adjustments to optimize crop growth. GIS-based soil sampling helps farmers collect data on soil composition and nutrient levels across their fields. This information guides precise nutrient application. Remote sensing can monitor nutrient uptake by crops, enabling farmers to adjust fertilization strategies accordingly.

GIS and remote sensing can be used to assess water availability, monitor water quality and plan irrigation systems effectively. Remote sensing can identify areas of the field experiencing water stress, allowing for targeted irrigation and water conservation efforts.

4.6.6 Pest and Disease Management

AI-powered sensors and cameras can continuously monitor livestock health and detect signs of disease, allowing for early intervention. AI models analyze environmental factors and historical data to predict the possibility of crop diseases, enabling proactive management. AI image recognition can recognize pests and diseases in crops based on images captured by cameras or drones, allowing for targeted treatment. Drones equipped with thermal and multispectral sensors can detect pest and disease outbreaks in crops. GIS and remote sensing technology helps farmers identify affected areas and take prompt action to control the spread. Machine learning can provide recommendations for integrated pest management strategies, including natural predators and pest-resistant crop varieties.

By using data analysis and remote sensing, IT applications can offer early warning systems for disease and pest outbreaks. This aids farmers in taking preventative action and lowers crop losses. Using photos taken by drones or cell phones, AI-driven image recognition can detect pests and plant illnesses.

4.6.7 Post-harvest Operations

AI can automate quality control processes by using computer vision to inspect and sort harvested crops based on size, colour and defects. AI models analyze environmental conditions and historical data to predict the shelf life of harvested produce and recommend storage conditions. AI algorithms optimize transportation routes for delivering farm produce to markets, reducing transportation costs and time. AI-driven communication platforms help farmers connect with buyers, distributors and other stakeholders more efficiently, facilitating the sale of their products.

Traceability: Product tracking from farm to table is made possible by IT systems, which improves food safety and transparency. **Blockchain Technology** reduces fraud and improves traceability in the agricultural supply chain by producing transparent, tamper-proof records of transactions and product origins.

Supply Chain Management: Information Technology (IT) tools assist farmers in coordinating their production procedures with the supply chain. AI can analyze market trends, historical sales data and consumer preferences to optimize production and distribution schedules. Produce movement tracking from farm to market, route optimization for transportation and maintaining product traceability and quality control are all included in this. Machine learning helps streamline transportation routes, reduce waste and manage inventory efficiently, ensuring timely delivery of products to market.

Waste Management: Machine learning techniques like Convolutional Neural Networks (CNN) and Support Vector Machines (SVM) have applicability in waste management tasks such as waste recognition, organization, isolation and categorization, among other functions. An autonomous system segregates the waste into bio-degradable and non-degradable items by using CNN algorithm in machine learning by detecting and classifying the waste according to the dataset provided to CNN. This system serves as an alternative to human involvement in locating and gathering refuse. It employs Convolutional Neural Networks (CNN) to detect garbage within images, generating bounding boxes around the identified garbage portions. Object detection involves recognizing instances of specific object classes (such as bottles, livestock, poultry or vehicles) within digital images and videos. Object detection and recognition can also be done by using Artificial Neural Networks (ANN) and drones. After completing the task of object detection, the nest step is to detect the distance and direction of the object from the base, which is essential for allowing pick up of the garbage. A robot is given the inputs about the location of the garbage, which moves in the desired direction and gets hold on the target object for subsequent disposal.

Financial Management and Documentation: For farmers and agricultural enterprises, IT solutions help with financial planning, cost analysis and budgeting. By streamlining these operations, IT solutions help farmers to better track their costs, revenue and profitability. Credit applications and investment decisions both benefit from this data.

4.6.8 E-Commerce and Marketing

Internet shopping has emerged as an alternative to traditional marketing systems due to the advent of IT tools. AI can analyze market data to predict crop prices, helping farmers make decisions on when and where to sell their produce. AI-powered e-commerce platforms connect farmers directly to consumers and facilitate online sales and marketing. By removing middlemen, online markets and digital platforms allow farmers and buyers to communicate directly. By using online marketplaces and e-commerce platforms, IT helps farmers reach a larger consumer base. This lessens the need for conventional middlemen and permits direct sales to customers.

The most revolutionary marketing channel is now online shopping based on the Internet. This channel is a perfect vehicle of marketing with global nature benefitted by shorter time, reduced cost, less crowd and queues than traditional markets, unlimited time and space with increased convenience. There are ample choice of products/services at a single platform. However, convincing the consumer and physical verification of the inputs/products to be purchased before buying is still a challenging task in this marketing system. Therefore, an effective digital marketing strategy needs to be adopted at the right time to make a purchase decision. Sufficient information should be gathered from authentic sources of websites, social networking sites, YouTube as well as from traditional sources (newspaper, TV, radio, family, friends etc.).

4.6.9 Education and Extension Services

Farmers can receive training, instructional materials and agricultural extension services through the use of IT. Farming best practises can be shared through mobile apps and online courses. Farmers are equipped with knowledge and best practices through the use of the Internet. Chatbots and smartphone applications driven by artificial intelligence provide farmers with information, counsel and direction regarding farming techniques. AI-powered chatbots and mobile apps provide farmers with training and information on best practices, helping them adopt new technologies effectively. AI can create virtual farm models that simulate different integrated farming scenarios. Farmers can interact with these models to see the potential outcomes of adopting specific practices or technologies. Augmented Reality (AR) and Virtual Reality (VR) technologies powered by AI can provide immersive demonstrations of farming techniques and technologies.

4.7 PROGRAMMING TOOLS FOR DETERMINING EFFICIENT IFS MODELS

For efficient use of factors of production, i.e., land, labour, capital and management practices, modelling/programming tools are available. Linear programming, goal programming and compromise programming tools/software may be used depending on the farming situation, agricultural resource allocation and objectives of the IFS model. The risks related to an enterprise can be assessed from the variance of return from that enterprise over the years. Generally, enterprises with high variance are highly risky and vice versa.

4.7.1 Linear Programming

This is one of the farm optimization techniques which involves modelling in a single objective framework. The objective is to maximize farm income by optimum utilization/allocation of available resources. Here, the constraints (maximum limit of allocation in a given situation) of land, labour and capital are defined. For example, the land availability for marginal, small, medium and large farmers can be to the extent of 1, 2, 3 and 4 ha, respectively. Similarly, capital limit is also defined, e.g., 1.0, 1.5, 3.0 and 6.0 lakhs for marginal, small, medium and large farmers, respectively. The labour availability in terms of person-days per annuum is also defined. The enterprise options and the respective area required are defined. One dairy unit of two cows may require land of 0.08 ha, one piggery unit of 20 pigs may occupy an area of 0.003 ha and a poultry unit of 2,000 birds may need an area of 0.025 ha. Taking all these factors into account, the system equations are solved to reach the maximum net profit.

4.7.2 Compromise Programming

This programming is adopted to determine the most suitable enterprises in a multi-objective approach (Behera et al. 2014). In this programming, determined weightage are given based on priorities. If capital minimization is the priority, highest weightage is given to this objective. Remaining weightage is given to other compromising objectives such as farm income maximization, labour cost minimization, etc.

4.7.3 Goal Programming

It is a tool for evaluation of the existing farming system and redesigning it for efficiency and profitability. Goal programming helps in handling multiple decision criteria. If a farmer wishes to make less investment on farming activities, but still wants to maximize the farm income and minimize the agricultural risks, this programming will be useful. In such types of situations, goal programming is applied to obtain a compromised solution based on the relative importance of each objective (Behera and Rana 2014).

4.7.4 Fuzzy Linear/Nonlinear Programming

This programming is a multi-objective optimization tool which incorporate additional constraints with the additional objective function and involves less sensitive analysis. This methodology has advantages over single objective planning as it can accommodate any number of objectives easily without much computational burden (Behera and Rana 2013). When all the parameters and variables are considered, it is called a fully fuzzy linear programming.

4.8 CONCLUSION

Information technology for advanced research and improvement of an integrated farming system and incorporation of modern tools, including artificial intelligence powered solutions in IFS, can facilitate better decision making, maximize resource use, boost output and support sustainable agriculture. To solve the difficulties of contemporary farming and meet the objectives of environmental sustainability and food security, these technologies are essential. Precision farming can assist in making the most of the land that is available, allowing for the production of more food without increasing the amount of land used for agriculture. These IT technologies enable farmers to optimize planting, fertilization and irrigation, resulting in higher crop yields and better-quality produce. By precisely applying inputs like water, fertilizers and pesticides where they are needed most, farmers reduce waste and minimize environmental impact, making the farming system most resource-efficient. These also contribute to increased profitability for farmers through cost savings. These technologies also help reduce agriculture's environmental footprint by conserving resources and minimizing pollution. In integrated farming systems, AI and machine learning are crucial instruments for raising profitability, sustainability and production. These technologies enable farmers to make data-driven decisions, reduce resource waste and respond effectively to changing environmental and market conditions, ultimately leading to more efficient and resilient agricultural practices. The farmers will be able to predict and minimize risks like drought, pests and diseases by keeping an eye on weather patterns, soil conditions and crop health. Precision farming technology will also improve food safety and traceability, making it possible to track products more accurately from the farm to the table. Adoption of automation in farming system by the use of sensors, drones and robots will enhance efficiency, increase production and profit while at the same time minimizing cost, drudgery and environmental footprints.

REFERENCES

Behera, U.K., Kaechele, H., France, J. 2014. Integrated animal and cropping systems in single and multi-objective frameworks for enhancing the livelihood security of farmers and agricultural sustainability in Northern India. *Animal Production Science* 55(10):1338–1346. DOI: https://dx.doi.org/10.1071/AN14526.

Behera, U.K., Rana, D.S. 2013. Fuzzy linear programming for integrated farming systems in multi-objective environment. *Indian Journal of Agronomy* 58(1):119–124.

Behera, U.K., Rana, D.S. 2014. Multi-objective optimization and design on integrated farming systems using goal programming. *Indian Journal of Agronomy* 59(3):460–467.

Bukhamsin, A., Khalil, M., Ran, T., Gillies, L., Ikram, B., Khaled, N.S., Jurgen, K. 2021. Robust, long-term, and exceptionally sensitive microneedle-based bioimpedance sensor for precision farming. *Advanced Science* 8(16):2101261. DOI: https://doi.org/10.1002/advs.202101261.

Kakamoukas, G., Sarigiannidis, P., Maropoulos, A., Lagkas, T., Zaralis, K., Karaiskou, C. 2021 February. Towards climate smart farming—a reference architecture for integrated farming systems. In *Telecom* 2(1): 52–74. MDPI.

Popović, T., Latinović, N., Pešić, A., Zečević, Ž., Krstajić, B., Djukanović, S. 2017. Architecting an IoT-enabled platform for precision agriculture and ecological monitoring: A case study. *Computers and Electronics in Agriculture* 140:255–265.

Yaduraju, N.T. 2021. Hand hoes to robotic weeders: The fight against weeds continues. In: *Lead Papers of 5th International Agronomy Congress,* November 23–27, 2021, Indian Agricultural Research Institute, New Delhi, India. pp 65–66.

Zhai, Z., Martínez, J.F., Beltran, V., Martínez, N.L. 2020. Decision support systems for agriculture 4.0: Survey and challenges. *Computers and Electronics in Agriculture* 170:105256. DOI: https://doi.org/10.1016/j.compag.2020.105256.

5 Modern Tools for Greenhouse Gas Emission Estimation and Mitigation in IFS

5.1 INTRODUCTION

Agriculture and allied activities contribute significantly (24%) to global greenhouse gas (GHG) emissions (IPCC 2014). Agricultural practices, waste management, energy consumption and organic matter burning, fertilizer use are the major contributors of methane (CH_4) and nitrous oxide (N_2O) emissions (https://www.epa.gov/ghgemissions/global-greenhouse-gas-emissions-data). Global warming is caused due to emissions of gases such as carbon dioxide (CO_2), CH_4, N_2O and fluorinated gases. These are called GHGs, having different effects on Earth's warming (https://www.epa.gov/ghgemissions/understanding-global-warming-potentials). Global warming potential (GWP) is a measure of the amount of energy the emission of one tonne of a gas will absorb over a time period of 100 years, relative to the emissions of one tonne of CO_2; therefore, CO_2 has GWP of 1, CH_4 has 27–30 and N_2O has GWP of 273 (https://www.epa.gov). The adverse impact of GHGs emission can be tackled by taking measures to reduce the emission of GHGs, a strategy called mitigation. Another strategy is to leave with emission, by changing the management options to reduce the adverse effects, called adaptation. These two strategies are applicable for agriculture with various options of adaptation and mitigation. Though both measures are essential to address the effect of changing climate, there are major differences between the two (Table 5.1). Mitigation measures addresses the root cause and the whole chain, while adaptation measures address the effect and form a part of the process (Figure 5.1).

5.2 TOOLS FOR ESTIMATION OF GREENHOUSE GAS (GHG)

Agricultural-based activities may lead to a source or sink for greenhouse gases. When the emission is higher compared to sequestration, it leads to source and vice versa leads to a sink for GHGs. Therefore, estimation of the GHGs is essential to take measures for adaptation and mitigation. The following software are available for estimation of GHGs in agricultural systems.

5.2.1 ALU Software

The Agricultural and Land Use GHG inventory (ALU) software is based on IPCC guidelines for estimating GHG emissions and removals. This software facilitates the compilation of activity data, assignment of emission factors and completion of the calculations with internal checks (https://www.nrel.colostate.edu/projects/alusoftware/home).

5.2.2 IPCC Inventory Software

This is an MS Windows-based software for estimation of GHGs by multiplying activity rate with emission factor. The Agriculture Forestry and Other Land Use (AFOLU) sheet may be used for estimation of GHGs in IFS. It uses IPCC 2006 emission guidelines and has adequate quality control

DOI: 10.1201/9781003239963-5

TABLE 5.1
Major Differences Between Mitigation and Adaptation Measures in Global Warming

Mitigation	Adaptation
It addresses the root cause, i.e., reducing GHG emissions by managing the source and sink.	It mainly focuses on how to adjust the situation by employing long-term strategies.
It emphasizes the use of renewable sources of energy to cut the sources of emissions.	It anticipates the adverse effects and strategies as to the control measures.
It enlarges the sink size such as forest, plantation, restricting deforestation, etc.	Location specific agronomic measures such as increasing the use efficiency of input use, e.g., fertilizers, irrigation water, pesticides, resistant crops to abiotic and biotic stresses are employed.
Mitigation needs planning for a larger area context and use of new technologies, therefore it is cost intensive.	Adaptation measures needs a thorough knowledge of local conditions/ risks and bringing in changes that give better results by reducing the risks.
Mitigation measures may have a long-term effect.	It gives short-term benefits.

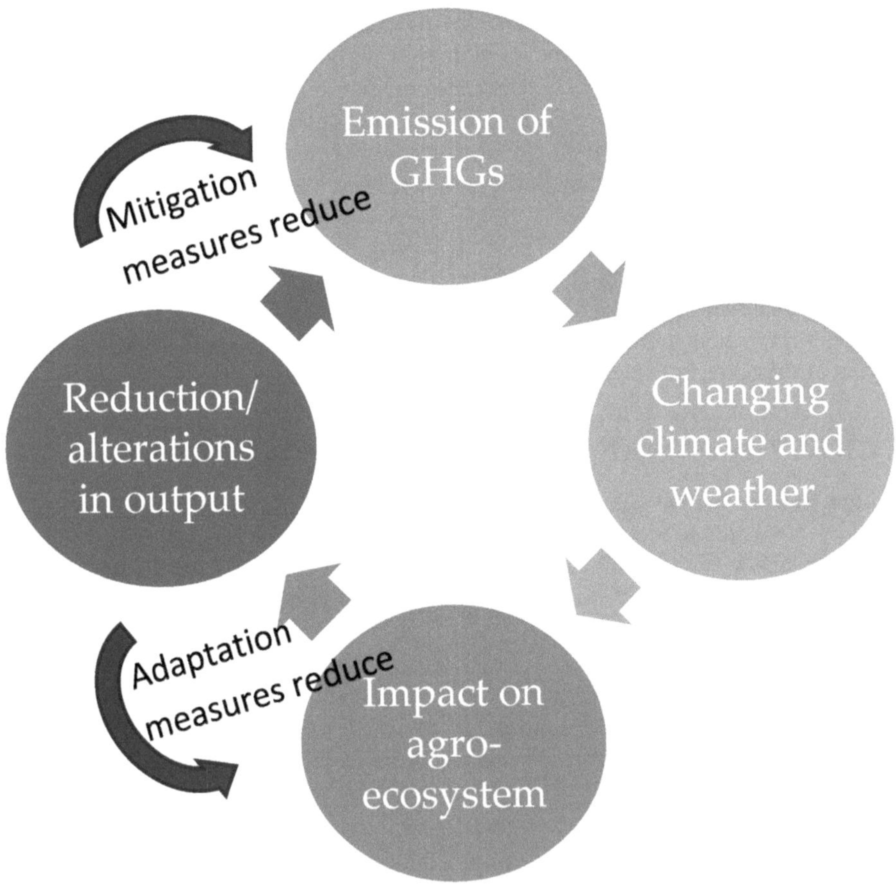

FIGURE 5.1 Mitigation measures reduce emissions of GHGs, while adaptation measures reduce susceptibility to climate change's consequences.

measures to ensure completeness and accuracy of data and values. It has major sections such as livestock, land, aggregate sources and non-CO_2 emission sources on land and others (https://www.ipcc-nggip.iges.or.jp/software/).

5.2.3 Cool Farm Tool (CFT)

CFT gives an opportunity for online estimation of GHG emission/sequestration (https://www.coolfarmtool.org). It uses crop as well as livestock data. For crops, data on yield, area, fertilizers, pesticides, energy use and transport are required. For livestock, data on herd size, feed, manure, energy and transport are required.

5.2.4 GHGFARM

This tool can be used to estimate and lower the net greenhouse gas emissions from farms. It can also be used to compare, analyze and quantify various farm management scenarios, which can lead to the adoption of more sustainable farming methods over the long run (Newlands 2007).

5.2.5 Agro-chain Greenhouse Gas Emissions (ACE) Calculator

The ACE calculator is a tool for estimation of total GHGs emissions for agricultural products bought by consumers. It takes various stages into consideration such as farm production, collection, transport, processing, distributions, retail, etc. It has a use interface for comparing GHG emission from two different scenarios (https://ccafs.cgiar.org/resources/tools/acge-calculator).

5.2.6 On-farm GHG Accounting Tools

This is from Agriculture Victoria; an Excel-based tool useful for estimating on-farm GHG emissions and carbon sequestration. The Excel-based GHG accounting framework (GAF) is freely available for dairy, sheep and beef, cropping, feedlot, sugar, cotton, horticulture, buffalo, deer and poultry industries (http://picc.org/au/resources/Tools.html). The GAF tools do not calculate soil organic carbon change, which is estimated by another simple model called Soil Organic Carbon Reserves and Transformations in Ecosystems (SOCRATES) model hosted by QUT.

5.2.7 The MiLA Tool

The Model for Integrative Life Cycle Assessment in Agriculture (MiLA) tool is a Microsoft Excel 2010 based multivariate empirical tool developed to estimate GHG emissions and cumulative energy demand (CED) of cropping systems (Peter et al. 2016). This tool is well suited for energy and food crops, which can be used by farmers, researchers and business organizations. Environmental aspects such as eutrophication, biodiversity and erosion potential along with socioeconomic factors should be taken into consideration.

5.3 MODELLING OF GHG EMISSION FROM DIFFERENT IFS COMPONENTS

Mitigation and adaptation measures in IFS models need information on the amount of emissions from different components. To measure the real-time emissions by installing infrastructure involves significant time and a huge cost. Therefore, alternative options for a modelling approach may be followed to fairly estimate the emission status. Appropriate mitigation adaptation strategies may then be implemented to reduce the extent of emissions.

5.3.1 Models for GHG Emission Study in Integrated Livestock Farming System

Empirical models use regression equations taking into account the nature of feed resources use, whereas, the dynamic models consider the interactions and other factors. The interactions among crop, dairy farm and weather are taken into consideration in the Integrated Farm System model (Rotz et al. 2009). For crop production, GHG simulation factors such as the amount of stubble removed by burning, amount of stubble retained and organic matter input is considered, while for the livestock system factors such as pasture/grasses fertilized with nitrogen, legume pasture not fertilized with N are considered (Meier et al. 2020). In addition, factors such as breed of animal, production stage, character of feeds, grazing and pasture management and manure handling should also be considered (Jose et al. 2016).

5.3.2 Models for GHG Emission Study in Fisheries Production System

A bio-economical model using biological information such as Schaefer's model of surplus production may be used (Schaefer 1957). In a maximum economic yield (MEY) model the economic factors are used (Dichmont et al. 2010). Modelling studies on GHG emission in the fisheries sector has given directives to the national government to consider strategies for modernizing old fishing vessels or a subsidy for installing emission abatement equipment (Kim et al. 2021).

5.3.3 Models for GHG Emission Study in Forestry

The Carbon Budget Model of the Canadian Forest Sector (CBM-CFS3) comprises carbon in living trees, dead organic matter and mineral soils. The environmental benefits of conserving old forests differ depending on the forest types, distributions of age-classes and threat of natural disturbances (https://nrcan.gc.ca/). Emissions from forest degradation in 74 developing nations with 2.2 billion hectares of forested area between 2005 and 2010 revealed total emission of 2.1 billion tonnes of CO_2 annually; of these, 53% came from the timber harvest, 30% from the harvest of fuel wood and 17% from forest fires (Pearson et al. 2017).

5.3.4 Models for GHG Emission Study from Crop Production

The GHG emissions from direct land usage should take into account the precise location of crop cultivation and historical land use (Lam et al. 2021). The Poisson regression model, inverse-probability-weighted regression adjustment (IPWRA) can be used to assess GHG emissions in crop production systems (Israel et al. 2020).

5.4 OPTIMIZING ENERGY USE EFFICIENCY IN IFS

An IFS generates various byproducts along with main products and the efficient use of these byproducts as resources/inputs for other enterprises holds the key for gaining efficiency in energy use. However, many times the appropriate treatment of byproducts is essential for their effective use. For an agri aquaculture system larger than 2 ha, a good farm configuration, including solutions of composting, vermicomposting, pelleting and anaerobic digestion, would be acceptable, providing the best earnings of up to USD 1,900 while releasing the least amount of greenhouse gases (Thao et al. 2022).

5.4.1 Energy Input, Output, Productivity and Ratio

Using the standard values (Table 5.2) provided by different researchers, the energy input from direct and indirect sources, such as seed, chemical and organic fertilizer, pesticides, human power, fuel

TABLE 5.2
Energy Conversion Factors for Agronomic Inputs, Yield of Crops and Other Enterprises

Agronomic Input/Output	MJ Unit^{-1}	Reference
Diesel (L^{-1})	47.7	Pimentel and Pimentel (2008)
Power tiller (h^{-1})	123.0	Calilung (1985)
Tractor (h^{-1})	542.0	Quilty et al. (2014)
Adult man (h^{-1})	1.96	Mittal et al. (1985)
Adult woman (h^{-1})	1.57	Mittal et al. (1985)
Electricity (KW h^{-1})	11.93	Mittal et al. (1985)
Nitrogen (kg^{-1})	63.5	Pimentel and Pimentel (2008)
Phosphorus (kg^{-1})	13.95	Pimentel and Pimentel (2008)
Potassium (kg^{-1})	6.69	Pimentel and Pimentel (2008)
Zinc sulphate (kg^{-1})	20.9	Singh and Mittal (1992)
Farmyard manure (kg^{-1})	0.30	Taki et al. (2012)
Vermicompost (kg^{-1})	0.50	Ram and Verma (2015)
Water for irrigation (m^3)	1.02	Acaroglu and Aksoy (2005)
Herbicide (kg^{-1})	303.8	Pimentel and Pimentel (2008)
Insecticide (kg^{-1})	418.4	Pimentel and Pimentel (2008)
Fungicide (kg^{-1})	115.0	Pishgar-Komleh et al. (2011)
Seed (kg^{-1})	15.5	Pimentel and Pimentel (2008)
Rice grain yield (kg^{-1})	15.2	Pimentel and Pimentel (2008)
Maize, wheat, lentil, moong, sorghum, cow pea, oat grain yield (kg^{-1})	14.7	Singh and Mittal (1992)
Maize stover yield (kg^{-1})	18.0	Mittal et al. (1985)
Rapeseed seed yield (kg^{-1})	25.0	Mittal et al. (1985)
Rapeseed stover yield (kg^{-1})	18.0	Mittal et al. (1985)
Okra, tomato, cauliflower, cabbage (kg^{-1})	0.80	Tuti et al. (2012)
Onion (kg^{-1})	1.6	Gopalan et al. (1971)
Banana (kg^{-1})	5.35	Singh and Mittal (1992)
Lemon (kg^{-1})	2.88	Singh and Mittal (1992)
Guava (kg^{-1})	2.60	Singh and Mittal (1992)
Berseem (kg^{-1})	10.0	Singh and Mittal (1992)
Rice and wheat straw yield (kg^{-1})	12.5	Singh and Mittal (1992)
Fuel wood (kg^{-1})	18.0	Singh and Mittal (1992)
Okra, Tomato, Cabbage, Cauliflower, Onion, Banana leaves and stem (kg^{-1})	10.0	Soni et al. (2013)
Lentil and moong stover (kg^{-1})	11.25	Soni et al. (2013)
Poultry (kg^{-1})	4.56	Gopalan et al. (1971)
Goat (kg^{-1})	8.12	Gopalan et al. (1971)
Mushroom (kg^{-1})	1.62	Salehi et al. (2014)

consumed, machinery used for various operations in the crop production, and the output, such as yields, were converted into an equivalent value of chemical energy (MJ ha^{-1}). The energy input, output, output-input ratio as well as the energy productivity in different treatments for rabi-season crops were calculated based on the following formulae given by Chamsing et al. (2006).

$$\text{Energy input (GJ ha}^{-1}\text{)} = E_f + E_s$$

where

E_f (GJ ha^{-1}) = energy input in farm operations (physical + chemical + biological energy inputs in farm operations)

E_s (GJ ha^{-1}) = energy sequestered in machinery (energy sequestered in manufacturing of machinery in GJ h^{-1} × machine working hour in h ha^{-1}

$$\text{Energy output (GJ ha}^{-1}) = (\text{Yield} \times E_{eqm}) + (\text{By-product} \times E_{eqb})$$

where

E_{eqm} = energy equivalent value of main product
E_{eqb} = energy equivalent value of by-product

$$\text{Energy productivity (kg GJ}^{-1}) = \frac{\text{Crop yield (kg ha}^{-1})}{\text{Energy inputs to crop production, GJ ha}^{-1}}$$

$$\text{Energy output-input ratio} = \frac{\text{Energy output (GJ ha}^{-1})}{\text{Energy inputs to crop production (GJ ha}^{-1})}$$

5.4.2 Indicators of Energy Use Efficiency in IFS

There are various indicators employed to evaluate the performance of IFS towards energy use efficiency (Table 5.3). In some of the indicators, higher values indicate better efficiency, whereas indicators which impact environment and investment negatively, lower values indicate better performance. For renewability index and ecosystem resource yield ratio, higher values indicate better performance, whereas for environmental loading of investment, investment-yield ratio, system transformity and non-renewable yield ratio, lower values indicate better performance (Hoang and Alauddin 2011).

5.5 ENERGY USE OPTIMIZATION IN CROP PRODUCTION SYSTEMS

Crop production needs various inputs such as seed, irrigation water, fertilizers, manures, herbicides, insecticides, fungicides, labour, machines, etc. Often these inputs are used without considering the environmental consequences as well as the amount of output it would likely generate. In a wheat production study in Turkey, it was observed that there is potential prospect to save about 14% (2.93 GJ ha^{-1}) of the total energy consumption leading to a 17% decrease in GHG emissions by the adoption of ideal energy consumption levels while maintaining a steady wheat yield (Imran and Ozcatalbas 2021). An optimization model in a rice production system revealed that water consumption should

TABLE 5.3
Indicators of Energy use Efficiency in IFS

Indicator	Formula
Renewability Index	Renewable resources/(Renewable resources+ Non-renewable resources + Purchased non-renewable resources)
Ecosystem resource yield ratio	(Renewable resources + Non-renewable resources)/yield
Environmental loading of investment	Purchased non-renewable resources/(Renewable resources + Non-renewable resources)
Investment-yield ratio	Purchased non-renewable resources/Yield
System transformity	(Renewable resources + Non-renewable resources + Purchased non-renewable resources)/Yield
Non-renewable yield ratio	(Non-renewable resources + Purchased non-renewable resources)/Yield

Source: Adapted from Hoang and Alauddin (2011).

come down by 25% for safeguarding the river ecosystem; this approach could balance the environmental implications of water-energy consumption and economic benefits of agriculture (crop production) under various climate change scenarios (Sedighkia and Abdoli 2022).

5.6 ENERGY USES OPTIMIZATION IN INTEGRATED FARMING SYSTEM

An integrated system poses a complicated situation for decision making in operations. The FarmDESIGN model seeks to address with complexity associated with mixed farm reconfiguration by fusing a bio-economic farm model and a multi-objective optimization algorithm (Groot et al. 2012). This model shows that small adjustments in the farm design result in significant enhancement of farm production efficiency. It was observed that by employing the FarmDESIGN model in North Western India, objectives of soil health buildup, nutritional security of farm household and reduction in synthetic inputs use could be fulfilled compared to the present negative organic matter balance of the majority of farms (Prusty et al. 2022). To optimize energy use, Data Envelopment Analysis (DEA) is also employed to assess energy efficiency, distinguish between efficient and inefficient production and compute wasteful use of energy (Amid et al. 2016).

5.7 BEST MANAGEMENT MEASURES TO REDUCE GHGS EMISSIONS

Agricultural operations/activities generate various amounts of GHGs. There are standard factors used to estimate such emissions (Table 5.4). Rice cultivation contributes to a portion of the total agricultural methane (CH_4) emissions. In low land rice, CH_4 emissions can be reduced by adoption of direct seeding, composted or partially decomposed organic materials (with low C:N ratio), a technology of alternate wetting and drying (AWD), incorporation or deep placement of fertilizers and cultivating early maturing rice varieties with lower GHGs emission potential.

Global warming, the rise of earth temperature due to deforestation, burning of fossil fuels releasing carbon dioxide (CO_2), emission of methane (CH_4) due to enteric fermentation and from rice paddies and nitrous oxide (N_2O) from fertilizer use, etc. is one of the major concerns of the twenty-first century at the international level. This global phenomenon is due to the increasing levels of CO_2, CH_4, N_2O and CFC (Chlorofluorocarbon) gases commonly called greenhouse gases (GHGs). These GHGs act as a blanket over the earth's atmosphere and trap the thermal radiations of the earth's

TABLE 5.4
Emission Factors Used for Estimating Greenhouse Gases Emissions from Agricultural Operations

Particulars	Unit	Emission Factor (CO_2-eq kg unit^{-1})	References
Machinery	MJ	0.071	Li et al. (2014)
Diesel	L	2.75	Li et al. (2018)
Urea	kg	2.51	Linquist et al. (2018)
Ammonium bicarbonate	kg	0.17	Linquist et al. (2012)
Superphosphate	kg	0.23	Lyu et al. (2021a)
Insecticide	kg	5.10	Lyu et al. (2021b)
Herbicide	kg	6.30	Lyu et al. (2021b)
Fungicide	kg	3.90	Lyu et al. (2021b)
Rice seed	kg	0.78	Mahajan et al. (2010)
Plastic tray and film	kg	1.47	Maraseni et al. (2015)
Electricity	kWh	1.12	Linquist et al. (2012)

surface. As a result, there is a substantial increase in the earth's temperature, causing melting of glaciers, ice caps, floods, droughts, reduction in the yields of crops, reduction in the cultivable land with the resulting adverse impact on human and animal health, etc.

The major contributor to global warming is the release of CO_2 due to anthropogenic factors. However, each molecule of CH_4 and N_2O has 62 and 275 times, respectively, greater global warming potential than a molecule of CO_2. Rice farming contributes significantly (23%) to the agricultural sources that may account for around 40% of the total annual emission of methane into the atmosphere.

The release of CH_4 from ruminants and vast areas of wetlands and rice paddies contributes to this global crisis. From these lands, CH_4 is emitted due to the action of certain anaerobic microorganisms called methanogens (e.g., Methanosarcina, Methanosaeta spp.) on the organic matter. However, out of the methane produced from bulk soil and organic plant debris, some amount is consumed by aerobic microorganisms called methanotrophs (e.g., Methylococcus, Methylocystis spp.) in the aerobic or oxic layer of the flooded soil. The balance amount of CH_4 is released to the atmosphere through ebullition, diffusion and rice plant system. Ebullition is the release of CH_4 as gas bubbles from flooded soil due to the influence of wind, water temperature, solar radiation, water level, atmospheric pressure, etc. This is the major process of methane emission during land preparation and early stages of rice growth. Diffusion of CH_4 due to the concentration gradient is generally slow and depends upon the supply of CH_4 and wind speed. During the active growth stage of rice plants, CH_4 is mostly transported through the aerenchymatous tissue of the rice plant and released to the atmosphere through the micro pores in the leaf sheath.

The emissions from different countries have been estimated by researchers (Table 5.5) along with the rice area and productivity of the respective country. It has been found that the co-relation between average CH_4 emission and rice productivity is positive.

5.7.1 Reducing CH_4 Emissions

The factors that affect the CH_4 production in rice paddies can be managed so as to reduce emission levels. Some of these factors are:

- Organic matter decomposition.
- Water level regulation.
- Rice cultivars.
- Fertilizer management.
- Residue management.

TABLE 5.5
Rice Area (mha), Productivity (t ha^{-1}) and CH_4 Emission (kg ha^{-1}) in Major Rice-Producing Countries

Country	Total Rice Area[a]	Average CH_4 Emission[b]	Rice Productivity[a]
India	43.08	182	2.00
China	28.67	256	6.20
Indonesia	11.64	161	4.47
Thailand	9.88	49	2.62
Vietnam	7.51	336	4.51
Philippines	4.06	175	3.31
Japan	1.71	182	6.44
Australia	0.12	810	8.74

[a]Rai, M. (2006). *Indian Farming* 56 (7): 4–9; [b]Jain et al. (2004). *Journal of Science and Industrial Research* 63: 101–115.

5.7.2 Organic Matter Decomposition

CH_4 is mainly released due to anaerobic decomposition of organic matter by the methanogens. Without compromising the food production, CH_4 emission can be reduced by incorporation of partially or fully composted organic matter rather than uncomposted material, such as rice straw, having a high C:N ratio. Use of biogas slurry instead of FYM/cow dung also reduces CH_4 emissions.

5.7.3 Water Level Regulation

Water management is the most efficient method of reducing CH_4 emissions from rice paddies. Rice is a semi-aquatic plant and therefore grows well in flooded soil. However, several experiments have sown that equivalent yields, higher water use efficiency and reduction in CH_4 emissions can be achieved through mid-season drainage or intermittent drainage or alternate wetting and drying. Standing water in rice during the ripening stage is not necessary and it only contributed to non-evapotranspiration losses giving rise to low water use efficiency. Aerobic rice cultivation is a new technique of producing rice in soils that are not puddled or flooded (maintaining an aerobic condition), but under irrigation with optimum external inputs, however, with application of less irrigation water compared to transplanted rice. The potential for emissions from rice is lower in rainfed and unevenly supplied rice environments than in fully irrigated rice environments.

5.7.4 Rice Cultivars

The cultivars differ in influencing CH_4 emissions due to the differences in maturity duration, capacity of the root to oxidize the rhizosphere and release of root exudates. The greater the duration of the variety, the greater will be the amount of CH_4 emissions. The rice variety that influences the methanogens and methanotrophs through the transport of O_2 into the rhizosphere also influences the CH_4 emissions.

5.7.5 Fertilizer Management

Type of fertilizer, dose, method and time of application affect CH_4 emissions. Ammonium sulphate and phosphogypsum producesless CH_4 in comparison to urea, similarly the higher the application rate the higher is the emission. Incorporation of fertilizers or deep placement of urea super granules reduces CH_4 emissions. The addition of oxidants such as ferric iron or sulphate to the soil also suppresses CH_4 production and reduces emissions. Use of nitrification inhibitors (which slow the conversion of NH_4^+ N into NO_3^- N) like N-Serve and AM also reduces GHG emissions. In the absence of synthetic inhibitors, neem products like neem cake or oil can be used as nitrification inhibitors.

5.7.6 Residue Management

Flooded rice fields with retained crop and weed residues are sources of CH_4 emissions. Huge amounts of residues are produced from rice-based cropping systems like rice-wheat and rice-rice systems. An efficient management strategy is to be followed for rapid decomposition of these residues and less emission of CH_4. Management options like residue return or removal, time of residue return, type of residue (high or low C:N) incorporation, etc. affects greatly the emission rate. The residues remaining in the field should be decomposed well before flooding. Therefore, from CH_4 emissions and sustainability of the rice-based cropping system point of view, the residues of the crop grown before rice should be removed, decomposed and incorporated into the soil.

Rice is grown in more than 43 million hectares in India, out of this a sizable proportion is lowland rice (irrigated 19.5 mha, rainfed lowland 15.5 mha and flood-prone 3.0 mha), which contribute to methane emission. However recent research findings point to the fact that the overall contribution of methane emission can be reduced with adoption of certain management practices as discussed in this chapter.

REFERENCES

Acaroglu, M., Aksoy, A.S. 2005. The cultivation and energy balance of *Miscanthus giganteus* production in Turkey. *Biomass and Bioenergy* 29(1):42–48. DOI: https://doi.org/10.1016/j.biombioe.2005.01.002.

Amid, S., Gundoshmian, T.M., Shahgoli, G., Rafiee, S. 2016. Energy use pattern and optimization of energy required for broiler production using data envelopment analysis. *Information Processing in Agriculture* 3(2):83–91. DOI: https://doi.org/10.1016/j.inpa.2016.03.003.

Calilung, E.J. 1985. Comarison of land preparation equipment used on small rice farms in the Philippines. In: *Thesis No. AE-85-2*. Asian Institute of Technology, Bangkok, Thailand.

Chamsing, A., Salokhe, V., Singh, G. 2006. Energy consumption analysis for selected crops in different regions of Thailand. *Agricultural Engineering International* 6(13):1–18. Available online: https://ecommons.cornell.edu/bitstream/handle/1813/10526/EE%2006%20013%20Salokhe%20final%206Nov2006.pdf?sequence=1. Accessed on 27 August 2023.

Dichmont, C.M., Pascoe, S., Kompas, T., Punt, A.E., Deng, R. 2010. On implementing maximum economic yield in commercial fisheries. *Proceedings of National Academy of Sciences USA* 107:16–21.

Gopalan, C., Shashtry, B.V.R., Balasubramanium, S.C. 1971. *Nutritive Value of Indian Foods*. National Institute of Nutrition (ICMR), Hyderabad. pp 47–58.

Groot, J.C.J, Oomen, G.J.M., Rossing, W.A.H. 2012. Multi-objective optimization and design of farming systems. *Agricultural Systems* 110:63–77. DOI: https://doi.org/10.1016/j.agsy.2012.03.012.

Hoang, V.N., Alauddin, M. 2011. Analysis of agricultural sustainability: A review of exergy methodologies and their application in OECD countries. *International Journal of Energy Research* 35:459–476. DOI: https://doi.org/10.1002/er.1713.

Imran, M., Ozcatalbas, O. 2021. Optimization of energy consumption and its effect on the energy use efficiency and greenhouse gas emissions of wheat production in Turkey. *Discover Sustainability* 2:28. DOI: https://doi.org/10.1007/s43621-021-00035-w.

IPCC. 2014. *Climate Change 2014: Mitigation of Climate Change. Contribution of Working Group III to the Fifth Assessment Report of the Intergovernmental Panel on Climate Change.* Edenhofer O., Pichs-Madruga R., Sokona Y., Farahani E., Kadner S., Seyboth K., Adler A., Baum I., Brunner S., Eickemeier P., Kriemann B., Savolainen J., Schlomer S., von Stechow C., Zwickel T., and Minx J.C. (eds) Cambridge University Press, Cambridge and New York, NY, USA.

Israel, M.A., Amikuzuno, J., Danso-Abbeam, G. 2020. Assessing farmer's contribution to greenhouse gas emission and the impact of adopting climate-smart agriculture on mitigation. *Ecological Processes* 9(51). DOI: https://doi.org/10.1186/s13717-020-00249-2.

Jose, V.S., Sejian, V., Bagath, M., Ratnakaran, A.P., Lees, A.M., Al-Hosni, Y.A.S., Sullivan, M., Bhatta, R., Gaughan, J.B. 2016. Modelling of greenhouse gas emission from livestock. *Frontiers in Environmental Sciences* 4(27). DOI: https://doi.org/10.3389/fenvs.2016.00027.

Kim, K., Kim, D.H., Kim, Y. 2021. Fisheries: A missing link in greenhouse gas emission policies in South Korea. *Sustainability* 13:5858. DOI: https://doi.org/10.3390/su13115858.

Lam, W.Y., Chatterton, J., Sim, S., Kulak, M., Beltran, A.M., Hujibregts, M.A.J. 2021. Estimating greenhouse gas emissions from direct land use change due to crop production in multiple countries. *Science of The Total Environment* 755:143338. DOI: https://doi.org/10.1016/j.scitotenv-2020.143338.

Li, X., Ma, J., Yao, Y., Liang, S., Zhang, G., Xu, H., Yagi, K. 2014. Methane and nitrous oxide emissions from irrigated lowland rice paddies after wheat straw application and midseason aeration. *Nutrient Cycling in Agroecosystems* 100:65–76. DOI: https://doi.org/10.1007/s10705-014-9627-8.

Li, Y., Jia, L., Wu, W., Yan, J., Liu, Y. 2018. Urbanization for rural sustainability–Rethinking China's urbanization strategy. *Journal of Cleaner Production* 178:580–586. DOI: https://doi.org/10.1016/j.jclepro.2017.12.273.

Linquist, B., Van Groenigen, K.J., Adviento- Borbe, M.A., Pittelkow, C., Van Kessel, C. 2012. An agronomic assessment of greenhouse gas emissions from major cereal crops. *Global Change Biology* 18(1):194–209. DOI: https://doi.org/10.1111/j.1365-2486.2011.02502.x.

Linquist, B.A., Marcos, M., Adviento-Borbe, M.A., Anders, M., Harrell, D., Linscombe, S., Reba, M.L., Runkle, B.R.K., Tarpley, L., Thomson, A. 2018. Greenhouse gas emissions and management practices that affect emissions in US rice systems. *Journal of Environmental Quality* 47(3):395–409. DOI: https://doi.org/10.2134/jeq2017.11.0445.

Lyu, T., Shen, J., Ma, J., Ma, P., Yang, Z., Dai, Z., Li, M., Zheng, C.G. 2021a. Hybrid rice yield response to potted-seedling machine transplanting and slow-release nitrogen fertilizer application combined with urea topdressing. *Crop Journal* 9(4):915–923. DOI: https://doi.org/10.1016/j.cj.2020.08.013.

Lyu, Y., Yang, X., Pan, H., Zhang, X., Cao, H., Ulgiati, S., Wu, J., Zhang, Y., Wang, G., Xiao, Y. 2021b. Impact of fertilization schemes with different ratios of urea to controlled release nitrogen fertilizer on environmental sustainability, nitrogen use efficiency and economic benefit of rice production: A study case from Southwest China. *Journal of Cleaner Production* 293:126198. DOI: https://doi.org/10.1016/j.jclepro.2021.126198.

Mahajan, G., Gill, M.S., Singh, K. 2010. Optimizing seed rate for weed suppression and higher yield in aerobic direct seeded rice in Northwestern Indo-Gangetic Plains. *Journal of New Seeds* 11:225–238. DOI: https://doi.org/10.1080/1522886X.2010.496109.

Maraseni, T., Chen, G., Banhazi, T., Bundschuh, J., Yusuf, T. 2015. An assessment of direct on-farm energy use for high value grain crops grown under different farming practices in Australia. *Energies* 8:13033–13046. DOI: http://dx.doi.org/10.3390/en81112353.

Meier, E.A., Thorburn, P.J., Bell, L.W., Harrison, M.T., Briggs, J.S. 2020. Greenhouse gas emissions from cropping and grazed pastures are similar: A simulation analysis in Australia. *Frontiers in Sustainable Food Systems* 3:121. DOI: https://doi.org/10.3389/fsufs.2019.00121.

Mittal, V.K., Mittal, J.P., Dhawan, K.C. 1985. *Research Digest on Energy Requirement in Agriculture Sector.* Co-ordinating cell, AICRP on energy requirements in agricultural sector, Punjab Agricultural University, Ludhiana.

Newlands, N.K. 2007. *GHGFARM – A Software Tool to Estimate and Reduce Net-Greenhouse Gas Emission from Farms in Canada.* Proceedings of the Third IASTED International Conference on Environmental Modelling and Simulation, August 20–22, 2007, Honolulu, Hawaii, USA.

Pearson, T.R.H., Brown, S., Murray, L., Sidman, G. 2017. Greenhouse gas emissions from tropical forest degradation: An underestimated source. *Carbon Balance and Management* 12(3). DOI: https://doi.org/10.1186/s13021-017-0072-2.

Peter, C., Specka, X., Aurbacher, J., Kornatz, P., Herrmann, C., Heiermann, M., Muller, J., Nendel, C. 2016. The MiLA tool: Modelling greenhouse gas emissions and cumulative energy demand of energy crop cultivation in rotation. *Agricultural Systems* 152:67–79. DOI: https://dx.doi.org/10.1016/j.agsy.2016.12.008.

Pimentel, D., Pimentel, M. 2008. *Food, Energy and Society, Third Ed.* CRC Press, Boca Raton, FL.

Pishgar-Komleh, S.H., Sefeedpari, P., Rafiee, S. 2011. Energy and economic analysis of rice production under different farm levels in Guilan Province of Iran. *Energy* 36:5824–5831.

Prusty, A.K., Natesan, R., Panwar, A.S., Jat, M.L., Tetarwal, J.P., Lopez-Ridaura, S., Adelhart Toorop, R., Akker, J.V.D., Kaur, J., Ghasal, P.C., Groot, J.C.J., Barba-Escoto, L., Kashyap, P., Ansari, M.A., Shamim, M. 2022. Redesigning of farming systems using a multi-criterion assessment tool for sustainable intensification and nutritional security in north western India. *Sustainability* 14:3892. DOI: https://doi.org/10.3390/su14073892.

Quilty, J.R., McKinley, J., Pede, V.O., Buresh, R.J., Correa Jr., T.Q., Sadro, J.M. 2014. Energy efficiency of rice production in farmers' fields and intensively cropped research fields in the Philippines. *Field Crops Research* 168:8–18. DOI: http://dx.doi.org/10.1016/j.fcr.2014.08.001.

Ram, R.A., Verma, A.K. 2015. Energy input, output and economic analysis in organic production of mango (*Mangifera indica*) cv. Dashehari. *Indian Journal of Agricultural Sciences* 85(6):827–832.

Rotz, C.A., Corson, M.S., Chaianese, D.S., Coiner, C.U. 2009. *The Integrated Farm System Model: Reference Manual.* USDA, ARS Pasture Systems and Watershed Management Research Unit, University Park, PA.

Salehi, M., Ebrahimi, R., Hassan, A.M., Mobtaker, G. 2014. An assessment of energy modelling and input costs for greenhouse button mushroom production in Iran. *Journal of Cleaner Production* 64:377–383.

Schaefer, M.B. 1957. Some considerations of population dynamics and economics in relation to the management of the commercial marine fisheries. *Journal of the Fisheries Research Board of Canada* 14:669–681. DOI: http://dx.doi.org/10.1139/f57-025.

Sedighkia, M., Abdoli, A. 2022. Balancing environmental impacts and economic benefits of agriculture under the climate change through an integrated optimization system. *International Journal of Energy and Environmental Engineering* 13:1053–1066. DOI: https://doi.org/10.1007/s40095-022-00482-9.

Singh, S., Mittal, J.P. 1992. *Energy in Production Agriculture.* Mittal Publications, New Delhi, India. pp 6–12.

Soni, P., Taewichit, C., Salokhe, V.M. 2013. Energy consumption and CO_2 emissions in rain-fed agricultural production systems of Northeast Thailand. *Agricultural Systems* 116:25–36. DOI: https://doi.org/10.1016/j.agsy.2012.12.006.

Taki, M., Ajabshirchi, Y., Mobtaker, H.G., Abdi, R. 2012. Energy consumption, input-output relationship and cost analysis for greenhouse productions in Esfahan Province of Iran. *American Journal of Experimental Agriculture* 2(3):485–501. DOI: 10.9734/AJEA/2012/1461.

Thao, N.T.T., LeThanh, S., Schnitzer, H., Thang, N.V., Hai, L.T. 2022. Development of decision support framework for optimizing resource recovery from a household-scale integrated agri-aquaculture system in the Mekong Delta, Vietnam. *Journal of Cleaner Production* 379(1):134643. DOI: https://doi.org/10.1016/j.jclepro.2022.134643.

Tuti, M.D., Vedprakash, B.M., Pandey, R., Bhattacharyya, D., Mahanta, J.K., Bisht, M.K., Mina, B.L., Kumar, N., Bhatt, J.C., Srivastava, A.K. 2012. Energy budgeting of Colocasia-based cropping systems in the Indian sub-Himalayas. *Energy* 45:986–993.

6 Organic Farming Systems

6.1 CONCEPT OF ORGANIC FARMING

Organic farming is a production management system that encourages and improves agro-ecosystem health, including biodiversity, biological cycles and soil biological activity, through the use of on-farm agronomic, biological and mechanical methods while prohibiting the use of all synthetic off-farm inputs. Before the invention of chemical fertilizers and pesticides, agriculture was purely organic basde. However, the importance of organic farming has been felt after witnessing the negative consequences of chemical-based modern farming (Mahalingam 2007). Organic farming solely relies on organic inputs, such as manures and organic matter for soil health, organic feed materials for animals/birds/fishes and biological/mechanical/cultural methods of pest/disease/weed control for crop/animals/birds/fishes. Therefore, organic farming is an in-built system that generates organic inputs from one enterprise/component to be used in another enterprise/component of the system (Rai 2010). Since organic byproducts are recycled in the system, there is negligible or no environmental footprints; therefore, it does not concern human civilization.

Organic farming, according to the United Nations Food and Agriculture Organization (FAO), is a system that begins to consider potential environmental and social impacts by eliminating the use of synthetic inputs such as synthetic fertilizers and pesticides, veterinary drugs, genetically modified seeds and breeds, preservatives, additives and irradiation. These are replaced with site-specific management practices that maintain and increase long-term soil fertility and prevent pests and diseases (https://fao.org/organicag/oa-faq/oa-faq1/en).

Organic agriculture was described by the FAO/WHO Codex Alimentarius Commission in 1999 as a holistic production management system that promotes and improves agro-ecosystem health, including biodiversity, biological cycles and soil biological activity. It emphasizes the use of management approaches over the use of off-farm inputs, recognizing that regional conditions necessitate regionally customized solutions. This is accomplished by employing agronomic, biological and mechanical processes, rather than synthetic materials, to perform any specific function within the system.

Organic farming is defined by the United States Department of Agriculture (USDA) as a system that avoids or largely excludes the use of synthetic inputs (such as fertilizers, pesticides, hormones, feed additives and so on) and relies on crop rotations, crop residues, animal manure, off-farm organic waste, mineral grade rock additives and biological systems of nutrient mobilization and plant protection to the greatest extent possible.

The International Federation of Organic Agriculture Movements (IFOAM), founded in 1972, emphasizes the importance of organic farming by stating that the world's food and farming systems are currently facing multiple challenges, including increased hunger, climate change and biodiversity loss as well as farmers and food workers not earning a living wage (https://ifoam.bio/about-us). Organic agriculture is defined by the IFOAM as an agricultural system that encourages environmentally, socially and economically sustainable production of food, fibre and timber. Soil fertility is viewed as the cornerstone to successful production in this system. Organic farmers strive to improve quality in all elements of agriculture and the environment by utilizing the natural properties of plants, animals and the landscape.

Organic farming is practiced based on a set of standards/guidelines and the process of production is certified by an authorized third party. This system of production encourages the *in situ* produced

DOI: 10.1201/9781003239963-6

inputs. However, certain external products are permitted and some are allowed for restricted use (https://apeda.gov.in).

External inputs permitted: Compost, slurry, cow pee, crop wastes, green manure, straw and various mulches are among the materials produced on an organic farm. Items produced outside the organic farm unit include wood shavings, sawdust, calcified seaweed, calcium chloride, natural calcium carbonate (limestone, gypsum, phosphate chalk), sodium chloride, magnesium sulfate (Epson salt), gypsum (calcium sulfate), silage and silage extract (apart from ammonium silage), clay (bentonite, perlite, zeolite), bacterial preparations (biofertilizers), biodynamic preparations, plant preparations and botanical extracts, vermiculate and peat.

Inputs for restricted use: e.g. meat meal, blood meal, bone meal, and feather meal without preservatives, compost made from animal excreta including poultry, fish and fish products without preservatives, guano, byproducts from the food and textile industries; biodegradable material of microbial, plant or animal origin without any synthetic additives, seaweed and sewage sludge and urban composts from separated sources, straw, vermicasts, animal charcoal, compost and spent mushroom and vermiculate substances, compost from organic household reference, byproducts from oil palm, coconut and cocoa (including empty fruit bunch, palm oil mill effluent, cocoa peat and empty cocoa pods), byproducts of industries processing ingredients from organic agriculture, basic slag, calcareous and magnesium rock, rock phosphates, pulverised rock, mineral potassium with low chlorine content (sulphate of potash, kainite, sylvinite), trace elements (boron, ferrous, manganese, molybdenum, zinc), wood ash from untreated wood, potassium sulphate, aluminium calcium phosphate, sulphur and stone meal.

Since almost all of the inputs used in organic farming are produced on-site, a significant push towards organic farming will necessitate several initiatives as well as infrastructure support. Engaging in proactive research on crop diseases, creating organic manure, developing natural insecticides, educating farmers and offering storage and connectivity are all crucial areas to consider. In contemporary agriculture, organic farming has gained significance owing to the growing concerns around the careless application of chemical pesticides and fertilizers. With fewer crop kinds available, there is a loss of genetic diversity and the genetic foundation of crops. The adoption of genetically modified (GM) crops may be important for short-term yield gains, but people are not yet ready to accept these foods and their long-term implications remain unknown. Globally, there has been a notable increase in demand for organic food in recent years. Thus, promoting organic farming practices in many of India's prospective regions as well as in developing nations reliant on agriculture could alleviate the problems associated with agricultural sustainability and satisfy the dietary preferences of an expanding number of urban customers.

6.2 BENEFITS OF ORGANIC FARMING SYSTEM

In India, the practice of organic farming is not new; it dates back thousands of years (Rajendra 2007). It is a farming system that focuses on cultivating the land and growing crops in a way that maintains the soil's life and health by using organic wastes (crop, animal, and farm wastes as well as aquatic wastes), other biological materials and helpful microbes (biofertilizers) to release nutrients to crops for increased sustainable production in an environment free of pollution. Organic farming in India provided higher (22%) net profit to farmers compared to conventional farming, despite a reduction (9.2%) in crop productivity, mainly due to the availability of premium price (20–40%) for the certified organic produce and reduction (11.7%) in cost of cultivation (Ramesh et al. 2010). During 2020–2021, the total area (cultivated + wild harvest) under organic certification in India was about 5.78 mha, of which the highest area of 2.43 mha was in Madhya Pradesh followed by Rajasthan (0.70 mha) and Maharashtra (0.53 mha). During 2020–2021, India exported organic products worth ₹7078.50 crores, of which the highest export was from Madhya Pradesh (₹ 2683.58 crores), followed by Maharashtra (₹913.74 crores) and Gujarat (₹723.20 crores).

Organic farming is becoming more and more popular; at present, it is practiced in nearly every country on the globe, and its proportion of farms and agricultural area is rising (Ramesh et al. 2005). According to a recent estimate by the International Federation of Organic Agriculture Movements (IFOAM), there are more than 24 million hectares of organically maintained land worldwide. Almost 130 countries worldwide have adopted organic farming, and the area under organic management is constantly expanding. While organic agricultural production is growing globally, the majority of sales occur in the more developed regions of the world. India is one of the countries that relies heavily on agriculture; out of its 1,400 million people, over 58% work in the agricultural industry. A study found that 52% of people are still in farming, producing just 13% of GDP, despite the fact that a significant number of farmers and farm labourers are leaving this sector (Bajaj 2008). This made it very evident that farmers' and farmworkers' incomes cannot increase. The current farming method, known as conventional farming or chemical farming (CF), accounts for 98% of all farming activities and uses chemical fertilizers, insecticides, herbicides, mechanized implements for numerous procedures and modern agricultural research. Prior to 1965, India practiced 100% natural farming (OF) without the use of artificial fertilizers or pesticides. According to the IFOAM in the Principles of Organic Agriculture (Aiyar 2007), 'the role of organic culture, whether in farming, processing, distribution, or consumption, is to sustain and improve the health of eco systems and organisms ranging from the smallest in the soil to human beings.' It is a remarkable thing to note that prior to Green Revolution agriculture, in India our forefathers followed organic farming with additional features like integrated farming system (IFS), mulching, green manure, cover cropping, crop rotation, multi-cropping and self-content with required infrastructure facilities in villages and value addition after harvesting (Zaman 2012).

6.2.1 Environmental Benefits

To preserve the natural balance and ensure the survival of life and property, different farming techniques are now required due to increased production, degradation of the foundational natural resources and environmental contamination. Today, when these agrochemicals are made from fossil fuels, are not renewable and are becoming harder to get, that seems like the more sensible option. It might eventually have a significant negative impact on our foreign exchange. The foundation of organic agriculture is the preservation of living ecological cycles and systems.

6.2.2 Advantages to Farm Family

A farmer can experience a threefold increase in turnover when they switch to organic farming from chemical farming. Water is not wasted and the land is not contaminated by this. Consequently, there is a significant boost in the farmer's income. Organic farming results in higher-quality food items and generally gives farmers, producers and marketers more clout in the marketplace. It restores the consumer's health. The ideal soil conditions for high-yield and high-quality crops are created by organic manures. It provides all of the nutrients (secondary and micronutrients for N, P and K) that the plant needs. It improves plant growth and physiological activities of plants and enhances the physical characteristics of the soil, such as its granulation and tilth, to allow for better water retention, easy root penetration and good aeration. Its capability to absorb moisture aids in the granulation of sandy soils and increases their water-holding capacity. The aeration of fibrous portions of the organic matter, with its high carbon content, encourages soil aggregation to improve the permeability and aeration of clay soil. Microbes that aid in aggregation get their energy from the carbon in organic substances. It boosts the beneficial chemical reactions that occur in the soil and its ability to deliver and retain nutrients. It reduces the need for purchased inputs. The majority of organic manures are leftovers or byproducts that, when they build up, can become contaminated. Pollution is decreased by using them in organic farming. Fertilizers that are organic are regarded as complete plant food. The constant use of chemical fertilizers can cause the pH of the soil to become acidic. Organic matter helps to restore this balance. It is thought that compared to crops grown with commercial fertilizers, organic crops offer humans and animals a healthier and more nutritious diet. Because organic plants are more resilient to pests and illnesses, handling of pesticides, which could otherwise be hazardous to farm families, is avoided. Compared to when farmers rely solely on one crop or business, this strategy offers the opportunity for crop diversification and lowers risks. Because organic farming requires more labour, there are more job opportunities available. For instance, the need for female labour has grown significantly in Karnataka for products like tea and spices. Increasing employment in high unemployment areas can boost local economies and slow the flow of people from rural to urban areas. Our tiny and marginal farmers might be able to escape poverty if they switch to organic farming. They must be made aware of the methods and advantages of organic farming. Organic farming is a market-oriented agriculture involving proper certification and farmers are advised to ascertain the marketability of their products and the likely profits before venturing into it.

6.2.3 Benefits to Soil System

Fertilizer has an efficiency of no more than 50%; the remainder is wasted. Measuring about 40 mt year^{-1}, the industrial nitrogen fixation (INF) accounts for just 15.3% of the overall nitrogen fixation. In contrast, 175 mt year^{-1} of biological nitrogen fixation (BNF) accounts for roughly 67.3% of the entire amount. Through mineralization, plants also obtain nutrients from organic sources, and the soil contains billions of microbes that are useful for these processes. The soil has already suffered greatly from the overuse and careless application of inorganic fertilizer, leading to a shortage of both macro- and micronutrients. Pesticides leave residues in soil and plants. In an organic farming system use of several bio-pesticides (*Bacillus thurengiensis, Trichoderma viridi,* GV, BT, NPV etc.) Neem-based botanical pesticides, bio-control agents (*Trichogramma*, *Cryptolaemus*, *Chrysoperla*, etc.) are capable of controlling pests and leave no residue for the soil systems.

6.3 THE KEY CHARACTERISTICS OF ORGANIC FARMING

Preserving soil organic matter levels, promoting soil biological activity and using cautious mechanical intervention to enhance soil fertility over the long term are key characteristics. They also include employing comparatively insoluble nutrient sources that soil microorganisms make available to the plant in order to indirectly supply nutrients to crops. Legumes, biological nitrogen fixation and efficient recycling of organic materials, such as crop leftovers and animal manures, can all contribute to nitrogen self-sufficiency. Weed, disease and pest control relying primarily on crop rotations, resistant varieties, natural predators, diversity, organic manuring and limited (preferably minimal) thermal, biological and chemical intervention. Comprehensive livestock management that takes into account evolutionary adaptations, behavioural needs, and animal welfare issues such as diet, housing, health, breeding and rearing. Pay close attention to the farming system's impact on the larger environment as well as the conservation of species and natural ecosystems.

6.3.1 Principles of Organic Farming

There are four principles of organic agriculture:

- **Principle of health**: Organic agriculture should sustain and enhance the health of soil, plants, animals, humans and the planet as one and indivisible. This idea emphasizes the interdependence of ecosystem health and human and community well-being, with good soils producing crops that, in turn, promote the health of both humans and animals. Living systems are healthy when they are complete and unbroken. Maintaining one's physical, mental, social and ecological well-being is more important than just being free from disease. Three essential aspects of health are regeneration, immunity and resilience. Maintaining and improving the health of ecosystems and creatures, from the tiniest in the soil to humans, is the purpose of organic agriculture, regardless of farming, processing, distribution or consumption. In particular, organic agriculture is intended to produce high-quality, nutritious food that contributes to preventive health care and well-being. In view of this it should avoid the use of fertilizers, pesticides, animal drugs and food additives that may have adverse health effects.

- **Principle of ecology**: The organic agriculture is rooted within living ecological systems. It states that production is to be based on ecological processes and recycling. The ecology of the particular production environment facilitates nourishment and well-being. For crops, this is the living soil; for animals, it's the farm ecosystem; and for fish and other marine life, it's the aquatic environment. Systems of wild harvesting, organic farming and pastoralistry should complement natural cycles and ecological equilibrium. These cycles work differently depending on the site, but they are ubiquitous. It is necessary to modify organic management to fit the ecosystem, scale, culture and local conditions. To preserve and enhance environmental quality and conserve resources, inputs should be minimized through reuse, recycling and effective material and energy management. Organic agriculture should attain ecological balance through the design of farming systems, establishment of habitats and maintenance of genetic and agricultural diversity. Those who produce, process, trade or consume organic products should protect and benefit the common environment, including landscapes, climate, habitats, biodiversity air and water.
- **Principle of fairness**: Organic agriculture should build on relationships that ensure fairness with regard to the common environment and life opportunities. Fairness is defined by justice, equity, respect and stewardship of the common good, both amongst individuals and in their interactions with other living things. This principle highlights the importance of treating everyone fairly in interpersonal interactions, including farmers, employees, processors, distributors, traders and consumers. It also applies to individuals engaged in organic agriculture. In addition to improving food sovereignty and lowering poverty, organic agriculture ought to enhance the standard of living for all parties concerned. Its goal is to generate enough high-quality food and other items to meet demand. This principle insists that animals should be provided with the conditions and opportunities of life that accord with their physiology, natural behavior and well-being. Ecologically and socially equitable management is necessary for natural and environmental resources used for production and consumption; these resources should be preserved for future generations. Systems of production, distribution and commerce that are open, equitable and take into consideration the true costs to the environment and society are necessary for justice.
- **Principle of care**: Organic agriculture should be managed in a precautionary and responsible manner to protect the health and well-being of current and future generations and the environment. A living and dynamic system, organic agriculture adapts to both internal and external conditions and needs. Organic agriculturalists can boost output and efficiency, but this should not come at the expense of people's health and well-being. As a result, it is necessary to analyze current practices and evaluate new technology. Care must be exercised, given our imperfect understanding of agriculture and ecosystems. According to this philosophy, the main considerations for management, development and technology selection in organic agriculture are responsibility and prudence. To guarantee that organic agriculture is safe, healthful and environmentally sound, science is required. But scientific understanding on its own is insufficient. Time-tested solutions are provided by accumulated wisdom, traditional and Indigenous knowledge and practical experience. Significant dangers should be avoided by organic agriculture by embracing suitable technologies and shunning unreliable ones, including genetic engineering. Making decisions should be

transparent and involve participation from all parties involved, taking into account their needs and values.

6.3.2 Organic Agriculture: Its Relevance to Indian Farming

Fertilizers are applied to only 45% of India's total cultivable area when irrigation facilities are available; the other 55% of arable land, which is primarily rain-fed, has very little fertilizer applied. These farmers frequently employ organic manure as a source of readily available nutrients from their own farm or location. Because it uses the fewest chemicals, the northeastern part of India offers a lot of room and potential for organic farming. An estimated 18 million hectares of similar land are available in the Northeast that might be used for organic farming. The year 2015 saw the establishment of Sikkim as the nation's first organic state. Given the vast area under naturally organic or default organic production, India has enormous potential to cultivate crops organically and become a significant global supplier of organic goods. It is necessary to establish a clear plan about organic farming and how it relates to markets. The way that people think about organic agriculture may have the biggest effect. To manage and expand diversity, integrate biological principles and resources into farming systems and increase agricultural productivity in an ecologically sustainable way, it draws on traditional and Indigenous farming knowledge combined with a few modern technologies. Traditions may become an intrinsic component of progress rather than an impediment to it. Farmers are challenged to take up new knowledge and viewpoints, as well as to innovate, by adopting organic agriculture (Gosavi 2009). This leads to increased farming participation, which can lead to more prospects for rural employment and economic upliftment. Thus, by emphasizing the utilization of local resources and self-sufficiency, organic agriculture contributes to the empowerment of farmers and local communities. Pest control in organic farming begins with making the correct decisions at the appropriate time, such as cultivating crops that are naturally resistant to diseases and pests or scheduling sowing dates to avoid pest and disease outbreaks. Careful management of planting time and area not only prevents pests but also enhances the number of natural predators with the potential to control insects and weeds. Other methods generally employed for the management of pests and diseases are: clean cultivation; refining soil health to resist soil pathogens and promotion of plant growth; crop rotation; encouraging natural biological agents for insects, weeds and disease control; using physical barriers for protection from birds and animals; modifying habitat to encourage pollinators and natural enemies of pests; and using pheromone attractants and trap pests. Soil-borne root infections are often less severe on organic farms than on conventional farms, but there are no persistent changes in foliar diseases between the two systems. The use of long and diversified crop rotations, crop combinations and regular application of organic amendments are likely to be associated with the successful control of root diseases in organic systems. Increased levels of soil microbial activity, which leads to increased competition and antagonism in the rhizosphere, the presence of beneficial root-colonizing bacteria and increased levels of vesicular-arbuscular mycorrhizal colonization of roots have all been identified as contributing factors in the control of root diseases. This is an unexplored area where native organisms provide protection against other harmful organisms.

6.3.3 Basic Steps of Organic Agriculture

There are five basic steps for following organic agriculture. The first step is to switch from conventional to organic land management. Next, the management of the overall environment to guarantee the sustainability and biodiversity of the system. Crop production using alternate nutrient sources, such as crop rotation, residue management, organic manures and biological inputs, constitutes the third fundamental phase. The fourth involves controlling weeds and pests by improved management techniques, cultural and physical methods and biological control systems. Lastly, keeping livestock in line with the organic idea and considering them an essential part of the system as a whole.

6.4 NUTRIENT MANAGEMENT IN ORGANIC FARMING

Many people define organic farming as a type of agriculture where pest and disease control, as well as nutrient delivery, are handled exclusively with organic inputs. It is actually a type of diversified agriculture that is specific to managing farming inputs solely with resources found in the area. The word 'organic' relates to the idea of farming as a business rather than specifically to the kind of inputs used. Frequently, organic farming has faced criticism on the grounds that, due to the extremely limited supply of organic supplies, farm productivity and profitability may increase only with organic inputs. Although there are few true organic resources available, using organic inputs has proven to be more profitable than using agrochemicals in situations when soil limitations and climatic fluctuations exist. Though the yields under organic farming during the initial years are less, gradually the productivity increases and by sixth year, it become at par with inorganic farming (Rajendran et al. 1999). In an organic production system, the availability of nitrogen from organic resources frequently prevents cereals from reaching their maximum yield potential. The yield differences between organic and conventional farming are limited to 10% to 15% when water availability is limited or when crops are rainfed. The FYM employed in these tests typically contains N, 0.5–0.8%, P, 0.2–0.4%, and K, 0.8–1.0%; alternate ways of efficient usage or the quality of organic matter/manure are not mentioned. When compared to the use of 15–20 t ha^{-1} FYM with and without chemical fertilizer, this nutrient-rich manure helps to increase crop output even at lower application rates (5–10 t ha^{-1}). Furthermore, it's usual practice to use liquid manures made by fermenting green leafy materials, animal urine, and other locally accessible resources. The significant discrepancies in crop yields found under the conventional and organic crop production systems are likely due to variations in the quality of the manures used. Nonetheless, a thorough comparison of the nutrient supply strategies used in conventional and organic systems is required. Future studies should look into how best to employ them. Farms have long had livestock as part of their operations. In organic agriculture, livestock are essential because they act as a bridge between using crop waste or farm-produced fodder and returning nutrients in the form of manure. Dairying, in particular, has helped a number of small and marginal farmers to improve their income. According to a field survey, marginal and small farmers have contributed to increased farm profitability and the availability of appropriate volumes of animal manure, even in advanced states like Punjab. Farmers rarely pay enough attention to how to store and use their resources, which causes 40–60% loss of nutrients, particularly nitrogen. Cattle dung pits are typically the source of NO3-N leaching that pollutes surface water and groundwater resources. Organic farmers and farming practices utilize technology such as composting and vermi-composting to minimize losses as much as possible. These not only improve the nutrients availability from organic sources but also prevent potential hazards of ground water pollution.

The potential of microorganisms in organic farming should be utilized to counteract these enemies in the soil. To improve the chemical, biological and physical qualities of the soil, organic farming systems depend on the management of soil organic matter. The utilization of 'biologically-derived nutrients' for plant nutrition, as opposed to readily soluble forms of nutrients, such as those found in bulky organic materials, is one of the fundamental tenets of soil fertility management in organic systems. Enhancement of soil biological activity is also recognized to be crucial in mitigating the growth of weeds, pests and diseases. The components of the system now need to be bound together harmoniously and complementarily on all sides. Possible organic farming nutrient sources include animal dung, crop leftovers, green manure, bio-fertilizers, bio-solids from agro-industries and food processing wastes. Animal excrement is often utilized as farmyard manure, even though it has competitive uses as fuel. By enriching composts with nutrient-bearing minerals and other additions, various compost production systems, such as N-enriched phospho composting, microbe mediated phospho composting, and vermi composting, have been developed. These manures have the ability to sufficiently meet the nutrient needs of crops and encourage the growth of advantageous macro- and microflora in the soil. Presently, only 30% of total cultivable areas in the country have irrigation facilities where agrochemicals use is higher compared to rain-fed zones. It is here that ingenuity and efforts are required to increase crop productivity and farm production despite recurrence of environmental constraints of drought and waster scarcity. At every stage of agricultural operations, increasing input use efficiency is a fundamental necessity of organic farming. This is accomplished, in part, by lowering losses and implementing innovative technology to increase manure's nutritional content.

6.4.1 Recycling of Crop and Animal Residues for Nutrient Supplementation

Organic farms use a variety of technologies to increase the amount of nutrients that may be obtained from manure, including recycling farmyard dung three or four times. A realistic estimate puts the amount of agricultural waste available in the nation annually between 600 and 700 million tonnes (mt), most of which is wasted. By turning this material into bioenergy, nutrients for malnourished soil and fuel for farmers, we must turn waste into wealth. About 1,800 mt of animal manure are produced in India annually. Even if two-thirds of the dung is used for biogas generation, it is expected to yield about 440 mt per year of manure, which is equivalent to 2.90 mt N, 2.75 mt P_2O_5 and 1.89 mt K_2O. Organic farms and food production systems are quite distinct from conventional farms in terms of nutrient management strategies. To create holistic farms that function like living organisms with balanced growth for both crops and cattle, organic systems embrace management alternatives. Thus, the nutrient cycle is as closed as it can be. Food is the only product that is exported from the farm, along with nutrients. It is forbidden to burn crop remnants and to apply animal manure to fields or store it in an improper manner. As a result, compared to the traditional method, it is thought to be more environmentally friendly and sustainable. Farm conversion from a high-input, chemical-based system to an organic system is designed after undertaking a constraint analysis for the farm with the primary aim to take advantage of local conditions and their interactions with farm activities, climate, soil and environment, so as to achieve (as far as possible) closed nutrient cycles with less dependence on off-farm inputs. Soil and water conservation practices are to be followed to minimize the loss of nutrients from soil. Crop varieties and crop rotations are chosen to fit the local environment and have the capacity to adequately balance the nitrogen requirements of crops. Micronutrients, potash, sulphur, phosphorus and other requirements are satisfied by local, ideally renewable resources. For this reason, organic farming is frequently referred to as knowledge-based farming as opposed to input-based farming. Additionally, the goal of organic farms is to maximize crop yield within a specific set of farming parameters. There are ample evidences to show that agrochemical-based, high input agriculture is not sustainable for long periods due to gradual decline in factor productivity, increased soil erosion and salinization with adverse impact on soil health and quality. Harnessing the varietal potential by appropriate biotechnology input is neglected area and needs adequate attention. The impact of organic agriculture on natural resources favours interactions within the agro-ecosystem that are vital for both agricultural production and nature conservation. Predation, pollination, habitats, waste recycling, carbon sequestration, nutrient cycling, soil stabilization through buffering and structural improvement and soil formation and conditioning are among the ecological services that are obtained. There are considerable environmental costs associated with conventional agriculture, and there is overwhelming evidence that switching to organic agriculture can significantly improve the environment. Conventional agriculture also carries significant pre-consumer health costs, especially when it comes to pesticide use. It is estimated that 25 million agricultural workers in developing countries suffer from pesticide poisoning each year.

6.4.2 Impacts of Different Organic Nutrient Management in Agriculture

Soil physical properties such as bulk density, capillary pore space, water-holding capacity and percentage of soil aggregate are significantly influenced with different organic packages. Among the organic packages, application of recommended nitrogen through FYM, vermicompost and neem cake by one-third each coupled with nitrogen containing biofertilizers (*Azospirillium* to paddy, *Azotobactor* to potato and *Rhizobium* to groundnut) and phosphorus solubilizing bacteria (PSB) with rock phosphate to all crops result in an increase of all physical parameters. An organic package with neem cake, FYM, vermicompost, rock phosphate, biofertilizer and PSB increases the biological qualities of the soil, its nutritional content and the population of bacteria that causes the least amount of weeds, diseases and insect pests.

6.5 PEST AND DISEASE MANAGEMENT IN ORGANIC FARMING

The first step in organic agricultural pest control is making the correct selections at the right times, such as selecting planting seasons to avoid pest and disease outbreaks or cultivating crops that are naturally resistant to diseases and pests. In addition to preventing pests, careful planting management of both time and space boosts the population of natural predators, which are naturally able to control weeds, insects and illnesses. Other methods generally employed for the management of pests and diseases are: clean cultivation, improving soil health to resist soil pathogens and promote plant; growth; rotating crops; encouraging natural biological agents for control of diseases, insects and weeds; using physical barriers for protection from insects, birds and animals; modifying habitat to encourage pollinators and natural enemies of pests; and using semi-chemicals such as pheromone attractants and trap pests. The quest for non-financial resources derived from natural resources should be prioritized in order to control biotic stressors. For an extended period, organic farmers have argued that industrial pesticides and fertilizers make crops more vulnerable to pest infestations. It has been demonstrated that organic crops are both more tolerant of and resistant to insect attack. Compared to ordinary rice, organic rice is said to have thicker cell walls and less free amino acid. Many studies have indicated that high plant N levels due to high inputs of soluble N fertilizers are associated with plant vulnerability to insect herbivory. Organic N is available slowly as the plant grows at thus acts as a self-control against the disease. Soil-borne root diseases are generally less severe on organic farms than conventional farms, while there have been no consistent differences in foliar diseases between the two systems. The successful control of root diseases in organic systems is likely to be related to the use of long and diverse crop rotations, crop mixtures and regular application of organic amendments. Root diseases can be prevented, in part, by increased levels of vesicular-arbuscular mycorrhizal colonization of roots, beneficial root colonizing bacteria and increased levels of soil microbial activity that cause increased competition and antagonism in the rhizosphere. Native creatures in this uncharted region offer defense against other dangerous critters.

6.5.1 Organic Methods of Insects Pest Management in Vegetables

The major insect pests of commonly grown vegetable crops in India are shoot borer, fruit borer, stem borer, hairy caterpillar (on drum stick) and army worm of vegetable crops. The organic management involves the use of *Andrographis paniculata* decoction 3 to 5% or *Sida spinosa* decoction 5%. For preparation of these decoctions, the plants are taken and cut into small pieces excluding roots. One kg of this is mixed with four litres of water and placed in a mud pot. This is boiled and reduced to one litre. On cooling, 500 ml of this extract is mixed with 100 ml of soap solution and 9.4 litres of water and sprayed on the top.

6.5.2 Use of Neem Kernel Extracts for Insects Pest Management

An acre needs roughly 3–5 kg of neem kernels to prepare neem kernel extracts. To use just the kernel, the outer seed coat must first be removed. Three kilograms of kernels should be plenty if the seeds are fresh. Five kilograms are needed if the seeds are old. Gently tap the kernel and secure it loosely with a cotton swab. Put this in a container with ten litres of water and soak it overnight. It is then filtered after that. 6–7 litres of extract can be obtained after filtering. Dilute 500–1000 ml of this extract with 9 litres of water. Before spraying, soap solution @ 10 ml $litre^{-1}$ should be added to help the extract stick well to the leaf surface. This concentration of the extract can be increased or decreased depending on the intensity of the pest attack.

6.5.3 Use of Garlic, Chili and Ginger Extract

Take around 18 grams of garlic, peel it and mash it into a paste. Nine grams each of green chili and ginger are combined to make a paste. One litre of water is used to dissolve each of the three pastes. Before spraying, this liquid is well mixed and filtered. A mixture of around 500 millilitres of this extract, 100 millilitres of soap solution and 9.4 litres of water is sprayed on top.

6.5.4 Management of Beetles and Bugs

Cow dung extract can be used to manage vegetable pests such as leaf beetles (also known as pumpkin beetles), pod sucking bugs and Epilachna beetles: Combine one kilogram of cow manure with ten litres of water. Use a gunny cloth to filter the extract. Refilter the filtrate using the same cloth after adding five litres of water. There will be a very clear solution in the filter. Spray the plants with the filtrate.

Diseases of vegetables such as tomato wilt, *Fusarium* wilt in chili, *Cercospora* leaf spot, yellow mosaic viral disease, *Alternaria* leaf spot and fruit rot of all vegetable crops can be managed by spraying 10% cow urine solution once in 10 days.

Half litre cow's urine along with ½ litre sour butter milk is mixed with 9 litres of water. This is sprayed once in seven days twice.

Cow's urine and water is mixed in the ratio 1:2. The seeds or roots of seedlings are soaked in this for half an hour before sowing or transplanting.

About forty kgs of neem cake per acre is applied as basal manure for vegetable crops to prevent diseases. If there is a disease attack in the nursery, then add 10% cow's urine extract along with the water that is used to irrigate the nursery.

6.5.5 Fumigation Combined with Other Organic Methods

It is advised to apply 10% cow urine extract via spraying to control crop diseases. Fumigation should be carried out in the evening on the same day or the following day. Embelica ribes powder is employed at a rate of 500 g ha^{-1}. Fumigation is carried out on the field in the opposite direction of the wind by placing the powder in a wide-mouthed kettle filled with burning charcoal. On the seventh day after fumigation, 300 ml of *Acorus calamus* extract along with one litre of cow's urine is mixed with 8.7 litres of water (measurement for one tank) and sprayed on the crop.

6.6 PHYSICAL METHODS OF PEST MANAGEMENT IN ORGANIC FARMING

6.6.1 Use of Light Traps

Light traps can be used to monitor and trap the adult insects to reduce the multiplication of pests and reduce the pest population. Light traps such as hurricane lamps, trap with electric bulbs and battery-operated bulbs may be used. The adult moths have a general tendency to be attracted to the light. This set up should be operated in the field during nighttime. A large plate or vessel fitted with kerosene mixed water is kept near the light trap, so that the attracted moths falls in this toxic liquid and die.

6.6.2 Yellow Sticky Trap

In this method, castor oil smeared with yellow-coloured sticky papers are kept in the field. Insect pests such as white flies get trapped on these sticky traps. After sufficient numbers of insects are trapped, the surface is cleaned and again castor oil is applied for fresh use.

6.6.3 Bird Perches

Many birds eat harmful insects, provided these birds find sitting places near the crop fields. For this purpose, different shapes of bird perches such as 'T' shaped nests or resting places made up of long dried twigs are installed in the crop field @40–50 numbers ha^{-1}. These installed structures attract the predatory birds for resting and the resting birds devour the larvae in the field.

6.6.4 Hand Picking of Insect Pests

When the cropped area is small, and monitored regularly, some of the insect pests at the first instance of incidence may be caught and killed. For killing, safer methods such as use of a small amount of kerosene in a polythene bag may be followed. The harmful larvae are picked up during evening hours and may be put in the bag. Clean cultivation methods are helpful for managing the pest populations. Wild grasses, weeds and other unwanted vegetation should be cleared from the fields, bunds and nearby areas, where many harmful insects lay eggs and proliferate their population.

6.6.5 Use of Neem as Pest Repellent

Various neem plant parts such as leaves, seed cake and kernels may be formulated as pest repellents. These are kept in a pot with an addition of a small volume of water and the mouth of the pot covered with a cloth. This mixture is kept as such for about three days. These are kept at different locations in the field with the mouth of the pot open during the evening. The foul smell that spreads from the pot acts as insect pest repellents.

6.7 IFS FOR MAINTAINING SAFETY AND QUALITY OF ORGANICALLY PRODUCED FOOD

Foods grown organically are in greater demand, especially because of customer opinions regarding the products' safety and quality. Organic farming methods have a significant positive influence on the ecosystem as well. It has been shown that foods grown organically have fewer pesticides, pharmaceutical and hormone residues and, in many cases, less nitrate. Foods contaminated with nitrates are dangerous and are usually linked to the heavy use of nitrogen fertilizers. Studies that compared nitrate contents of organic and conventional products found significantly higher nitrates in conventional products. Quality after storage has been reported to be better in organic produce relative to chemical-based produce after comparative tests. 'Organic' in organic agriculture is a labelling term that denotes products that have been produced in accordance with certain predefined parameters and certified by a duly constituted certification agency or authority. The organic label is therefore a process claim rather than a product claim. Organic standard will not exempt producers and processors from compliance with general regularity requirements such as food safety regulation, pesticide registration, general food and nutrition labelling rules.

6.7.1 IFS Steps to Switch from Conventional to Organic Farming

Many adjustments are needed when switching from conventional to organic farming. The farmer's mentality is one of the largest shifts. Sadly, conventional methods sometimes rely on band-aid solutions that don't deal with the root of the issue. Transitioning farmers typically focus too much about substituting permissible organic products for synthetic inputs, rather than thinking about management techniques based on preventive strategies. Here are a few steps new entrants should follow when making the transition to organic farming:

A. **Understand the basics of organic agriculture and the organic farming standards**

As knowledge-based systems, organic production requires newcomers and transitional producers to familiarize themselves with good and sustainable agricultural practices. In addition to conducting their own experiments and reading pertinent material, transitional producers should be ready to take part in both official and informal training sessions. As previously indicated, moving to an organic farming system involves more than just replacing synthetic products with permitted organic ones. Organic farming is a holistic system that relies on sound practices focused on preventative strategies. Since there are often few organic remedies available to organic producers for certain problems, prevention is the key element in organic production.

B. **Identify resources that will help you**

In general, experienced organic farmers are a great resource for sharing important technical knowledge. A competent mentor should be able to offer expertise, real-world experience and recommended reading lists to transitional producers. Some of the most significant difficulties that transitional farmers may face can be identified by mentors. Additionally, mentors can assist in locating production components that would otherwise be hard to locate. Producers should also contact agrologists, veterinarians and other agricultural and financial consultants in order to learn ways to improve their current farming practices.

The Internet is a valuable source of information, especially to new organic farmers. Numerous organic and ecological organizations, including the Organic Agriculture Centre of Canada (OACC), the Atlantic Canadian Organic Regional Network (ACORN), the Canadian Organic Growers (COG), the Certified Organic Associations of British Columbia (COABC), the Sustainable Agriculture Research and Education (SARE), the National Sustainable Agriculture Information Services/Appropriate Technology Transfer for Rural Areas (ATTRA) and the Agri-réseau/agriculture biologique-Quebec, offer a wide variety of reading materials. Consider joining an organic organization or network to access these valuable resources and establish good working contacts.

C. **Plan your transition carefully**

Develop a transitional plan with clear and realistic goals. The plan should clearly identify various steps to be taken in making the transition to organic and be sure to include realistic timeframes. Determine your advantages and disadvantages. Think about how to improve on your strengths while addressing any deficiencies. The business portion of the transitional plan ought to include a multi-year budget as well as a successful but practical marketing campaign. Make sure you have included all of your expenses on the list. Provide all necessary equipment to start the transition, including processing equipment, specialized applicators and equipment for composting, mechanical weeding equipment and extra handling equipment for organic goods. Although the demand for organic products is continually growing, growers need to make sure they have a reliable market for the organic products they plan to produce.

Careful planning is very important. Yields are frequently low in the early stages of the transitional phase, and premium prices for certified organic products are typically not yet achievable. Consider reasonable prices and yields while assessing the project's viability. In certain cases, early in the changeover process, it is better to stick with conventional tactics to prevent a sharp decline in yield, which could endanger the operation's financial stability. It is advisable for farmers who intend to transform their livestock activity to first certify their fields. This allows time to learn more about organic livestock management requirements while, at the same time, starting to produce organic feeds. Although organic certifiers generally want to see the entire farm become organic, certifiers generally allow new entrants several years of transition time before the whole farm is fully certified.

The concurrent production, processing, or handling of livestock, crops and other comparable products, both organic and nonorganic, is known as parallel production. While certifiers strongly discourage this kind of activity, some do permit it, particularly in the transitional phase. Producers that are allowed to engage in parallel production must be ready to handle a substantial amount of record-keeping in order guaranteeing organic purity and traceability.

D. **Understand your soils and ways to improve them**

Given that soil is the foundation of the organic farming system, it is essential that newcomers comprehend the diverse qualities and constraints of the soils on their farms. The appropriateness of the soil might differ greatly between fields. Excellent assets are fields with good drainage, a high content of legumes, a good level of fertility and organic matter, an appropriate pH, biological health and reduced pressure from weeds and pests. These fields are frequently the first to be prepared for certification and transition.

Many tools exist to assess soils. Soil chemical, physical and biological analyses, soil survey and legume composition field assessments and field yield histories are very important and should be considered early in the transition. Unhealthy soils require particular attention. Farmers may need to find acceptable soil additives, such as composted manure, limestone, rock dust and additional sources of nitrogen, phosphorous, potassium and micronutrients, if they intend to cultivate crops without keeping any livestock. Transitional growers may face difficulties obtaining optimal yields without extra livestock manure, compost and/or other off-farm soil inputs, even with the best of crop rotations that include green manure crops like legumes (nitrogen fixing crops). When these inputs are scarce or expensive, producers may benefit from integrating livestock on their farm.

Remember that in organic farming, farmers must be able to recycle nutrients through appropriate nutrient management practices. These practices include crop rotations, cover crops (such as green manure, catch crops, and nitrogen-fixing crops), and minimizing nutrient losses from leaching, overfertilization and improper handling, spreading and storage of manure and compost.

E. **Identify the crops or livestock suited for your situation**

The degree of difficulty to grow or raise the product organically; land and soil suitability; climate suitability; level of demand for the product; marketing challenges; capital required; current prices for conventional, transitional and organic products; and profitability over additional workload should all be taken into account before planting a crop or rearing any livestock.

F. **Design good crop rotations**

Plan the crop rotation(s) carefully and choose the best cover crops (green manure, winter cover crops, catch crops, smother crops, etc.) after the crops have been selected. In organic farming, crop rotation is one of the most crucial management techniques. They can break the life cycles of pests, control weeds, replenish and recycle fertility and enhance the tilth and structure of the soil. Some rotational crops may also be cash crops, generating supplemental income.

On some farms, land base availability may be a limiting factor when planning your crop rotations. The transitional plan should, therefore, include crop rotation strategies. Responding to external influences such as new market opportunities may also have a substantial impact on crop rotations; therefore, farmers must assess the effects of producing new crops on crop rotations and land base availability.

G. **Identify pest challenges and methods of control**

It is important to know the crop's most common pests, their life cycles and adequate control measures. For instance, Colorado potato beetle may be a pest of significant importance when growing potatoes; cucumber beetles in cucurbitaceous crops (cucumber, squash, and melons); flea beetle in many seedlings crops; clipper weevil and tarnish plant bug in strawberry crops.

Crop rotation, variety selection, sanitation, floating row coverings, catch crops, flamers, beneficial insect introduction, bio pesticides and inorganic pesticides are all options for reducing pest load. Transitional growers should be prepared to employ and test some of these possibilities. To increase your chances of success, discuss pest challenges with your agrologists, IPM specialists and/or other established organic producers when exploring a new style of cultivation. The availability of organic goods has greatly improved in recent years. New pest control products containing Bt, spinosad, kaolin clay are effective and currently available to organic growers. It is often reported that the types of weeds found on the farm evolve with time as growers change the way they grow their crops and control their weeds. By keeping track of the weed population, growers will be able to refine their crop rotations and improve their control measures.

Cattlemen are required to give careful care that supports health and satisfies the behavioural needs of diverse livestock under organic livestock management. Farmers rarely need to use traditional medicine when they implement strong herd health measures. Nonetheless, organic livestock farmers want to make an effort to become knowledgeable about complementary medicines, including homeopathy, herbal/fragrance cures and immune system boosters.

H. **Be ready to conduct your own on-farm trials**

Prosperous organic farmers consistently experiment with novel and/or inventive approaches to management. Organic farmers must always assess techniques like interplanting, cover crops and the application of different soil and pest control agents. Be willing to try out novel strategies.

I. **Be ready to keep good records**

One of the most crucial needs for preserving organic integrity is record-keeping. It is expected of farmers to maintain thorough records of their harvest, processing and marketing. Everything that comes into and goes out of the farm is included in this data. When conducting an inspection of a farm enterprise, impartial third-party inspectors demand that farmers produce the aforementioned documents. Once the record-keeping requirements are understood and the reporting procedure established, paperwork becomes routine.

6.8 COMMON DIFFERENCES IN CONVENTIONAL AND ORGANIC FARMING

Organic and conventional agriculture belong to two different paradigms. The key distinction between the two opposing agricultural paradigms is that organic farming strives to preserve natural balance through decentralization, diversity, community perspectives and preservation rather than exploitation of natural resources. As a result, using organic practices appears to bring about a new dimension in the functioning of the agro-ecosystem. The operation of this system cannot be understood by adding up individual ecological parameters. Organic farming seems to improve soil fertility in a way and to an extent which cannot be achieved by conventional farming even if the later consistently respects some ecologically principles.

Organic agriculture is one of several to sustainable agriculture and many of the techniques used (e.g., inter-cropping, rotation of crops, mulching, integration of crops and livestock) are practiced under various agricultural systems. The distinctive feature of organic agriculture, as outlined by numerous regulations and certification schemes, is the compulsory use of soil-building crop rotations and the near total prohibition of synthetic inputs. Although there are several exceptions in both situations, the fundamental guidelines of organic production are that natural inputs are permitted and synthetic inputs are not. Some natural inputs, like arsenic, are forbidden because different certification programmes have established that they are bad for the environment or human health. Additionally, some artificial inputs that have been found necessary and in line with the principles of organic farming are permitted (insect pheromones, for example). All certification programmes keep lists of particular, permitted synthetic inputs and forbidden natural inputs; the creation of such a list is currently being discussed in Codex. In addition to these two prerequisites, many certification programmes also include standards for extra environmental protection measures. Even while a large number of farmers in developing nations do not utilize synthetic inputs, this does not mean that their practices may be considered organic on their own.

Chemical farms now have little use for the skillful husbandry that was previously the guiding concept of land operations. Today, the emphasis is exclusively on productivity – high input in exchange for high returns and productivity (which is largely diminishing for farmers worldwide). Four critical concerns are ignored: what happens to the land, the food it provides, the people who eat it and the communities that suffer. Because of the frequent use of synthetic fertilizer and a lack of crop rotation, modern farming depletes the soil's fertility year after year. About half of the nitrate in the artificial fertilizer used on crops is dissolved by rain. The dissolved nitrate runs off the fields to contaminate water courses. Heavy rains can sweep away topsoil and render the area unfit for farming where repetitive deep ploughing is utilized to turn over the ground. Compression damage to soil structure is a severe problem in intensively cultivated areas. A tractor may run over the soil six or seven times during conventional tillage, and the wheelings can cover up to 90% of a field. Even a single tractor pass can compress the surface sufficiently to reduce soil porosity by 70%, increasing surface run-off and, hence, water erosion. In the worst-case scenario, surface run-off may exceed 100% – no water penetrates the surface. Large amounts of potentially valuable organic materials disappear into the sky in clouds of polluted smoke in nations where stubble is burned.

6.9 GENERAL REQUIREMENT FOR ORGANIC CERTIFICATION

6.9.1 General Information

A registered cultivator shall comply with National Programme for Organic Production (NPOP) norms and shall adhere to the National Standards for Organic Production (NSOP) and general standards for organic agricultural production, animal husbandry production, honey, wild collection, processing, packaging, storage, labelling and transport standards. Prepare, implement and revise an organic production plan on an annual basis, and submit it to a certification body. Permit on-site inspections with full access to the production and handling operation, including non-certified production and handling operations, areas, structures and offices by organic certification Inspectors and other higher officials of certification, as well as APEDA officials, as needed. Maintain all records relevant to the organic operation for at least five years after their creation, and grant authorized representatives of the state or central government officials of the accrediting agency access to such records during normal working hours for review and copying in order to determine compliance with NPOP norms and standards. Pay the appropriate fees within the time frame specified. The operator shall inform in case of any application, including drift, of a prohibited substances to any, production unit, site, facility, livestock or product that is part of an operation and changes in certified operations or any portion of a certified operation that may affect the organic integrity in compliance with standards.

6.9.2 Application for Organic Certification

A person seeking organic certification of production or handling operation shall submit an application for registration in the prescribed format in triplicate. The application must include information on an organic production or handling system plan, as well as any information requested in the application, such as name, address, contact information, telephone number of the authorized person, and so on. The names of the organic certifying bodies to whom applications have previously been made, the outcome, any non-compliance indicated, a copy of such records and the reason for applying must be provided. Any further information required to determine compliance with the given standards must be provided. The operator must pay the stipulated registration charge, one-time inspection fee and one-time trip cost along with the application form. The other prescribed fees shall be paid by the operator as notified during the course of the certification process.

6.9.3 Review of Application

After the submission of the application, it will be scrutinized. Any required information must be given to the operator, and the operator must submit the relevant information quickly. Applications that do not include the required fee will be rejected. After reviewing the application, a decision on acceptance or rejection will be made. The application will be returned to the applicant with the reasons for rejection and the payments contained. The application fee will not be reimbursed under any circumstances. After registration, an initial on-site inspection must be scheduled and reported to the operator, or it must be noted in the registered copy of the application itself. An applicant can withdraw the application at any time but the fees paid shall not be refunded.

6.9.4 Inspection

Initial field inspection shall be fixed at a reasonable time so that the operator can demonstrate compliance or capacity to comply with the standards while conducting inspection of land, facilities and activities. Such initial on-site inspection shall be delayed up to six months from the date of registration so as to give time for the operator to comply with required standards, including record-keeping. All on-site inspections must be performed in the presence of the operator or an authorized representative of the operator who is familiar with the operation. This obligation, however, does not apply to unannounced/surprise inspections. There will be one annual inspection, with subsequent inspections scheduled based on the risk assessment performed at the initial inspection.

6.9.5 Verification during Inspection

During the field inspection, the inspector must ensure that the NPOP standards and state standards are met or may be met. The verification is based on the information about the operator's organic production plan and the real implementation of the standards. To ensure that forbidden substances/materials are not used, samples of soil, water, wastes, seeds, plant tissues, plant, animal and processed goods are gathered for examination. NABL approved ISO 17025 laboratories will test the samples. The operator is responsible for the expense of sending samples for analysis. During the onsite inspection, an interview is conducted with the person in charge of the organic production system to ensure the accuracy of the information acquired during the inspection as well as the thoroughness of the inspection, observations gathered during the on-site inspection. The inspector must also gather additional necessary information as well as areas of concern. After the inspection checklist and inspection report are completed, the operator or his representative's signature is collected. A copy of the check list and inspection report must be submitted to the operator and evaluator in question. Inspection reports shall be evaluated by the evaluator within a reasonable time and any additional information required shall be addressed to the operator. In case of any non-compliance to the prescribed standards an explanation shall be obtained from the operator and sanctions shall be imposed if required.

6.9.6 Group Certification Standards

This system applies to farmer groups, cooperatives, producer groups, contract production and small-scale processing units. The producer group shall have similar production system and within the same geographical proximity. Farmers with four hectares or more can join the group, but they must be inspected individually. Such a farm's entire area must be less than 50% of the total area of the group. Although processors and exporters may be in the same group, they must be inspected annually. A legal status or organized organization is required for the group. The group must keep a written internal cControl system (ICS). Individual members/committees will be allocated the group's duties for carrying out specified activities. The group shall develop an internal quality system manual comprising of implementation of the internal control system and assessment of risk.

Internal standards shall be prepared in the local language under the framework of NPOP standards. The internal standards shall include the definition of the production unit, method of dealing with part conversion, parallel production, period of conversion, production norms for entire production unit and harvest and post-harvest procedures. The IQS shall include buying procedure, trading procedure and processing procedure for the group.

6.10 CONTINUATION OF CERTIFICATION

To maintain certification, the operator must renew registration by paying renewal fees. The operator must submit an updated annual report for the production or handling operation. The operator must submit an updated corrective action for previously discovered minor nonconformities. After receiving the renewal application for certification continuation, it will be scrutinized and the facts will be verified before making a decision. There are multiple organic certification agencies in various Indian states (Table 6.1).

6.11 CONSTRAINTS OF ORGANIC FARMING IN INDIA

The major issue for organic farmers is marketing their produce. The primary barrier to marketing organic products is a lack of understanding among individuals (customers/buyers). Furthermore, organic items are expensive, and only the wealthy and foreigners can buy them. Furthermore, before purchasing organically certified produce, consumers should confirm its authenticity. Organic growers cannot afford to invest money on organic certification. There are not enough organic sales units. The organic marketing in most countries is still relatively small, and, on an average, it is less than half a percent of the total agricultural sector except in Germany and Austria, where 2–3% of their agricultural area is under organic production. There are several companies in India that grow organic vegetables, fruits, plantation crops, spices and tea and export them to other countries. Farmers linked with large exporters typically do not have to worry about the sale of their products or their certification; however, small and marginal farmers face numerous challenges. Several institutions and movements are working together to promote organic agriculture in India and to reform policies to favor ecological agriculture. In June 2001, the director general of foreign trade issued a notification declaring the export of an agricultural product as organic would be permitted only if it was produced, processed and packed under a valid organic certificate issued by a certifying agency duly accredited by either APEDA oor the coffee board, the spices board or the tea board.

To strongly promote organic farming, it is necessary to disseminate information about organic farming and its advantages over conventional agricultural methods. The use of bio-fertilizers and bio insecticides necessitates understanding and willingness by the farmer. Knowledge of the availability and utility of extra nutrients for soil enrichment is also essential for increasing productivity. Attention to the application of composts/organic manure is also needed. The organic matter is spread during the months when the right moisture level is absent on the soil. The whole manure turns into wastes in the process. The required operation is, of course, labour intensive and costly, but it is necessary to obtain the desired results.

It has been discovered that before the production of organic crops can begin, their marketability, particularly at a premium above conventional supply, must be ensured. Inability to acquire a premium price, at least for the period required to achieve traditional crop yield levels, will be a setback. Many specialists and knowledgeable farmers are unsure whether organic materials can provide all

TABLE 6.1
Organic Certification Agencies in Different States of India

IMO Control Pvt. Ltd. Mr. Umesh Chandrasekhar, Director No. 1314, Double Road, Indiranagar 2nd Stage Bangalore-560 038.(Karnataka) Phone No.: 080-25285883, 25201546 Fax: 080-25272185 Email: imoind@vsnl.com	**Indian Organic Certification Agency** (INDOCERT) Mr. Mathew Sebastian Executive Director Thottumugham P.O. Aluva-683 105 Cochin, (Kerala) Telefax:0484-2630908-09/2620943 Email:Mathew.Sebastian@indocert.org
Lacon Quality Certification Pvt. Ltd Mr. Bobby Issac Director Chenathra, Theepany, Thiruvalla – 689 101. (Kerala) Telefax: 0469 2606447 Email: laconindia@sancharnet.in	**Natural Organic Certification Agency** Mr. Sanjay Deshmukh, CEO Chhatrapati House Ground Floor Near P. N. Gadgil Showroom Pune-411 038 (Maharashtra) Phone No.: 020-25457869, 56218063 Fax: 020-2539-0096 Email: contact@nocaindia.com
OneCert Asia Agri Certification Pvt. Ltd. Mr. Sandeep Bhargava Chief Executive Officer Agrasen Farm, Vatika Road, Vatika P.O., Off Tonk, Jaipur-303 905, (Rajasthan) Phone No.: 0141-2770342 Telefax No: 0141-2771101 Email: info@onecertasia.in	**Bureau Veritas Certification India Pvt.** Ltd. (Formerly known as BVQI (India) Pvt. Ltd.) Mr. R. K. Sharma Director Marwah Centre, 6th Floor Opp. Ansa Industrial Estate Krishanlal Marwah Marg Off Saki-Vihar Road Andheri (East) Mumbai-400 072 (Maharashtra) Phone No.: 022- 56956300, 56956311 Fax No. 022-56956302/10 Email: scsinfo@in.bureauveritas.com
ECOCERT India Pvt. Ltd Dr. Selvam Daniel(C.R.) Sector-3, S-6/3 & 4, Gut No. 102 Hindustan Awas Ltd. Walmi-Waluj Road Nakshatrawadi Aurangabad – 431 002 (Maharashtra) Phone No.: 0240-2377120, 2376949 Fax No.: 0240-2376866 Email: ecocert@sancharnet.in	**SGS India Pvt. Ltd.** Dr Manish Pande Divisional Manager – Food, Retail & CSRS250 Udyog Vihar Phase – IV Gurgaon – 122 015 (Haryana) Phone No.: 0124-2399990-98 Fax No.: 0124-2399764 Email: namit_mutreja@sgs.com
Control Union Certifications **(Formerly known as Skal International (India))** Mr. Dirk Teichert Managing Director "Summer Ville", 8th Floor 33rd – 14th Road Junction Off Linking Road, Khar (West) Mumbai –400052 (Maharasthra) Phone 022-67255396/97/98/99 Fax 022-67255394/95 Email: cuc@controlunion.in cucindia@controlunion.com controlunion@vsnl.com	**Uttaranchal State Organic Certification Agency (USOCA)** Director 12/II Vasant Vihar Dehradun-248 006 (Uttaranchal) Phone No.: 0135-2760861 Fax: 0135-2760734 Email: uss_opca@rediffmail.com Tel. No.: 0422-2544199; 0422-6586060 E-mail: profdrkkk@yahoo.com; iscopcbe@yahoo.co.in Website: www.iscoporganiccertification.org

(*Continued*)

TABLE 6.1 (CONTINUED)
Organic Certification Agencies in Different States of India

APOF Organic Certification Agency (AOCA) Mr. K. Dorairaj Chief Operating Officer #3, 1st floor, 9th cross, 5th main, Jayamahal Extn, Bangalore – 560046 Phone No: 080-55369888 Fax: 080-23430155 Email: aocabangalore@yahoo.co.in	**Food Cert India Pvt. Ltd** Quality House, H. No. 8- 2- 601/P/6, Road No. 10, Banjara Hills, Panchavati Colony, Hyderabad – 500 034 Tel. No.: +91- 40-23301618, 23301554, 23301582 Fax: +91-40-23301583 Email: foodcert@foodcert.in
Rajasthan Organic Certification Agency (ROCA) 3rd Floor, Pant Krishi Bhawan, Janpath, Jaipur 302 005 (Rajasthan) Phone No: 0141-2227104, Tele Fax: 0141-2227456 Email: dir_rssopca@rediffmail.com	**Aditi Organic Certifications Pvt. Ltd** No. 531/A, Priya Chambers Dr. Rajkumar Road, Rajajinagar, 1st Block, Bangalore – 560010 Tel.: +91-80-32537879 Fax: +91-80-23373083 Mobile: +91-9845064286 Email: aditiorganic@gmail.com Website: www.aditicert,net
Vedic Organic Certification Agency Plot No. 55, Ushodaya Enclave, Mythrinagar, Miyanagar, Hyderabad – 500 050 Mobile No.: 09290450666, Tel. No.: 040-65276784, Fax: 040-23045338 Email: voca_org@yahoo.com; usha_preetham@yahoo.co.in	**Chhattisgarh Certification Society** India (CGCERT), Raipur A-25, VIP Estate, Khamhardih, Shankar Nagar, Raipur-492007, Chhattisgarh (India) Telefax: +91-771-2283249 Email: cgcert@gmail.com 1424 A,
ISCOP (Indian Society for Certification of Organic Products) Rasi building, 162/163, Ponnaiyarajapuram Coimbatore – 641 001 Tamil Nadu Mob. No.: 094432 43119	**Tamil Nadu Organic Certification Department (TNOCD)** Coimbatore Thadagam Road, Coimbatore-641013, Tamil Nadu (India) Tel.: +91-422-2405080, Fax: +91-422-2457554. Email: tnocd@yahoo.co.in

Source: http://ncof.dacnet.nic.in/Training_manuals/Training_manuals_in_English/Organic_Agriculture_in_India.p

of the nutrients in the appropriate proportions. Even if this problem can be surmounted, they are of the view that the available organic matter is insufficient to meet the requirements.

In spite of the adoption of the NPOP during 2000, the state governments have yet to formulate policies and a credible mechanism to implement them. There are just four accreditation agencies, and their knowledge is confined to fruits and vegetables, tea, coffee and spices. The certification agencies are insufficient. In India, small and marginal farmers have been practicing a form of organic farming using the traditional farming system. They use renewable resources from their own farms and carry out agricultural activities in an environmentally sustainable manner. However, now the costs of the organic inputs are higher than those of industrially produced chemical fertilizers and pesticides, including other inputs used in the conventional farming system. Bio-fertilizers and bio-pesticides have yet to become popular in the country. There is a lack of marketing and distribution networks for them since stores are unwilling to engage with these products due to low demand. The situation is exacerbated by unpredictable supplies and growers' lack of awareness.

Higher profit margins for chemical fertilizers and pesticides for retailing, as well as strong advertising campaigns by manufacturers and dealers, are other important issues hurting India's organic input markets. In many cases, farmers face some yield loss when switching from conventional to organic farming methods and eliminating synthetic inputs. Restoration of full biological activity in terms of growth of beneficial insect populations, nitrogen fixation from legumes, pest suppression and fertility problems will take some time and the reduction in the yield rates is the result in the interregnum. It may also take years to make organic production possible on the farm.

REFERENCES

Aiyar, P. 2007. *Agriculture: Where India and China Stand.* Published in The Hindu (Coimbatore) 3 September 2007.

Bajaj, K. 2008. *Improving Agricultural Productivity is Crucial.* The Economic Times (Chennai) 21 July 2008.

Gosavi, P. 2009. *Organic Farming – Some Economic Aspects; A Micro Study.* Southern Economist 15 August 2009. pp 11–14.

http://ncof.dacnet.nic.in/Training_manuals/Training_manuals_in_English/Organic_Agriculture_in_India.p.

https://apeda.gov.in/apedawebsite/organic/organic_contents/Appendix_1_Crop%20Production.pdf.

https://fao.org/organicag/oa-faq/oa-faq1/en.

Mahalingam. 2007. *Agriculture in India: Paradoxes Galore.* KISAN WORLD 7 May. pp 3–7.

Rai, M. 2010. *Organic Farming: Potentials and Strategies.* Net. Wikipedia "OF". http://en. Wikipedia.org.

Rajendra, P. 2007. *Organic Farming: A Fresh Thrust Needed.* The Analyst Special Issue, The ICFAI University Press. pp 47–48.

Rajendran, T.R., Venugopalan, M.V., Tarhalkar, P.P. 1999. *Technical Bulletin No. 1/2000.* CICR, Nagpur. p 37.

Ramesh, P., Panwar, N., Singh, A.B., Ramana, S., Yadav, S.K., Shrivastava, R., Subba Rao, A. 2010. Status of organic farming in India. *Current Science* 98(9):1190–1194. Available online: https://www.researchgate.net/publication/281767191_Status_of_Organic_Farming_in_India. Accessed on 8 December 2022.

Ramesh, P., Singh, M., Rao, A.S. 2005. Organic farming: Its relevance to the Indian context. *Current Science* 88(4):561–568.

Staff Report. *"Farmers Urged to Eliminate Input Costs Zero Budget Farming Can Make Cultivation a Profitable Occupation."* The Hindu (Coimbatore) 30 December 2007.

Zaman, A. 2012. More crop and income with less water under sustainable agriculture. *Green Farming Strategic Vision* 3(4):10.

7 *In Situ* Resource Generation, Nutrient Management, ITK and Use for IFS

7.1 ORGANIC RESOURCE GENERATION AND RECYCLING

An organic farming system necessitates several organic products for meeting the on-farm needs. The most important inputs for organic farming are organic sources of nutrients, such as manures, composts, vermicompost, *Azolla*, blue green algae (BGA), oil cakes, slurry and organic products for pest control such as neem products and bio-dynamic preparations. Scientific procedures are required to be followed for generation of these valuable inputs. In this chapter, detailed descriptions on how these inputs are produced on organic farms are discussed.

7.2 PREPARATION OF VERMICOMPOST

Earthworms feed on large quantities of organic material and produce vermicompost (also called worm compost, vermicast, worm castings, worm humus or worm manure). The process of producing vermicompost is called vermicomposting. Exotic species of earthworms produce the vermicompost in a short period of time compared to local species. Vermicompost are rich in essential nutrients as well as several soil enzymes, beneficial bacteria and mycorrhizae. Therefore, these are used as a source of plant nutrients as well as a soil conditioner. It improves the physical as well as biological properties of the soil.

The earthworms species utilized for composting purposes include *Eisenia foetida*, also known as brandling worms, and *Lumbricus rubellus*, also referred to as red wigglers. The common earthworm species in India are *Perionyx excavatus* and *Lampito mauritii*.

7.2.1 SITE SELECTION FOR VERMICOMPOSTING

Vermicomposting unit should be installed at a cool, moist and shady place near to availability raw materials, cattle shed and water source. The site must be free from harmful insects and predators which harm the worms. Fresh organic waste should be kept in shade for 15–20 days for partial decomposition before mixing with waste materials. The waste materials/farm waste should be dried and chopped before mixing with cow dung. Lengthwise, a thatched roof shelter with an unpaved (*katcha*) floor is constructed in the east-west direction that protects the site from direct sunlight. The roof of the shed should ideally be open on all sides. With three vermibeds measuring 10' x 3' each and a 1' space in between, a shed area of 12' x 12' can adequately accommodate the treatment of 9–12 quintals of refuse over a cycle of 40–45 days. The length of the shed may be modified in accordance with the amount of refuse that needs to be treated and the spatial constraints. The thatched roof is maintained at a height of 8 ft in the centre and 6 ft on all sides. In anticipation of precipitation, the foundation of the site is elevated by a minimum of 6 in. above the surface. Following the procedure outlined below, the elevated ground is covered with vermibeds.

DOI: 10.1201/9781003239963-7

7.2.2 Materials Used for Vermicomposting

The process of converting waste into compost through the activity of earthworms, as well as the increase in the number of earthworms, are simple procedures that may be easily controlled by farmers. It is advisable to employ organic material only after it has undergone pre-processing. To ensure effective decomposition and to preserve the quality of vermicompost, it is necessary to employ an adequate quantity of earthworms (1 kg of worms for every 2.0–2.5 quintals of raw material). At the onset of the enterprise, cow dung can be utilized to cultivate a substantial quantity of earthworms. Once these earthworms have proliferated into sizable populations, they can be employed to process a significant amount of organic raw materials. Earthworms require a period of seven weeks to reach the stage of adulthood. Starting from the eighth week, they begin to deposit cocoons. A single fully grown worm has the ability to generate two cocoons every week. The incubation period for each cocoon varies between 5 and 10 days, depending on the species of worms, quality of nutrition, and general conditions. During this period, each cocoon gives birth to 3–7 offspring. The consequent augmentation amounts to approximately 1200–1500 worms annually. The population undergoes exponential growth, doubling within a month.

Carbon is needed by microorganisms to form their cell structures, and nitrogen is needed for the production of their cellular proteins. It was found that the best C:N ratio for effectively composting raw materials was 30:1 or less. Low nitrogen organic materials decompose slowly, taking longer to break down and producing less compost overall. These materials are identified by a wide carbon-to-nitrogen ratio. Under these conditions, biological activity declines and numerous organism successions are required to decompose the carbonaceous materials. Narrowing the C:N ratio of low-nitrogen organic compounds is required to enhance the breakdown process. Nitrogen can be supplied in the form of nitrogenous fertilizer to accomplish this. Superphosphate is frequently added to compost to increase its phosphorus content. It also evaluates the NH_3^- evaporation loss from the decomposable materials during the compost formation process. Ten parts cow dung to three parts organic waste is the ideal ratio for combining cow dung and organic wastes. Before adding earthworms, materials like polythene, glass and stone must be removed from the raw material. To allow the earthworms to breathe and to encourage aerobic decomposition, the bedding should be loose and have a moist feel like that of a wrung-out sponge. Vermicompost can also be produced from domestic waste, non-toxic solid and liquid waste from industries and municipal rubbish through this technique. Earthworms are beneficial to the ecology because they improve the general health of the ecosystem and transform trash into useful fertilizer. Earthworm activity converts garbage into compost, and earthworm populations grow as a result of basic processes that farmers can easily manage. The best course of action is to use organic material only after preprocessing. To maintain the quality of vermicompost and facilitate efficient decomposition, a sufficient number of earthworms should be used (1 kg of worms for every 2.0–2.5 quintals of raw material). Earthworms can be raised in large quantities using cow dung in the beginning of the business. These earthworms can be used to digest large amounts of organic raw materials once they have procreated and grown into sizable populations. It takes seven weeks for earthworms to mature into adulthood. During the eighth week, they begin to lay cocoons. Twigs that are completely grown are capable of producing two cocoons per week. Each cocoon produces three to seven young during the five- to ten-day incubation period, which varies according on the kind of worms, the quality of the food and other environmental factors. The resulting increase comes up to about 1,200–1,500 worms annually. The population doubles in a month due to exponential expansion.

7.2.3 Size and Dimensions of Beds

There are different dimensions used depending upon the quantity of material and requirement of vermicomposts. For small-scale units for backyard or garden or in a field, the most convenient pit of easily manageable size is 2m x 1m x 0.75m. When permanent brick construction unit is to be

FIGURE 7.1 (a) Brick-cemented vermicomposting unit with thatching to protect from sunlight and rainfall; (b) Low-cost HDPE vermicomposting unit covered with green net.

prepared, a tank of dimension 4m x 3m x 1m constructed under the shade of tree (Figure 7.1a). This is further divided into three parts with 22.5 cm brick walls with vents to accommodate 20 to 30 kg of cattle or farm waste per day. Individual tanks of dimensions 1 m × 1 m × 0.75 m has been recommended (Sarangi and Lama 2013). Low-cost high-density polyethylene (HDPE) woven beds with a ready-made shed may be used for vermicomposting. The units of length 12 m, width 4 m and height 2 m are fixed on the selected site and covered with a green net (Figure 7.1b).

For convenience of handling the fresh organic matter and harvesting of matured vermicompost, the length may be divided into two or three parts. Small-scale household level vermicomposting can be done by use of bins. Bins should have holes in the sides to allow air to flow, and a spout that can be opened or closed or holes in the bottom to drain into a collection tray. Plastic bins require more drainage than wooden ones because they are non-absorbent. The design of a small bin usually depends on where an individual wishes to store the bin and how they wish to feed the worms. Most small bins can be grouped into three categories:

- *Non-continuous* – an undivided container. The bottom of the bin is lined with a layer of bedding materials. Above the bedding, a layer of organic materials for composting is put along with worms. The worms will begin to decompose the organic waste and bedding when another layer is added on top of it. Because it is compact and simple to assemble, this kind of bin is frequently utilized. However, because all the materials and worms need to be removed, harvesting is somewhat challenging.
- *Continuous vertical flow* – a series of trays stacked vertically. Similar to any other bin, the bottom-most tray gets loaded first and isn't picked until it's full. Rather, a thick layer of bedding is placed on top, and organic material is added using the tray above. After the lower tray is completely composted, the worms go to the upper tray. The bottom tray can be retrieved once a sufficient number of worms have migrated and should be largely worm-free. Harvesting is made simpler with the help of these containers.
- *Continuous horizontal flow* – a series of trays lined horizontally. The earthworms' migration in the direction of a food source is also necessary for this technique to facilitate harvesting. Typically, the bin is made to resemble a non-continuous bin but be longer horizontally. Usually, a large gauge chicken wire screen divides it in two. When one half is full, it is used; bedding and organic materials are then placed into the other half. The compost can be gathered when the worms eventually move to the side containing the food. Not only are these bins easier to harvest, but they are also larger than a non-continuous system while being small enough to be used indoors.

7.2.4 Procedure of Vermicomposting

The step-wise procedure as suggested by Sarangi and Lama (2013) is given below.

Bottom layer consists of a thin layer (1 in.) coarse sand, on this a layer (2 in.) of garden soil, over this a layer of cow dung or vermicompost. Measured amount of paddy straw as initial substrate is then put in the pits (1 m × 1 m × 0.75 m) under a shed and spread layer after layer, each having a thickness of 15 cm and mixed with lime (2.5–7.5%). Cow dung slurry is applied to each layer of paddy straw until a point is reached where a few droplets of water are released when the straw is pressed. After a period of one week, 400 earthworms (*E. eugeniae*) are introduced into each pit. Additionally, 300 g of phosphate-solubilizing fungi (*Trichoderma viridae*) are added per tonne of raw material. The pits are then covered with wet gunny sacks. Regularly examine the compost for moisture levels, if necessary, water should be added until the material releases a few drops of water when squeezed, indicating a moisture level of 60–70%. Aeration is given by turning the material fortnightly. The compost is mature when it turn black to brown in colour with tea leaf dust-like appearance and feed material is converted to loose granular mounds due to feeding and defecation by the worms. The prepared compost is passed through a 3 mm sieve for separating the earthworms present after composting and loosening the compost mass for final use.

7.2.5 Management during Preparation of Vermicompost

The organic material used should be chopped as the smaller the size the faster will be the rate of decomposition. Fresh organic materials should not be used for vermicomposting. It may cause harm to worms due to excessive heat produced by the raw materials. The unit should be covered to avoid direct sunlight and rainfall on the beds. Protect the worms from ant, rat and bird. Maintaining optimum temperature in the vermi-bed is one of the most important factors for successful production of quality vermicompost. Earthworms prefer optimum temperature of 25–45°C, best being 24°C. Watering should be done to keep the bed moist (40–50%), but should not be too wet, which is not congenial for worms. To maintain moisture level, spray water on the pit as and when required. The surface of the bed is covered with a gunny bag to maintain moisture content and temperature. The pH level should be near neutral or slightly alkaline. The surface of the bed is sealed with 25 mm thick layer of soil paste, and at every 300 mm distance, holes of 50 mm diameter are made and sticks are introduced into them for two days and are later removed. This provides the track for air circulation. The tanks can also have holes of 15 mm diameter all-round (walls) at 300 mm apart to provide cross-ventilation.

Turning should be done twice at an interval of 20 days to facilitate proper aeration. Application of water should be stopped, 2–3 days before harvesting to facilitate worm movement into lower layers and easy separation of vermicompost. For easy separation of worms, the harvested vermicompost is kept over a heap of wet undecomposed raw material for 24 hours. Sieving should be done properly with the help of a sieve; drying of the sieved material should be done under shade. The worms are collected for new culture beds. The cocoons and young worms are used for seeding the new culture beds. The vermicompost recovered is rich in macro-nutrients, microbes such as actinomycetes and nitrogen fixers, and it is used as a manure. Vermicompost can be applied at 3 to 5 tha^{-1} and for plotted plants at 250 g per pot.

7.2.6 Vermiwash Preparation and Use

Vermiwash is a dark brown-coloured liquid collected after saturating the actively vermicomposting substrate. It is rich in essential plant nutrients, plant growth-promoting hormones (auxins and gibberellins), enzymes and symbiotic microorganisms. It is used as a source of nutrients as well as bio-pesticide. The vermiwash can be prepared by the use of small containers or barrels of 200 litres capacity. A hole is made at the bottom of the container and fitted with a tap for collection of vermiwash. At the bottom of the container gravel or small pieces of stones/bricks are filled up to a height of 10–15 cm or 10–15% volume of the container. Over this coarse sand is added of the same amount. Pre-decomposed organic wastes or cow dung are then added and water is given to maintain optimum moisture. Then 1000–1500 number of earthworms are released in to it. Watering is done continuously for one week and after 10 days vermiwash is collected from the fitted tap. It is used as a spray with 20–30% dilution with water, which develops resistance of plants to diseases and pests.

7.3 COMPOST PREPARATION

Composting is a process by which organic wastes are biologically decomposed under controlled conditions to a state in which it can be handled, stored and applied to the land without adverse effect on the environment. It is a self-heating, thermophilic and aerobic biological process which occurs naturally in heaps of bio-degradable organic matter such as manure, moist hay and straw. This bio-degradation process is carried out by different kinds of heterophytic microorganisms, bacteria, fungi, actinomycetes and protozoa, which derive their energy and carbon requirements from the decomposition of carbonaceous materials. Effective microorganisms (EM) may be used for preparation of compost. EM is a mixed culture of beneficial microorganisms that can be applied as inoculants to increase the microbial diversity of soils, which, in turn, can improve soil quality and health, enhance the growth, yield and quality of crops. In EM composting, activated EM (AEM) is applied @ 4 litres heap-1 of compost along with molasses or jaggery syrup to hasten the process of decomposition. Mixing of AEM, molasses/jaggery syrup and water could be in the ratio of 1:1:100 or 1:1:50. The different composting methods are available for preparing various types of composts.

7.3.1 Indore Method

The waste materials such as plant residues, animal wastes, vegetable wastes and weeds can be composted by the Indore method. The waste materials are first chopped into small pieces of 5–10 cm size and dried to 40–50% moisture level. A compost pit of dimensions of 30 × 14 × 3 ft with sloping sides (narrow at bottom and at wide surface) is prepared and the raw material is spread in layers of 3" thickness. A mixture of urine, earth, and wood ashes is sprinkled and this is followed by a second layer of dung. The pit is filled up this way until the material occupies a height of 3 ft above the ground level. The heap is properly moistened with dung, using earth or night soil. Sufficient quantity of water is sprinkled over the heap to wet the composing materials to the level of 50% moisture. Periodical turnings at 15, 30 and 60 days after piling are given to aerate and then the materials are covered with a thin layer of soil of about 2–3 cm thickness. As air can conveniently penetrate only to a depth of 1.5–2.0 ft extra aeration has to be provided, which is done by means of artificial vents (holes) of 4" diameter pipe for every 4 ft length of the pit. The pit is watered twice a day i.e., morning and evening with rose can. The compost obtained by the Indore method will have a composition of 0.8% N, 0.3% P and 1.5% K_2O. This method of compost preparation involves considerable labour in the preparation of heap and periodical turnings, therefore it can be expensive and impracticable when large quantities of materials are to be processed.

Activated compost is prepared by incorporation of fresh organic materials to an already fermenting heap so that quicker decomposition is achieved with the established microbial population.

7.3.2 Bangalore Method

The basic raw material is spread in a pit of 20 x 4 x 3 ft dimensions to a depth of 6" layer, moistened with 20–30 gallons of water if the material is dry. Over this FYM or preferably a mixture of dung, urine and litter (un-decomposed) from the cattle shed is placed as a layer of 2" thickness. It is again covered on the top with a layer of earth to a thickness of 6". It is beneficial to mix the earth with bone meal or oil cakes, wood ash, etc., to improve the manurial value of the compost. The piling of layers is continued until the heap raises above the ground level to a height of 2 ft. Then the heap is kept open for one week to facilitate aerobic decomposition. Later the heap is plastered with a layer of moist clay for anaerobic fermentation to occur. Fissures, or cleavages (cracks) that occur in the clay layer, have to be sealed off periodically. The compost will be ready in four to five months period starting from the day of preparation. This process is called as aerobic and anaerobic decomposition of compost. In this process the basic raw material is not so well decomposed as in the other methods. But organic matter and N contents are well conserved. The number of turnings is reduced. The out turn of the compost is relatively greater and is a cheaper process. When pits or trenches are not available for composting, town refuse and night soils can be composted in above ground heaps of 1 m height, 1m width and of any convenient length by placing refuse and night soil in alternative layers as in trenches and adding the final fuse on the top. In this method, the materials decompose more quickly than in pits and can be used after three to four months. The compost obtained by this method will contain 1.5% N, 1.0% P and 1.5% K_2O.

7.3.3 NADEP Compost

The NADEP method of organic composting was developed by a Gandhian worker called Narayan Deorao Pandharipande of Maharashtra (Pusad). The process is similar to the heap method of composting, but it is done in brick-lined enclosures provided with air holes on all sides. Compost can be prepared from a wide range of organic materials, including dead plant material such as crop residues, weeds, forest litter and kitchen waste. This method of making compost involves the construction of a simple, rectangular brick tank with enough spaces maintained between the bricks for necessary aeration. The recommended size of the tank is 10 ft (length) × 5 ft (breadth) × 3 ft (height). All the four walls of NADEP tank are provided with vents by removing every alternate brick after the height of 1 ft. from bottom for aeration. Tank can be constructed of mud mortar or cement mortar. Before filling, the tank is plastered by dilute cattle dung slurry to facilitate bacterial activity from all four sides. It is also filled in definite layers each layer consisting of the following sub layers. Sub-layer-1 consists of 4–6 in. thick layer of fine sticks, stems (to facilitate aeration) followed by 4 to 6 in. layer of dry and green biomass. Sub-layer-2 contains 4 kg of cow dung and is mixed with 100 litres of water and sprinkled thoroughly on the agricultural waste to facilitate microbial activity.' Sub-layer-3 consists of 60 kgs. of fine dry soil and is spread uniformly over the soaked biomass for moisture retention and acts as a buffer during biodegradation. In this way, approximately 10–12 layers are filled in each tank. After filling the tank, the biomass is covered with 3 in. of the thick layer of soil and sealed with cow dung and mud plaster. Thus, materials required are: agricultural waste (dry and green) – 1350–1400 kg; cattle dung or biogas slurry – 98–100 kg; fine sieved soil – 1675 kg; water – 1350–1400 litres. After 15–30 days of filling the organic biomass in the tank gets automatically pressed down to 2 ft. The tank is refilled by placing two to three layers over it and is resealed. After this filling the tank is not disturbed for three months except that it is moistened at intervals of every 6–15 days. The entire tank is covered with a thatched roof to prevent excessive evaporation of moisture. Under no circumstances should any cracks be allowed to develop. If they do, they should be promptly filled up with slurry. This method requires large proportions of soil, which may not be feasible in all situations.

7.3.4 The Coimbatore Method

This method involves anaerobic degradation followed by aerobic process. The basic raw materials are loosely spread in a pit of size 12 × 6 × 3 ft and to a depth of 9 in. and water is sprinkled until the entire material is moist. Cow dung slurry of 2 in. thick is added to enhance the rate of biodegradation. Then about one kg of powdered bone meal or rock phosphate is broadcasted uniformly above the layer and above this an emulsion of 5–10 kg of fresh cattle dung in 5–10 litres of water is applied. Repeat this process until a heap 2 ft above ground level is formed. Then the entire exposed surface area of the heap is plastered with mud to facilitate semi-aerobic fermentation, which will takes place for above four to six weeks depending upon the nature of the raw material. After four to six weeks, the mud plaster is removed to permit aerobic fermentation. If the heap has sunk unevenly which is a sign of defective fermentation, the material is heaped again after forking and moistened. The decomposition is complete in three to four months and is fit for application to the field.

7.3.5 Phospho-compost Preparation

In this preparation, farm wastes/crop residues, cow dung are mixed with rock phosphate about 30% of the compostable material (Singh 1985). This mixture is made into slurry so as to provide adequate moisture and, after uniform mixing, the slurry is allowed to decompose in a compost pit for about two to three months. N-enriched phosphor-compost is produced by using phosphate solubilizing microorganisms, namely, *Aspergillus awamori*, *Pseudomonas straita* and *Bacillus megaterium*; phosphate rock, pyrite and bio-solids to increase the manurial value compared to ordinary FYM and compost.

7.3.6 Spent Mushroom Substrate Compost

The straw or substrate used for production of mushroom is considered as spent substrate when one crop of mushroom has been obtained and further extension becomes unremunerative. Each kilogram (kg) of mushroom produced leaves behind 5–6 kg of byproduct called spent mushroom substrate (Ma et al. 2014). After about two to three weeks of mushroom harvest, the growing material is considered spent (Rinker 2017). This spent mushroom substrate (SMS) can be composted for further use as a good organic manure for raising healthy crops. Compost preparation from SMS can be done either by natural weathering process by filling in a *Kachcha* pit of size 5 ft × 5 ft × 5 ft or by aerobic or anaerobic decomposition process. For aerobic decomposition, a perforated base is first prepared in the bottom of the above pit with the help of waste wooden materials. The perforated bottom of the pit is connected perpendicularly with 2 in. diameter hollow plastic pipes placed at a distance of 1 sq. ft. Holes of 0.5 in. diameter are provided in the plastic pipes at a distance of 15 cm each in a circular fashion. The pit is filled with SMS and left for re-composting for a period of two years. In the anaerobic process, the pit is filled with SMS and covered with 1 ft thick layer of normal soil to create anaerobic conditions and left as such for two years. The spent mushroom compost is a good source of nutrients such as N (1.3–4.2%), P (0.1–0.4%), K (0.5–1.8%), Na (0.05–0.2%), Mg (0.2–0.4%), and several trace elements such as Fe, Cu, Zn, Mn, Mo, and B (values in ppm: 1000–2500; 4–12; 50–200; 100–300; 1–2; 6–15) and soluble salts and no pests or weed seeds (Jasinska 2018).

7.3.7 General Principles of Composting

Farm bio-wastes generated from animals, plants and weeds are collected and mixed to a heterogeneous mass after cutting the plant byproducts to small pieces of about 5 cm size. The composting materials are mixed with cow dung slurry and soil in the ratio of 70:20:10 (agricultural waste: animal dung: soil) and to improve the quality, rock phosphate is added @ 5% of the weight. Moisture is maintained at 50–60% by sprinkling water over the materials and then the top is plastered with red earth or mud. Periodic turning may be given to facilitate aeration as per the method of composting. Generally, compost is ready for use in the field within six months of preparation.

7.4 GREEN MANURE PREPARATIONS

Plant species such as Dhaincha (*Sesbania* spp.), Sunnhemp (*Crotalaria juncea*), Alfalfa/Lucerne (*Medicago sativa*), Cowpea (*Vigna unguiculata*), and Pillipesera (*Phaseolus trilobus*) are used as green manure and species such as Gliricidia (*Gliricidia* spp.), Karanj (*Pongamia pinnata*), Neem (*Azadirachta indica*), Mahua (*Madhuca longifolia*), Subabul (*Leucaena leucocephala*) are used as green leaf manuring purposes. *Dhaincha* provides a cheaper and renewable source of N, especially when inadequate infrastructure and transport means is a constraint for use of other source of nutrients. Common species used in Asia are *Sesbania cannabina* (former name *acculeata*) as well as *Sesbania rostrata*. This green manuring plant is adapted to poorly drained and flooded soils of at least moderate fertility, with pH 5.5 to slightly alkaline. It can grow in more acid conditions (down to pH 4.3), but N fixation is less efficient. Establishment on heavy clays can be difficult. It can tolerant low to moderate salinity. Seed germination percentage and seedling growth decline sequentially as salinity rises to 16 dS m^{-1}. High levels of hard seededness require scarification to achieve rapid germination. This can be achieved by various means of physical abrasion of the seed coat, hot water treatment at 80°C for about 5 minutes or 100° C for 3 seconds or treatment with concentrated sulphuric acid for 15 minutes, followed by thorough washing with freshwater, and sowing immediately.

The cultivation practices of *Dhaincha* as given by Sarangi et al. 2012 are described here. The seed rate of 30–35 kg ha^{-1} is used for green manuring of *Dhaincha*. The land is ploughed and the seeds are sown in the dry soil, when winter rain is received germination starts and this crop gets sufficient time for green manuring as well as for seed production. Alternatively, the seeds can be sown on a previously ploughed land after receiving winter rain sufficient for germination of seeds. If grown as an intercrop with rice, planting should be delayed 30 days to avoid excessive competition with the rice. For green manuring, the green biomass is incorporated into wet soil at 8–10 weeks along with puddling. The seed rate is less (20–25 kg ha^{-1}) when it is grown for seed purposes as the spacing is higher. *Dhaincha* is a photoperiod sensitive, short-day plant and the critical photoperiod is 12–12.5 hrs. It flowers in 30–35 days when the day length is shorter than 11 hrs and bears seed in 65 days with a seed yield 1.8-3.0 t ha^{-1}. A plant spacing of 2.0x 1.5 m is used, and nipping of terminal buds carried out two months after sowing to encourage branching and improve seed yield. Pods ripen over a long period, and they are normally hand-harvested. The harvesting of seed crop should be done before the pods turn full brown, otherwise there will be shattering of pods, thereby loss of seed. Morning harvest is preferable to avoid pod-shattering and seed loss. Seed harvested during the rainy season is often of lower quality due to pod borer infestation. Blister beetles (Coleoptera, Meloidae) attack *S. rostrata* when it is in flower, eating leaves and growing tips. The gGreen vegetable bug (*Nezara viridula*) reduces yields in seed crops through sucking developing pods. Adults feed on the crop, but it is not suitable for nymphal development, making it an ideal trap crop for this insect. Pod borers also are damaging to seed crops. *S. rostrata* is susceptible to damping off caused by *Pythium* and *Rhizoctonia, Cercospora* leafspot, and leaf mosaic virus. Root knot nematodes (*Meloidogyne* spp.) are damaging in better drained soils. However, it can act as a plant trap for the nematodes, *Hirschmanniella oryzae* and *H. spinicaudata*, that damage rice crops.

7.5 OIL CAKES

Oil cakes are prepared after extraction of oil from oilseeds or fruits. These are compacted mixture of the meal remaining after vegetable oils are extracted from oil containing seeds. These are edible oil cakes such as groundnut, mustard, sesame, coconut and non-edible oil cakes (neem, karanj, mahua, castor). The de-oiled cakes are important source of organic nutrients. Edible oil cakes are used for feeding to animals, whereas non-edible oil cakes are mainly used as manure. Non-edible oil cakes contain harmful toxic substances, which makes these unsuitable for feeding the animals. *Pongamia* oilcake obtained from *Pongamia pinnata* (Karanj) a leguminous nitrogen fixing tree contain about 4.28% nitrogen (Osman et al. 2009). Oil cakes are used for preparation of EM Bokashi (Bokashi is a Japanese term, which means fermented organic matter) by fermenting organic matter (rice bran, oilcake, fish meal etc.) with EM within a period of just 8 days. Bokashi acts as a store house of EM microbes.

Direct use of oil cakes may be avoided; instead mixed with composts or soaked with water and then applied to plants. Direct use may attract insects and ants. One kg of groundnut oil cake may be soaked with 7–8 litres of water for 10–15 days and to prevent unpleasant odour, mixed with probiotics EM as deodorant. If immediate application is required, soak the coconut oil cake with enough water for 5–6 hours so that the oil cakes absorb water and swell, then the dissolved content is applied to trees. Oilcakes are also effective against phyto-nematodes, seeds can be treated with extracts of oil cakes before sowing in order to protect the crop from nematodes.

7.6 POND SILT/SEDIMENT

The pond bottoms gradually get filled with soil eroded from upper riches, accumulation of leaf biomass and other plant products as well as from left over organic matter applied to the pond as feed for aquatic animals. Deposition of pond silt beyond a certain limit reduces the pond volume, restricts the growth of fish and reduces the stocking density. Therefore, there is a need to desilt the ponds periodically and utilize the removed silts. Silt deposition is often a sign of poor health for the pond. Often, weeds on the bank of the pond grow and produce a huge biomass, these should be removed along with mud. The removed pond silt or mud along with the weed biomass are, however, an important organic source of nutrients for plants. These can be removed manually or by the use of pond vacuums. This can be done by drying the pond bottom or in the wet conditions by flushing by pressure. In the drying method, the pond is dried under the sun for two to four weeks, and then the mud is removed to the composting yard for use in the compost preparation or directly used for application to trees/plants near the pond. Areas where drying of the pond bottom is not possible and where the pond bottom contains acid sulphate soils, removal is done by flushing under wet conditions. The oxidation of the matter is avoided and care should be taken to prevent contamination. After removal, the pH should be measured and a liming plan is adopted. Liming of the pond in acid sulphate areas should be appropriately followed for achieving higher yield from aquaculture (Sarangi et al. 2022).

Pond sediments are used as soil conditioners and therefore provide an opportunity for a successful integration between aquaculture and agriculture. These sediments are rich in humic acid, fulvic acid and humin released through decomposition of organic matter in sediment (Rahman et al. 2004). Agricultural byproducts are used as feed for fish and, in turn, the pond sediments are valuable organic sources of nutrients for the plants. Fish pond sediments are rich in clay content (50–60%), organic matter (49.8 to 62.0 g kg^{-1}), having 47–50% water-holding capacity and neutral to slightly alkaline in pH (7.2-8.2) and are good for soil aggregate stability and microbial activity (Avnimelech et al. 1999; Rahman et al. 2004).

7.7 *IN SITU* PRODUCTION OF *AZOLLA*

Azolla, a free-floating freshwater fern, is one of the most important biofertilizers in agriculture, particularly in rice farming. It is used as an organic nitrogen source as well as for supply of organic matter. This fern always forms symbiosis with nitrogen-fixing blue green algae (BGA), *Anabaena azollae*. This BGA remains associated with meristem and groves with *Azolla* and are never in direct contact with the external environment. The BGA (*Anabaena azollae*) filament gets into the developing cavities in the dorsal lobes of *Azolla* leaf. *Azolla* provides shelter and nutrients to BGA, the cyanobacterium, while *Anabaena* fixes the atmospheric N in the form of ammonia. Under ideal conditions, *Azolla* contribute 40–60 kg N ha^{-1} and a biomass of 20–25 t ha^{-1} (Kannaiyan 1993). *Azolla* can be produced in a nursery or multiplication area and then applied to a rice field or it can be simultaneously cultivated in a paddy field as dual cropping and later incorporated after covering the entire area.

7.7.1 *Azolla* Multiplication in Nursery

There are several species of *Azolla* such as *A. caroliniana, A. maxicana, A. pinnata, A. nilotica, A. rubra, A. microphylla* and *A. filiculoides*. Among these, *A. pinnata* is the most common species found in India. The multiplication can be done directly in small plots of 10–40 m^2 area or in

FIGURE 7.2 Dual culture of *Azolla* and rice in the waterlogged fields.

concrete tanks of about 10 m^2 area containing soil and water. When grown in fields, it should have bunds and irrigation facility and under the shade of trees. The bunds are plastered with mud to avoid leakage of water from the plots. Over the soil, about 10 kg cow dung mixed with 20 litres of water is spread and water is applied to a depth of 10 cm. Water level should be maintained throughout the growth period. *Azolla* inoculums of about 200 g m^{-2} is applied. Rock phosphate @ 2 g m^{-2} is applied in three split doses at 4 days interval as top dressing. For insect pest control, a neem cake @ 2.5 g m^{-2} is applied 7 days after inoculation of *Azolla*. After two to four weeks, the *Azolla* growth can be harvested for further multiplication in large plots.

7.7.2 *Azolla* Cultivation in Paddy Fields

Azolla can be directly multiplied in rice fields and incorporated *in situ* during the rice crop growing period (Figure 7.2). This method of cultivation adds fresh biomass of *Azolla* as well as helping to suppress weed growth in the paddy field. A fresh basic stock of *Azolla* is obtained from nursery, applied @ 2 t ha^{-1} to the main field one week after transplanting, allowed to grow for about one month and then incorporated into the soil by the use of a cono weeder (Sarangi et al. 2014). Even after incorporation, the left-over *Azolla* develops again, which are trampled in as second crop. *Azolla* incorporation in rice fields produces increased grain yield by 32.5%, increased CH_4 emission by 31.5% and decreased N_2O emission by 3.4-fold compared to the application of NPK fertilizers only (Kimani et al. 2020).

7.8 SILAGE

Grass or other green fodder are preserved in the form of silage for use during the lean period of fodder availability. Silage is the green succulent roughage preserved more or less in its original condition, with minimum deterioration and minimum loss of nutritive constituents of fodders. When the weather conditions do not permit drying or preparation of hay or straw, this procedure may be followed. The green biomass without first being dried is compacted and stored in airtight conditions, in pits specially made for the preservation that are called a silo, and used as animal feed during the summer/winter season, when the production is low. Silos are pits dug 2.4 to 3.0 m depth, with variable sizes and one m^3 of space accommodates about 400–500 kg fodder. The location of a pit should be free from water stagnation. The pit should be surrounded on all sides with thick plastic sheets. A pit can also be constructed using bricks and cement. The size of a silo should be calculated based on the number of cows to be fed and the length of the feeding period. The process of conserving green fodder is called ensilage. Crops rich in soluble sugars or carbohydrates such as maize (corn), jowar, bajra, oat, hybrid Napier are most suitable for ensiling. Cultivated and natural grasses can be ensiled with addition of 3–3.5% molasses. Crop should be harvested between the flowering and milk stage. In general, crops with thick stems are conserved in the form of silage while thin stemmed crops are conserved as hay.

Fodder crops are harvested when it has 30–35% dry matter and just the ears are emerging. It is always better to chop the fodder first (since packing is better and loss of nutrients is minimized) followed by the addition of 0.5% salt, 1% urea to cereals and grasses to improve palatability and nitrogen content. Fodder should be evenly distributed throughout the pit. At the top of the silo, fodder should be packed 3–4 ft above ground level. It should be covered with long paddy straw or poor-quality grasses on all sides and then covered with wet mud and dung to seal and prevent entry of air and water. The layer may be about 4–5 in. thick. Temperature rises to about 27 to 38°C and fermentation starts to convert green crops into silage. In anaerobic conditions, with the help of microorganisms, sugar contained in green fodder is converted in to lactic acid, which helps to preserve the green fodder for a longer duration. Silage will be ready in 45–60 days with a golden yellowish green colour and with good aroma.

7.9 MUSHROOM SPAWN PRODUCTION

The most important requirement for mushroom production is the initial seed called spawn, which is the mycelia of the mushroom fungus multiplied on substrates like cereal grains. However, the production of spawn requires special technical skill as well as a well-equipped laboratory set up for quality and economic production of spawn. Major equipment required are autoclave, laminar flow, BOD incubator, weighing machine, pressure cooker, boiler, boiling vessels, steel racks, other minor instruments and chemicals. There are three steps for quality spawn production. The first step is pure culture preparation by tissue culture in potato dextrose ager (PDA) media. The second step is mother spawn production by the use of jowar, bajra, wheat, rice or maize grains. These substrates are cleaned, washed and cooked for 30 minutes. After draining excess water, 20 g pharmaceutical grade calcium carbonate ($CaCO_3$) is mixed with one kg of cooked substrates. Coating of lime on grains raise the pH above 7, so that the growth of spawn is accelerated. Then these substrates are filled in polypropylene bags to hold 200 g per bag of mother spawn. The open end of the bags is plugged with cotton plugs and covered with butter paper. The filled bags are then sterilized in an autoclave at 121°C (15 lb $in.^{-2}$ pressure) for about one to one and half hours. Then the mother culture is further multiplied by following the above procedure and filled in saline bottles or polypropylene bags for use as commercial grade spawn.

Maintenance of cleanliness inside the spawn production unit is very essential. Ultra Violet (UV) light may be used for 30 minutes before inoculation by keeping sterilized substrate, forceps inside the chamber. The UV light should be switched off 15 minutes before entering inside the inoculation chamber. Inoculation should always be done near the spirit lamp flame to avoid contamination. The fully prepared spawn bags of bottom mushroom can be stored in cold room (4–6°C) for one to two months for future use. The shelf life of spawn varies with the type of mushroom (Table 7.1).

7.10 SPAWN REARING AND PRODUCTION OF CARP FRY

Crap fry (*Catla catla, Labeo rohita, Cirrhinus mrigala*) are generally produced within a short period of 15–30 days in small earthen ponds of 0.02-0.5 ha with water depth of 1.0–1.2 m. Cemented nurseries of 50–100 m^2 with 1.0–1.2 m water depth are also used as nurseries for high-density rearing. While seasonal ponds can be used effectively with minimum pre-stocking pond preparation, perennial ponds require specific steps for preparation prior to seed stocking. The steps involved for

TABLE 7.1
Duration and Storage Conditions Requirement for Mushroom Spawn of Various Types

Mushroom type	Days for Complete Colonization of Mother Spawn	Days for Complete Colonization of Commercial Spawn	Incubation Temperature (°C) During Colonization	Storage Temperature (°C)	Shelf Life of Spawn (months)
Agaricus	20–21	12–14	25	4	2
Pleurotus	8–12	8–10	25	4	1
Lentinula	20–22	15–16	25	4	3
Volveriella	6–8	5–7	32	>15	<15 days
Calocybe	15–17	12–14	25	>16	15 days

preparation of seasonal ponds include removal of terrestrial grasses, if any, followed by ploughing, liming, water filling, fertilization, water culture and control of aquatic insects. The perennial ponds, on the other hand, involve removal of aquatic weed and control of predatory and weed fish prior to pond preparation. The details of the above processes are discussed as follows.

7.10.1 Management of Aquatic Weeds

Perennial ponds, if not managed properly, lead to luxuriant growth of several types of aquatic weeds, which are floating, submerged, emergent and marginal in nature (Figure 7.3). These aquatic weeds pose several problems as they absorb nutrients, arresting pond productivity, harbouring predatory and weed fishes/insects and hindering free movement of fish and netting operations Though a wide range of manual, mechanical, chemical and biological methods are available for control of these weeds, the manual method is commonly advocated for weed clearance in nursery and rearing ponds.

FIGURE 7.3 Freshwater aquatic weeds.

FIGURE 7.4 Common freshwater predatory fish.

7.10.2 Eradication of Predatory and Weed Fish

The presence of predatory and weed fish (Figure 7.4) in the ponds severely affects the seed survival through devouring on the stocked seed as well as competing with them for space and oxygen. While the commonly found predatory fishes in the fish ponds are *Murrels, Magur, Singhi, Pabda, Wallago,* etc., the weed-fishes include *Puntius, Barbus, Oxygaster, Anabas, Amblypharyngodon, Colisa,* etc. Most of these fish normally breed in the pond prior to the onset of carp spawning season, increasing their population. Thus, eradication of these species is essential for successful seed rearing. Though a wide range of pesticides is used for the purpose, dewatering followed by sun drying the pond is the most effective methods adopted for eradication of predatory and weed fish. Mahua oil cake is most commonly used for eradication (applied @ 2.5 t $ha^{-1}m^{-1}$, three weeks before stocking). The oil cake also serves as organic manure after its decomposition Application of commercial bleaching powder (30% chlorine) @ 350 kg $ha^{-1}m^{-1}$ of water is another effective method for killing the fish. Alternatively, application of urea @ 100 kg $ha^{-1}m^{-1}$ followed by bleaching powder @ 175 kg $ha^{-1}m^{-1}$ after 18–24 hours also effectively controls these fish.

7.10.3 Pond Fertilization

Fertilization is carried out in the pond with both organic manures and inorganic fertilizers to increase the availability of plankton, which are the preferred food for the early stages of fish. The ponds are first applied with lime depending on the soil pH. Use of cow dung @ 5–6 t ha^{-1} or poultry dropping @ 2–3 t ha^{-1}, a fortnight before stocking, helps in improving the natural productivity. However, phased manuring schedule with combination of 750 kg groundnut/mustard oil cake, 200 kg cow dung and 50 kg single super phosphate per hectare in split doses helps in maintaining the desired plankton population. As basal dose, half of the above amounts is made into a thick paste and is spread throughout the nursery two to three days prior to stocking. Similarly, the rest of the amount is applied in two to three split doses depending on the plankton level of the pond.

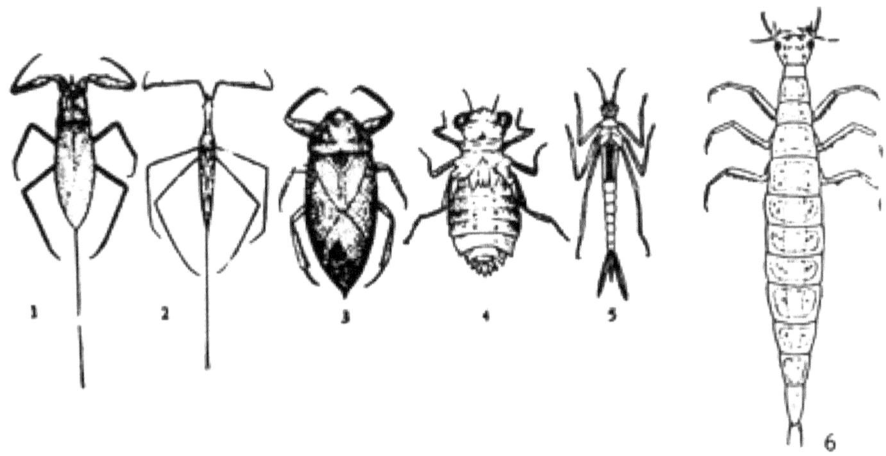

FIGURE 7.5 *Common freshwater pond insects: 1. Lacotrephes japonesis, 2. Ranatra uniedor, 3.* Fish killer (*Kirkaldyia deyrolloi*), 4. Dragon fly larva (*Zygoptera*), 5. Dragon fly larva (*Anisoptera*), 6. Water centipede (*Cybister chinenses* Shar).

7.10.4 Control of Aquatic Insects

Aquatic insects and their larvae (Figure 7.5) often prey upon the carp larvae and sometimes kill them through pricking their body. They also compete for food with the fish seed leading to poor survival and growth. The population of these insects increases tremendously just after pond fertilization. Common insects having a potential threat to the fry include back-swimmers (*Notonecta*), water stick insects (*Ranatra*), pond skaters (*Gerris*), water scorpions (*Nepa*), water measurers (*Hydrometra*) and diving beetles (Dytiscidae) etc. The shorter multiplication time and flying ability of these insects often pose problem for their control. Application of soap oil emulsion (cheap vegetable oil @ 56 kg ha^{-1} with one-third its weight any cheap soap) is a simple and effective method to kill these aquatic air-breathing insects. Emulsion of kerosene @ 100–2001 or diesel @ 751 and liquid soap @ 560 ml or detergent powder @ 2–3 kg ha^{-1} water area can he used as an effective substitute.

7.10.5 Stocking Procedure of Spawn

The spawn (Figure 7.6) is transferred from the hatchery and stocked in nurseries during cool hours, preferably in the morning, after due acclimatization to the new environment. The optimum stocking density of spawn recommended for an earthen nursery is 3–5 million ha^{-1}. However, higher densities of 10–20 million ha^{-1} can be followed for nursery rearing in cement tanks. While monospecies culture of carp is usually recommended, multi-species rearing can be taken up in case of non-availability of a sufficient number of ponds.

7.10.6 Supplemental Feeding

The stocked seeds are provided with a mixture of finely powdered groundnut oil cake and rice bran in equal proportions by weight as supplementary feed @ 600 g $lakh^{-1}$ spawn for the first five days and 1.2 kg $lakh^{-1}$ spawn per day for the subsequent days in two equal instalments during morning and evening hours.

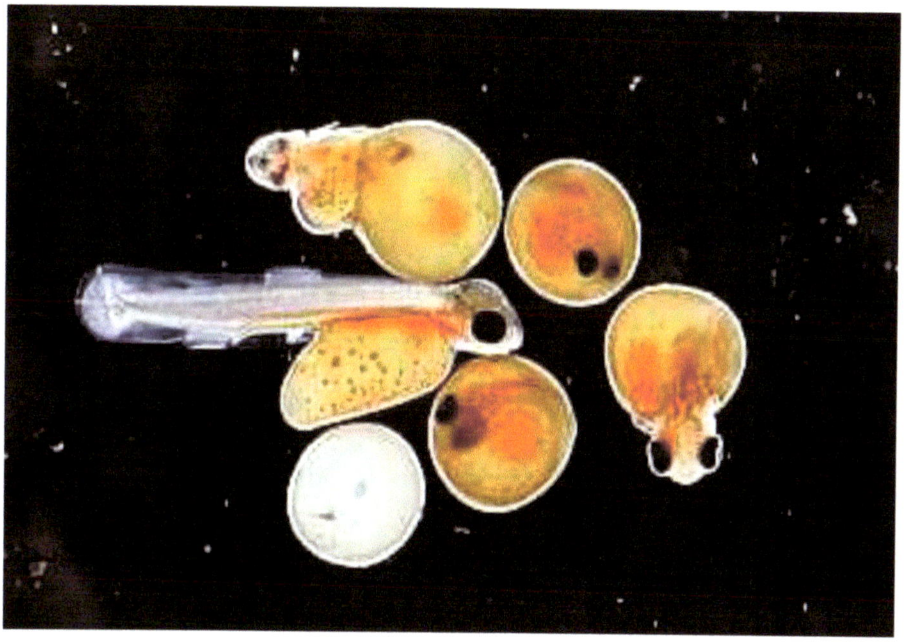

FIGURE 7.6 Fish spawn immediately after hatching.

7.10.7 Harvesting

During 15 days of rearing period the fry (Figure 7.7) attain the size of about 25–30 mm which is the ideal size for transferring to the rearing pond. Harvesting is done by repeated netting with adrag-net of 1/8" mesh. Survival levels of 50% are normally achieved in well-managed ponds. During a

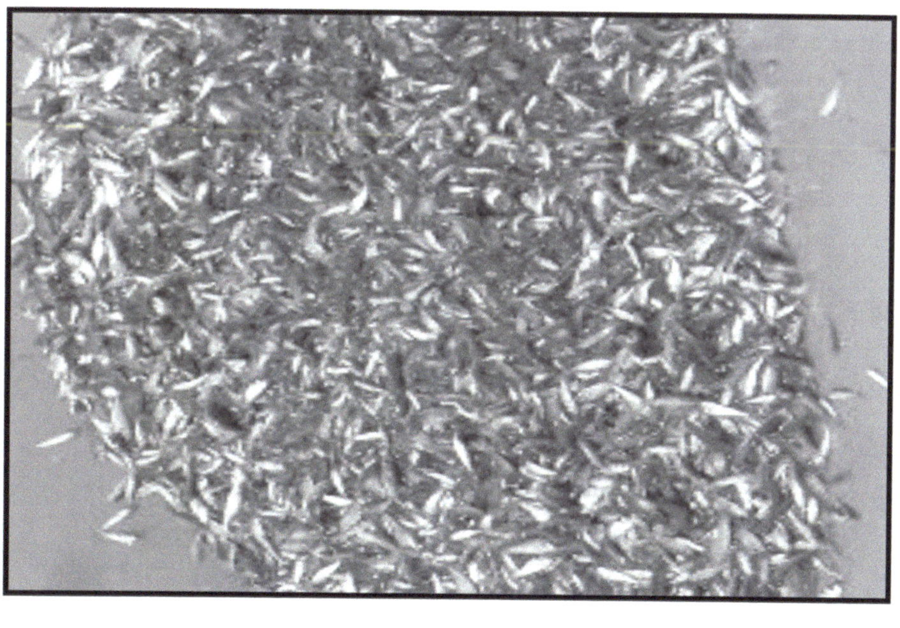

FIGURE 7.7 Harvested fish fry.

season from June to September at least 2–3 crops of fry can be raised in earthen ponds and 4–5 crops in concrete tanks. Net income of ₹150,000–175,000 per ha per two crops in a year can be achieved.

7.10.8 Water Quality Management

Proper management of the rearing environment offers optimum environmental conditions for the growth and better health of the cultivated fish. It also strengthens the defence mechanism of the fish to fight against invading disease-producing organisms. Some of the physico-chemical parameters of water have a direct influence upon the spawn/ fry health. Any abrupt and wider fluctuations of such values often cause a state of stress in fish resulting sometimes in widespread disease outbreaks. Dissolved oxygen content, pH, turbidity, temperature, introduction of some chemicals, detergents, pesticides and naturally produced toxic products like hydrogen sulfide, ammonia, dinoflagellate toxins, etc. are potential stress-related parameters. Overfeeding, overstocking, excessive application of inorganic fertilizers and accumulation of organic matter in ponds can deteriorate the water quality.

Therefore, water quality parameters (Table 7.2) in a pond should be monitored regularly, so that conditions that can adversely affect the growth of fish/prawns can be avoided. The population of phytoplankton and microorganisms are major determinants of the level of oxygen and metabolites in the pond. The dial fluctuation of DO (including its vertical profile), pH and CO_2 serve as indicators of their population. Since CO_2 is the major factor affecting the dial fluctuation of pH, monitoring pH fluctuation may be adequate. Also, $CO_{....2}$ is more difficult to measure. Daily measurements are conducted at early hours i.e. 5–6 am and after noon measurements in the i.e. 2–3 pm. This represents the period before the start of photosynthesis and the peak of photosynthesis, respectively. Thus, the maximal and the minimal of these parameters occur during this period. The other parameters do not have a distinct dial pattern and therefore can be monitored only once a day, preferable at a common time. Feed and growth data need to be presented with water quality parameters, side by side. This is because algal blooms are consequences of nutrients from feeds and excess feed can cause the rapid deterioration of water quality. Careful monitoring and data collection will remain useless unless it influences decisions regarding water management. Most of the water quality problems can be solved with adequate water exchange. Thus, if large quantities of water suitable for aquaculture is available, monitoring would not be as critical and high production levels can be targeted. If water

TABLE 7.2
Optimum Aquaculture Water Quality Parameters for Indian Major Carps

Water Parameter	Optimum Level
Water depth	1.0–1.2 m
Temperature	26–32°C
Dissolved oxygen (DO)	>5.0 ppm
pH	7.5–8.5
Total ammonia nitrogen	<1.0 ppm
Total nitrate nitrogen	<5.0 ppm
Nitrite nitrogen	<0.01 ppm
Sulphide	<0.03 ppm
Biological oxygen demand (BOD)	<2.0 ppm
Chemical oxygen demand (COD)	<70 ppm
Sacchi disc visibility (transparency)	25–45 cm
Ratio of phytoplankton and zooplankton	10:1
Water colour	Light green/light brown

is limited, the risk of encountering water quality and disease problems increases as one goes for more intensive culture.

7.11 EFFECTIVE MICROORGANISM -BASED BIO-PESTICIDE

Effective microorganisms (EM) contain photosynthetic (phototrophic) bacteria, lactic acid bacteria, yeasts, actinomycetes and fermenting fungi (*Aspergillus* and *Penicillium*). These microorganisms can coexist in mixed cultures and are physiologically compatible with one another. EM 1 is used as base culture for preparing various EM-based formulations. It is a yellow-brown liquid with a pleasant odour and sweet-sour taste and dormant. Activated EM (AEM) or rice wash AEM is prepared by the use of EM 1, molasses or jaggery and water. The molasses are dissolved with lukewarm water and mixed with EM 1 in a container. For one litre of molasses, one litre of EM 1 and 18 litres of water are required. These are mixed and filled in the container so that no air gap is left. Then stored in a place in ambient temperature away from direct sunlight for 7–10 days. Then open the cap for a few seconds to let off the gas formed during the fermentation period. After the end of fermentation, a layer of yeasts starts to appear on the surface of the solution with a pleasant smell. The product is ready when pH drops below 4.0. Rice wash AEM is prepared, when molasses/jaggery are not available, and instead water used for initial rinsing of rice could be used for making AEM. EM fermented plant extract (EM FPE) is prepared from AEM by the use of chopped plant parts from neem, calotropis, marigold, papaya, aloe vera, turmeric and garlic leaves. These are chopped into small pieces and are added at the time of preparing AEM or rice wash AEM. The container is made airtight and kept in a warm place away from direct sunlight. Every day the container is opened and the liquid is stirred for few minutes. EM FPE is ready to use when pH is <4.0 and there is hardly any gas formation. Pour the prepared EM FPE through a cloth to filter out the un-decomposed leaves and use the clear solution as a bio-pesticide. It is applied @ 2 litres per hectare after dilution with 700 litres of water and the spraying is done during the morning and evenings, at an interval of 15 days.

7.12 PREPARATION OF NEEM PRODUCTS FOR INSECT PEST CONTROL

Neem (*Azadirachta indica*) trees in IFS not only provide products for pest control but also have environmental benefits. The neem has inherently high efficacy of CO_2 fixation (more than 14 micro mole of CO_2 per m^2 per second and release oxygen by absorbing CO_2 even during the night time). Neem trees having a thick foliage cover with a very high leaf surface area, carry out maximum CO_2 fixation and also protect against other environmental pollution by trapping dust particles and absorbing toxic fumes, NO_2 and SO_2.

The leaves and seed of the neem are very useful ingredients for preparation of bio-pesticides (Table 7.3). Azadirachtin is one of the most important bitter compounds present in neem plant parts for use as pesticide, antifeedant and repellent. Neem seed cake can be used along with nitrogen fertilizers (urea) to increase uptake and to make slow release of nitrogen by regulating the rate of nitrification. Neem-coated urea ensures slow and continuous availability of nitrogen matching the crop requirement during the life cycle. For insect pest control, the simplest way is the use of water extract of neem leaves. For 5 litres of water, 1 kg of green neem leaf is required. Since the quantity of leaves required for preparation of this extract is quite high (nearly 80 kg are required for 1 hectare) this can be used for nursery and kitchen gardens. Spraying of 1.25% to 5% of neem kernel aqueous extract is recommended on the crops. For managing stored grain pests, dried neem leaves are mixed with the grains. A layer (half inch) of neem leaves is placed at the bottom of the storage container over which a layer (about a foot) of sun-dried grains is placed, over which another layer of neem leaves is placed and this is repeated up to the top of the storage structure. Instead of leaves, neem leaf powder can be used mixed with grains @ 1% of weight. Bio-pesticides are prepared from neem seed kernel to control stem borers, cutworms and soil borne diseases. The pulp from collected neem fruits is removed and the clean seeds are dried under the sun for few days. The seed shells are broken by

TABLE 7.3
Neem Products for Use as Bio-Pesticides

Name of the Product	Amount (Litres) for 1 ha Area	Amount (ml) to be Mixed with 10 Litres of Water
Margocide O.K. (80% EC)	2	100
Margocide O.K. (20% EC)	1	50
Achuk 0.15% EC	1	50
Nimbicidin 0.03% EC	0.2	100
Bioneem (EC)	1	50
Nimbasol	0.5	25
Repelin 93% EC	2	100
Nimajal F (EC)	0.4	20
Rakshyak (EC)	2	100
Neemgold (EC)	0.5	25
Neem marg (EC)	1	50
Multineem 90% EC	0.6	30
Neemax (powder)	4 kg	200 g
Econeem (liquid)	1	50
Fortune (liquid)	1	50
Neevar (liquid)	1	50

using stones or mortar and the kernels are collected and then crushed for use. About 500 g of seed kernels are mixed thoroughly with 10 litres of water, then filtered and sprayed on crop foliage to control the insect pests. The neem seed kernel oil can also be extracted as it contains about 45% oil. For preparing neem cake extract, 100 g of neem cake is required for 1 litre of water. The neem cake is put in a muslin pouch and soaked in water. It is soaked overnight before use in the morning. It is then filtered and emulsifier is added (1 ml $litre^{-1}$ of water). It can then be used for spraying. Neem oil can be used to control root knot nematode and at the same time encourage the multiplication of earthworms in soil. Neem oil controls fungal diseases such as powdery mildew, black spot, scab, rust, leaf spot, anthracnose and tip blight. Neem oil (15–30 ml) is added to 1 litre of water, stirred well and to this emulsifier is added (1 ml $litre^{-1}$). It is very essential to add the emulsifier and mix properly. This should be used immediately before the oil droplets start floating. A knapsack sprayer is better for neem oil spraying in preference to a hand sprayer.

7.13 INDIGENOUS TECHNICAL KNOWLEDGE (ITK)

Indigenous knowledge (IK) can be defined as a systematic body of knowledge acquired by local people through the accumulation of experiences, informal experiments and an intimate understanding of the environment in a given culture (Gope et al. 2017). Before the invention of modern inputs (fertilizers, pesticides, growth hormones etc.), farmers used ancestral knowledge in agriculture for protection of crops and maintenance of soil and crop health.

7.13.1 ITK for Seed Treatment

Pulse seeds are treated with milkweeds (*Calotropis*), karanja (*Pongamia pinnata*) and neem leaf extracts. The leaves of these plants are crushed, dissolved in water and the extract is mixed with pulse seeds. The seeds are kept overnight for uniform coating of extract on the seed coat, then the solution is drained out and seeds are dried under the sun for two hours and then sown. This treatment offers resistance to fungal and bacterial diseases. Wheat seeds are soaked in the milk before sowing to avoid rust diseases. Maize seeds are soaked in cow urine for 10–12 hours before sowing. Rice seedlings raised from seed treated with neem kernel extract offer resistance against leaf hoppers.

7.13.2 ITK for Management of Insect Pest of Crops

In rice, the gundhi bug or ear head bug (*Leptocorisa acuta*) is a serious insect pest that sucks the milky sap of the developing spikelet and makes it chaffy. ITK using locally available snails, frogs, fishes collected from ponds, ditches, paddy fields very well manage this pest. These dead creatures are made into small pieces and kept in cloth packets for 2–3 days for rotting. Each small cloth packets containing rotten snails/frogs/fishes are tied on the tip of a stick. About 25–35 sticks per hectare are placed over the paddy field. This act as a repellent for the gundhi bugs and the paddy crop is saved. Case worm and leaf folder in rice crop are managed by applying kerosene-mixed sand over the area and then a rope is pulled over the rice field to dislodge the insect pest over the standing water and then this water is drained out. If it is not possible to drain out the water, wild colocassia plants are chopped and applied to the field along with raw cow dung to kill the larvae of caseworm. Gram pod borer (*Helioths armigera*) is managed by using *karanja* (*Pongamia pinnata*) leaf extract. One kg of *karanja* leaves and barks are soaked in five litres of water and boiled for one hour. It is then strained and diluted with 15 litres of water. This solution is sprayed over the affected area. The hairy caterpillar (*Amsacta albistriga*), an insect pest of blackgram, groundnut, maize, castor, brinjal, moringa and other crops, are controlled by ITK involving tamarid (*Tamarindus indica*) and lemon (*Citrus* sp.) juice. In this practice, 150 ml each of lemon juice and tamarind extract are mixed in 15 litres of water and sprayed on the hairy caterpillar–affected field of area 4000 m^2. White fly in brinjal crop could be controlled by the use of hing (*Ferula asafoetida*) seeds. About 10 g hing seeds are taken in a pouch and kept in irrigation channel, so that it is immersed in irrigation water which flows to the brinjal crop field. This practice reduces the white fly menace.

Termites can be effectively managed by employing ITK, such as in fruit crops, sugar and flour, which is spread near the base of plants to attract the ants, which feed on the termites. Application of a mixture of 200 g of turmeric powder in one litre of mustard oil by making a paste and applying on tree trunks attract ants for control of termites. In field crops, termites are controlled by placing chopped leaves of milkweeds (*Calotropis* sp.) in gunny bags in the irrigation channel. The exudates from these leaves are lethal to the termites. Aphids are managed by spraying tobacco leaf extracts. Tobacco leaves and twigs (4 kg) are boiled in 40 litres of water for 40 minutes, after cooling one kg soap powder is mixed and the solution is diluted 7–8 times and sprayed to control aphids and white flies in citrus crops. Cow dung slurry can protect crops from aphid attack and acts as a repellent to cutting and biting insects. Cattle urine may be used against aphids, thrips, mites and caterpillars. Application of extracts prepared from the parts of nettle (*Urtica* spp.) and fruits of prickly ash (*Zanthoxylum armatum*) plants are used to control chewing, biting and cutting insects. Leaf and leaf extracts of pride of India plant (*Melia azadirachta*) controls various kinds of insects. In northeastern regions of India, the borer insects of citrus plants are controlled by putting cotton-soaked kerosene or petrol in holes and blocking the holes with mud or cow dung.

7.13.3 ITK for Management of Diseases of Crops

Blast and bacterial blight in paddy are controlled by spraying a solution of cow dung with an equal volume of water. Application of excess dose of nitrogenous fertilizers cause bacterial leaf blight in a paddy. Diluted cow dung slurry (10 kg cow dung in 100 litres of water) is filter through a cotton cloth and sprayed over the paddy field. To manage bunchy top disease in chilies, dusting of ash, spraying of sour butter milk/liquid waste of tanned leather/cow or goat urine is followed. Fresh cow dung is applied near the collar region of the chili plant to control fungal diseases (damping off and dieback).

Karanj or neem or castor cake application controls the soil-borne diseases such as root rot, collar rot, etc. Diluted milk (one litre milk mixed with 9 litres of water) is effective to control powdery mildew and viral diseases. Brown spot disease in rice can be controlled by papaya leaf extract. Two kg fresh papaya leaf is soaked in 3–4 litres of water overnight, filtered, diluted with 50–60 litres of water and to this 250 ml of soap solution is added before spraying. Leaves of *Khair* (*Acacia catechu*) can be put into an irrigation channel to control brown spot disease in a paddy. Cultivation of marigold (*Tagetes* spp.) in rotation with solanaceous vegetables could help manage bacterial diseases.

7.13.4 ITK for Storage of Agricultural Produce

Proper storage and preservation of agricultural produce reduces the loss and maintains the quality of produce. Several ITK are involved in storage of produce. In northeastern hill regions of India, farmers store the dried ears/cobs of cereal crops above the kitchen, so that the smoke generated during the cooking process keeps the insect pests away (Sarangi et al. 2009). Indigenous rodent-proof grain storage structures are also built, locally called *Nahu* by the *Adi* tribes of Arunachal Pradesh (Sarangi et al. 2009).

7.13.5 ITK for Crop, Soil and Water Management

Maintaining soil fertility and productivity for long-term use by the use of organic sources of nutrients, restricting soil erosion and practicing crop diversity can be considered knowledge drawn from Indigenous knowledge (Sarangi 2009). The planting of vetiver grass (*Vetiveria zizanoides*) across the slope restricts soil erosion.

In Odisha, the Indigenous method of management of direct seeded rice in rainfed lowlands is known as '*Beushening*'. This ITK maintains optimum plant population, aerates the crop root zone and controls weeds. In this method a narrow bottom country plough is run over the rice crop area about one month after sowing with standing water, followed by laddering and seedling redistribution. This practice loosens the soil, suppresses weeds and controls insect pests, thereby the growth and tillering of rice are increased with better nutrient uptake.

India has rich traditional knowledge of water management in different agro-ecological regions (Table 7.4). *Khadin* is an Indigenous method of rainwater harvesting, first designed by the Paliwal Brahmins of Jaisalmer, western Rajasthan in the 15th century, is a very long (100–300 m) earthen embankment built across the lower hill slopes lying below gravelly uplands (http://www.rainwaterharvesting.org/Rural/Traditional2.htm). The *Khadin* system is based on the principle of harvesting rainwater on farmland and subsequent use of this water-saturated land for crop production with provision of sluices and spillways to drain off excess water. Rainwater harvesting by farm ponds in Assam, Odisha and West Bengal, locally called *Pokhari/Pukhri/Gadia,* is a traditional system, for multipurpose use of harvested rainwater for crop-fish integration and domestic use of water. Rich traditional knowledge on creation, management and sustainable use of irrigation water, flood and drought control can be learnt from the 5000-year-old *Ahar* and *Pyne* system in Bihar. *Ahars* are reservoirs with an embankment on three sides while *Pynes* are diversion channels laid from the river or the catchment area for impounding water in the *Ahars* and channels. During the flood period, excess water is harvested in *Ahars* through the *Pynes* and during the drought period, one *Pyne* can irrigate up to 400 acres. Thus, this ITK system helps control flood and drought and acts as a protecting mechanism for the villages (https://paramparaproject.org/traditions_ahar_pyane.html). *Johad*, also known as a *pokhar* or a percolation pond, is a small earthen check dam that captures and conserves rainwater. It is a community-owned traditional harvested rainwater storage wetland primarily used for storage of water resources for future use. The *Johad* water harvesting system is widely used in states like Haryana, Rajasthan, Punjab and western Uttar Pradesh of North India. The collected/stored water is available year-round. *Johads* help recharge groundwater in the nearby wells (http://www.navrangindia.in/2020/04/johads-age-old-simple-traditional-water.html). The *Paar* system of

TABLE 7.4
Indigenous Rainwater Harvesting and Management Systems of India

ITK on Water Management	Features	Agro-Ecological Region
Ahar and Pyne	*Ahar* is a reservoir for collection of flood/rain water and *Pyne* are channels connected to *Ahar* for impounding harvested rainwater.	Flood and drought affected region of southern Bihar
Johad	Small earthen check dam to capture and conserve rainwater	Water scarce region of North India
Khadin	Very long (100–300 m) earthen embankment at lower hill slopes for rainwater harvesting	Arid desert region (Jaisalmer, Rajasthan)
Paar	Percolation water from catchment area are conserved by a serios of dug out storage structures	Western Rajasthan
Pat	Water from swift-flowing hill streams is diverted to crop field by irrigation channels by the use of stone and mud diversion bunds and aqueducts.	Jhabua district of Madhya Pradesh
Pokhari/Pukhri/Gadia	Dug out farm ponds for water harvesting, storage and multiple water use for crop-fish and domestic purpose.	Assam, Odisha and West Bengal
Naada/Bandha	A stone check dam constructed across a stream or gully to capture mosoon runoff on a stretch of land. Silt deposit makes the land fertile.	Mewar region

water harvesting constructed through traditional masonary technology is a common place where the rainwater flows from the *agar* (catchment) and in the process percolates into the sandy soil. To access the *rajani pani* (percolated water) *kuis* or *beris* are dug in the *agor* (storage area). *Kuis* or *beris* are normally 5 metres (m) to 12 m deep. About six to ten *kuis* are constructed in a *paar*, depending on the size of the *paar*. The *pat* system of irrigation developed from Jhabua district of Madhya Pradesh. This system was devised according to the peculiarities of the terrain to divert water from swift-flowing hill streams into irrigation channels called pats. Diversion bunds are made across the stream by piling up stones and then lining them with teak leaves and mud to make them leakproof. The pat channel passes through small nullahs by the construction of a stone and mud wall aqueduct that joins the stream off and on, and also sheer cliffs before reaching the fields. This system of water management helps the farmers for early winter crop establishment and getting good crop yields. *Naada/bandha* is a stone check dam for water harvesting and storage that is found in the Mewar region of the Thar desert. It is constructed across a stream or gully to capture monsoon runoff on a stretch of land. Submerged in water, the land becomes fertile as silt deposits on it and the soil retains substantial amounts of water.

There are several Indigenous ways of irrigation water application such as *Rehat or Rahat* (Persian wheel) system, in which water is lifted manually using a pulley, rope and bucket by using animal power. Water is lifted by a series of small buckets fixed on a circular belt or chain moving on a vertical wheel or drum. Bullocks or camels walking in a circle turn the drum through a set of bevel gears and shafts, which moves Rehat. Revolving movement of the belt rises the bucket full of water on one side and empties it when it reaches the top. *Tenda* is an Indigenous method of irrigation water application in Odisha. This system consists of a bamboo platform made with four stout bamboos at one corner of the farm pond to lift the water by a bucket connected with a long bamboo tied vertically with a horizontal bamboo/wooden lever with balancing weight made by stones/bricks and mud at the other end. By pulling down the vertical bamboo tied with bucket towards the farm pond, the weight of the lever goes up. After filling the bucket with water and gently releasing the vertical bamboo, the bucket comes up as the balancing weight comes down. Then the bucket filled with water is emptied to a half cylinder shape hollow channel made from palm tree trunk. From this channel the water flows through the field channels to the cropped area. *Sena* is another Indigenous method of lifting irrigation water to the crop field. It consists of a basket made of split bamboo and operated by two persons by holding the ropes attached to either side of the sena. By the swinging it backward, it collects the water and forward movement delivers the water.

7.13.6 ITK for Management of Animal Diseases and Fishery Production

Foot and mouth disease (FMD) is a predominant disease among cattle with symptoms of fever, blisters inside the mouth, leading to excessive secretion of stringy or foamy saliva and drooling, and blisters on the feet that may rupture and cause lameness. Farmers in the coastal regions of India treat this disease by feeding bamboo leaves mixing with semi-ripened fruit of sapota or fig or coriander. Feeding a mixture of leaves of tamarind and fruit of black pepper are also practiced. Cattle fever is also treated by feeding rhizome of local ginger (*Aamada*) with seeds of black pepper. Mites in cattle are controlled by mixing, grinding and making a paste of leaf dust of tobacco with lime and applied externally over the skin. Arthritis in cattle restricts the movement and reduces the efficiency. Bark of arjun (*Terminalia arjuna*) tree mixed with ginger and lentil to make a paste and applied on the affected location. Oral feeding of mixture of bark of date palm, arjun tree and honey is also employed to treat arthritis. Diarrhoea and dysentery in cattle is controlled by feeding yellow dodder (*Cuscuta campestris*), ginger and black pepper. For easy delivery of calves, farmers feed the animal mixture of *Cynodon dactylon*, root of *Amaranthus* sp, rice washed water and molasses.

The water quality in pond water is improved with application of tamarind/acacia branches/leaves and also pseudo stem of the banana. Feeding of fish with broken rice mixed with cow dung by making ball is a traditional practice. These are kept nearer to the water surface in the four corners of the pond. Wave action slowly wash away portion of the feed and it get mixed with water and fish feed on these. To control diseases of fish, farmers catch the fish and put on saltwater containers for 10–15 minutes and then release. Farmers apply garlic paste to control disease, by the antimicrobial action of allicin, present in garlic. Farmers also apply a mixture of garlic paste and common salt in the pond to control fish disease as it has a disinfecting ability. Neem (*Azadirachta indica*) leaf paste and lime in fish pond is used to protect fish against external injuries causing reddening and lesions. About 1 kg of garlic (*Allium satium*) is ground to make a paste, then it is thoroughly mixed with about 20 litres of fresh water. This solution is then sprayed over the infected pond. During the period of disease outbreak in aquatic species, as a prophylactic measure, neem leaves are ground to make a paste. Turmeric is added (250 g) in to the paste and mixed properly. The paste is then mixed with 25–30 litres of water and sprayed in the cultured fish pond evenly.

7.13.7 ITKs in Aquaculture

ITKs are treasure troves of ancient wisdom, beliefs and traditional knowledge passed on from generation to generation for preservation, effective utilisation and conservation of natural resources, soil, plant, animal and other organisms. Indigenous knowledge is effective, locally available, relatively cheap and less destructive to local environments. Most rural/tribal people survive on a local knowledge base (Kalita et al. 2010). Low-input–based small-scale aquaculture (sometime referred to as 'rural aquaculture') is poised to play a significant role in increasing freshwater fish production in the country. This low-input aquaculture system is closely associated with ITKs (Yadava and Choudhury 1986; Bhagabati and Kalita 1987; De and Saha 2001; Kalita et al. 2004; Saha and Nath 2013). Some of the popular ITKs widely used in aquaculture in India are as follows.

- *Garlic and fenugreek are used as fish attractants*

Garlic and fenugreek seed are employed by certain farmers as fish attractants to facilitate the process of harvesting. A mixture is created by combining 1 kg of semi-dried garlic with 250 g of roasted and ground fenugreek seed. After being mixed, both compounds are applied in a pond. After one hour, a group of fish is drawn to it due to the emission of perfume from this compound, making it easier for the fisherman to catch the fish. This mixture is sufficient to treat a water body covering an area of one hectare. This combination is also employed as bait or lure for angling.

- *Application of cattle urine*

In certain regions of India, a practice has been observed where fish farmers utilise cow urine to combat algal bloom by applying it to the surface of fish ponds. However, it is worth noting that this particular method has been prohibited in other nations.

- *Dispersing ash in a fish pond*

In this technology, paddy straw or hay is incinerated and the resulting ash is dispersed around the fish pond. Farmers perceive it as a preventive method to manage the fish disease.

- *Asafoetida (Hing) are used for attracting Magur and Singhi*

The majority of fish farmers in India engage in the cultivation of Magur and Singhi due to their high market demand. However, harvesting these species poses significant challenges as they tend to burrow deep into the mud. To address this issue, one can place twigs, leaves, or branches from a tree in the pond prior to harvesting. Next, the hing is wrapped in a towel and placed near the leaf or twigs to attract the fish with its perfume. After a duration of 5 to 10 minutes, the farmer proceeds to enclose the twigs with a net and gather the fish.

- *Utilization of gunny bag and bamboo pole for the management of Argulus parasite*

Certain farmers employ the practice of placing aged gunny bags at the bottom of their ponds. This is done with the intention of attracting and capturing the larvae of Argulus, a type of parasite. The collected larvae are then dried and exterminated, effectively eliminating any eggs deposited by the Argulus. Alternatively, some farmers opt to install bamboo poles within their ponds. The purpose of this is to provide a surface for the fish to rub against, causing the parasites to detach from the body surface of fish.

- *Utilizing turmeric as an antimicrobial agent*

Turmeric possesses antibacterial and anticancer effects, with curcumin being the main component responsible for its diverse medicinal benefits. The fish farmers disperse the paste around the water in order to address the Aeromonas hydrophilla infection. It is occasionally employed as an exogenous insecticidal and antifungal agent.

- *Applications of Mahua as a piscicide*

Mahua (*Madhuca longifolia*) is a member of the Sapotaceae family, renowned for its fragrant blossoms and significant cultural significance among Indigenous communities. It is widely used by tribal people in the production of both fermented and non-fermented food items. In the fisheries sector, farmers utilize it to regulate the growth of weeds and reduce predatory fish populations at concentrations of 200–250 parts per million (ppm). The Mahua plant contains the active compounds saponin and mourin, which are hematotoxins that have the ability to kill red blood cells. These compounds are responsible for causing the death of fish.

- *Utilization of segments of banana trunk*

Rural fish farmers utilize sections of the banana trunk, which are placed on the water's surface to preserve alkalinity and act as a supplementary source of food. The presence of oxygen-containing pores in the banana trunk ultimately contributes to an increase in oxygen levels. Nevertheless, it is extracted from the pond before reaching a state of decomposition.

- *Utilization of jackfruit rachis for the purpose of controlling tadpoles*

The presence of an excessive number of tadpoles poses a significant issue, mainly because they compete with fish for sustenance and also create complications during the process of harvesting and sorting and grading after the harvest. The Jackfruit rachis is utilized to control this issue. Because of the adhesive properties of the Jackfruit rachis, the tadpole becomes attached to it, and later the tadpoles are detached from the pond.

- *Use of matured palm fruit for crab management*

Rural fish farmers employ ripe palm fruit as a means to manage or eliminate crab populations in fish ponds. The aroma of the mature fruit lures the crab, causing them to gather around the fruit for feeding. Farmers may effortlessly remove the crabs by scooping them up. Occasionally, vacant clay containers with mature palm fruit are also employed for the extraction of crabs. After a period of 3–4 days, the pots are extracted from the pond, which has been populated with crabs.

- *Utilization of paddy straw to mitigate turbidity*

Turbidity, which refers to the presence of suspended particles in water, is a frequent issue that occurs during periods of rainfall. To manage the ongoing issue of clay turbidity, farmers in Assam utilize rice straw or sections of banana stem. According to Kalita et al. (2004), farmers regularly extract the rotted hay from the pond after a period of 4–5 days.

- *Akashi fruit as a tranquilizer for fish.*

The Indigenous inhabitants of Jharkhand utilize the fruit of the Akashi plant as a piscine sedative. This fruit has a partially granulated texture and is combined with water at a ratio of 1 kg per 100 litres. It is then evenly distributed over the surface of the water. After a duration of 30 minutes, fish begin to rise to the surface of the water and they are collected either by using a net or by manually selecting them.

- *Regulate the pH level of water*

Farmers employ the branches of tamarind and moringa trees to lower the pH level of water. The branches are submerged in the pond for a duration of 3–4 days. Shrimp producers utilize toddy, which is palm sap, to regulate the pH levels (Savji et al. 2011). In their study, Goswami et al. (2006) found that the pH of pond water may be raised by submerging chopped sections of banana pseudo stem, which secrete alkaline substances. West Bengal farmers have adopted this approach to reduce the occurrence of protozoan illnesses and Argulus.

- *Flock of piscivorous birds is an indicator of the presence of fish shoals.*

People obtain information regarding the existence of fish shoals or aggregations when a significant number of fish-eating birds fly over bodies of water such as rivers and reservoirs. This facilitates the local population in efficiently seeking and capturing fish from natural aquatic environments.

- *Coconut spikes are used as a fish aggregating device*

In Kerala, fishermen employ a distinct fishing technique known as *kolachal*, which serves as an artificial fish aggregating device (FAD). This method creates breeding and hiding spots for fish. *Kolachal* functions as an artificial substrate that supports the abundant growth of periphyton, a very nutritious natural food source for fish. Consequently, it acts as a feeding ground for fish.

- *Tannin for preservation of fishing nets*

Traditional farmers utilize natural fibres such as cotton, sisal, hemp, manila and coir in the production of fishing nets. Historically, tannin has been employed for the conservation of fishing nets. To prepare tannin, the bark is broken into small pieces, thoroughly crushed and heated with water in metal containers until a concentrated solution is achieved. The residual portion is compressed, and the liquid is then gathered and filtered. The net is submerged in the solution for a period of one to two days and then dried. To inhibit the leaching of tannins, immerse the material in a 1% solution of copper sulfate and ammonia.

- *Utilizing a petromax and sickle for fishing*

In northern Kerala, farmers employ petromax lamps with sickles to catch snakehead fish (*Channa* spp.) for fishing purposes. This technique is employed during the rainy season in rice fields, as well as in connected canals and trenches. This procedure necessitates the presence of 3–4 individuals. A single individual grasps the petromax and proceeds in a forward direction. Another adherent of the light carrier possesses a bag crafted from palm leaves and a hand net. A different individual wields a keen-edged sickle. Fish are drawn to and rendered disoriented by the powerful illumination of the petromax. The fisherman effortlessly captures fish by using a sickle, hand net or sack.

- *Singhi (Heteropneustes fossilis) for treating anemia*

Coastal farmer/people use 'Singhi *(Heteropneustes fossilis)*' as an excellent food for pregnant women and children to get rid of anemia. High iron content of this fish helps in synthesis of hemoglobin for treating the anemia disease. The Asian stinging catfish, *Heteropneustes fossilis*, is a species of air-breathing catfish found in India, Bangladesh, Pakistan, Nepal, Sri Lanka, Thailand and Myanmar.

- *Transporting fish seed using a mixture of rice beer and silt.*

Fish seed traders in various regions of India and Bangladesh commonly transport fish seed using metal hundies or containers. They incorporate 50 ml of Indigenous rice beer for every 10 litres of water in these containers. According to the vendor, the rice beer functions as an anesthetic during transit, leading to improved survival rates for the fry. Saha and Chowdhary (1956) have documented the conventional technique of transporting young fish and small fish in hundis, a historic practice in Bengal. The hundis are filled with water from the same source as the fry and are supplied at a quantity of 50,000 in the smaller hundis and 75,000 in the larger ones. Approximately 50–60 g of finely ground sediment is evenly distributed over the surface of the water in the hundi. While being transported, the hundis are sometimes rattled. Basu (1951) found that the inclusion of silt causes the suspended organic contaminants to coagulate. Management measures such as silt addition, sediment removal using a rough cloth rope and partial water exchange allowed for transportation for up to 30 hours.

- *Transporting prawns by the use of plastic tubes*

Reports have shown that in Tamil Nadu, there is a practice of using plastic tubes to cover the rostrum of brood stock of freshwater prawn. This is done to prevent any potential harm to the packaging material during transportation.

- *Collection of seeds*

According to Kalita et al. (2004), it has been observed that during the monsoon season in Assam, there is a tradition of collecting the offspring of murrells, specifically *Channa punctatus*. These fish migrate in groups and periodically come to the surface of natural waters to breathe. This collection is done using bamboo sieves, mosquito nets, or cloths.

- *Avian deterrent techniques*

The utilization of audio/videotapes and coloured plastics for bird deterrence has been seen as a straightforward technique in numerous fish farms located in Karnataka, Andhra Pradesh and Odisha.

- *Evaluating the viability and vigour of the seed*

In order to assess the quality of seeds, farmers place a small number of them in a plate filled with water and agitate the water with their finger. The seeds that exhibit counter-current swimming are regarded as healthy seeds. Individuals that choose to settle or gather at the centre are determined to be unhealthy individuals.

- *Control of leaches*

It has been stated that farmers in Assam make use of cucumber peels or bitter gourd leaves by converting them into a paste. These bitter plant elements are thought to aid in the elimination of leaches (Kalita et al. 2004).

- *Management of Epizootic ulcerative syndrome (EUS) disease*

Some farmers are utilizing neem plant branches in fish ponds to manage ulceration on the bodies of the fish.

- *Conducting a site appropriateness analysis to determine the feasibility of excavating a pond and maintaining the dike.*

To assess the soil's water retention capacity, farmers create a mud ball using the soil from the intended pond construction site. If the ball remains intact, it indicates that the soil possesses sufficient water retention capacity, making it acceptable for the construction of a pond. In regions with significant precipitation, the majority of dykes are susceptible to erosion and harm. Agriculturalists cultivate plantation crops such as coconut, papaya, and vegetable crops like cucumber, gherkin, etc., or cultivate turf with fodder grass. In Manipur, farmers cultivate *Colocasia* on pond dykes at a density of 3000 plants per hectare. *Colocasia* is a vegetable widely consumed in several regions of India (Singh et al. 2011).

- *Breeding fish*

To harden the eggshell, West Bengal farmers use the appropriate amounts of water-immersed catechu (*Acacia catechu*) and myrobalan (*Myrobolus indica*) nut extract in the hatching pool (Goswami et al. 2006). Higher hatching rates are known to be made possible by this, which is known to aid in shell hardening and avoid immature hatchling release.

- *Using banana peels and leaves in the breeding of ornamental fish*

Carrot or potato slices and banana peels are placed in ornamental fish larval rearing tanks to improve the production of live feed Infusorians. Some fish breeders employ banana leaves for egg deposition and teak leaves to maintain an acidic pH in the fish tanks during breeding egg layers. Fish feeding using gunny bags and salt supplement

Farmers in Andhra Pradesh, India, have devised a basic feeding technique known as 'the bag feeding technique,' in which the feed is stored in feed bags with tiny holes that can be organized in two or three rows. Major carps in India have a habit of browsing and sucking feed via holes. Due to the ability of NaCl to promote growth, these farmers also employ common salt in their feed (Nandeesha et al. 2000; Keshavanath et al. 2003; Gangadhara et al. 2004).

7.14 TYPES OF MANURES AND FERTILIZERS

Application of balanced manures and fertilizers is essential to ensure sustainable yield as well as soil health under integrated farming system (Figure 7.8). Under ideal conditions, the major sources of nutrients for IFS should be from organic sources and minor amounts may be met from chemical fertilizer sources (Figure 7.9). Crops remove nutrients from soil and these nutrients should be replenished to the soil. Nitrogen, phosphorus and potassium are taken up by plants in large quantities, these are called 'major' or 'primary nutrients.' Calcium, magnesium and sulphur are required in relatively smaller but in appreciable quantities, these are called 'secondary nutrients.' Micronutrients (iron, manganese, zinc, copper, chlorine, molybdenum, boron) are required by plants in small quantities for their growth and development. Apart from these carbon, hydrogen and oxygen are essential plant nutrients obtained by plants from air and water.

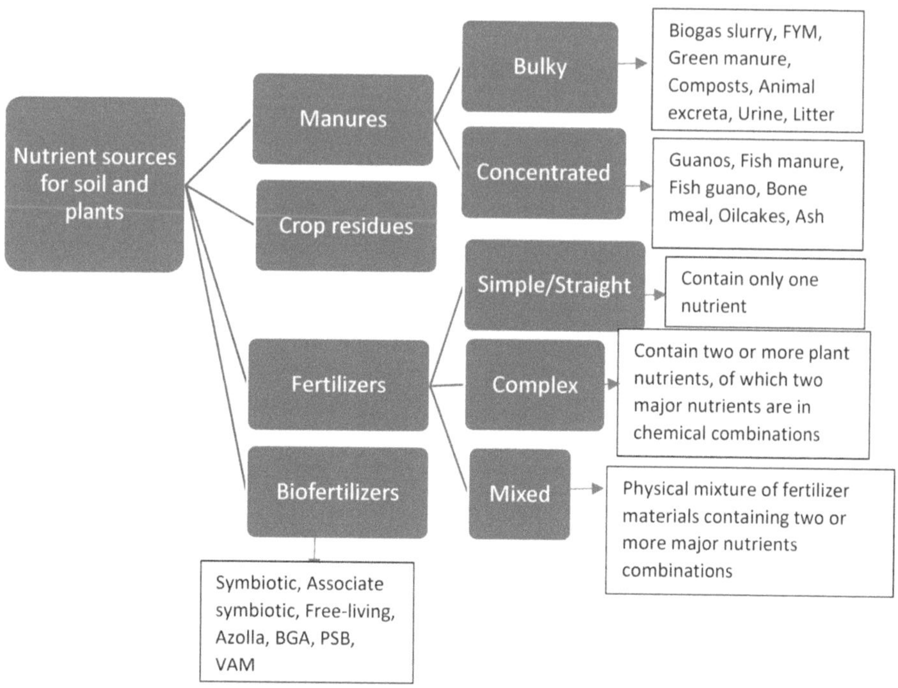

FIGURE 7.8 Types of fertilizers and manures used for an integrated farming system.

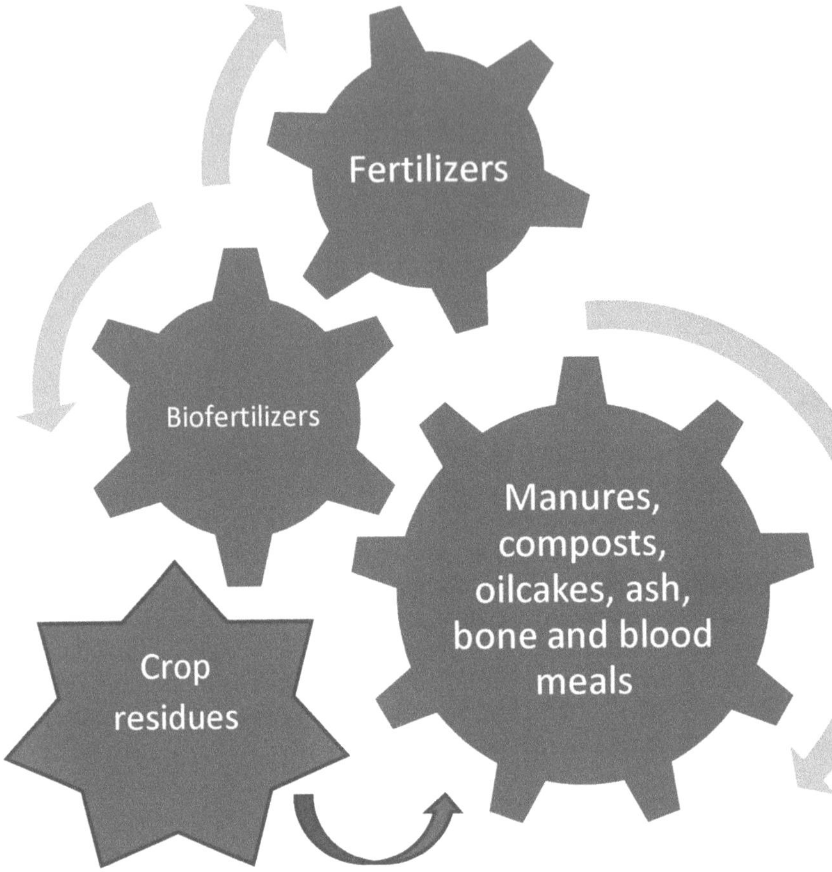

FIGURE 7.9 Major sources of nutrients for soil and plants for integrated nutrient management.

7.14.1 Fertilizers and Calculation of Dose

Crop plants uptake 13 essential nutrients (N, P, K, Ca, Mg, S, Fe, Mn, Zn, Cu, Cl, Mo and B) from soil and three (C, H and O) from air/water. Intensive cropping results depletion of these nutrients from soil. Therefore, to achieve higher crop productivity and soil health on a sustainable basis, these nutrients are to be supplemented through manures and fertilizers. Manures and composts are organic in origin, whereas fertilizers are either manufactured or mined. Chemical fertilizers are inorganic or synthetic organic materials in concentrated form. Straight fertilizers contain only one primary nutrient, whereas compound fertilizers contain at least two of the nutrients nitrogen, phosphorus and potassium, obtained chemically or by blending or both. When compound fertilizers contain two or more plant nutrients of which two major nutrients are in chemical combination, these are called complex fertilizers. Mixed fertilizers are physical/granulated mixtures of fertilizer materials containing two or more major plant nutrients. In this preparation, no chemical reaction should take place and they are prepared by physically mixing the ingredients thoroughly either manually or mechanically. The problem of segregation by physical mixing can be overcome by the process of granulation. Granular fertilizers are composed of particles of roughly the same composition and about one-tenth of an inch in diameter. This kind of fertilizer is superior in efficiency compared to the fine or powdery fertilizer, due to ease in handling and less fixation in soil. However, granulated mixed fertilizers are costlier compared to physical mixtures due to the additional cost involved in the process of granulation.

Fertilizer grade refers to the guaranteed minimum percentage of nitrogen (N), phosphate (P_2O_5) and potash (K_2O) contained in the compound fertilizer. The fertilizer grade of a fertilizer containing 10 kg N, 26 kg and 26 kg K_2O in 100 kg product is expressed as 10-26-26. The numbers representing the grade are separated by hyphens and are always stated in the order of N, P_2O_5 and K_2O. Fertilizer ratio refers to the ratios of the percentage of N, P_2O_5 and K_2O in the fertilizer. It designates the relative proportion of three major nutrients, keeping the percentage of N as 1 in the ratio. Thus, a 5-10-5 fertilizer mixture which contains 5% N, 10% P_2O_5 and 5% K_2O has a fertiliser ratio of 1:2:1. Conditioners are materials added during preparation of fertilizers for reducing hygroscopicity, to prevent caking and to keep it free-flowing. Since these materials are low grade organic materials, they therefore improve the physical condition of the fertilizers. Fillers (sand, soil, coal powder, dolomite, china clay) are added to make up the weight, to bring the percentage of nutrients to desired values and to produce a mixture of the desired grade.

For preparing a mixed fertilizer, ingredients are selected such as N source (ammonium sulphate, ammonium chloride, calcium ammonium nitrate and urea), P source (single super phosphate, bone meal and rock phosphate) and K source (muriate or sulphate of potash). Care should be taken to select fertilizers which do not hamper the quality of the crop. Potassium chloride and ammonium chloride fertilizers should not be selected for tobacco and potato crops. For acidic soils, rock phosphate may be preferred. Compatibility of mixing is one of the most important factors for preparing a mixed fertilizer (Table 7.5).

Fertilizer recommendations or dose for crops are generally given in elemental or oxidized form of nutrients. This is given to choose the fertilizer material as per the availability of fertilizer or manures, type of soil and crop. The dose is given in the form of N:P_2O_5:K_2O for major nutrients and if other nutrients are to be applied, this is added in the sequence. For example, the fertilizer dose of rice during *Kharif* season in coastal areas is given as 80:60:40 kg N:P_2O_5:K_2O ha^{-1}. Therefore, these are to be converted into the required fertilizer amount to be applied. For this, the fertilizer grade and the area of crop for which the fertilizer to be applied should be known.

The fertilizer amount is converted from nutrient content of the fertilizer by multiplying the conversion factor, which is found as

$$\text{Conversion factor to fertilizer} = 100/\text{nutrient content in the fertilizer (\%)}$$

TABLE 7.5

Compatibility of Fertiliser Materials Important for Preparing Mixed Fertilizers

Fertilisers	KCl	K_2SO_4	SPM	AS	CAN	Urea	SSP/TSP	AP	BS	RP	Lime
KCl											
K_2SO_4											
SPM											
AS											
CAN											
Urea											
SSP/TSP											
AP											
BS											
RP											
Lime											

Note: Green colour indicate fertilizers which can be mixed; Orange colour indicate fertilizers which may only be mixed shortly before use and red colour indicate fertilizers which cannot be mixed; KCl: Potassium chloride; K_2SO_4: Potassium sulphate; SPM: Sulphate of potash magnesia (contained in potassium schoenite); AS: Ammonium sulphate; CAN: Calcium ammonium nitrate; SSP: Single super phosphate; TSP: Triple super phosphate; AP: Ammonium phosphate; BS: Basic slag; RP: Rock phosphate.

For converting from N to urea, the factor is 100/46=2.17. Similarly, the factor to convert P_2O_5 to single super phosphate (SSP) is 100/16=6.25. The factor to convert K_2O to muriate of potash (MOP) is 100/60=1.67. The factors to convert from N and P_2O_5 to Diammonium phosphate (DAP) are 5.55 and 2.17 respectively, as DAP contains 18% N and 46% P_2O_5. For straight fertilisers, the fertilisers are calculated individually. For example, to calculate how much urea, SSP and MOP will be required for a crop with recommended dose of fertilizer (RDF) of 120:60:40 kg $N:P_2O_5:K_2O$ ha^{-1}, the amounts will be 120×2.17, 60×6.25 and 40×1.67 i.e. 260.4, 375 and 66.8 kg of urea, SSP and MOP required per ha area. However, for compound fertilisers, first, the amount of compound fertilizer required to supply the highest nutrient content in the compound fertilizer is to be calculated. Then, the amount found out is to be checked for another nutrient it supplies. The rest of the nutrient is provided by straight fertilizer. To find out the amount of DAP, urea and MOP required for a crop with RDF of 120:60:40 kg $N:P_2O_5:K_2O$ ha^{-1}, first the amount of DAP required to supply 60 kg P_2O_5 is to be found out, as, DAP contains, higher P_2O_5 (46%) compared to N (18%). Therefore, the DAP requirement will be 60×2.17 = 130.2 kg ha^{-1}. Now, this amount is to be checked for how much of N it supplies. As 100 kg DAP provides 18 kg N, 130.2 kg DAP provides [(18/100) ×130.2]=23.44≈23 kg N. Therefore, the rest amount (120 kg – 23 kg=97 kg) of N is to be fulfilled from urea. Urea required will be 97×2.17=210.49 kg ha^{-1}. MOP requirement will be 40×1.67=66.8 kg ha^{-1}. After calculation of fertilizers per hectare basis, it is then converted into the actual crop area basis by multiplying the area conversion factor. Under integrated nutrient management (INM), manures and biofertilizers are to be given priority, apart from chemical fertilizers. Therefore, the amounts of essential plant nutrients provided from these sources are to be determined and the rest amounts are applied from chemical sources.

7.14.2 Application Methods of Fertilizers

Fertilizers should be applied in such a way that it involves least cost, less loss and nutrients are easily available to the crop plants. The suitability of application method is determined by having a thorough knowledge of the nature of the soil, crop and fertilizer. Soil with higher pH (>8) increases the volatilization losses of ammoniacal fertilisers. Band placement of P fertilizers in soils with low available P, reduces fixation. In soils with medium available P, incorporation is better for higher availability to crops. To utilize the residual soil moisture, deep placement in the moist soil zone is essential. Broadcasting method of application is suitable when the crop has a fibrous root system and is planted in close spacing. The nature of fertilizer also determines the method of application. Placement method is adopted for mud-ball urea, urea pellets and briquetts. Drilling is the method for granules or prills. Liquid fertilizer may be applied with irrigation water as fertigation. Soluble fertilizers are applied as foliar spray or as root dip. The roots of seedlings are dipped in nutrient solutions and then transplanted. Hill placement refers to applying fertilizer either in bands or near the plants. This method is employed when relatively small quantities are to be applied. N-fertilizer is applied in pinches at the base of each plant of cotton, maize, cauliflower, cabbage, brinjal, tomato and other vegetables by this method. Band placement of fertilizers refer to application in narrow bands on one side or both the sides of the row about 5 cm below the seed and 5 cm to one side of the seed or plant grown or around each plant. In low fertile soils, band placement helps an initial good start of the crop. This method reduces fixation and volatilization losses. Point placement refers to the application in a hole or in a depression followed by covering the fertilizer with soil. This method is suitable for top dressing of fertilizer in widely spaced crops, e.g., sugarcane, tobacco and fruit crops. Sub-soil placement of fertilizers is done with the help of heavy power machinery when the subsoils are acidic in nature in humid and sub-humid regions.

7.14.3 Increasing Use Efficiency of Applied Nutrients

- *Adopting proper method of fertilizer application*: In a particular crop different application method may be followed as per the stage of the crop. During the initial stage, when the root as well as foliage growth is very limited, targeted soil application may be followed. When the crop canopy is developed foliar application will increase the nutrient absorption by the crops. Foliar spray of water-soluble fertilizers (19-19-19 and 13-0-45) increased the protein (2%), fat and fibre content of potato tubers (Sarangi et al. 2021).
- *Proper time of application*: Matching the time of application of fertilizers with the crop demand, increases the nutrient absorption by roots and reduces the excess availability and leaching losses. During the vegetative growth stage of crop, most of the nitrogen is absorbed and for cereals top dressing of rest nitrogen is done. Whereas for pulses, nitrogen is applied only as a starter dose as basal and it should not be applied after root nodulation. In very short duration crops such as leafy vegetables or seedlings in the nursery, the entire dose of N may be applied as basal dressing to the soil. Phosphorous promotes root development and thus gives a good start for vigorous growth of plants. Unlike nitrogen, phosphorus is relatively immobile in the soil; therefore, the entire recommended quantity is applied as basal. Potassium is needed gradually by the crop throughout the growth and development stages. Generally, these fertilizers are also applied as basal doses. However, recent studies reveal that split doses of potassium fertilizers application achieve better yield and potassium use efficiency (Sharma and Singh 2021). Application of potassic fertilizers during the later period also increases the tolerance to insect pests and diseases.
- *Selection of proper fertilizer material*: Ammoniacal forms of N (Ammonium Sulphate and Ammonium Chloride) are more efficient for waterlogged paddy fields. Ammoniacal fertilizers should be applied in the reduced zone and nitrate fertilizers should be applied in the oxidized zone in wet land rice as top-dressing in splits. Ammonium sulphate is more effective in sodic and S deficient soils. Sulphur containing fertilizers such as single super phosphate, ammonium sulphate and ammonium phosphate sulphate are suitable for oilseed crops, particularly when grown in S deficient soils. Rock phosphate is suitable for long duration crops in acid soils. Zinc sulphate is more effective in sodic soils. Instead of potassium chloride (KCl), potassium sulphate is recommended for high value crops such as tobacco, potato and grapes in S deficient soils. Fertilization with KCl (MOP) reduces the starch content of the potatoes by about 2%, leading to a starch yield that was about 1 t ha^{-1} lower than that after K application in the sulphate form (Naumann et al. 2020).
- *Application as per nature of crop*: In fodder crops to get succulent green fodder, with higher protein content, application of N fertilizer in several splits is necessary. However, when the fodder crop is grown for silage purpose, to increase the carbohydrate content, N application should be stopped sufficiently before the harvest. Similarly in sugarcane, sugar beet, potato and tobacco crops, N fertilizer should not be applied during the maturity stage. Fertilizer application method and time also depend on the root growth pattern of the crop. In cotton and tobacco crops, the root growth during the initial stage is very limited; therefore, for better absorption, the placement of nutrients very near to the plant is essential. The root systems of maize and soybean are extensive and exploit the most soil volume, whereas in the case of carrots, cotton, potatoes and other shallow-rooted crops, the root volume is very limited. Potatoes have a sparse and weak root system, concentrated in the top soil (0–30 cm). In contrast, alfalfa and sweet clovers have a deep-tap root system and due to the root penetration of these crops there is loosening of compact subsoils. In irrigated crops, phosphates may be applied to legumes in rotation and N to cereals or other crops. In rainfed or dryland cropping P fertilizers should be applied to the *Kharif* crop while N and K to each crop of *Kharif* and *Rabi* if none of the crops are legumes.
- *Modification of fertilizer to reduce losses and sustain release*: Urea is the predominant form of N fertilizer; however, it is subjected to various losses. To reduce the losses, coating may be done with different materials such as neem, lac, rock phosphate, sulphur and gypsum to produce coated fertilizers. The coated urea contains about 30–34% N. Urea super granule (USG) or urea pellets of 1–4 g each are placed in the reduced zone of rice soils to reduce the losses. The urea fertilizer availability may also be extended by application of nitrification inhibitors

such as N Serve [2 chloro-6 (trichloromethyl) pyridine], AM (2-amino-4 chloro-6 methyl pyrimidine), ASU (Guanyl thio urea), Nitrapyrin, TU (Thiourea), DD (Dicyanadiamide), ST (Sulphathiazole), DCS (N-2,5-dichloro phenyl sucinamic acid) and ATC (4-amino-1, 2,4-triazole hydrochloride). Slow or controlled release form of urea also increase the use efficiency. The solubility of Isobutylidene diurea (IBDU) is 1g per litre. It is manufactured by the condensation reaction of urea and isobutyraldehyde. IBDU contains 31–32% N. Crotonylidene diurea (CDU) is the reaction product of urea with acetaldehyde with 30% N. Guanyl Urea Sulphate (GUS) is the sulphate salt of Guanyl urea produced from calcium cyanamide by acid treatment. It is watersoluble and absorbed by the soil colloids as a result, the leaching losses are reduced. Urea Formaldehyde (UF) or ureaform is produced by condensation of urea with formaldehyde, containing 38–42% N. UF is a coarse granular material and less hygroscopic than urea. Oxamide $(CONH_2)_2$ is the diamide of oxalic acid and is white crystalline solid. Its solubility is about 0.4 g per litre with 31.8% N. For micronutrients, chelated forms are recommended for better nutrient use efficiency. The chelated compounds are available in the market as 'sequestrenes' or 'versenols'. Iron and zinc fertilizers are generally available as chelated organic compounds of Diethyl-ethylene-triamino-penta acetic acid (DTPA), Ethylene-diamino-tetra-acetic acid (EDTA) and Cyclohexane-1-2-diamino-tetra acetic acid (CDTA).

- *Management of residual effect of manures and fertilizers*: Manures are not completely utilized by the crops during the year, these are applied. Similarly, when high doses of fertilizers are applied, there are residual availability during the next season. However, the residual effect of chemical fertilizers may be beneficial or harmful. Application of some fertilizers cause soil acidic or alkaline conditions. Whereas rock phosphate reduces the soil acidity, ammonium chloride and ammonium sulphate create soil acidity.
- *Maintaining optimum soil moisture for better absorption of applied nutrients*: The movement of applied nutrients towards plant roots by the process of mass flow and diffusion is facilitated by maintaining optimum soil moisture. Soil moisture stress (deficit or excess) negatively impacts the efficiency of applied nutrients. Under rainfed/dryland conditions, conservation of soil moisture improves nutrient use efficiency. In irrigated agriculture, it is advisable to apply irrigation water during the period of fertilizer application. However, excessive irrigation often leads to fertilizer leaching, necessitating additional fertilizer applications (Bayer et al. 2014).
- *Integrated nutrient management*: It is desirable to properly conserve and use all organic waste materials in the farm to build soil fertility and its water holding capacity, and for supplying the micronutrients. The organic residues are to be incorporated three to four weeks before sowing the crop. When using composts of high C:N ratio, it is advisable to add small quantity of N fertilizer to it, so that the N availability to the crop is not hampered due to immobilization by the microorganisms. The inclusion of legumes in the crop rotation and intercropping systems, use of bacterial and algal cultures, biofertilizers and green manures increases the use efficiency of applied nutrients in the cropping system.

7.14.4 Manures

The important components of manures are excreta of farm animals (dung) and birds (guano) along with crop residues. These are of two types, bulky and concentrated. Bulky organic manures contain relatively small quantities of plant nutrients and large quantities of organic matter. Bulky organic manures are applied in large quantities to supply plant nutrients, including micronutrients. These manures also improve soil physical properties such as soil structure, bulk density, water holding capacity, etc. The important bulky organic manures are farmyard manure (FYM), composts, vermicomposts and green manures. FYM contains 0.5–1.5% N, 0.4-0.8% P_2O_5 and 0.5–1.9% K_2O. The nutritional value of urban compost (1.0–2.0% N, 1.0% P_2O_5 and 1.5% K_2O) is better than rural compost (0.4–0.8% N, 0.3–0.6% P_2O_5 and 0.7–1.0% K_2O). Green manures on average contain contains 2.8% N, 0.5% P_2O_5 and 1.7% K_2O. Concentrated organic manures contain relatively higher quantities of plant nutrients compared to bulky organic manures. These are guano, fish manure, fish guano, oil cakes, bonemeal, bloodmeal, horn and hoof meal, etc. These are also called organic nitrogen fertilizer. Bloodmeal contain 10–12% N, 1–2% P_2O_5 and 1% K_2O, whereas bone meal

TABLE 7.6
Nutrient Value of Some Bulky and Concentrated Organic Manures

	Nutrient Content (%)		
Organic Manures	**Nitrogen (N)**	**Phosphate (P_2O_5)**	**Potash (K_2O)**
Green manures	2.8	0.5	1.7
Compost (rural)	0.4–0.8	0.3–0.6	0.7–1.0
Compost (urban)	1.0–1.2	1.0	1.5
Farm yard manure	0.5–1.5	0.4–0.8	0.5–1.9
Castor cake	4.3–5.8	1.8–1.9	1.0–1.3
Cotton decorticated seed cake	6.4–6.5	2.8–2.9	2.1–2.2
Mahua cake	2.5–2.6	0.1–0.9	1.2–1.9
Karanj cake	3.9–4.0	0.9–1.0	1.2–1.4
Coconut cake	3.0–3.2	1.8–1.9	1.7–1.8
Neem cake	5.2–5.3	1.0–1.1	1.4–1.5
Mustard cake	5.1–5.2	1.8–1.9	1.2–1.3
Sesamum cake	6.2–6.3	2.0–2.1	1.2–1.3
Bloodmeal	10.0–12.0	1.0–2.0	1.0
Bonemeal (raw)	3.0–4.0	20.0–25.0	–
Bonemeal (steamed)	1.0–2.0	25.0–30.0	–
Fishmeal	4.0–10.0	3.0–9.0	0.3–1.5
Horn and hoof meal	13.0	–	–
Activated sludge	5.0–6.5	3.0–3.5	0.5–0.7
Night soil	1.2–1.3	0.8–1.0	0.4–0.5

contains 1–4% N and 20–30% P_2O_5. Neem cake contains about 5% N, 1% P_2O_5 and 1.5% K_2O. Groundnut cake contains about 7% N, 1.5% P_2O_5 and 1.3% K_2O. Sesamum cake contains about 6% N, 2% P_2O_5 and 1.3% K_2O (Table 7.6).

Apart from organic manures, ash obtained from domestic fuel use and factories is a rich source of potash for crops, particularly when hard woods are burnt. It is applied to crops such as chili, onion, garlic, tobacco, banana crops which require large amounts of potash. It is also used in nursery beds to prepare a loose soil bed, so that seedlings can be easily uprooted without damaging the roots.

7.14.5 Biofertilizers

These are important component of INM and increase soil and crop productivity with less cost. These are the preparations containing live or latent cells of efficient strains of nitrogen fixing, phosphate solubilizing and cellulolytic micro-organisms (bacteria, fungi and algae) which help crop plants' uptake of nutrients by their interactions in the rhizosphere when applied to seed or soil. Symbiotic nitrogen fixation is the biological N fixation by the enzymatic reduction of dinitrogen (N_2) from the atmosphere to ammonia. The ability to fix N is only found in a variety of Prokaryotic organisms due to the presence of an enzyme system, nitrogenase, which provides the biochemical machinery for N fixation. Symbiotic N fixation are the soil bacteria of the genera *Rhizobium, Bradyrhizobium* and *Azorhizobium*, collectively referred as *Rhizobia. Rhizobia* have the ability to induce the formation of nodules on the roots of leguminous plants. Some free-living bacteria such as *Klebsiella, Azotobacter, Azospirillum, Rhodopseudomonas, Rhosdospirillum, Diplococcus, Micrococcus, Beierinckia, Drexia* and *Azomonas* also fix atmospheric N. *Azospirillum, Phosphobacterium* and *Acetobacter diazotrophicus* are recommended for cash crop like sugarcane for a nitrogen economy. *Azospirillum* is applied to transplanted paddy and vegetable crops as a root dipping treatment. The required quantity (10 kg ha^{-1}) of *Azospirillum* is mixed with about 10 litres of water at one corner of the field and all the roots of the seedlings are dipped in this for a minimum period of half an hour before transplanting.

7.14.5.1 *Rhizobium* Bacteria for N Fixation in Soil

Rhizobium bacteria form nodules on the roots of leguminous plants. In response to root exudates or flavonoids, the *Rhizobium* species produce lipo-Oligosaccharides, or Nod factors, that are responsible of eliciting meristems and nodule-like structures in roots. The rod-shaped bacteria present in root nodules as pleomorphic forms (bacteroids) which are normally involved in fixing atmospheric N into a combined form (ammonia) utilizable by the host plant. These bacteria are generally applied as seed treatment of *Rhizobium* culture for better root nodulation and atmospheric nitrogen fixation. Materials required for seed treatment are *Rhizobium* culture, sugarcane molasses/jaggery/gur, rice starch and water. For small seeded pulses like green gram and black gram, 300 g of *Rhizobium* culture, 300 ml of rice starch, 50 g of jaggery and 700 ml of water are required for 10 kg seeds. First the 300 g of *Rhizobium* culture is put in a bowl, to this 700 ml of water, 300 ml of rice starch and 50 g of jaggery are added. These components are thoroughly mixed to make a slurry. Then the 10 kg seeds are poured on to this slurry and mixed thoroughly so that a uniform coating of the *Rhizobium* culture is made all over the seeds. The coated seeds are kept under shade for about 30 minutes; avoid drying of the treated seeds with direct sunlight. It is advisable to do the seed treatment during evening hours and sowing in the field during morning hours when there is no scorching heat of the sun. For large seeded pulse crops like chickpea for 10 kg seeds 200 g of *Rhizobium* culture is required. First 100 g of jaggery, 500 ml of rice starch and 500 ml of water are mixed and boiled and then cooled. The *Rhizobium* culture is mixed well with this. Then the chickpea seeds are poured on to this and mixed thoroughly to make a uniform coating over the seeds. Then the treated seeds are dried under shade and sown during cool hours. The N fixation rate varies from crop to crop but generally in the range of 40–300 kg N ha^{-1} $year^{-1}$.

7.14.5.2 Phosphorus Solubilizing Microorganisms (PSM)

Phosphorus plays a key role in balanced plant nutrition for increasing crop productivity next to nitrogen. Though most soils contain insoluble inorganic phosphate, these are of no use to crops unless they are solubilized in the form of phosphorus pentoxide (P_2O_5). Phosphate exist in several forms, the most important for plant nutrition is mono calcium phosphate [$Ca(H_2PO_4)2H_2O$], which is water soluble and the dominant form in superphosphate. Dicalcium phosphate ($CaHPO_4$ $2H_2O$) is another form which is sparingly soluble in water and present in many phosphatic fertilizers. The other forms are in the form of complexes like tricalcium phosphate [$Ca_3(PO_4)_2$] in alkaline soils and hydroxyapatite [$Ca_{10}(PO_4)_6$ $(OH)_2$] in bones and rocks. In acidic soils it forms complexes with iron and aluminium. Most of the applied phosphatic fertilizers are fixed in the soil by iron, aluminium and calcium. It is also present in organic form such as phytins, phospholipids, nucleic acids, phosphorylated sugars and coenzymes after the decay of plant materials. The fixed soil phosphorus can be mobilized by the action of certain bacteria (*Pseudomonas, Bacillus, Micrococcus*) and fungi (*Penicillium, Aspergillus, Fusarium*). These are called phosphate solubilizing microorganisms (PSM). These organisms convert insoluble inorganic phosphate compounds into soluble form by the action of organic acids (malic, gluconic, glutaric, glyoxalic, succinic, fumaric, malonic, citric, tartaric, lactic, ketoglutaric, ketobutyric acids) produced by them. These efficient microorganisms have a consistency capability to solubilize the fixed phosphorus in the rock phosphates (Mussoorie, Udaipur, Matoon, Morocco, Jordan) with the action of their enzymes. PSM is applied as soil application @ 5 kg ha^{-1} mixed with 1,000 kg of cow dung/FYM and about 1 bag of rock phosphate. The mixture of PSM, FYM and rock phosphate are kept under shade or ceiling overnight to maintain 50% moisture. Soil application is done in rows or during the levelling operation of the field.

7.14.5.3 Potash Mobilizing Microorganisms

Potassium is one of the most significant major nutrients for crop production and is present in soil as readily, slowly and difficultly available forms. The difficultly available portion or fixed potassium is slightly soluble and often makes up to 90–98% of the total soil potassium. This is present in the crystalline structure of primary minerals, such as orthoclase feldspars and muscovite mica. The slowly available potassium constitutes about 2–10% of the total in mineral soils in the form of KCl and NaCl sylvinite and longbeinite is K_2SO_4. $2MgSO_4$. The mixtures are widely used for potassic fertilizers. There are bacteria such as *Frateuria aurantia* belonging to Pseudomonadaceae family, which solubilize the fixed potassium in soil. Application of these bacteria can save up to 50% of the cost of chemical potassic fertilizers. These bacteria can be used as a co-inoculation with *Rhizobium, Azospirillum, Azotobacter* and PSM. The bacteria are applied as soil application @ 2.5 kg ha^{-1} mixed with 500 kg of mature cow dung or FYM before sowing of the crops.

7.14.5.4 Sulphur Mobilizing Microorganisms

Sulphur is an important secondary plant nutrient, essential for synthesis of several amino acids including methionine, crystine and cysteine, which are important components of plant and animal proteins. The nitrogen to sulphur ratio in good quality protein should be around 15:1. Sulphur acts more like nitrogen than any of the other essential plant nutrient and its deficiency symptoms in plants closely resemble nitrogen deficiency symptoms. Leaching loss of sulphur form the soils are very much like nitrogen, particularly through percolating waters in light-textured sandy and salty soils. Therefore, for better plant nutrition, sulphur should be available in an optimum quantity for plants. Bacteria such as *Acetobacter pasturianus* secretes the sulphur and make it available to the plants in usable form. It is recommended for crops such as oilseeds, cotton, fruit crops, cabbage, onion, turnip and other vegetables @ 625 g ha^{-1} as soil application.

7.15 GROWING FODDER CROPS FOR INTEGRATION OF ANIMAL-BASED ENTERPRISES

Fodder production is one of the challenges to improve upon the productivity of livestock enterprise in an IFS. Among the various production factors, use of biofertilizers play a key role in N-economy of cereal fodders. Use of *Azotobactor* and *Azospirillum* is beneficial to increase the forage yield along with saving of 20–25 kg N ha^{-1}. Integration of cereal and legume fodders, not only give better nutrition to the animals but also improve soil health. The green biomass required for animals can be met from different sources (Table 7.7). Maintaining a combination of annual-perennial and cereal-legume combinations and staggered planting dates of fodder crops is beneficial for continuous fodder production and for maintaining the soil fertility. Hybrid Napier Bajra, being a perennial fodder crop, holds promise for year-round production in all seasons. Sorghum, being tolerant to moisture stress and moderately tolerant to soil salinity, performs better compared to other fodder crops. Para grass is a promising fodder crop for lowlands as it can tolerate waterlogging. For saline soils, Rhodes grass (*Chloris gayana*) and Karnal grass (*Leptochloa fusca*) are ideal options. Karnal grass inter-cropping with mesquite (*Prosopis juliflora*) in an agroforestry system in an extremely alkaline soil (pH 10.4, exchangeable sodium percentage 90) produced 47 t ha^{-1} green forage in 15 cuttings without any amendment in a growth period of about 50 months, proving its potential as a primary colonizer of abandoned alkali lands (Singh et al. 1991). In coconut gardens, guinea grass can be grown as an intercrop.

Some fodder crops are grown during both *Kharif* and *Rabi* seasons (e.g., maize), whereas some are grown only during *Rabi* (e.g., oat). Planting materials are also different for different fodder crops. Alfalfa, berseem, rice bean, oat and sorghum are grown from seeds, whereas, guinea grass, hybrid napier bajra (HNB), humidicola grass and para grass are propagated from stem cuttings with

TABLE 7.7
Green Biomass Sources for Animal Component in IFS

Terminology	Definition/Examples
Forage/pasturage	It embraces all vegetations consumed by grazing animals, including fruits and twigs of trees and shrubs.
Grass	Botanically, any plant of the family Poaceae is called grass. The term 'grass' in the context of feeding livestock and grassland agriculture is not limited to the narrow botanical sense alone; it refers to all true grasses, grass like plants and plants grazed by livestock.
Grassland	It refers to an ecosystem in which the dominant component is comprised of herbaceous species.
Rangeland	It refers to large, naturally vegetated mostly unfenced land of low rainfall area that are grazed by livestock. Rangeland vegetation includes shrub lands, grass lands, open forest where soil is dry, sandy, saline or alkaline or even wet soils with steep topography. The rocks preclude growing of commercial farm and timber crops. Grassland is component of rangeland.
Pasture	It includes improved, irrigated grazing lands that receive abundant rainfall, generally it is seeded and fertilized to domesticate forage crops and ordinarily fenced.
Permanent pasture	It consists chiefly of perennial grasses and legumes in humid and subhumid region, which are grazed year after year. The fields have high carrying capacity (the maximum number of organisms or animals or plants that can be supported in a given area or habitat over a period of year). Carrying capacity generally refers to the number of grazing livestock that can be maintained on a given grassland over a period of time.
Temporary pasture	It is also called annual pasture. It consists of annual forage plants that are grazed during their period of growing and maturity. These are savanna, steppe and prairie.
Savanna	It is a tract of level land covered with tall grasses and trees growing apart. The grasses grow under the scattered tree canopy. This is seen under hot and seasonally dry climatic conditions. Trees are sufficiently wide spaced to allow sunlight to fall on the ground for growth of grasses. Found near the equator.
Steppe	It is grassy, levelled and without tree species, develops farther away from the equator and with colder climatic conditions.
Prairie	It is dominated by taller grasses under high rainfall conditions during the months from April to September.
Cereal fodders	Maize, guinea grass, pearl millet (bajra), sorghum (jowar), napier grass (elephant grass), para grass, Dinanath grass, oat, Rhodes grass, marvel grass, sabi grass, signal grass, congo signal, dharaf grass, anjan (buffel) grass, yellow anjan, sudan grass, barley, teosinte, setaria grass
Legume fodders	Alfalfa (lucerne), berseem (Egyptian clover), Sweet clover (Senji), cowpea (lobia), guar, rice bean, horse gram, stylo, shaftal, butterfly pea, subabul
Non-legume and non-cereal fodders	Sunflower, mustard, turnip, carrot, beet, amaranthus

two buds called rooted slips (Photo). HNB is a perennial grass. The detailed package of practices of some of the fodder crops are given in Table 7.8.

7.15.1 Fodder Maize (*Zea mays*)

Maize being a C4 plant requires abundant moisture, fertilization and temperature, it produces a huge biomass (Figure 7.10a) under good management practices, hence it is called 'the queen of cereals.' It can be grown all year. The seeds should be sown at a minimum depth of 5 cm to protect from birds, parrots and also for better stand establishment and preventing lodging of plants. To protect the

TABLE 7.8
Package of Practices of Different Fodder Crops

Crop	Varieties	Spacing (cm × cm)	Seed Rate (kg ha−1)	FYM (t ha−1)	Fertilizer (N-P-K kg ha−1)	Yield (tha−1)
Maize	African tall, Vijay composite, Moti composite, Jawahar	30 × 10	40–50	5–10	80-40-40	35
Sorghum	CoFS-29, SSG-59-3, Raseela MFSH-4, Sudan	30 × 10	30–40	5–10	60-30-30 (apply 30 kg N ha^{-1} after each cut)	70
Bajra	Giant Bajra	30 (Solid row)	7–10	5–10	40-20-20	25
Hybrid Napier Bajra	CO-3, CO-4, CO-5, Smart Napier, Super Napier, Australian Red Napier	50 × 50	3000 kg setts or 40,000 rooted slips ha^{-1}	10	50-50-50 (apply 40 kg N ha^{-1} after each cut)	20 per cut (6-7 cuts)
Oat	Kent, JHO-822, JHO-99-2, OL-1193	20 × 10	80–100	5–10	80-40-40	30
Guinea grass	CO-1, PGG-9, PGG-14, Hamil, Maccuimi, Reversedale	50 × 50	40,000 rooted slips ha^{-1}	5–10	80-40-40 (once in a year)	20 per cut
Para grass	Indian Para grass (IPG)	50 × 50	40,000 rooted slips ha^{-1}	5–10	40-40-40 (apply 40 kg N ha^{-1} after each cut)	15 per cut
Nandi grass	Kazungula	100 × 45	2–5	5–10	80-40-40	40-50
Rice bean	K1, Bidhan Rice bean 1, Bidhan Rice bean 2	30 × 15	30–40	5–10	20-40-40	15
Alfalfa	CO-1, Sirsa Type 9, Chetak (S-244), Anand Lucerne-3 (AL-3)	25 (Solid row)	25–40	5–10	20-80-10	20-35
Berseem	BL1, Mescave, Wardan	25 (Solid row)	25-40	5–10	20-80-10	40-50
Cowpea	Russian giant, EC4216, UPC4200, Bundellobia-1, UPC-5287	30 × 15	30–40	5–10	20-40-20	20
Butterfly pea	IGFRI-1530-1, IGFRI-173-1, IGFRI-94-1, IGFRI-S-23-1, IGFRI-S-12	50 × 15	20–30	5–10	20-60-20	20
Stylo	*S. hamata*, *S. scabra*, *S. guyanensis*	30 × 15	6–10	5–10	20-60-20	20

FIGURE 7.10 (a) Fodder maize (*Zea mays*); (b) Fodder sorghum (*Sorghum bicolor*).

crop from seed-borne diseases like collar rot and seedling rot, the seeds are treated with Bavistin @ 2 g kg^{-1} seed. Biofertilizer treatment with *Azotobacter* or *Azospirillum* @ 500 g per 10 kg seeds is done seven days after treating with seed- treating chemicals. A paste of *Azotobacter* is made in a seed-treating drum, then seeds are put in it and mixed. After treatment, the seeds are dried under shade for a few hours. Maize can be harvested for feeding as green fodder or for making silage and hay. Harvesting is done at cob formation to milking stage (50% flowering) for green fodder and at dough stage for silage.

7.15.2 Fodder Sorghum (Sorghum bicolor)

Sorghum (jowar) is grown during both *Kharif* and *Rabi* seasons as a fodder crop (Figure 7.10b). It remains green and palatable over a longer period than maize and bajra fodders. It grows well in a hot and dry climate, also it is drought resistant and moderately tolerant to salinity. To economize the nitrogen nutrition, biofertilizers such as *Azotobacter* or *Azospirillum* can be applied @ 250 g per 10 kg seeds. Harvesting is done at 50% flowering stage and not before 50 days after sowing as, prior to these, plants contain a poisonous substance, Hydrocyamic acid. In the case of multi-cut varieties, a second harvest is done at 45 days after the first harvest. Sorghum can be intercropped with cowpea, rice bean and guar in 1:1 or 2:1 row ratio.

7.15.3 Hybrid Napier Bajra

It is an inter-specific hybrid between bajra/pearl millet (*Pennisetum americanum*) and napier grass/elephant grass (*Pennisetum purpureum*). It can tolerate high temperature and can resist drought. It is perennial in nature and propagated by stem cuttings (rooted slips). Stem cutting with two buds are planted inside the soil. At planting only P and K are applied (N fertilizers are not to be applied). After crop establishment (10–15 days after planting) 50 kg N ha^{-1} is applied. First harvesting of fodder (15 cm above surface) is done 75 days after planting. After harvesting, hoeing, weeding, application of 40 kg N ha^{-1} is done and then earthing up is necessary as HNB has a bushy growth. Earthing up helps better tillering and hence higher fodder yield. In a year, 6–7 cuttings may result about 120 t ha^{-1} of fodder (Figure 7.11a).

7.15.4 Oat (Avena sativa)

It is grown during the *Rabi* or winter season as an irrigated crop (Figure 7.11b). It requires about 20–25°C for germination and still lower temperature for its better growth and development. Light winter rains followed by bright sunshine and lower humidity are very useful for the crop, as high humidity and cloudy weather invite many fungal diseases. Varieties with smaller seeds (Kent) require less seed (80 kg ha^{-1}), whereas bold-seeded varieties (OL-9, JHO 822) require a higher seed

FIGURE 7.11 (**a**) Hybrid Napier Bajra; (**b**) Oat (*Avena sativa*).

rate (100 kg ha^{-1}). Oat responds well to the treatment of bio-fertilizers (*Azotobacter* @ 250 g per 10 kg seeds). First harvesting is done at 50% flowering stage (50-60 days after sowing) with fodder yield of 25–30 t ha^{-1}. After the first cut, 20 kg N ha^{-1} is top dressed. Multi-cut varieties (JHO-822) yield 3–4 cuts.

7.15.5 Guinea Grass (*Panicum maximum*)

It can tolerate shade under trees and bushes (Figure 7.12a). Since germination takes 10–12 days, the seeds should be soaked or sown under pre-sowing irrigation conditions. While sowing, fine sand should be mixed with seed for uniform distribution of seeds. The crop can also be grown by raising seedlings in a nursery bed. Seedlings are transplanted at one month age. Transplanting is done with one seedling per hill, so about 40,000 seedlings are required per hectare with spacing of 50×50 cm. Manures and fertilizers (50 kg N, all P and K) are applied during sowing. After first cut, hoeing, weeding and rest fertilizer is applied and earthing up is done. First cut at 70 days after planting and subsequent cuts at 45 days interval. Total 4–5 cuts with stubble height of 15 cm.

7.15.6 Para Grass (*Brachiaria mutica*)

It is also called buffalo grass or water grass. In marshy land the growth of the crop is excellent (Figure 7.12b). It is grown round the year and highly tolerant to saline or sodic conditions. It can also tolerate water stress in the month of April and May. The first harvesting is done at 60–75 days after planting and subsequent cut at monthly interval is undertaken for getting a higher yield. Average fodder yield is 60–80 t ha^{-1} per hectare per annuum.

FIGURE 7.12 (**a**) Guinea grass (*Panicum maximum*); (**b**) Para grass (*Brachiaria mutica*).

FIGURE 7.13 **(a)** Nandi grass (*Setaria sphacelata*); **(b)** Rice bean (*Vigna umbellata*).

7.15.7 Nandi Grass (*Setaria sphacelata*)

It is a bunch-type perennial grass producing many tillers (Figure 7.13a) belonging to the Poaceae family. It has a flattened (tussock) growth habit (fan-shaped) with hairless (absence of foliar bristles) leaves. It is a multi-cut type of fodder crop. It is also planted for soil conservation. Good management results in a higher green fodder yield (40–50 t ha^{-1}).

7.15.8 Rice Bean (*Vigna umbellata*)

It provides high quality green as well as dry fodder. It can be grown as an intercrop with maize, as maize supports the climbing habit of rice bean for its growth. Rice bean is a N-fixing legume that improves the N status of the soil, thus providing nitrogen to the succeeding crop. Seeds are sown at 4–5 cm soil depth. Rice bean is an ideal catch crop, suitable for intercropping and multiple cropping system and also serves as an excellent cover crop and green manure crop (Figure 7.13b). It is harvested at 50% flowering stage (70–80 days after sowing) to obtain green fodder yield of about 20 t ha^{-1}.

7.15.9 Alfalfa/Lucerne (*Medicago sativa*)

It is called the 'Queen of Fodders.' It is a legume fodder (Figure 7.14a) belonging to Fabaceae family. It is cultivated for harvesting green fodder as well as for grazing and for preparing hay and silage. It is a climate resilient fodder crop as it can tolerate drought, hot and cold weather conditions. It is grown as a cover crop to control soil erosion. The quality of fodder is good as well as this crop improves soil fertility. Due to a hard seed coat, the seeds should be soaked overnight in water before sowing. The first harvesting is done at about 60–75 days after sowing and subsequently 5–10

FIGURE 7.14 **(a)** Alfalfa (*Medicago sativa*); **(b)** Berseem (*Trifolium alexandrinum*).

cuttings are done at 30–40 days interval depending upon the management practices. The green fodder yield of 20–35 t ha^{-1} could be obtained in 5–6 cuts.

7.15.10 Berseem/Egyptian Clover (Trifolium alexandrinum)

It is called the 'king of fodders.' It is a fast-growing legume fodder crop (Figure 7.14b) with high-quality forage, mainly cut and fed as green chopped fodder or mixed with oats for preparation of silage. It is suitable for heavy textured soil and improves the physical, chemical and biological properties of soil. Soaked seeds are sown and mixed with soil by running a brush or branch. It is harvested at 50–60 days after sowing with stubble height of 6–10 cm and subsequently after 40–45 days of previous harvest. Total fodder yield of 40–50 t ha^{-1} obtained from 5–6 cuttings.

7.15.11 Cowpea/Lobia (Vigna unguiculata)

It is grown suitably both in tropical and in subtropical zones and an important legume fodder both for *Kharif* and summer seasons. It conseves soil in hilly tracts and is a soil cover crop in coastal regions. Seeds are treated with *Rhizobium* culture @ 200 g per 10 kg seed. Application of sulphur and calcium is important for forage quality and palatability, therefore a single super phosphate (SSP) containing 12% S and 19.5% Ca should be applied. First cut at 60 days after sowing and another cut at 50 days of first cut gives total fodder yield of 20 t ha^{-1}.

7.15.12 Butterfly Pea (Clitoria ternatea)

It is a deep-rooted, tall and climbing type legume fodder crop. It can tolerate high rainfall as well as drought. However, it is difficult to eradicate due to deep root system and heavy self-seeding habits. First cut at 70–75 days after sowing and subsequent cuts at 35 days interval. Stubble height is kept at 10 cm. Green fodder yield of 20–30 t ha^{-1} can be obtained.

7.15.13 Stylo (Stylosanthes hamata)

It is a tropical legume shrub belonging to Fabaceae family, having several very significant pasture and forage species. In addition to providing high protein feed stock, improves soil fertility by nitrogen fixation. First cut at 70–80 days after sowing and subsequent cut after 60–70 days.

REFERENCES

Avnimelech, Y., Kochva, M., Hargreaves, J.A. 1999. Sedimentation and resuspension in earthen fish ponds. *Journal of the World Aquaculture Society* 30(4):401–409. DOI: https://doi.org/10.1111/j.1749-7345.1999.tb00988.x.

Basu, S.P. 1951. Effect of addition of soil in transport on larvae of Indian carps. *Proceedings of Indian Science Congress* 38(3):227.

Bayer, A., Whitaker, K., Chappell, M., Ruter, J., van lersel, M.W. 2014. Effect of irrigation duration and fertiliser rate on plant growth, substrate solution EC and leaching volume. *Acta Horticulturae* 1034:477–484. DOI: https://doi.org/10.17660/ActaHortic.2014.1034.59.

Bhagabati, A.K., Kalita, B. 1987. Studies on traditional fishing in some beels in Kamrup, Assam. In: *Proceedings of the Workshop on Development of Beel Fisheries in Assam*, 21–22 April, 1987. Guwahati, Assam. p 47.

De, H.K., Saha, G.S. 2001. Indigenous technical knowledge in feed and nutrition. *Aquaculture Asia* VI(2):20–21.

Gangadhara, B., Nandeesha, M.C., Keshavanath, P., Varghese, T.J. 2004. Growth response, biochemical composition and digestive enzyme activity of *Labeo rohita* (Ham.) fed salt incorporated diets. *Journal of Applied Aquaculture* 16(1/2):169–176.

Gope, L., Behera, S.K., Roy, R. 2017. Identification of indigenous knowledge components for sustainable development among the Santhal community. *American Journal of Educational Research* 5(8):887–893. http://pubs.sciepub.com/education/5/8/8.

Goswami, B., Mondal, S., Dana, S.S. 2006. Indigenous technological knowledge in fish farming. *Indian Journal of Traditional Knowledge* 5(1):60–63.

http://www.navrangindia.in/2020/04/johads-age-old-simple-traditional-water.html. Available online. Accessed on 29 January 2023.

http://www.rainwaterharvesting.org/Rural/Traditional2.htm. Available online. Accessed on 29 January 2023.

https://paramparaproject.org/traditions_ahar_pyane.html. Available online. Accessed on 29 January 2023.

Jasinska, A. 2018. Spent mushroom compost (SMC) – retrieved added value product closing loop in agricultural production. *Acta Agraria Debreceniensis* (150):185–202. DOI: https://doi.org/10.34101/actaagrar/150/1715.

Kalita, B., Choudhury, M., Ojha, S.N. 2004. Indigenous technical knowledge on pond construction and maintenance, fish seed transportation, and fish health management in Assam hills. *Indian Journal of Traditional Knowledge* 3(2):192–197.

Kalita, B., Dutta, A., Bhagwati, S.K., Sharma, A. 2010. Indigenous technical knowledge for fish harvesting in Karbi-Anglong District of Assam. *Indian Journal of Traditional Knowledge* 9(2):252–255.

Kannaiyan, S. 1993. Nitrogen contribution by *Azolla* to rice crop. *Proceedings of Indian National Science Academy* B59(3&4):309–314. Available online: https://www.researchgate.net/publication/285251497_Nitrogen_contribution-by_Azolla_to_rice_crop. Accessed on 16 December 2022.

Keshavanath, P., Gangadhar, B., Khadri, S., 2003. Growth enhancement of carp and prawn through dietary sodium chloride supplementation. *Aquaculture Asia* 8(4):4–8.

Kimani, S.M., Bimantara, P.O., Hattori, S., Tawaraya, K., Sudo, S., Cheng, W. 2020. Azolla incorporation and dual cropping influences CH_4 and N_2O emissions from flooded paddy ecosystems. *Soil Science and Plant Nutrition* 66(1):152–162. DOI: https://doi.org/10.1080/00380768.2019.1705736.

Ma, Y., Wang, Q., Sun, X., Wang, X., Su, W., Song, N. 2014. A study on recycling of spent mushroom substrate to prepare chars and activated carbon. *Bioresources* 9(3):3939–3954.

Nandeesha, M.C., Gangadhar, B., Varghese, T.J., Keshavanath, P. 2000. Effect of dietary sodium chloride supplementation on growth, biochemical composition and digestive enzyme activity of *Cyprinus carpio* (Linn.) and *Cirrhinus mrigala* (Ham.). *Journal of Aquaculture in the Tropics* 15:135–144.

Naumann, M., Koch, M., Thiel, H., Gransee, A., Pawelzik, E. 2020. The importance of nutrient management for potato production part II: Plant nutrition and tuber quality. *Potato Research* 63:121–137. DOI: https://doi.org/10.1007/s11540-019-09430-3.

Osman, M., Wani, S.P., Balloli, S.S., Sreedevi, T.K., Rao C.S., D'silva, E. 2009. *Pongamia* seed cake as a valuable source of plant nutrients for sustainable agriculture. *Indian Journal of Fertilisers* 5(2):25–26; 29–32. Available online at http://oar.icrisat.org/424/1/IndJourFer5_2_25-26_29-32_2009.pdf. Accessed on 30 December 2022.

Rahman, M.M., Yakupitiyage, A., Ranamukhaarachchi, S.L. 2004. Agricultural use of fishpond sediment for environmental amelioration. *Thammasat International Journal of Science and Technology* 9(4):1–10. Available online at: https://www.researchgate.net/publication/288520919_Agricultural_use_of_fish-pond_sediment_for_environmental_amelioration#fullTextFileContent. Accessed on 14 January 2023.

Rinker, D.L. 2017. Spent mushroom substrate uses. In: Diego C.Z., Pardo-Gimenez A. (eds) *Edible and Medicinal Mushrooms*. John Wiley & Sons Ltd., Inc. Hoboken, NJ, USA. pp 427–454. DOI: https://doi.org/10.1002/9781119149446.ch20.

Saha, K.C., Chowdhury, N.K. 1956. Studies on the mortality in spawn and fry of (redsoil), water medium in transporting carp spawn (Dimpona) in the existing trade to check mortality during transit. *Indian Journal of Fishing* 3(1):119–126.

Saha, R., Nath, D. 2013. Indigenous technical knowledge of fish farmers at Dhalai Distict of Tripura, India. *Indian Journal of Traditional Knowledge* 12(1):80–84.

Sarangi, S.K. 2009. Indigenous farming system of Adi tribes: Source of food, nutrition and medicine. *LEISA INDIA* 11(1):22–23.

Sarangi, S.K., Lama, T.D. 2013. Straw composting using earthworm (*Eudrilus eugeniae*) and fungal inoculant (*Trichoderma viridae*) and its utilization in rice (*Oryza sativa*)- groundnut (*Arachis hypogaea*) cropping system. *Indian Journal of Agricultural Sciences* 83(4):420–425.

Sarangi, S.K., Mahanta, K.K., Mandal, S., Maji, B. 2012. Dhaincha (*Sesbania* spp.) cultivation for nitrogen fixation and non-timber fuel wood production in coastal soils. In: Saha C., Nayak L.K., Karmakar S. (eds) *Proceedings of 25th National Convention of Agricultural Engineers on Advances in use of*

Non-conventional Energy Sources for Agriculture, Fisheries and Rural Development. 19–20 January, 2012, Agricultural Engineering Division, West Bengal State Centre, Published by The Institution of Engineers (India), 8, Gokhale Road, Kolkata-700 020, India. pp 41–45; 123.

Sarangi, S.K., Mainuddin, M., Maji, B. 2022. Problems, management, and prospects of acid sulphate soils in the Ganges Delta. *Soil Systems* 6:95. DOI: https://doi.org/10.3390/soilsystems6040095.

Sarangi, S.K., Maji, B., Sharma, P.C., Digar, S., Mahanta, K.K., Burman, D., Mandal, U.K., Mandal, S., Mainuddin, M. 2021. Potato (*Solanum tuberosum* L.) cultivation by zero tillage and paddy straw mulching in the saline soils of the Ganges Delta. *Potato Research* 64(2): 277–305. DOI: https://doi.org/10.1007/s11540-020-09478-6.

Sarangi, S.K., Maji, B., Singh, S., Sharma, D.K., Burman, D., Mandal, S., Ismail, A.M., Haefele, S.M. 2014. Crop establishment and nutrient management for dry season (boro) rice in coastal areas. *Agronomy Journal* 106(6):2013–2023. DOI: https://doi.org/10.2134/agronj14.0182.

Sarangi, S.K., Singh, R., Singh, K.A. 2009. Indigenous method of rat proof grain storage by Adi tribes of Arunachal Pradesh. *Indian Journal of Traditional Knowledge* 8(2):230–233. http://nopr.niscpr.res.in/bitstream/123456789/3956/1/IJTK%208%282%29%20230-233.pdf.

Savji, K.A., Gunalan, B., Solanki, H. 2011. Comparative evaluation of toddy (*Borassus flabellifer* SAP) on pH reduction and control of plankton bloom in two brackish water shrimp culture ponds. *International Journal of ChemTech Research* 3(4):2047–2057.

Sharma, S., Singh, J. 2021. Split application of potassium improves yield and potassium uptake of rice under deficient soils. *Journal of Soil and Water Conservation* 20(2):213–220. DOI: https://doi.org/10.5958/2455-7145.2021.00027.8.

Singh, C.P. 1985. Preparation of Phospho-Compost and its effect on the yield of moong bean and wheat. *Biological Agriculture & Horticulture* 2(3):223–229. DOI: https://doi.org/10.1080/01448765.1985.9754435.

Singh, G., Gill, H.S., Abrol, I.P., Cheema, S.S. 1991. Forage yield, mineral composition, nutrient cycling and ameliorating effects of Karnal grass (*Leptochloa fusca*) grown with mesquite (*Prosopis juliflora*) in a highly alkaline soil. *Field Crops Research* 26(1):45–55. DOI: https://doi.org/10.1016/0378-4290(91)90056-2.

Singh, Y.B., Singh, S.K., Saha, H., Tiwari, V.K., Das, P., Waikhom, G. 2011. *Eurale ferox* Salib integration into fish ponds in Manipur, India: A sustainable approach. *African Journal of Basic & Applied Sciences* 3(5):198–200.

Yadava, Y.S., Choudhury, M. 1986. Banas fishing in beels of Assam. *Journal of Bombay Natural History Society* 83(2):452–456.

8 Impact of Climate Change on Agriculture and Adaptations through IFS

Globally the impact of climate change on agriculture has been prominent in all agro-ecological regions and more so in the coastal regions because of the nearness to the ocean. There are ample evidences that the climate on Earth is changing and further changes cannot be avoided (IPCC 2007c). The impact is due to aberrant rainfall events, such as increased frequency of cyclonic storms and rise in temperatures and sea levels. Agricultural crops are very sensitive to weather elements and increased soil as well as water salinity are due to sea water ingress, dry spells, drought, floods and high-speed storms. Recent climate models project a significant change in the behaviour of major monsoon systems due to an increase in greenhouse gases (GHG) emissions. Variations in monsoonal rainfall and air temperature could adversely impact agricultural production particularly in Asian countries. In most of the rainfed areas, both excessive rainfall and drought had a negative influence on rice yield between 1966 and 2002, with drought having a significantly greater impact than extreme rainfall (Auffhammer et al. 2011). Fishery production is affected due to changes in hydrological scenario and salinity levels. Animal production is hampered due to the direct impact of increased stress caused by temperature rise and the increased frequency and intensity of heat waves. The indirect effects are due to reduction in the quantity and quality of fodder/feed materials, drinking water and increased activity of pathogens and vectors. Another but even more uncertain climate-dependent factor in agricultural production is crop pests and diseases. A significant part of crop yield losses is due to pests and diseases. In rice, 37% yield loss is attributed to insects, diseases and weeds together (Oerke 2006). These phenomena adversely affect food production; therefore, food security is compromised. Hence, it is essential to assess the risks due to climate change, devise suitable technologies to reduce the impact and implement those strategies in the agriculture sector. All possible impacts and outcomes of climate change on the farming system (Figure 8.1) should be considered for devising adaptation and mitigation plans.

8.1 BENCHMARK INFORMATION

Benchmarking is the collection of data regarding the existing status of impact of climate change and readiness to tackle the impact. As per the World Bank, from the standpoint of climate policy, the effects can be quantified in terms of greenhouse gas emissions or energy consumption, and the activities linked to these can be process inputs like fuel or electricity used, or process outputs like products manufactured, services rendered or heat produced (Partnership for Market Readiness 2017). The following formula can be used to calculate benchmarks, which are often indices of environmental performance used in climate policies.

Environmental performance = impact (GHG or CO_2 emissions, energy use, etc.)/activity [units of output (product, heat, service) produced or inputs (fuel, electricity) consumed]

Benchmarking involves planning, data collection, analysis, integration, monitoring and improvement. Benchmark surveys are carried out to obtain data on the present status of farming communities with respect to their resources, constraints and possibilities. Data on socioeconomic aspects (household parameters, knowledge, attitude and practices), natural resources (soil, water and crop resources, quality and utilization under climate change situations) and gender aspects

DOI: 10.1201/9781003239963-8

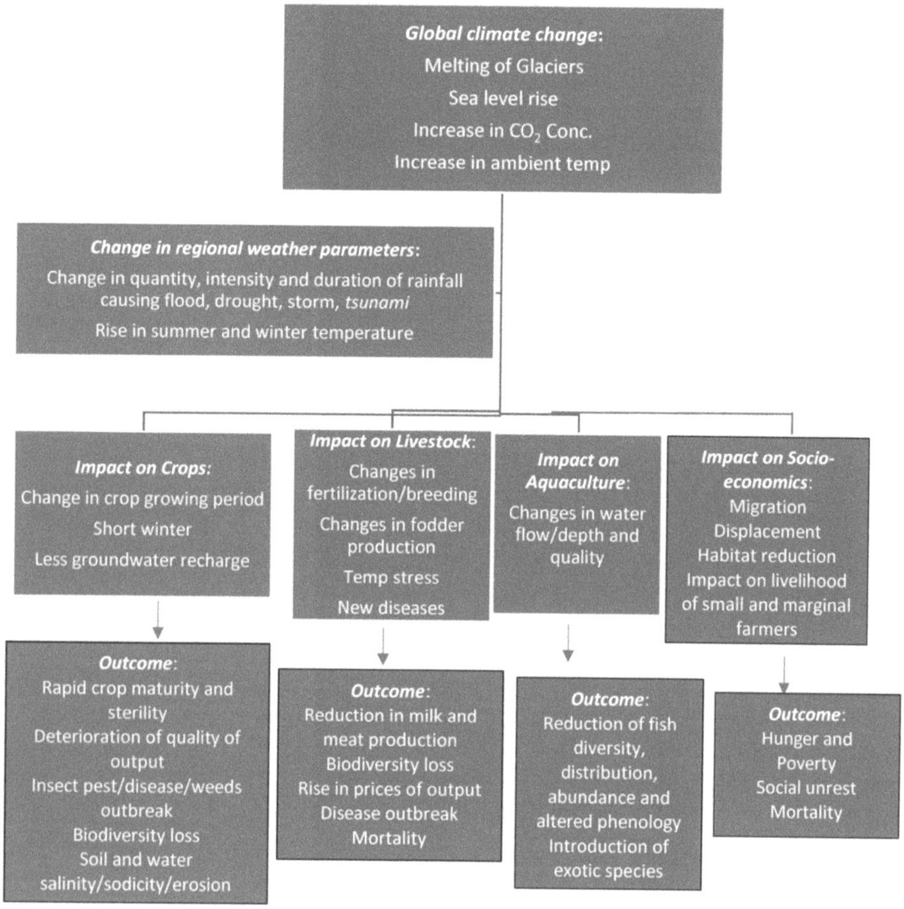

FIGURE 8.1 General assessment of climate change impact and outcome on farming system.

(role in managing household and farm activities) are collected. Along with data on GHG emissions, water pollutions, soil salinization, acidification, erosion and weather aberrations, the benchmarking should also consider farm resources availability, technology adoption rate and socioeconomic vulnerability as a result of climate change. The data analysis is done to identify, diagnose and prioritize hinderances to farm productivity. Based on the resource's availability and feasibility, future opportunities are identified for site-specific resource management under climate change situations. As per IPCC (2007a), risk assessment in agriculture involves assessment of hazards, vulnerability and adaptive capacity. Hazard is physical phenomenon that pose a threat to people, properties or financial resources. Natural hazards are floods, droughts, cyclones, whereas man-made hazards are accidents due to fire, building collapse, injudicious use of natural resources, etc. Vulnerability is the extent to which a community/population, structures, service or geographic area is likely to be damaged. Physical vulnerability depends on location and infrastructure. Economic vulnerability depends on economic status. Social vulnerability includes gender, age, physical fitness, etc. Environmental vulnerability is the extent of degradation, pollution, contamination, which covers exposure (rate and variation of environmental change), sensitivity (availability and importance of the affected resource) and adaptive capability (ability to respond to environmental change). A disaster happens when a hazard impacts a vulnerable population and causes damage, casualties and disruption. Therefore, a disaster is the product of hazard and vulnerability. Adaptive capacity is the knowledge, skills, resources, abilities and strength which enable an individual, household and community to prevent, prepare, stand against, survive and recover from a disaster. Therefore, risk is assessed from hazard, vulnerability and adaptive capacity. Risk is the likelihood of detrimental consequences or expected losses (to life, property, livelihoods, economic activity or environment) resulting from interaction between natural- or human-induced hazards, vulnerable conditions and adaptive capacity. Reducing anyone or more of the three contributing variables (hazard, vulnerability and exposure) will reduce risk, and increasing the adaptive capacity will lessen the risk. However, the natural hazard component is beyond the limit of human beings; therefore, reducing vulnerability and increasing adaptive capacity are practical strategies to manage risk in the agriculture sector.

8.2 FUTURE SCENARIO OF CLIMATE CHANGE

The global path of future development to reach an outcome of climate change is defined by different scenarios. As per the third and fourth assessment report of the Intergovernmental Panel on Climate Change (IPCC), there are six families (A1F1, A1B, A1T, A2, B1 and B2) of climate change scenarios (IPCC 2001, IPCC 2007b). IPCC in its fifth assessment report (AR5) gave new development scenarios called Representative Concentration Pathways (RCPs) based on anticipated radiative forcings (IPCC 2014). Based on model (MPI-ESM – German Climate Computing Centre's Earth System Model) calculations of these radiative forcings (from low to very high), the impact of climate change on global temperatures, sea ice melt, sea level rise, etc. are estimated. There are four RCPs (RCP 2.6, RCP 4.5, RCP 6.0 and RCP 8.5) as per AR5. The numerical values (2.6, 4.5, 6.0 and 8.5) indicate the radiative forcings in W m^{-2} in 2100. The Shared Socioeconomic Pathways (SSPs) have been put forth as community scenarios covering different climate futures as well as different possible and internally consistent socioeconomic developments (Riahi et al. 2017). Apart from these, there are high-end scenarios particularly for coastal regions. A variety of unlikely, but not impossible, high-end sea-level scenarios (HESs) based on expert elicitation assessment for Greenland and Antarctic ice sheets and high-end physical-based model projections for glaciers, ocean sterodynamic effects, and glacial isostatic adjustment and contributions from land-water (Dayan et al. 2021). Regional drivers of change such as use of groundwater, changes in cropping pattern and population density changes are also incorporated to devise regional scale-based climate change scenarios.

The A1F1 scenario assumes the energy use is fossil fuel based with rapid economic growth. The world population is estimated to be 9 billion by 2050 and gradually decline thereafter. There are extensive social and cultural interactions at the global level. Income and way of life converge between regions. Scenario A1B give emphasis on all energy sectors, whereas in the case of the A1T scenario, the energy use is non-fossil-based sources, in both A1B and A1T, other conditions are the same as A1F1. Scenario A2 assumes a more divided world with continuous increasing global population. Countries operate independently and tend to be self-reliant. In this scenario, the economic developments are regionally oriented with high degree of emissions. Scenario B1 is more integrated and envisages an eco-friendlier world with emphasis on economic, social and environmental stability. B1 hypothesises rapid economic growth with fast changes towards a service and information economy. Here, the population projections are as per A1 scenario. There will be an introduction of clean, green and resource-efficient technologies and a reduction in material use intensity. Unlike the B1 scenario, the B2 scenario is a divided world, however with eco-friendly approaches. The population growth will be increasing but the pace is slower than A2 scenario. It stretches regional-based emphasis towards solutions to economic, social and environmental stability. Economic developments are intermediate with less rapid and more fragmented technological change than that of A1 and B1.

The salient features of RCP 2.6 include average global temperature rise during 2081–2100 is 1°C compared to 1986–2005. Sea level rise is 0.4 m with a small increase in extreme weather events. It envisages renewable energy generations and use, new technologies to capture emissions, use of bicycles and public transport. These conditions require high efforts to achieve global CO_2 emissions decline to about 400 ppm by 2100. If these conditions are met, there will be requirement of low level of adaptations. RCP 2.6 is favourable to the particular sensitivity requirement of tropical coral reef. If the temperature rise is less than 1.2°C, about 50% of coral reefs can be protected.

RCP 4.5 envisages mean radiative forcing by greenhouse gas emissions stabilizing at 4.5 Wm^{-2} by 2100 by employing a range of technologies and strategies for reducing GHG emissions. Average temperature rise of 1.8°C (2081–2100) compared to 1986–2005. Sea level rise of 0.47 m (2081–2100) is compared to 1986–2005. Moderate increase in extreme weather events. Gradual increase in CO_2 concentration to 538 ppm in 2100. It envisages a mixed transport system and emphasis on renewable energy sources. It requires a medium level of adaptations and all nations to undertake CO_2 emission regulations simultaneously and effectively. Global temperature rises between 2°C and 3°C by 2100 with mean sea level rise 35% higher than that of RCP2.6. Many plant and animal species will become incapable of adapting to the effects of RCP 4.5 and higher RCPs.

RCP 6.0 foresees high radiative forcing level of 6 Wm^{-2} by the end of the current century (2100). The global average temperature rise is 2.2°C by 2081–2100 relative to average temperature of 1986-2005. Sea level rise of 0.48 m and moderate increase of extreme weather events. CO_2 concentration of 670 ppm by 2100. Continuous global warming through 2100 will make the global temperature rise by 3–4°C by 2100. Mixed use of energy sources and transport system contributes to rise in CO_2 concentration.

RCP 8.5 is the worst-case scenario of highest level of GHG emissions. It is also the business-as-usual scenario or high emission scenario, leading to a radiative forcing of 8.5 Wm^{-2} by 2100. The average temperature rise during 2081–2100 is 3.7°C compared to the average temperature of 1986–2005. Sea level rise of 0.63 m in 2081–2100 (average rise relative to 1986–2005). Large increase in extreme weather events during 2081–2100. This scenario envisages coal-based power use, and transport systems more dependent on personal cars and use of trucks. This scenario requires high cost and level of adaptation to climate change. The cumulative CO_2 emissions in RCP 8.5 amount to about 7,300 Gt CO_2 over the course of the entire century (Riahi et al. 2011).

The Shared Socioeconomic Pathways (SSPs) have been put forth as community scenarios covering different climate futures as well as different possible and internally consistent socioeconomic drivers (Lehtonen et al. 2021). There are five alternative socioeconomic futures (Table 8.1). SSP1 assumes sustainable development with low challenges to mitigation and adaptation. SSP2 takes a middle-of-the road development path with medium challenges to mitigation and adaptation. SSP3

TABLE 8.1
Climate Change Scenarios – Shared Socioeconomic Pathways (SSPs)

SSP	Scenario	Estimated Warming (2041–2060)	Estimated Warming (2081–2100)	Very Likely Range in °C (2081–2100)
SSP1-1.9	Very low GHG emissions: CO_2 emissions cut to net zero around 2050	1.6 °C	1.4 °C	1.0 – 1.8
SSP1-2.6	Low GHG emissions: CO_2 emissions cut to net zero around 2075	1.7 °C	1.8 °C	1.3 – 2.4
SSP2-4.5	Intermediate GHG emissions: CO_2 emissions around current levels until 2050, then falling but not reaching net zero by 2100	2.0 °C	2.7 °C	2.1 – 3.5
SSP3-7.0	High GHG emissions: high GHG emissions: CO_2 emissions double by 2100	2.1 °C	3.6 °C	2.8 – 4.6
SSP5-8.5	Very high GHG emissions: CO_2 emissions triple by 2075	2.4 °C	4.4 °C	3.3 – 5.7

covers high challenges to mitigation and adaptation due to regional rivalry. SSP4 is an inequality pathway with low challenges to mitigation and high challenges to adaptation. SSP5 envisages fossil fuel–based development with high challenges to mitigation and low challenges to adaptation.

Apart from the conventional scenarios, there are cases of high-end scenarios (HES). The information on upper extreme of the impact of the climate change is vital for the planners, decision-makers and managers of vital infrastructures (coastal cities, ports, coastal cultural heritages, chemical/biochemical industries, nuclear plants) as well as for other risk-averse stakeholders in the coastal regions (Dayan et al. 2021). High-end scenarios such as global total sea level rise (SLR) exceeding 2 m by 2100 (lies within the 90% uncertainty bounds for a high emission scenario) have been put forth, which is more than twice the upper value given by IPCC in AR5 (Bamber et al. 2019). The HES (SLR >2 m by 2100 for a temperature rise of 4°C or more) will result a land loss of 1.79 M Km^2, which will significantly reduce the most important food production regions of the world and force population displacement of 187 million (Nicholls et al. 2011). The impact of the HES will be severe in small islands, Africa and parts of Asia, as the likelihood of protection in these regions is lowest (Nicholls et al. 2011). The largest ice sheet of the northern hemisphere (the Greenland Ice Sheet), has a potential to raise global mean sea level (GMSL) by 7.4 m, while the largest ice sheet at the South Pole (the Antarctic Ice Sheet), which is the largest ice mass on Earth, could increase GMSL by 58 m (https://sealevel.nasa.gov).

The regional scale climate change scenarios are guided by population growth, resource use, local-level land and water management strategies, etc. In the coastal regions, polder/island level water and salt balance scenarios affect the agricultural system. Models involving crop management options, salt management and climate change scenarios employed for the Ganges Delta region revealed significant changes in salt concentrations in the soil, pond and canals (Mainuddin et al. 2021). An integrated population-gross domestic product (GDP)–food-water model to examine water demand to 2100 in Bangladesh revealed that urban water demand is likely to grow proportionately significantly more than irrigation water demand, necessitating urgent alternative sources of water (Kirby and Mainuddin 2022). Therefore, studies on regional scale scenarios of climate change are very important for crop planning and resource use. Regional comparisons of water footprints will guide in shifting crops of high-water footprint in a region to a region where they have a lower water footprint by comparing management systems in different regions.

8.3 IMPACT OF CLIMATE CHANGE ON CROP PRODUCTION

8.3.1 Impact of Changes in Rainfall on Crop Production

Climate change will significantly impact rainfall patterns, soil moisture storage, evaporation and runoff (Skendzic et al. 2021). Globally 72% of the land is expected to suffer aridification, with an increase of about 10% arid and semi-arid regions, while humid and semi-humid regions would decrease by 23% and 4.9%, respectively (Tabari 2020). However, due to climate change, the frequency and intensity of extreme rainfall events will increase because more water vapour is held in a warmer environment and it can occasionally rain quickly (Blenkinsop et al. 2021). Heavy rain during the sensitive stages of crop development may cause crop failure or significantly impact the yield and quality of produce. Heavy rain causes floods and waterlogging (reduces O_2 in the crop root zone), damages young plants and inhibits aerobic process in roots, resulting in rotting of roots and wilting of plants, chlorosis, leaf epinasty and ultimate the death of plants. The length of the wet seasons and air humidity will both rise with more rainfall. When coupled with elevated temperatures, these may facilitate the growth of fungal infections. High humidity, temperature and moisture increases the surviving and breeding populations of insect pests, diseases and weeds. This could also create more climate-resilient populations and lead to the appearance of new pests, diseases and weeds.

8.3.2 Impact of Rising Temperatures on Crop Production

The cultivated crops have optimum temperature requirements during different growth and developmental stages. Rice crop can tolerate relatively high temperatures (35/25°C day/night); however, they are more sensitive to heat during the reproductive stage, causing spikelet sterility under high temperatures (Wassmann et al. 2009a). The production potential of rice in high temperature regions may decrease under continued warming since temperatures are often already approaching critical levels (Wassmann et al. 2009b). Rising temperatures may exacerbate heat stress on the dry season crop in Bangladesh and the Indian states of West Bengal, Bihar, Jharkhand, Odisha, Tamil Nadu, Kerala and Karnataka (Wassmann et al. 2009b). For the wheat crop, high temperature–induced heat stress adversely affects the yield and quality due to a significant effect on several metabolic processes such as photosynthesis, respiration, reproduction and grain development (Yadav et al. 2022). The photosynthetic rate of wheat decreased by 17–25% with yield penalty of 29–44% due to heat stress during anthesis and grain-filling periods (Djanaguiraman et al. 2020). High temperatures lead to heat stress in drier regions, increases transpiration, and ultimately causes yield reduction. In fruit and vegetable crops, high temperature on fruit surfaces caused by prolonged exposure to sunlight speed up the ripening process, therefore impairing the quality as well as the yield. Winter loving vegetable crops (potato, cabbage, cauliflower, tomato, etc.) are severely affected directly due to increases in winter temperatures and indirectly by the incidence of insect pests and diseases in tropical regions. Due to a rise in atmospheric temperature, water storage is hampered and fuel cost for pumping of irrigation water increases significantly.

8.3.3 Impact of Increased Concentration of CO_2 on Crop Production

Rising CO_2 concentrations will have both detrimental as well as beneficial effects on crop production. It leads to an increase in plant growth as the rate of photosynthesis increases but reduce the nitrogen uptake by plants, resulting in harvests with less nutritious content (Bhargava and Mitra 2020). Concurrent elevated CO_2 concentration and temperature will lead to exceeding the optimal growth range, decreased stomatal conductance, canopy net photosynthesis rate and subsequently reduced biomass and tuber yield in the potato (Lee et al. 2020). At a higher CO_2 level, C3 plants will profit more than C4 plants; but, with continuing global warming, the benefits of higher CO_2 on crop productivity will be countered by an increase in temperature (Streck 2005). Most of the weeds are C3 plant and are likely to benefit with increased CO_2 level. Increase in ambient CO_2 (50% higher) resulted in higher tuber malformation (in about 63%), increased glucose (22%), fructose (21%) and reducing sugars (23%), a trend towards lower tuber greening (about 12%), reduced tuber quality due to increased browning and acryl amide formation in French fries (Hogy and Fangmeier 2009). In grapes, tartaric acid and total sugar contents increased by about 8% and 14%, respectively, by rising CO_2 levels up to a maximum increase in the middle of the ripening season; however, at the maturity stage, the CO_2 effect on both quality parameters almost completely disappeared (Bindi et al. 2001). Increased CO_2 in coconut, areca nut and cocoa resulted in a higher biomass output but, at elevated air temperatures, a slight decline in biomass output was observed (Datta 2013).

8.4 IMPACT OF CLIMATE CHANGE ON ANIMAL HUSBANDRY

Animal husbandry affects climate change and vice versa. The livestock sector is accountable for generating 14.5% of total anthropogenic greenhouse gas emissions and at the same time climate change affects growth, production, reproduction and feed availability of livestock (Cheng et al. 2022). To adapt livestock systems to climate change, it is imperative to enhance the stress coping ability of livestock species by improved management and selection, along with matching soil and water management to achieve optimum crop and forage productivity (Nardone et al. 2010).

8.4.1 Effect of High/Low Temperatures on Livestock

Heat stress on domestic animals may result in lowering milk/meat output, reproductive capacity and health (Sejian et al. 2016). It is projected that livestock-based farming systems will be increasingly negatively impacted by global warming (Bernabucci 2019). Heat stress lowers fertility and production and increases vulnerability to diseases (Singh 2021). In humid tropical conditions, under the cumulative effect of high atmospheric temperature and humidity, the process of conduction, convection and radiation (non-evaporative cooling) methods of animals becomes less effective; therefore, animals adapt by evaporative cooling methods (sweating and panting) to alleviate heat stress (Bal 2021). Heat stress also impact the immune system of animals; however, management practices like provision of shade, sprinklers, nutritional supplements can ameliorate the negative effects of heat stress (Gupta et al. 2023). The combined effect of temperature and humidity in terms of the temperature humidity index (THI) suggest that in humid areas, the limiting element for heat stress is humidity; in dry climates, the limiting factor for heat stress is dry bulb temperature (Bohmanova et al. 2007). In North India, when the THI is greater than 72, animals suffer from mild heat stress, at the value of 80 moderate heat stress occurs and beyond 90, there is incidence of severe heat stress (Bal 2021). Poultry birds reduce the intake of feed by 5% per 1°C rise in temperature in the temperature range of 32–38°C, to reduce the heat generated from metabolism (Sohail et al. 2012). Under low temperature (cold stress), the need for higher dietary nutrient and energy intake in dairy cattle is due to an increased metabolism rate (FAO 2016a). In poultry birds cold stress affects liver function and thyroid hormone production, which are key factors in energy expenditure and body temperature homeostasis (Nguyen et al. 2016). To cope with the heat stress on livestock systems, there is a need for adaptation strategy involving physical modifications in housing systems with cooling devices along with other dietary measures and breed management. Supplementation of antioxidants (enzymatic and non-enzymatic) boost the defence mechanism against oxidative stress (Rahal et al. 2014). Under high temperature and humidity conditions, poultry houses should be open style for ventilation with proper shading and grass cover on the ground surface to balance sunlight reflection and a shiny roof surface to reflect more solar radiation (Bal 2021). Coping strategies involving lower-protein diets along with limiting amino acids may be followed instead of high-protein diets during hot weather (Pawar et al. 2016). In cold-stress prone regions, provision of windbreaks helps dairy animals to take shelter in the leeward side of the cold wave and keeping the animals in a herd helps them to stay warm (Bal 2021). In poultry houses an optimum night temperature should be maintained by designing with low-heat conducting materials (woods and bamboo), in addition to the provision of lighting with electric bulbs. The orientation of the poultry house should be in the east-west direction of solar energy. Energy-rich feeds with oil/fat should be provided to maintain optimum body temperature.

8.4.2 Impact of Climate Change on Livestock in Different Agro-ecological Zones

The Food and Agricultural Organization (FAO) of the United Nations has identified the broad classifications of agro-ecological zones (AEZ) of the world: tropics, sub-tropics, temperate, boreal and artic (FAO 1996). The majority of the world's ruminants are found in the tropical and sub-tropical AEZs (arid and semi-arid regions), whereas the highest productivity is found in the temperate AEZ (Herrero et al. 2012). In temperate AEZ, the impact of climate change on livestock performance can be minimal, even the crop productivity can be increased due to increases in average temperatures (Seguin 2008). According to Rojas-Downing et al. (2017), a 2°C increase will have a favourable effect on livestock productivity and grazing in wet temperate regions and a detrimental one in arid and semiarid regions. An increase in the length of the growing period will positively affect forage quality, quantity and duration of availability. Climate change may lead to positive effects for agriculture in the boreal zones by significantly increasing the crop feasible growing degree days (GDD) to about 76% by 2099 compared to the current level of 32% (King et al. 2018). In this zone, there will be increased crop and fodder production with scope for the introduction of new crops and the reduction in housing cost for livestock due to longer growing seasons (Iglesias et al. 2007).

8.4.3 Impact of Rainfall Variability on Livestock

The negative effect of erratic rainfall on livestock population, particularly drought in the drier regions of the world is observed. In Tanzania (East Africa), due to drought there are cases of mortality of cattle (Leweri et al. 2021). In African conditions, for farmers, the probability of choosing goats and chickens increased, whereas beef cattle, dairy cattle and sheep options decreased with the increase in rainfall (Rust and Rust 2013). The productivity efficiency of animals is decreased and mortality increased after the incidence of floods due to various epizootic diseases, poor nutrition and sub-clinical diseases (Rasool et al. 2021). The flooding risks on animals can be long term, such as effects on reproduction, nutrition and carcass quality, as well as short-term, including comingling of unfamiliar cattle, nutritional restricts and disease transmission (Crist et al. 2020). The climate change effect on flooding has been reported from various parts of the world, including from the continental United States covering North Carolina, Nebraska and Illinois (Crist et al. 2020) and from India (Rasool et al. 2021), Bangladesh (Biswas et al. 2019) and South Africa (Ngarava et al. 2021). In Australia, particularly in Queensland and New South Wales, floods led to the death of cattle and also forced farmers to sell animals due to a shortage of feed and infrastructure to treat animals (https://www.nutrienagsolutions.com.au).

8.5 IMPACT OF LIVESTOCK SYSTEMS ON CLIMATE CHANGE, ADAPTATION AND MITIGATION

Animal farming also influences climate change through enteric fermentation, land use change, feed production, manure production and processing and transport by emitting CO_2, N_2O and CH_4 (Rojas-Downing et al. 2017). Livestock enterprises are natural resource exhaustive and are important sources of CH_4 and N_2O emissions, having 28 and 265 times higher global warming potential, respectively, than CO_2 (Grossi et al. 2019). The GHGs are produced during the enteric fermentation, manure storage, fertilizer/manure use and feed production. On a global scale, the livestock sector is responsible for 44% of anthropogenic CH_4, 53% of anthropogenic N_2O and 5% of anthropogenic CO_2 emissions (Rojas-Downing et al. 2017). Land use changes due to exhaustive animal farming not only contribute to increased GHGs emission (Rojas-Downing et al. 2017) but also favour the emergence of zoonotic diseases (Rulli et al. 2021).

Adaptation measures such as integrated systems of livestock farming with forestry, crop production and diversification of animal and crop components in addition to production and management strategies modifications and changing the timing and locations of farm operations will reduce the impact of livestock farming on climate change (IFAD 2010). Scientific interventions to reduce emission by technologies and practices with respect to aspects of enteric fermentation, manure excretion, storage and application to soil (direct emissions) as well as land use change (deforestation for pasture and crop lands), fertilizers, chemicals, agricultural operations, feed production, processing, and transportation (indirect emissions) contribute to mitigation measures. Carbon sequestration by reforestation, improvement of animal nutrition and genetics to reduce enteric fermentation, management measures for manures, fertilizers and feeds to increase use efficiency add significantly to livestock mitigation measures.

8.6 IMPACT OF CLIMATE CHANGE ON AQUACULTURE

Aquaculture is the fastest-growing agricultural sector in the world, with serious issues for sustainability due to climate change effects (Maulu et al. 2021). Various factors of climate change (temperature rise, sea level rise, rainfall patterns, salinity and extreme weather events) adversely affect this sector and the severity may increase further if efforts to prepare for consequences, adaptation and mitigation options are compromised (Elsheikh 2021).

8.6.1 Impact of Rising Temperatures on Aquaculture

Fish being ectotherms (cold-blooded) are bound to be affected by rising ecosystem temperatures. However, the degree of the effect of warming will vary with the species (Portner and Farrell 2008). Rising temperatures can enhance population of apex fish predators such as brown trout (*Salmo trutta*), when ambient temperature are suboptimal for overall physiological performance (Ogorman et al. 2016). Coral reefs in marine ecosystems are extremely sensitive to rising water temperatures, prolonged high temperatures lead to bleaching, expulsion of symbiotic algae and thereby degradation and death of corals (Dao et al. 2021). Aquaculture organisms such as fish, zooplankton and phytoplankton need optimum ranges of water temperature for growth and development. Fluctuations of water temperatures below or above the optimum temperature range lead to a decline in population and, under extreme conditions, there may be few or none (Jain et al. 2013). In warm water, there is less dissolved oxygen, therefore the survivability is affected. Continued global warming and increases in water temperatures will lead to loss of fish habitats (17% loss in trout and salmon by 2030 and 42% by 2090) resulting in a threat to the global food supply (Jain et al. 2013). Sea surface temperatures affect the juvenile fish densities, with strong response to warmer temperatures for tropical and subtropical range-expanding fish, while juveniles of temperate species responded more strongly to benthic habitats (McCosker et al. 2022). There is also evidence of changes in distribution of fish species due to global warming, a 1°C rise affecting the distribution of 72% of all fish species with a tendency to move to deep waters (Campana et al. 2020).

8.6.2 Impact of Floods and Cyclones on Aquaculture

Floods and cyclones cause damage to fisheries infrastructure such as floating nets/gears, fishing vessels and boats (Febrianty et al. 2021; Muhala et al. 2021). In pond-based aquaculture, floods result in loss of fish stock to the tune of about 54% and, due to breach of embankment or high-water levels, there is also the introduction of unwanted/invasive species into the pond leading to management issues (Rutkayova et al. 2017). Extreme floods negatively impact the aquatic ecosystem services (Talbot et al. 2018) such as supporting services (primary production, soil formation), regulating services (water quality, diseases, climate regulation), provisioning services (food and drinking water supply) and cultural services (aesthetic value, recreation and tourism). In a farm pond-based farming system, floods result in output loss and the loss is further aggravated if management strategies based on climate information are ignored (Olutumise 2023). In coastal areas, flooding, particularly with saline river water, negatively affects the water quality in terms of higher salinity and turbidity and decreasing the chemical oxygen demand (Chukwu Okeah et al. 2018).

8.6.3 Impact of Rising CO_2 on Aquaculture

Fish exposed to high CO_2 concentration showed significantly less growth rates with a thinner dermis layer; however, it did not have significant effect on mortality, eye cataract, visible external damage on skin and mineral deposits in kidneys (Mota et al. 2019). In recirculating aquaculture systems (RAS), even extreme CO_2 levels (40 mg CO_2 L^{-1}) did not affect the conductivity and toxic elements deposition except for Fe, which increased with higher CO_2 level due to an inverse relation with pH (Aslam et al. 2019). However, it is advised to monitor the CO_2 level in water as increased CO_2 level reduces the release rate of CO_2 from fish's own metabolism, resulting in increased blood CO_2 (hypercapnia), corresponding decrease in blood pH (acidosis) and reduction in the oxygen-carrying capacity of haemoglobin (Southgate 2005). In marine ecosystems, increases in CO_2 are accompanied by environmental hypoxia (low O_2), this induces increased O_2 uptake by the blood, enhancing hypoxia tolerance in European sea bass (*Dicentrarchus labrax*), this is due to increased haemoglobin-O_2 affinity (Montgomery et al. 2019). Increasing concentration of CO_2 (from 15 mg L^{-1} to 30 mg L^{-1}) in a recirculating aquaculture respirometer system with pikeperch (*Sander lucioperca*) resulted in 6% decrease in final body weight; however, with habituation of the fish to hypercapnia conditions when other water quality parameters are in acceptable limits (Steinberg et al. 2017).

8.7 CLIMATE CHANGE AND OUTBREAK OF INSPECTS-PESTS

The sensitivity of insects to temperature variations is high, with climate change–related temperatures rise and temperate regions are more likely to face more pest outbreaks in comparison to tropical regions (Schneider et al. 2022). Due to global warming, among the 31 globally important plant-eating (phytophagous) insect pests, 41% pests are expected to be increasingly severe as insect pests, whereas only 4% will have reduced pest severity and 55% will show mixed responses (Lehmann et al. 2020). Temperature rise influences the insect migration, especially the lepidopteran insects, whereas rainy conditions favour migrating insects (desert locusts) due to availability of vegetation (Sujayanand and Karuppaiah 2016). Global warming may trigger expansion of geographic range of insect pests with increased number of generations and transmission of plant diseases as vectors with reduction in natural enemies of insect pests; as a result, less effectiveness of biological control measures (Skendzic et al. 2021).

Climate change may also negatively affect the insect pests, e.g., soil inhibiting insects directly affected by heavy rainfall by the way of washing away of eggs, larvae as well as disrupting their diapause (Skendzic et al. 2021). However, migratory and invasive insect pests such as desert locust (*Schistocerca gregaria*) are favoured by climate change effects (increased cyclone frequency and more extreme climate variability) as a warm, sandy and wet soil environment increases the egg laying, hatching and breeding (Salih et al. 2020). Climate change may also alter the traditional route of migrating pests to new areas of breeding and propagation (Salih et al. 2020). However, under extreme climate change scenarios (SSP585), the areas suitable for desert locusts will contract by the 2070s with a decrease of 0.88×10^6 km^2 moderately suitable area and a reduction of 1.55×10^6 km^2 highly suitable area (Guan et al. 2021). Rising temperatures may reduce the incubation and breeding period of desert locusts, thereby a greater number of generations in a year (FAO 2016b). Favourable weather conditions for locusts such as high rainfall during the period of cyclones in one region leads to migration to other regions. Migration of locusts from Africa (Ethiopia, Kenya, Sudan, Somalia) to India and Pakistan has been due to heavy rainfall from consecutive cyclones (2018 and 2019) in these countries, thus helping locust breeding and propagation (Salik and Suleri 2020).

8.8 ADAPTATION TO CLIMATE CHANGE THROUGH IFS

An integrated farming system, being an independent and self-generating system, has less dependence on external inputs by *in situ* production and recycling of resources. This system is resilient to climate change, however, to adapt to the extreme climate change scenarios, adaptation (Figure 8.2) and mitigation (Figure 8.3) measures are essential. There is a reduction in risk as failure of one component is compensated by success of another. This may be within or among the enterprises. In coastal regions, failure of vegetable crops due to waterlogging by high rainfall during the dry season is compensated by better performance of *Boro* rice. The failure of fishery component is compensated by duckery or poultry components. Provision of water harvesting in IFS, reduces the waterlogging, soil erosion and crop failure; enables integration of water dependent components (fishery, poultry) and reduces the pumping of groundwater. In less rainfall dry land areas, livestock is an integral component of the IFS to augment income. Trees, plantations and orchards sequester carbon in the soil and biomass, thus reducing CO_2 emission. The combination of perennial fodder crops under agro-forestry systems, intercropping, alley cropping reduces the impact of climate change

Crop	Aquaculture	Forestry	Livestock
•Change in cropping pattern & crop establishment methods. Rubber dams for coastal watersheds. Biotic & abiotic stress tolerant crops and varieties. Micro-irrigation systems. Multiple-stress tolerant crop varieties. Land shaping and rainwater harvesting. Agroforestry and alley cropping. Provision of drainage. In situ management of crop residues by happy seeder, zero-till machines, double disc-coulters, straw choppers. Laser land levelling, raised bed/zero tillage planting. Precision nutrient management. ICT-based agro-met services. Index-based insurance.	•Community participation in water management. Adoption of climate resilient species and technology. Shifting to most suitable sites/region. Integration of ITKs. Promotion of insurance to natural calamity. Reclamation measures for soil and water. Maintaing sanitatin and hygiene. Integration with crops, livestock e.g. paddy-cum-fish, pig-cum-fish, poultry-cum-fish, duck-cum-fish.	•Agro-forestry and aforestation. Soil and water conservation measures. Vegetative shelter belts. Bio-energy based plantation. Adjustment in planting practices, schedule and species distribution. Inclusion of fast-growing, reclamative and productive species for degraded lands. Bio-drainage plants in waterlogged fields. Measures to minimize the forest fire, insects and diseases. Monitoring and early warning.	•Muti-purpose tree species based livestock farming. Forage-livestock system. Diversification of livestock species. Adjustment of stocking rates. Feed and pest management. Adoption of vaccination and insurance measures. Integration of farm animals with thermo-tolerance. Adjustment and redesigning of shelters/sheds. Disease control and surveillance.

FIGURE 8.2 Adaptive strategies to climate change in IFS.

Crop Production	Aquaculture	Forestry	Livestock
•Aerobic rice culture. Site-specific nitrogen application. Increasing N use efficiency and reducing fertilizer run-off. Modified application method and timing of fertilizers and pesticides. Integrated nutrient management. Conservation agriculture for soil carbon sequestration. Use of renewable energy, cover crops, legumes in rotation. In situ generation of resources.	•Use of feedstocks and products with low emission potential. Use of solar and wind energy. Waste water management to minimize water pollution. Scientific use of inputs, equipments and methods of fishing and fish processing. Fuel-efficient fishing. Use of electricity in fishing operation. Best operating procedure for cold storages and ice plants. Building design for efficient energy use and low C foot-print.	•Halting deforestation and promotion of afforestation. Use of forest products for bioenergy generation. Selection of tress species with higher productivity and carbon sequestration. Agroforestry options to reduce green house gas emission, reduced fossil fuel and energy use. Sequestering carbon in biomass and soil. Silvo-pastoral system to add trees to pastures.	•Scientific waste, manure and urine management. Improved diet, feed and fodder management for increased digestibility. Feed supplement to reduce methane generation. Improved health and herd management. Integration of non-ruminant animals (swine, poultry, rabbit) in IFS. Strategies to reduce enteric fermentation.

FIGURE 8.3 Mitigation measures to climate change in IFS.

by carbon sequestration and nutrient cycling (Bell and Moore 2012). Use of biochar (produced by pyrolysis of biomass) has the potential for climate change mitigation. These biomasses are produced in IFS, such as from various herbaceous residues (Miscanthus and Switchgrass), agro-industrial residues (sugarcane bagasse, corn stover) and crop residues (Mulabagal et al. 2021).

Review of the resilience of IFS to climate change revealed suggestive evidence of a greater farm level resilience to climate change by adoption of IFS models (Gil et al. 2017). The following adaptation strategies may be followed by smallholder farmers under a climate change scenario (Jack et al. 2021).

- Replacing cattle with sheep, goat, camel and buffaloes.
- Following agroforestry – integrated land use systems.
- Developing high-quality fodder crop varieties.
- Selection and cross-breeding of heat tolerant species.
- Practicing migratory grazing practices.
- Efficient resource conservation and utilization by adoption of scientific practices.
- Water harvesting, conservation and recycling.
- Lowering stocking rate and feeding during nighttime.
- Feeding green fodder to cattle instead of dry fodder.
- Improved health programmes, including vaccination.

8.9 CLIMATE-SMART TECHNOLOGIES FOR COASTAL REGIONS OF INDIA

8.9.1 Enhancing Productivity of Acid Sulphate/Saline Soil in the Coastal Areas

Crop yields in acid saline/acid sulphate soils (pH < 4.0) are extremely low and are greatly underutilized. Good yield can be obtained from such soils by adoption of liming/rock phosphate application along with green manuring. Green manuring is to be done in the preceding kharif rice and application of amendments before the start of the rabi cropping. These packages of practices increase the soil pH and crop yield

8.9.2 Reducing Salinity Buildup in Coastal Saline Soil through Mulching/Cover Crop

The salinity build up in soil can be controlled by growing crops with large foliage cover rather than leaving fallow. Mulching of soil during dry season; rice husk and straw, organic farm waste or any other suitable materials can be used as mulch. When mulching material is not available, the soil is to be kept ploughed instead of leaving unploughed. Ploughing of soil on initiation of pre-monsoon shower and sowing of green manuring like *Sesbania*.

8.9.3 Efficient Nutrient Management of Coastal Saline Soils

Nitrogen fertilizer use efficiency is better when combined (organic + inorganic) sources of nutrients are applied. Addition of *Sesbania*, *Gliricidia*, city compost, *Azolla* and leaves of locally available trees along with inorganic N-fertilizer improves crop yield and soil health.

8.9.4 Farm Pond Land Shaping Technique for Enhancement of Productivity of Salt-Affected Coastal Land

About 20% of the farm area is converted into on-farm reservoir (OFR) to harvest excess rainwater. The dug-out soil is used to raise the land to form high and medium land situations besides the original lowland situation in the farm for growing multiple and diversified crops throughout the year instead of mono-cropping with rice in the *Kharif* season. The pond is used for rainwater harvesting for irrigation and pisciculture. Poultry/livestock farming can also be practiced in the farm along with crops and fish with the use of pond water. The high land free from water logging in *Kharif* with less salinity build up in dry seasons and thus can be used for multi- and diversified crop cultivation throughout the year.

8.9.5 Deep Durrow and High Ridgeland Shaping Technique for Enhancement of Productivity of Salt-Affected Coastal Land

About 50% of degraded farm land is shaped into alternate high ridges (1.5 m top width x 1.0 m height x 3m bottom width) and deep furrows (3m top width x 1.5 m bottom width x 1.0 m depth). The rainwater is harvested in the deep furrows and the harvested rainwater is used for initial irrigation during *Rabi*. The water stored in furrows is also used for fish cultivation and supplementary irrigation in *Kharif*, if required. The ridges are used for cultivation of vegetables and other horticultural crops/multi-purpose tree species (MPTs) year-round instead of mono-cropping with rice in *Kharif.*

8.9.6 Paddy-cum-Fish Cultivation Land Shaping Technique for Enhancement of Productivity of Salt-Affected Coastal Land

Trenches (3m top width x 1.5 m bottom width x 1.0 m depth) are dug around the periphery of the farm land leaving about 3.5m-wide outer boundary and the dugout soil is used for making dikes (about 1.5 m top width x 1.0 m height x 3m bottom width). The dikes are used for growing vegetables and/or fruit crops/multi-purpose tree species (MPTS) year-round. The remaining portion of the farm land, including the trenches is used for more profitable paddy + fish cultivation in *Kharif.* The farm land (non-trench and non-dike area) is used for low water requiring crops during dry (*Rabi*/summer) season with the rain water harvested in trenches.

8.9.7 Shallow Furrow and Medium Ridgeland Shaping Technique for Enhancement of Productivity of Salt-Affected Coastal Land

About 75% of the farm land is shaped into medium ridges (1.0 m top width ×0.75 m height × 2.0m bottom width) and shallow furrows (2.0 m top width × 1.0 m bottom width × 0.75 m depth) with a gap of 3.5m between two consecutive ridges and furrows. The furrows are used for rainwater harvesting and paddy-fish-cultivation during *Kharif.* The cropping schedule is similar to that followed in deep furrow and high ridge except rice can be grown in furrows in *Rabi*/summer with lesser supplementary irrigation.

8.9.8 Efficient Water Utilization through Drip Irrigation System

Fresh water scarcity is the main problem for cropping in the non-monsoon period in coastal region. The drip system can be efficiently used for cropping with the limited available freshwater in the rabi season. For cultivation of vegetable crops like okra, lateral spacing will be 70 cm and crop spacing will be 30 cm. The system can be operated with a one HP pump. Salinity of water may be less than 1.5 dSm^{-1}.

8.9.9 Summer Rice Crop Establishment through Drum Seeding of Pre-Germinated Seeds in Coastal Salt-Affected Soils

Direct wet seeding of rice using a drum seeder is effective and could enhance yield and net returns for farmers in coastal areas. Integrating the use of salt-tolerant varieties like Canning 7, or others with higher yield and better grain quality, with drum seeding and nutrient management using chemical and organic fertilizers could enhance the productivity of boro rice in coastal saline soils. If the supply of FYM is limited, *Azolla* can be used as a substitute for FYM.

8.9.10 Improved Nursery Management for Wet Season (*Kharif*) Rice in Coastal Rainfed Lowlands

Farmers in the coastal regions often practice suboptimal nursery management, which leads to reduced plant population due to seedling mortality and, hence, poor yields (≈ 1.5–2 t ha^{-1}), below the national average. A lower seed density of 25 g m^{-2} with application of 50-30-15 kg N-P2O5-K2O + 5 t FYM ha^{-1} produces healthy and robust seedlings with high seedling vigour. The full dose of phosphorus and potassium and 50% of nitrogen are applied as basal at sowing. The rest of the nitrogen is applied in two equal splits at 15d of sowing and 7d before uprooting. Forty-d-old seedlings of the improved rice variety Amal-Mana produce a higher yield than 25-d-old seedlings.

8.9.11 Improved Main Field Management of Kharif Rice in Submergence-Prone Coastal Rainfed Lowlands

Optimum main field management practices like balanced fertilizer dose of (50-20-10 kg N-P_2O_5-K_2O + 5 t FYM ha^{-1}) two seedlings per hill, spacing of 15 cm x 15 cm and rice variety Amal-Mana has the capability to withstand abiotic stresses prevailing in the coastal agro-ecosystems of rice-based lowlands, and also provide higher and stable yields. These cost-effective approaches on management options have great potential for enhancing rice yield in stress-prone rainfed coastal areas.

8.9.12 Water Productive Cropping System for Coastal Areas

The suitable cropping system for coastal areas were identified based on water productivity (WP), which was calculated for potential cropping systems for these areas. The WP on net return basis (₹ m^{-3}) was 25.01, 24.54, 23.45, 22.61, 10.78 and 9.18 for rice-tomato, rice-potato, rice-brinjal, rice-potato, rice-sunflower and rice-chili, respectively.

8.9.13 DSR Technology for Wet Season Rice in Coastal Rainfed Lowlands

The problem of stagnant flooding in coastal areas generally coincides with transplanting of rice, and damage occurs as the crop is still in the transplanting shock stage. To solve this problem, DSR technology fine-tuned. Before the onset of monsoon during last week of May, lines are made 20 cm apart after ploughing the field thoroghly and rice seeds are sown dry in the field. With the first shower of monsoon, the seeds with the heat of the soil start germinating. About 15 days after sowing, thinning and gapfilling is done to maintain an optimum plant population. Selective herbicide is used to reduce the weed population in the direct seeded plot.

8.9.14 Raised Bed Planting of Rabi Maize

Technologies has been developed for successful growing of rabi maize with raised beds to withstand the soil salinity and waterlogging caused by unseasonal rain. This climate resilient package of practices increases the yield of rice-based systems and increases the net income.

8.9.15 Zero Tillage Sown Rapeseed

Zero tillage sowing of rapeseed after harvest of *Kharif* rice provides a scope for cropping system intensification in salt-affected coastal soils. The developed package of practices helps to raise an early crop by avoiding salinity development.

8.9.16 Zero Tillage Planted Potato with Paddy Straw Mulching

This technology enables efficient use of residual soil moisture; conserve applied irrigation water and prevent soil salinity development. Early sowing and harvesting achieved by this practice help in taking another rabi pulse crop like green gram. This system increases the cropping intensity, yield and profit of farmers.

8.9.17 Modified System of Rice Intensification (MSRI) for Summer Season Rice Crop in Coastal Salt-Affected Soils

The entire set of management practices of SRI as evolved in other rice production environments are not suitable under coastal salt-affected areas. The MSRI includes nursery in raised beds, with irrigation facility to wash out salts, optimum FYM application, adjusted seedling age, application of biofertilizers, etc. Under the salinity stress condition, the seedling stage of rice is very sensitive; therefore, very young seedlings as suggested did not sustain this stress; similarly, the problem of mortality of single seedling transplanting per hill needs to be addressed. A package of practices has been developed for successful cultivation of rice with MSRI.

8.10 CONTINGENCY MEASURES TO ADAPT TO CLIMATE CHANGE–INDUCED CALAMITIES

A contingency plan is an action plan prepared before a natural calamity/disaster/weather aberration, which gives alternate options for managing the situation than the usual expected practices. The plan gives a step-by-step action to be done before, during and after the event. It may be called as a backup plan or plan B to cope with the untoward incidents in agriculture. A district-wise agricultural contingency plan for 20 different stages/union territories of India (Andhra Pradesh, Assam, Bihar, Chhattisgarh, Gujarat, Haryana, Himachal Pradesh, Jammu and Kashmir, Jharkhand, Karnataka, Kerala, Madhya Pradesh, Maharashtra, Odisha, Punjab, Rajasthan, Sikkim, Tamil Nādu, Uttar Pradesh and West Bengal) is available at the Indian Council of Agricultural Research-Central Research Institute for Dryland Agriculture website (http://www.icar-crida.res.in). Having an alternative plan before hand for the emergency situation, one could avoid the complete failure or loss, tackle the panic situations and adapt to the changing scenario. Therefore, it is a proposition to unexpected and unintended situations and emergencies. The plan is based on the risk and possible scenarios as well as the availability of alternate inputs and resources. This plan also involves the preparations for the mitigation of the consequences and the cost-benefit analysis.

In agriculture, contingency cropping is the growing of an alternative crop in place of a normally sown highly profitable crop of the region due to aberrant weather conditions. In dryland agriculture, contingency of growing another crop arises due to delay in the onset of the monsoon. Mid-season correction is a contingency management practice to overcome unexpected and unfavourable weather conditions. Thinning or population adjustment is an example of mid-season correction to overcome the soil moisture deficit during the dry spell. Rainfall aberration is one of the climate change–induced phenomena during the wet season. The possible effects of the abnormal rainfall are changes in the length of the growing period, moisture stress, poor germination and sub-optimum plant population (Table 8.2).

In crop production, contingency plan generally involves changes in sowing or planting time, change in crop for the use of shorter duration and stress tolerant crops and varieties, change in seed rate, change in fertilizer dose and schedule of application, change in plant population, use of growth regulators and change in harvesting time. These plans are prepared to cope with weather aberrations and natural calamities (drought, floods, unusual hailstorm, frost, cyclone, tsunami, sea water inundation due to damage of river bank, etc.). For preparing the contingency plan, the weather forecasts (short, medium and long range) of the India Meteorological Department (IMD) should be thoroughly taken into consideration. Interdisciplinary consultation of subject matter specialists (agriculture, animal husbandry, fisheries, irrigation and drainage, soil conservation, etc.) will strengthen the plan. Arrangement of inputs such as seeds, fertilizers, medicines and service providers ensures the timely execution of the contingency plan. The contingency measures for a rice-based cropping system mainly revolves around the agronomic management practices to reduce the damage caused by the unfavourable events (Table 8.3).

TABLE 8.2
Adverse Impact of Rainfall Aberrations on Crop Performance During Wet Season

Rainfall Aberration	Possible Impact on Crops
Delay in onset of the monsoon	Length of cropping season or cropping duration is reduced. Crop sowing is delayed.
Early withdrawal or cessation of monsoon	Moisture stress at maturity. Grain filling is affected.
Dry spell immediately after sowing	Germination affected; plant population reduced.
Dry spell at vegetative stage	Adversely affect stem elongation, leaf area expansion, branching or tillering.
Dry spell at flowering stage	Adversely affect anthesis, pollination, grain filling. Number of seeds/grains per inflorescence is reduced.
Dry spell at ripening	Grain filling is adversely affected. Grain size is reduced.

TABLE 8.3
Contingency Measures for Rice-Based Cropping System in Coastal Areas

Aberrant Weather Situation	Contingency Measures
Delay in onset of the monsoon by 2–6 weeks	Select short duration rice varieties (Shatabdi, Khitish, Swarna Mahsuri)
Delay in onset of the monsoon by >8 weeks	Kharif rice may be replaced with maize or vegetable crops (Brinjal, Chili, Okra)
Timely onset of the monsoon followed by a dry spell for 2–3 weeks resulting in poor crop stand	Gap filling if mortality is <20% by the use of reserve seedlings from nursery or by splitting of tillers from the existing health hills, re-sowing if the damage is >20%, straw mulching, dust mulching. Postpone top dressing of N and apply N as foliar spray (2% urea)
Mid-season drought due to a long dry spell of 2–3 weeks resulting in seedling mortality	Gap fill the damaged hills by the use of extra seedlings, mulching, postponement of top dressing of N fertilizer, spraying N as urea (2%), anti-transpirant, and KCl, thinning 33–50%, lifesaving irrigation.
Drought at flowering or reproductive stage	If damage is severe, harvest the crop for fodder purpose. Prepare for sowing for pulse crops (lentil, lathyrus, black gram, green gram) or mustard.
Heavy rainfall causing submergence during the nursery stage of rice	Adopt direct sowing of rice (DSR); if nursery is to be grown, select upland area, grow extra seedlings than the requirement, follow dapog method of nursery raising, choose submergence tolerant varieties (Swarna-Sub 1, CR 1009-Sub 1), if submerged gradually drain the excess water and apply K nutrient.
Heavy rainfall during the vegetative stage of rice causing seedling mortality	Utilize the extra nursery seedlings for gap filling, spray $ZnSO_4$ @ 0.2%. If there is problem of algal growth, apply $CuSO_4$ @ 2 g m^{-2} area.
Heavy rainfall during reproductive to maturity stage of rice	Postpone harvesting till cessation of rain and withdrawal of standing water. Create provision for drainage of excess water. Spray 2% brine solution or 5% common salt solution on panicles to prevent viviparous germination. After harvesting and threshing, dry the grains properly before storage.
High intensity rainfall during the rabi season causing waterlogging	Raised bed planting, provision of drainage by changing the field layout for all the rabi crops. Use sticker cum spreader (e.g., Ankur Gold @ 5 ml with 15 litres of spray volume).
Flooding in vegetable crops	Raised planting. Manage damping off by spraying Dithane M 45 @ 2g L^{-1} of water.

TABLE 8.4
Contingency Measures for Livestock Enterprise in Case of Drought and Flood

Aberrant Weather Situation	Contingency Measures
Drought	Utilization of river bank, pond dyke or periphery of other water harvesting structures for growing fodder crops. Select perennial fodder crops (Hybrid Napier Bajra, Paragrass, Nandigrass) or multi-cut fodder crops such as sorghum, Sudan grass, oats. Preparation of hay and silage from left over fodder. Water harvesting and water conservation measures, desilting of pond. Vaccination of animals, stocking emergency medicines, deworming of all animals and quarantine of sick animals. During the drought, for poultry birds mix Vitamins (A, B-Complex, C, D, E, K) @ 5 ml L^{-1} of drinking water.
Flood	Construction of cattle sheds or houses of animals/birds in elevated areas to prevent entry of flood water. Insurance for animals. On the basis of the weather forecast, harvest the fodder crops for safe storage and use during the flood. Provision for hay and silage making. Shift animals to a safer place in case of cyclone. Untie the animals during the storm period. Spray Butax for management of external parasites. Spray fly repellents in cattle sheds. Use clean drinking water for animals. Deworming and insurance claim after the event. During the flood, for poultry birds mix anti-biotic powder (Terramycin) @ 10 g L^{-1} of drinking water as a precaution to avoid disease outbreak. In damp areas use lime powder to prevent ammonia accumulation.
Heat wave	Shelter should be designed to protect from heat with proper ventilation. Provision of plenty of drinking water. Select heat-stress tolerant breeds. Ensure adequate space for each animal, if required reduce the stocking density. Allow resting (stop movement) of the animals during the hot period of the day. Observe for heat stress symptoms of animals and immediately treat the affected one. Sprinkle water on the cattle to wet the skin.

TABLE 8.5
Contingency Measures for Aquaculture During Drought and Floods

Aberrant Weather Situation	Contingency Measures
Drought	Desilting of ponds, raising the embankment height to increase the volume of harvested water, if water level recedes the critical level supply water from other sources (tube well, canal, river), create an environment for resting of fish by placing tree branches, old pipes and broken tiles. If water depth is low, partially harvest the fish, reduce feed supplement, check dissolved oxygen level, test water for quality parameters (salinity and pH) and take reclamation measures. Monitor growth, disinfect water, if needed do liming, maintain plankton and provide feed supplements.
Floods	Seek government insurance schemes. Redesign pond dyke height based on the historical extreme rainfall events. Fix nets around the pond/culture area to prevent the escape of fishes. Harvest the fishes of commercial sizes. After the event, repair the damaged infrastructure, including the bonds/ embankments. Drain the excess water. Check the stocking rate and water quality. Maintain optimum pH and salinity of water. If pH of water is low, apply lime and monitor the health of fish. Take preventive measures against outbreak of diseases. Insurance claim settlement.

A contingency plan for all the components of the IFS needs to be planned thoroughly. The key risks for livestock enterprise are to be identified including the diseases, so that the health of the animals is maintained and economic loss is minimized. The contact numbers of important service providers such as veterinarian, feed supplier, local emergency services (fire, power, gas, police) should be kept ready. Alternative strategies are to be adopted in case of drought and flood (Table 8.4).

Scientific aquaculture practices are essential to protect this enterprise during climate change–induced natural calamities. Drought, flood, low/high temperature, cyclones and tsunami affect the aquaculture enterprise adversely, hence contingency measures to handle such disasters are essential. Disease outbreak is another important hazard to the aquaculture, which needs utmost attention in all stages of production to processing. Maintenance of water quality and hygiene are essential to protect from diseases. Contingency measures are essential for the prevention, preparedness, early warning, impact/need assessment and sustainable recovery of aquaculture from major disasters (Table 8.5). The farmer/farming community should gain knowledge and expertise on preparedness for climate change by attending government programmes/trainings by different agencies and getting linked with sources of information on weather forecast and disaster-related bulletins.

REFERENCES

Aslam, S.N., Navada, S., Bye, G.R., Mota, V.C., Terjesen, B.F., Mikkelsen, O. 2019. Effect of CO_2 on elemental concentrations in recirculating aquaculture system tanks. *Aquaculture* 511:734254. DOI: https://doi.org/10.1016/j.aquaculture.2019.734254.

Auffhammer, M., Ramanathan, V., Vincent, J.R. 2011. Climate change, the monsoon and rice yield in India. *Climate Change* 111:411–424. https://link.springer.com/article/10.1007/s10584-011-0208-4.

Bal, S.K. 2021. Agro-meteorological basis of extremes of temperature with special perspective to livestock and poultry. In: Pankaj P.K., Phand S., Nirmala G., Kumar R.N., Pushpanjali (eds) *Climate Resilient Animal Husbandry*. ICAR-Central Research Institute for Dryland Agriculture and National Institute of Agricultural Extension Management, Hyderabad, India. pp 23–31; 137. Available online: https://www.manage.gov.in/publications/eBooks/Climate%20Resilient%20Animal%20Husbandry.pdf. Accessed on 24 March 2023.

Bamber, J.L., Oppenheimer, M., Kopp, R.E., Aspinall, W.P., Cooke, R.M. 2019. Ice sheet contributions to future sea level rise from structured expert judgement. *The Proceedings of the National Academy of Sciences, USA* 116:11195–11200. DOI: www.pnas.org/cgi/doi/10.1073/pnas.1817205116.

Bell, L.W., Moore, A.D. 2012. Integrated crop-livestock systems in Australian agriculture: Trends, drivers and implications. *Agricultural Systems* 111:1–12. DOI: https://doi.org/10.1016/j.agsy.2012.04.003.

Bernabucci, U. 2019. Climate change: Impact on livestock and how can we adapt. *Animal Frontiers* 9(1):3–5. DOI: https://doi.org/10.1093/af/vfy039.

Bhargava, S., Mitra, S. 2020. Elevated atmospheric CO_2 and the future of crop plants. *Plant Breeding* 140(1):1–11. DOI: https://www.onlinelibrary.wiley.com/doi/epdf/10.1111/pbr.12871.

Bindi, M., Fibbi, L., Miglietta, F. 2001. Free air CO_2 enrichment (FACE) of grapevine (*Vitis vinifera* L.): II. Growth and quality of grape and wine in response to elevated CO_2 concentrations. *European Journal of Agronomy* 14(2):145–155. DOI: https://doi.org/10.1016/S1161-0301(00)00093-9.

Biswas, J.C., Haque, M.M., Maniruzzaman, M., Ali, M.H., Kabir, W., Kalra, N. 2019. Natural hazards and livestock damage in Bangladesh. *Natural Hazards* 99:705–714. DOI: https://doi.org/10.1007/s11069-019-03768-0.

Blenkinsop, S., Muniz Alves, L., Smith, A.J.P. 2021. Science brief review: Climate change increases extreme rainfall and the chance of floods. In: Le Quere C., Liss P., Foster P. (eds) *Critical Issues in Climate Change Science*: 1–5. DOI: https://doi.org/10.5281/zenodo.4779119.

Bohmanova, J., Misztal, I., Cole, J.B. 2007. Temperature-humidity indices as indicators of milk production losses due to heat stress. *Journal of Dairy Science* 90(4):1947–1956. DOI: https://doi.org/10.3168/jds.2006-513.

Campana, S.E., Stefansdottir, R.B., Jakobsdottir, K., Solmundsson, J. 2020. Shifting fish distributions in warming sub-Arctic Oceans. *Scientific Reports* 10:16448. DOI: https://doi.org/10.1038/s41598-020-73444-y.

Cheng, M., McCarl, B., Fei, C. 2022. Climate change and livestock production: A literature review. *Atmosphere* 13(1):140. DOI: https://doi.org/10.3390/atmos13010140.

Chukwu Okeah, G.O., Wekpe, V.E., Ikebude, C.F. 2018. Impact of coastal flooding on fish production in Brass, Niger Delta Nigeria, implications for coastal resource management. *Oceanography & Fisheries* 6(1):555678. DOI: https://www.10.19080/OFOAJ.2018.06.555678.

Crist, S., Mori, J., Smith, R.L. 2020. Flooding on beef and swine farms: A scoping review on effects in the midwestern United States. *Preventive Veterinary Medicine* 184:105158. DOI: https://doi.org/10.1016/j.preventmed.2020.105158.

Dao, H.N., Vu, H.T., Kay, S., Sailley, S. 2021. Impact of seawater temperature on coral reefs in the context of climate change – a case study of Cu Lao Cham – Hoi An biosphere reserve. *Frontiers in Marine Science* 8:704682. DOI: https://doi.org/10.3389/fmars.2021.704682.

Datta, S. 2013. Impact of climate change in Indian horticulture - a review. *International Journal of Science, Environment and Technology* 2(4):661–671. https://www.ijset.net/journal/139.pdf.

Dayan, H., Cozannet, G.L., Speich, S., Thieblemont, R. 2021. High-end scenarios of sea-level rise for coastal risk-averse stakeholders. *Frontiers in Marine Science* 8:569992. https://www.frontiersin.org/articles/10.3389/fmars.2021.569992/full.

Djanaguiraman, M., Narayanan, S., Eradayani, E., Prasad, P.V.V. 2020. Effects of high temperature stress during anthesis and grain filling periods on photosynthesis, lipids and grain yield in wheat. *BMC Plant Biology* 20(268). DOI: https://doi.org/10.1186/s12870-020-02479-0.

Elsheikh, W. 2021. Effects of climate change on aquaculture production. *Eurasian Journal of Food Science and Technology* 5(2):167–173. https://dergipark.org.tr/en/download/article-file/2122988.

FAO (Food and Agricultural Organization of the United Nations). 1996. *Agro-Ecological Zoning Guidelines*. Rome, Italy. Available online at https://www.fao.org/3/w2962e/w2962e00.htm. Accessed on 27 March 2023.

FAO (Food and Agricultural Organization of the United Nations). 2016a. *Climate Change and Food Security: Risks and Responses*. Rome, Italy. Available online: https://www.fao.org/3/i5188e/I5188E.pdf. Accessed on 24 March 2023.

FAO (Food and Agricultural Organization of the United Nations). 2016b. *Weather and Desert Locusts*. Rome, Italy. Available online: https://www.fao.org/3/i6152en/i6152en.pdf. Accessed on 03 April 2023.

Febrianty, I., Mahyudin, I., Camelila, D., Amalia, A., Rahmi, N.S. 2021. The impact of floods on fish cultivation business in floating net cages in Sungai Alang village, Karang Intan district, Banjar regency, South Kalimantan. *Journal of Economics, Finance and Management Studies* 4(8):1352–1356. https://ijefm.co.in/v4i8/11.php.

Gil, J.D.B., Cohn, A.S., Duncan, J., Newton, P., Vermeulen, S. 2017. The resilience of integrated agricultural systems to climate change. *WIREs Climate Change* 8:e461. DOI: https://doi.org/10.1002/wcc.461.

Grossi, G., Goglio, P., Vitali, A., Williams, A.G. 2019. Livestock and climate change: Impact of livestock on climate and mitigation strategies. *Animal Frontiers* 9(1):69–76. DOI: https://doi.org/10.1093/af/vfy034.

Guan, J., Li, M., Ju, X., Lin, J., Wu, J., Zheng, J. 2021. The potential habitat of desert locusts is contracting: Predictions under climate change scenarios. *PeerJ* 9:e12311. DOI: https://doi.org/10.7717/peerj.12311.

Gupta, S., Sharma, A., Joy, A., Dunshea, F.R., Chauhan, S.S. 2023. The impact of heat stress on immune status of dairy cattle and strategies to ameliorate the negative effects. *Animals* 13(1):107. DOI: https://doi.org/10.3390/ani13010107.

Herrero, M., Thornton, P.K., Notenbaert, A., Msangi, S., Wood, S., Kruska, R., Dixon, J., Bossio, D., van de Steeg, J., Freeman, H.A., Li, X., Rao, P.P. 2012. *Drivers of Change in Crop-Livestock Systems and Their Potential Impacts on Agro-Ecosystems Services and Human Wellbeing to 2030: A Study Commissioned by the CGIAR Systemwide Livestock Programme*. International Livestock Research Institute, Nairobi, Kenya. Available online at https://cgspace.cgiar.org/bitstream/handle/10568/3020/SLP%20drivers%20study%20final%20draft.pdf.

Hogy, P., Fangmeier, A. 2009. Atmospheric CO_2 enrichment affects potatoes: 2. Tuber quality traits. *European Journal of Agronomy* 30:85–94. DOI: https://doi.org/10.1016/j.eja.2008.07.006.

https://sealevel.nasa.gov/understanding-sea-level/regional-sea-level/ice-sheets#:~:text=The%20Greenland%20Ice%20Sheet%2C%20the sea%20level%20by2058%20meters. Accessed on 28 March 2023.

https://www.icar-crida.res.in/CP-2012/index.html. Accessed on 10 April 2023.

https://www.nutrienagsolutions.com/au/farmers-facing-long-term-impact-floods. Accessed on 28 March 2023.

IFAD (International Fund for Agricultural Development). 2010. *Livestock and Climate Change*. http://www.ifad.org/lrkm/events/crops/papers/climate.pdf.

Iglesias, A., Avis, K., Benzie, M., Fisher, P., Harley, M., Hodgson, N., Horrocks, L., Moneo, M., Webb, J. 2007. *Adaptation to Climate Change in the Agricultural Sector.* AEA Energy and Environment and Universidad de Pditecnica de Madrid. Available online at https://agriculture.ec.europa.eu/system/files/2020-02/ext-study-adapt-climate-change-full-text_2007_en_0.pdf. Accessed on 27 March 2023.

IPCC. 2014. *Climate Change 2014. Synthesis Report.* Contribution of Working Groups I, II and III to the Fifth Assessment Report of the Intergovernmental Panel on Climate Change. Core Writing Team. Pachauri R.K., Meyer L.A. (eds) IPCC, Geneva, Switzerland. p 151. Available online: https://www.ipcc.ch/site/assets/uploads/2018/02/SYR_AR5_FINAL_full.pdf. Accessed on 13 March 2023.

IPCC. 2001. *Climate Change 2001. Synthesis Report.* A contribution of Working Groups I, II and III to the Third Assessment Report of the Intergovernmental Panel on Climate Change. Watson, R.T., and the Core Writing Team (eds) Cambridge University Press, Cambridge, United Kingdom and New York, NY, USA. pp 398. Available online: https://www.ipcc.ch/site/assets/uploads/2018/05/SYR_TAR_full_report.pdf. Accessed on 13 March 2023.

IPCC. 2007a. *Climate Change 2007. Climate Change Impacts, Adaptation and Vulnerability.* Contribution of working group II to the fourth assessment report of the Intergovernmental Panel on Climate Change (IPCC). Parry M.L., Canziani O.F., Palutikof J.P., van der Linden P.J., Hanson C.E. (eds) Cambridge University Press, Cambridge, UK. pp 976. Available online: https://www.ipcc.ch/site/assets/uploads/2018/03/ar4_wg2_full_report.pdf. Accessed on 07 March 2023.

IPCC. 2007b. *Climate Change 2007. Synthesis Report.* Contribution of Working Groups I, II and III to the Fourth Assessment Report of the Intergovernmental Panel on Climate Change. Core Writing Team. Pachauri R.K., Reisinger A. (eds) IPCC, Geneva, Switzerland. pp 104. Available online: https://www.ipcc.ch/site/assets/uploads/2018/02/ar4_syr_full_report.pdf. Accessed on 13 March 2023.

IPCC. 2007c. *Climate Change 2007. The Physical Science Basis.* Contribution of working group I to the fourth assessment report of the Intergovernmental Panel on Climate Change (IPCC). Solomon S., Qin D., Manning M., Chen Z., Marquis M., Averyt K.B., Tignor M., Miller H.L. (eds) Cambridge University Press, Cambridge, and New York, NY. pp 996. Available online: https://www.ipcc.ch/site/assets/uploads/2018/05/ar4_wg1_full_report-1.pdf. Accessed on 07 March 2023.

Jack, A.A., Adegbeye, M.J., Reddy, P.R.K., Elghandour, M.M.M.Y., Salem, A.Z.M., Adewumi, M.K. 2021. Ruminant productivity among smallholders in a changing climate: Adaptation strategies. In: Lackner M., Sajjadi B., Chen W.Y. (eds) *Handbook of Climate Change Mitigation and Adaptation.* Springer, New York, NY. pp 1–41. DOI: https://doi.org/10.1007/978-1-4614-6431-0-148-1.

Jain, S., Sharma, G., Mathur, Y.P. 2013. Effects of temperature variations on fish in lakes. *International Journal of Engineering Research & Technology* 2(10):2516–2523. https://www.ijert.org/research/effects-of-temperature-variations-on-fish-in-lakes-IJERTV2IS100744.pdf.

King, M., Altdorff, D., Li, P., Galagedara, L., Holden, J., Unc, A. 2018. Northward shift of the agricultural climate zone under 21st century global climate change. *Scientific Reports* 8:7904. DOI: https://doi.org/10.1038/s41598-018-26321-8.

Kirby, M., Mainuddin, M. 2022. The impact of climate change, population growth and development on sustainable water security in Bangladesh to 2100. *Scientific Reports* 12:22344. DOI: https://doi.org/10.1038/s41598-022-26807-6.

Lee, Y.H., Sang, W.G., Baek, J.K., Kim, J.H., Shin, P., Seo, M.C., Cho, J.I. 2020. The effect of concurrent elevation in CO_2 and temperature on the growth, photosynthesis, and yield of potato crops. *PLoS One* 15(10):e0241081. https://www.10.1371/journal.pone.0241081.ecollection2020.

Lehmann, P., Ammunet, T., Barton, M., Battisti, A., Eigenbrode, S.D., Jepsen, J.U., Kalinkat, G., Neuvonen, S., Niemela, P., Terblanche, J.S., Okland, B., Bjorkman, C. 2020. Complex responses of global insect pests to climate warming. *Frontiers in Ecology and the Environment* 18(3):141–150. DOI: https://esajournals.onlinelibrary.wiley.com/doi/full/10.1002/fee.2160.

Lehtonen, H., Aakkula, J., Fronzek, S., Helin, J., Hilden, M., Huttunen, S., Kaljonen, M., Niemi, J., Palosuo, T., Pirttioja, N., Rikkonen, P., Varho, V., Carter, T.R. 2021. Shared socioeconomic pathways for climate change research in Finland: Co-developing extended SSP narratives for agriculture. *Regional Environmental Change* 21(7). DOI: https://doi.org/10.1007/s10113-020-01734-2.

Leweri, C.M., Msuha, M.J., Treydte, A.C. 2021. Rainfall variability and socio-economic constraints on livestock production in the Ngorongoro conservation area, Tanzania. *SN Applied Sciences* 3(123). DOI: https://doi.org/10.1007/s42452-020-04111-0.

Mainuddin, M., Karim, F., Gaydon, D.S., Kirby, J.M. 2021. Impact of climate change and management strategies on water and salt balance of the polders and islands in the Ganges delta. *Scientific Reports* 11:7041. DOI: https://doi.org/10.1038/s41598-021-86206-1.

Maulu, S., Hasimuna, O.J., Haambiya, L.H., Monde, C., Musuka, C.G., Makorwa, T.H., Munganga, B.P., Phiri, K.J., Nsekanabo, J.D. 2021. Climate change effects on aquaculture production: Sustainability implications, mitigation and adaptations. *Frontiers in Sustainable Food Systems* 5:609097. https://www.10.3389/fsufs.2021.609097.

McCosker, E., Stuart-Smith, R.D., Edgar, G.J., Steinberg, P.D., Verges, A. 2022. Sea temperature and habitat effects on juvenile reef fishes along a tropicalizing coastline. *Diversity and Distributions* 28:1154–1170. DOI: https://doi.org/10.1111/ddi.13484.

Montgomery, D.W., Simpson, S.D., Engelhard, G.H., Birchenough, S.N.R., Wilson, R.W. 2019. Rising CO_2 enhances hypoxia tolerance in a marine fish. *Scientific Reports* 9:15152. https://www.10.1038/s41598-019-51572–51574.

Mota, V.C., Nilsen, T.O., Gerwins, J., Gallo, M., Ytteborg, E., Baeverfjord, G., Kolarevic, J., Summerfelt, S.T., Terjesen, B.F. 2019. The effects of carbon dioxide on growth performance, welfare, and health of Atlantic salmon post-smolt (*Salmo salar*) in recirculating aquaculture systems. *Aquaculture* 498:578–586. DOI: https://doi.org/10.1016/j.aquaculture.2018.08.075.

Muhala, V., Chicombo, T.F., Macate, I.E., Guimaraes-Costa, A., Gundana, H., Malichocho, C., Hasimuna, O.J., Remedio, A., Maulu, S., Cuamba, L., Bessa-Silva, A.R., Sampaio, I. 2021. Climate change in fisheries and aquaculture: Analysis of the impact caused by Idai and Kenneth cyclones in Mozambique. *Frontiers in Sustainable Food Systems* 5:714187. DOI: https://doi.org/10.3389/fsufs.2021.714187.

Mulabagal, V., Baach, D.A., Egiebor, N.O., Sajjadi, B., Chen, W., Viticoski, R.L., Hayworth, J.S. 2021. Biochar from biomass: A strategy for carbon dioxide sequestration, soil amendment, power generation, CO_2 utilization and removal of Perfluoroalkyl and Polyfluoroalkyl Substances (PFAS) in the Environment. In: Lackner M., Sajjadi B., Chen W.Y. (eds) *Handbook of Climate Change Mitigation and Adaptation.* Springer, New York, NY. pp 1023–1085. DOI: https://doi.org/10.1007/978-3-030-72579-2_80.

Nardone, A., Ronchi, B., Lacetera, N., Ranieri, M.S., Bernabucci, U. 2010. Effects of climate changes on animal production and sustainability of livestock systems. *Livestock Science* 130(1–3):57–69. DOI: https://doi.org/10.1016/j.livsci.2010.02.011.

Ngarava, S., Zhou, L., Mushunje, A., Chaminuka, P. 2021. Impacts of floods on livestock production in Port St Johns, South Africa. In: Nhamo G., Chapungu L. (eds) *The Increasing Risk of Floods and Tornanodes in Southern Africa.* Sustainable Developments Goals Series. Springer, Cham. pp 221–237. DOI: https://doi.org/10.1007/978-3-030-74192-1_13.

Nguyen, P.H., Greene, E., Donoghue, A., Huff, G., Clark, F.D., Dridi, S. 2016. A new insight into cold stress in poultry production. *Advances in Food Technology and Nutritional Sciences* 2(1):1–2. DOI: http://dx.doi.org/10.17140/AFTNSOJ-2-124.

Nicholls, R.J., Marinova, N., Lowe, J.A., Brown, S., Vellinga, P., de Gusmao, D., Hinkel, J., Tol, S.J. 2011. Sea-level rise and its possible impacts given a 'beyond 4°C world' in the twenty first century. *Philosophical Transactions of the Royal Society A. Mathematical, Physical and Engineering Sciences* 369:161–181. DOI: https://doi.org/10.1098/rsta.2010.0291.

Oerke, E.C. 2006. Crop losses to pests. *The Journal of Agricultural Science* 144(1):31–43. DOI: https://doi.org/10.1017/S0021859605005708.

Ogorman, E.J., Olafsson, O.P., Demars, B.O.L., Friberg, N., Gudbergsson, G., Hannesdottir, E.R., Jackson, M.C., Johansson, L.S., Mclau-ghlin, O.B., Olafsson, J.S., Woodward, G., Gislason, G.M. 2016. Temperature effects on fish production across a natural thermal gradient. *Global Change Biology* 22:3206–3220. DOI: https://onlinelibrary.wiley.com/doi/full/10.1111/gcb.13233.

Olutumise, A.I. 2023. Impact of relaxing flood policy interventions on fish production: Lessons from earthen pond-based farmers in Southwest Nigeria. *Aquaculture International* 31:1855–1878. DOI: https://doi.org/10.1007/s10499-023-01062-2.

Partnership for Market Readiness (PMR). 2017. *A Guide to Greenhouse Gas Benchmarking for Climate Policy Instruments.* World Bank, Washington DC. License: Creative Commons Attribution CC BY 3.0 IGO. Available online at: https://openknowledge.worldbank.org/bitstream/handle/10986/26848/115503-WP-PUBLIC-ADD-SERIES-PMRTechnicalNoteBenchmarkweb.pdf?sequence=1&isAllowed=y. Accessed on 07 March 2023.

Pawar, S.S., Sajjanar, B., Lonkar, V.D., Kurade, N.P., Kadam, A.S., Nirmal, A.V., Brahamane, M.P., Bal, S.K. 2016. Assessing and mitigating the impact of heat stress on poultry. *Advances in Animal and Veterinary Sciences* 4(6):332–341. DOI: http://dx.doi.org/10.14737/journal.aavs/2016/4.6.332.341.

Portner, H.O., Farrell, A.P. 2008. Ecology, physiology and climate change. *Science* 322(5902):690–692. DOI: https://www.science.org/doi/10.1126/science.1163156.

Rahal, A., Kumar, A., Singh, V., Yadav, B., Tiwari, R., Chakraborty, S., Dhama, K. 2014. Oxidative stress, prooxidants, and antioxidants: The interplay. *BioMed Research International* 2014:761264. https://www.10.1155/2014/761264.

Rasool, S., Hamdani, S.A., Ayman, N., Fayaz, A., Shubeena, S., Thahaby, N., Nabi, B., Hai, A., Akand, A.H. 2021. The impact of natural disasters on livestock sector: A review. *Journal of Biomedical Research & Environmental Sciences* 2(8):669–674. https://www.jelsciences.com/articles/jbres1295.pdf.

Riahi, K., Rao, S., Krey, V., Cho, C., Chirkov, V., Fischer, G., Kindermann, G., Nakicenovic, N., Rafaj, P. 2011. RCP 8.5 – A scenario of comparatively high greenhouse gas emissions. *Climate Change* 109(33). DOI: https://doi.org/10.1007/s10584-011-0149-y.

Riahi, K., van Vuuren, D.P., Kriegler, E., Edmonds, J., O'Neill, B.C., Fujimori, S., Bauer, N., Calvin, K., Dellink, R., Fricko, O., Lutz, W., Popp, A., Cuaresma, J.C., Samir, K.C., Leimbach, M., Jiang, L., Kram, T., Rao, S., Emmerling, J., Ebi, K., Hasegawa, T., Havlik, P., Humpenöder, F., Da Silva, L.A., Smith, S., Stehfest, E., Bosetti, V., Eom, J., Gernaat, D., Masui, T., Rogelj, J., Strefler, J., Drouet, L., Krey, V., Luderer, G., Harmsen, M., Takahashi, K., Baumstark, L., Doelman, J.C., Kainuma, M., Klimont, J., Marangoni, G., Lotze-Campen, H., Obersteiner, M., Tabeau, A., Tavoni, M. 2017. The shared socioeconomic pathways and their energy, land use, and greenhouse gas emissions implications: An overview. *Global Environmental Change* 42:153–168. DOI: https://doi.org/10.1016/j.gloenvcha.2016.05.009.

Rojas-Downing, M.M., Nejadhashemi, A.P., Harrigan, T., Woznicki, S.A. 2017. Climate change and livestock: Impacts, adaptation and mitigation. *Climate Risk Management* 16:145–163. DOI: http://dx.doi.org/10.1016/j.crm.2017.02.001.

Rulli, M.C., Dodorico, P., Galli, N., Hayman, D.T.S. 2021. Land-use change and the livestock revolution increase the risk of zoonotic corona virus transmission from rhinolophid bats. *Nature Food* 2:409–416. DOI: https://doi.org/10.1038/s43016-021-00285-x.

Rust, J.M., Rust, T. 2013. Climate change and livestock production: A review with emphasis on Africa. *South African Journal of Animal Science* 43(3):255–267. DOI: https://dx.doi.org/10.4314/sajas.v43i3.3.

Rutkayova, J., Vacha, F., Marsalek, M., Benes, K., Civisova, H., Horka, P., Petraskova, E., Rost, M., Sulista, M. 2017. Fish stock losses due to extreme floods-findings from pond-based aquaculture in the Czech Republic. *Journal of Flood Risk Management* 11(2):351–359. DOI: https://www.onlinelibrary.wiley.com/doi/epdf/10.1111/jfr3.12332.

Salih, A.A.M., Baraibar, M., Mwangi, K.K., Artan, G. 2020. Climate change and locust out-break in East Africa. *Nature Climate Change* 10:584–585. DOI: https://doi.org/10.1038/s41558-020-0835-8.

Salik, K.M., Suleri, A.Q. 2020. *Understanding Climate Change and Rising Locust Outbreak Risks in Pakistan.* Sustainable Development Policy Institute. Available online: https://sdpi.org/assets/lib/uploads/uploads/2020/08/Desert-Locust-final.pdf.

Schneider, L., Rebetez, M., Rasmann, S. 2022. The effect of climate change on invasive crop pests across biomes. *Current Opinion in Insect Science* 50:100895. DOI: https://doi.org/10.1016/j.cois.2022.100895.

Seguin, B. 2008. The consequences of global warming for agriculture and food production. In: Rowlinson P., Steele M., Nefzaou A. (eds) *Livestock and Global Climate Change.* Cambridge University Press, Hammamet, Tunisia. pp 9–11.

Sejian, V., Gaughan, J.B., Bhatta, R., Naqvi, S.M.K. 2016. *Impact of Climate Change on Livestock Productivity.* Available online: https://www.feedipedia.org/sites/default/files/public/BH_026_climate_change_livestock.pdf. Accessed on 24 March 2023.

Singh, V.K. 2021. Dryland agriculture and animal husbandry under climate change scenario. In: Pankaj P.K., Phand S., Nirmala G., Kumar R.N., Pushpanjali (eds) *Climate Resilient Animal Husbandry.* ICAR-Central Research Institute for Dryland Agriculture and National Institute of Agricultural Extension Management, Hyderabad, India. pp 1–5; 137. Available online: https://www.manage.gov.in/publications/eBooks/Climate%20Resilient%20Animal%20Husbandry.pdf. Accessed on 24 March 2023.

Skendzic, S., Zovko, M., Zivkovic, I.P., Lesic, V., Lemic, D. 2021. The impact of climate change on agricultural insect pests. *Insects* 12(5):440. DOI: https://doi.org/10.3390/insects12050440.

Sohail, M.U., Hume, M.E., Byrd, J.A., Nisbet, D.J., Ijaz, A., Sohail, A., Shabbir, M.Z., Rehman, H. 2012. Effect of supplementation of prebiotic mannanoligo saccharides and probiotic mixture on growth performance of broilers subjected to chronic heat stress. *Poultry Science* 91(9):2235–2240. DOI: https://doi.org/10.3382/ps.2012-02182.

Southgate, P. 2005. *The Importance of Measuring Carbon Dioxide in Aquaculture.* Available online at https://www.thefishsite.com/articles/the-importance-of-measuring-carbon-dioxide-in-aquaculture. Accessed on 03 April 2023.

Steinberg, K., Zimmermann, J., Stiller, K.T., Meyer, S., Schulz, C. 2017. The effect of carbon dioxide on growth and energy metabolism in pikeperch (*Sander lucioperca*). *Aquaculture* 481:162–168. DOI: https://doi.org/10.1016/j.aquaculture.2017.09.003.

Streck, N.A. 2005. Climate change and agroecosystems: The effect of elevated atmospheric CO_2 and temperature on crop growth, development, and yield. *Crop Production* 35(3):730–740. DOI: https://doi.org/10.1590/S0103-84782005000300041.

Sujayanand, G.K., Karuppaiah, V. 2016. Aftermath of climate change on insect migration: A review. *Agricultural Reviews* 37(3):221–227. DOI: https://doi.org/10.18805/ag.v37i3.3537.

Tabari, H. 2020. Climate change impact on flood and extreme precipitation increases with water availability. *Scientific Reports* 10:13768. DOI: https://doi.org/10.1038/s41598-020-70816-2.

Talbot, C.J., Bennett, E.M., Cassell, K., Hanes, D.M., Minor, E.C., Paerl, H., Raymond, P.A., Vargas, R., Vidon, P.G., Wollheim, W., Xenopoulos, M.A. 2018. The impact of flooding on aquatic ecosystem services. *Biogeochemistry* 141(3):439–461. https://www.10.1007/s10533-018-0449-7.

Wassmann, R., Jagadish, S.V.K., Heuer, S., Ismail, A., Redona, E., Serraj, R., Singh, R.K., Howell, G., Pathak, H., Sumfleth, K. 2009a. Climate change affecting rice production: The physiological and agronomic basis of possible adaptation strategies. *Advances in Agronomy* 101:59–122. https://www.10.1016/S0065-2113(08)00802-X.

Wassmann, R., Jagadish, S.V.K., Sumfleth, K., Pathak, H., Howell, G., Ismail, A., Serraj, R., Redona, E., Singh, R.K., Heuer, S. 2009b. Regional vulnerability of climate change impacts on Asian rice production and scope for adaptation. *Advances in Agronomy* 102:91–133. https://www/10.1016/S0065-2113(09)01003-7.

Yadav, M.R., Choudhary, M., Singh, J., Lal, M.K., Jha, P.K., Udawat, P., Gupta, N.K., Rajput, V.D., Garg, N.K., Maheshwari, C., Hasan, M., Gupta, S., Jatwa, T.K., Kumar, R., Yadav, A.K., Prasad, P.V.V. 2022. Impacts, tolerance, adaptation, and mitigation of heat stress on wheat under changing climates. *International Journal of Molecular Science* 23:2838. DOI: https://doi.org/10.3390/ijms23052838.

9 IFS Models for Different Agro-climatic Zones

India is divided into 15 agroclimatic zones by the Planning Commission on the basis of climatic as well as physiographic characteristics. The Himalayan zone consists of Western Himalaya (Ladakh, Kashmir, Punjab, Jammu; brown soils and silty loam, steep slopes) and Eastern Himalaya (Arunachal Pradesh, Sikkim and Darjeeling, Manipur; high rainfall and high forest covers heavy soil erosion, floods). The Gangetic region is divided into four zones. The upper Gangetic plains (North region of U.P.) are irrigated by canal and tube wells with rich groundwater resources. The middle Gangetic plains covering Bihar and Uttar Pradesh is a high rainfall region with 39% irrigate area and 142% cropping intensity. The lower Gangetic plains include West Bengal, where the soils are mostly alluvial and this region is flood-prone. The trans-Gangetic plains cover Punjab, Haryana, Union territory of Delhi and Chandigarh and Sriganganagar district of Rajasthan. This zone has the highest net sown area, cropping intensity, and it is irrigated with ground water. The plateau region of India is divided into four agro-climatic zones. The Eastern plateaus and hills region consists of Chota Nagpur, Gujrat hills, Madhya Pradesh, W. Baghelkhand plateau and Odisha. The soils are shallow to medium sloping, undulating, and irrigation is from tank and tube wells. Major parts of Madhya Pradesh, Uttar Pradesh and Rajasthan constitute the Central plateau and hills region. Western plateau and hills region covers Sahyadry, Maharashtra and Madhya Pradesh. Rainfall in this region is 904 mm with sown area 65%, forest 11% and irrigation coverage 12.4%. Southern plateau and hills is typically a semi-arid zone covering Tamil Nadu, Andhra Pradesh and Karnataka. In this zone, dry land farming is practiced in 81% of the area. India's coastal region constitutes East coast plains and hills zone (Odisha, Andhra Pradesh, Tamil Nadu and Pondicherry; soils of this zone are alluvial and coastal sand with irrigation sources from canals and tanks), West coast plains and hills region (Sourashtra, Maharashtra, Goa, Karnataka and Tamil Nadu with variety of cropping pattern, rainfall and soil types) and Gujarat plains and hills region (19 districts of Gujarat, Dadra & Nagar Haveli and Daman and Diu). Western dry region covers nine districts of Rajasthan (hot, sandy desert, rainfall erratic and high evaporation; scanty vegetation, famine and droughts). The Island region comprises of Eastern Andaman, Nicobar, Western Lakshadweep (typical equatorial, rainfall 3,000 mm in nine months, forest zone and undulating).

9.1 IFS MODEL FOR MEDIUM AND LARGE HOLDINGS OF WESTERN DRY ZONE AND TRANS-GANGETIC PLAINS

As per the Agricultural Census Division, Government of India, farmers having 10 ha and above land are classified as large farmers (Table 9.1), and the average size of holdings for large farmers during 2015–2016 was 17.07 ha (Agriculture Census 2015–2016, 2019).

Highest number (35,000) and area (562,000 ha) under large category of holding is found in Rajasthan. The western dry agroclimatic zone comprises nine districts of Rajasthan. In this dry zone, agro-forestry and livestock-based IFS consisting of trees, crops, pasture, cows, goats and sheep are recommended. Crops (pearl millet, green gram, dew gram, cluster bean) constitute 50% of the area, the livestock (4 cows and 15 goats/sheep) share 30% while trees such as khejri (*Prosopis cineraria*), ber (*Ziziphus mauritiana*), Israeli babool (*Acacia tortilis*) and anjan grass (*Cenchrus ciliaris*) cover 20% of the area of the 7 ha IFS model (Panwar et al. 2019). This IFS model generates a

DOI: 10.1201/9781003239963-9

TABLE 9.1
Classification of Farmers Based on Holding Size and Average Land Holding for Each Category During 2015–2016

Category	Classification (ha)	Avg. (ha)
Marginal	<1	0.38
Small	1–<2	1.40
Semi-medium	2–<4	2.69
Medium	4–<10	5.72
Large	10 and above	17.07

net income of ₹2.6 lakh yr^{-1} and employment of 920 man-days. The highest (46%) income is from trees and grass, followed by crops (32%) and livestock (22%).

The Trans-Gangetic Plains zone is an irrigated zone with crops as the major component of IFS model. In a model of crops and fruits farming system covering 34.86 ha area under crops and 14.41 ha area under fruits resulted net income of about ₹20.56 lakh yr^{-1} from crops and about ₹0.69 lakh yr^{-1} from fruits (Singh et al. 2017a). In a study in Punjab (Singh et al. 2017b), it was observed that the majority (63%) of the respondents adopted crops + dairy based IFS, followed by crops + fruits (13.7%), crops + beekeeping (9.6%), crops + floriculture (6.9%) and crops + poultry (4.1%). Three major IFS models (crop+fishery, crop+dairy and crop+piggery+forestry) are also reported from Punjab with farmers interest to expand area under the IFS model as well as to increase the number of the enterprises under each IFS model (Singh et al. 2021).

9.2 COMMERCIAL IFS MODELS

This is a market-oriented system of farming, where a large area is diverted to enterprises for bulk production meant for export. Commercial agriculture/agribusiness involves raising crops and livestock with the purpose to wholesale the harvests on the market to make cash. Tea, coffee, rubber, tobacco, sugarcane, cotton, jute and banana are some of the crops which form the major component of commercial IFS. These models involve a large area and investment. Modern technologies, machines, irrigation methods and improved processing facilities are the requirements for commercial IFS. Livestock is one of the components of this model. Plantation-based (tea, coffee, rubber) commercial IFS is a blend of industry and agriculture. Here, the high-value products are meant for sale in demand driven distant markets instead of local sale. In the plantations, generally a single crop is grown in a large area (sometimes with widely planted trees). Tea is grown on a commercial basis in India like coffee in Brazil and rubber in Malaysia. Commercial farms in Europe operatingd on an integrated system recorded 5–20% increase in farm income (Edwards 1987).

9.3 IFS MODEL FOR EAST COAST PLAINS AND HILLS ZONE

In this zone, rice-based cropping system and fishery are the important components of IFS along with horticulture and livestock. The IFS model for 1.25 ha area covers 0.5 ha crops (rice-green gram), 0.475 ha fish pond, 0.15 ha livestock and 0.125 ha area under horticulture (coconut, guava, intercropping of vegetables and fodder crops). Livestock consists of three cows, one buffalo and 500 poultry birds in five batches. Composite fish culture (catla, rohu and mrigal) is done in a pond with coconut plants on the pond embankment. This model generated net income of ₹1.58 lakh yr^{-1}

TABLE 9.2
Scientific Aspects of Rice-Fish Farming System for Higher Productivity and Profitability

Sl. No.	Aspects of Rice-Fish System	Particulars
1	Rice varieties	Andhra Pradesh: Badava Mahsuri; Odisha: Gayatri, Sarala, Durga, Tulasi, Panidhan, Kalashree, Nalini; Assam: Manohar Sali, Prafulla, Joymati, Rupsundari; Bihar: Kishori, Satyam, Santosh, Janaki, Sudha, Varadhi, Vaidehi; Chhattisgarh: Jaldubhi; Tamil Nadu: TNR1, TNR2; Uttar Pradesh: Chakia-59, Jal-Lahari, Jalapriya, Madhukar, Jalamagna, Jalnidhi; West Bengal: Sasi, Giri, Mahananda, Sabita, Purnendu, Suresh, Mandira, Jogen, Nalini, Matangini, Golak and for salt-affected area: SR 26B, Lunishree and Sonamani
2	Fish species	The weed fishes such as *Oxygaster bacaila, Amblyopharynodon mola, Gambusia affinis, Esomus danricus* and *Ambasis* because of their prolific breeding habit grow within a short period and having high market demand. Suitable live fishes are *Clarias batrachus, Heteropneustes fossilis, Channa* sp., *Anabas testudinieus* and *Oreochromis* sp. Carps such as *Puntius* spp., *Labeo bata* may also be considered. Grass carp should be prohibited for this system as it will eat away the rice crop. Catla and Silver carps may be introduced only after good rice crop establishment. Rohu is not suitable due to shallow water depth. Suitable freshwater prawns (*Macrobrachium rosenbergii* and *M. malcolmsonii*) grow well in this system, significantly contributing to the net income from the system.
3	Fish stocking	The stocking density depend on the stage of the fish. For nursery, hatchlings of 2.5–3.0 million ha^{-1}; for rearing, fry of 0.1–0.2 million ha^{-1} and for grow out system, fingerlings of 5,000–10,000 ha^{-1} are recommended. For single-stocking and single harvesting system, stocking density of 5,000–6,000 ha^{-1} is followed, however for single stocking-multiple harvesting system the density can be increased to 10,000 ha^{-1}.
4	Feeding	In natural rice-fish system, rice insects, the detritus matter, phytoplankton and zooplankton provide the nutrients to the fish component. However, for better growth of fish, feed supplementation with oil cake (groundnut or mustard) and rice bran (1:1 ratio) is recommended. For prawns, additional protein in the form of fish powder along with oil cake and rice bran should be applied.
5	Shelter	Freshwater prawns undergo moulting; therefore, shelter should be provided by placing pipes, palm leaves, broken tiles, asbestos sheets, etc.
6	Water quality	Maintenance of water quality is a prerequisite for growth of fish as well as rice. In this system, saline water should be restricted as rice crop will be affected. Water pH should be monitored and optimum level maintained by application of lime.
7	Harvesting	After the rice crop harvest, water from all the area concentrates in the pond refuge/trenches. Periodic harvesting of fish may be done before the water dries out.

Source: Ayyapan et al. 2004; Singh et al. 2001

and employment of 493 man-days with 61% income from fishery, 20% from horticulture, 12% from livestock and 7% from cropping system (Panwar et al. 2019).

Rice-fish is a remunerative farming system (Table 9.2) in the medium deep (25–50 cm water depth) and deep water (50–100 cm water depth) lands, where water is available at least for eight to nine months. This system has the potential for annual production of 16–18 t of food crops, 0.6 t of fish and prawn, besides 5.2 t of fodder in the form of straw from one hectare of land (Singh et al. 2001). In this zone, the majority (>75%) of farm holdings are found in a marginal and small category with an average holding size of 1.03 ha (Annamalai 2014).

9.4 IFS MODEL FOR LOWER GANGETIC PLAINS OR DELTA REGION

This zone covers most parts of West Bengal, where alluvial soils are predominant and flood is a common natural calamity. Rice, potato, jute and vegetables are major crops. Horticulture is also an important avocation comprising of floriculture, banana, coconut, mango and papaya. Fishery is an important component of IFS of Deltaic region, where pisciculture is practiced either in ponds or as a dual culture with rice. Farmers of this zone conserve the Black Bengal goat, a small breed of goat (Singh 2012a) and Garole breed of sheep (Arora et al. 2005). Garole sheep have a high reproductive performance, breed year-round, are resistant to diseases and have a grazing habitat in knee deep marshy saline areas (Arora et al. 2005). Goat farming with about 15 goats by using Black Bengal breeds provide additional net income to the farmers to the tune of ₹28,000 yr^{-1} (Singh and Dige 2015). Though fishery is an important component of IFS in this zone, with an increase in population, during 2019–2020, the total demand (19.91 lakh tons) for fish exceeded the production (16.19 lakh tons) in West Bengal (Paul 2022). To meet the demand, fish are imported from Andhra Pradesh and Odisha. In this zone, in high salinity (5–40 ppm) situations, shrimp-rice rotation system could result in 50% more profits than the double cropping system of rice-rice (Ismail et al. 2022). The rice-fish-horticulture (vegetables, tubers and fruit crops)-duck farming is also a profitable option for this zone, which can generate net income to the extent of ₹2.14 lakh ha^{-1} yr^{-1} with stable income for small holder farmers (Sinhababu and Poonam 2022). Homestead farming system involving diverse crops (vegetables, fruits, ornamental crops, betel vine, medicinal, spices and condiments), fishery in pond, livestock, poultry, mushroom cultivation and apiculture generate income of ₹1.5 lakh yr^{-1} as well as balanced nutrition for the family (Sarangi 2022). The model homestead farming system in a small area of 1,000 m^2 in coastal West Bengal provides diverse vegetables all year and fruit crops with water harvesting and pisciculture in ponds (Figure 9.1). There is production of manures from livestock and poultry for fertilizing the soils.

9.5 IFS MODEL FOR WESTERN HIMALAYAN ZONE

This zone covers Jammu and Kashmir (Srinagar, Anantnag, Udhampur, Riasi and Kathua), Himachal Pradesh, Siwalik hills of Punjab and Uttarakhand. This is an agricultural biodiversity heritage site

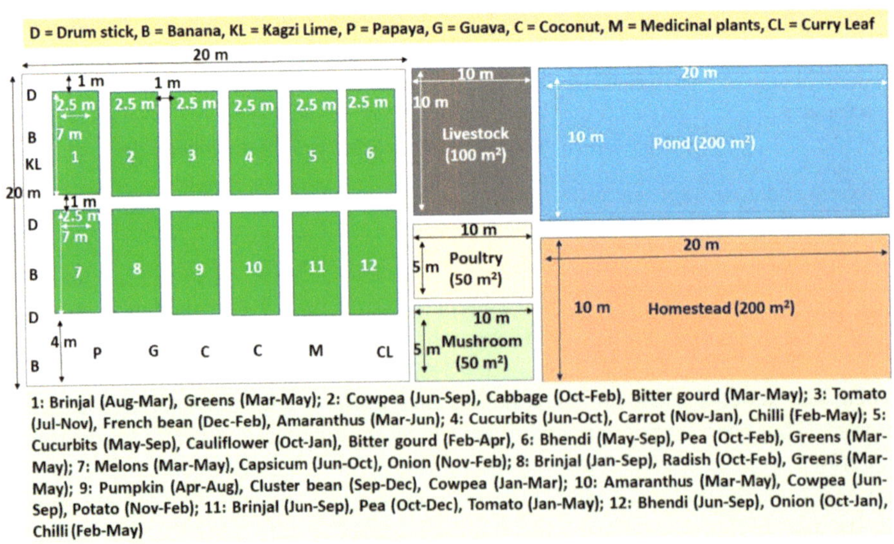

FIGURE 9.1 Layout (not to scale) of a model homestead farming system (for 1000 m^2 area) suitable for coastal area of West Bengal.

TABLE 9.3
Mountain Farming Systems of Western Himalayan Zone Based on Altitude (adapted from Roy and Kumar 2018)

Altitude	Height (m above msl)	Farming System Components
Low	1000–1200	Field crops, Livestock, Fruits, Secondary Agriculture
Mid-low	1200–1500	Food crops, Fruits, Vegetables, Livestock, Secondary Agriculture
Middle	1500–1750	Food crops and Vegetable based IFS, Fruits based IFS, Livestock based IFS, Protected cultivation of vegetables, medicinal and aromatic plants, Secondary Agriculture
Mid-high	1750–2500	Fruits and Vegetables based IFS, Livestock based IFS, Food crops, Pseudo-cereal-based IFS, Protected cultivation of vegetables, medicinal and aromatic plants, Secondary Agriculture
High	2500–3500	Livestock based IFS, Fruits and Vegetable based IFS, Food crops and Pseudo-cereal-based IFS, Protected cultivation of vegetables, medicinal and aromatic plants.

(one of the global biodiversity hotspots) with rich diversity of temperate fruits and herbs; horticulture-based and multiple cropping-based IFS is recommended (Singh 2009). The Himalayan agroecosystem has a sub-tropical to temperate climate suitable for fruit crops like apple, citrus, walnut, plum, peach, banana, mango and pineapple; vegetables such as chilli, tomato, potato, radish, pea, cabbage and cauliflower; cash crops such as ginger, cardamom and saffron; flowers such as orchids, gladiolus, marigold and chrysanthemum. Livestock component consisting of sheep, goat, cattle and buffaloes are very important as per suitability to different altitudes. However, in this zone there is a necessity to determine the maximum population of livestock which can be kept without endangering the wildlife and livestock productivity (Syed and Khan 2017).

Agricultural issues include male out-migration, fragmentation of land holdings and environmental uncertainty, necessitating efficient execution of reforms pertaining to land consolidation to enhance production (Shukla et al. 2018). There is opportunity for expansion of certified organic farming in this zone, as presently majority of the farming community (~ 54%) are practicing organic farming by default (Raina and Sharma 2021). Economically viable diversified IFS models for marginal farmers could overcome the problems of out-migration and land abandonment (Singh and Negi 2014). Diversified model for 0.5 ha land includes fruit trees on field margins, use of terraced land for growing cereals, pulses, vegetables and spices, 100 bushes of tea plants, floriculture, fish ponds (2 nos.), one unit of Angora rabbit, mushroom cultivation, backyard poultry, goats and milch cattle. However, for this mountain region, IFS models should consider the altitude level (Table 9.3). Protected cultivation of carnation in small area (1000 m^2) may result in significant net return (₹30 Lakhs in domestic market and ₹46 Lakhs in export-oriented market) with cost-benefit ratio of 1.43–1.60, payback period of 2–3 years and an internal rate of return of 73% (Roy and Kumar 2018).

9.6 IFS MODEL FOR EASTERN HIMALAYAN ZONE

This zone comprises seven sister states of the northeastern hills region (Arunachal Pradesh, Manipur, Meghalaya, Mizoram, Nagaland, Sikkim and Tripura), Darjeeling hills, Assam and two districts of West Bengal (Jalpaiguri and Cooch Bihar). In spite of high natural forest cover, due to the practice of shifting cultivation, steep slope and high rainfall, soils are prone to erosion caused by runoff and floods. The IFS strategy for this zone involves a watershed-based farming system, suitable soil and water conservation methods, efficient land use by agri-horti-silvi-pastoral system, additional income generation through livestock farming and the construction of water harvesting

FIGURE 9.2 (**a**) Grassed water way for natural drainage; (**b**) Terracing and contour bunding.

structures (Satapathy 2004). This zone requires watershed treatment measures such as excavation of staggered contour trenches (3 m × 0.45 m × 0.05 m) and mini percolation tank (3 m × 2 m × 1.5 m) to intercept and harvest rainwater, plantation of fruit and forest species at hill tops, stone fencing and loose boulder check dam and grassed waterways for natural drainage system (Figure 9.2a), terracing and contour bunding (Figure 9.2b) in the middle and lower portions of the hills (Verma et al. 2003).

Sustainable IFS in this zone requires adoption of the regenerative agriculture to restore the degradation of natural resources base and soil health through the maximum use of the *in-situ* resources, bio-terracing in place of cost intensive mechanical terracing, agroforestry and livestock farming. Bio-terracing involves planting of contour-aligned hedgerows of fast-growing trees with permanent cultivation of crops in alleys formed by hedgerows. The sloping land leads to gradual formation of terrace through accretion of soil and stabilization of risers by the proliferation of strong root systems of the hedgerow plants (Patiram et al. 2003). Vegetative barriers are provided by leguminous hedges (*Erythrina indica*) and broom grass [*Thysanolaena maxima* (Roxb.) Kuntzc (Family – Poaceae)] as the soil is excavated during every ploughing in the alleys, gradually developing into bench terraces. However, caution should be taken where *Erythrina* is used for bio-terracing, since in high rainfall areas it becomes a biological menace and needs to be removed by uprooting. Planting of broom grass (Figure 9.3a) in contour lines has multiple benefits for soil erosion control, yielding fodder for animals and the sale of inflorescence for making brooms (Tiwari et al. 2012). The terrace is used for field and horticultural crops, and risers provide space for cultivating suitable fodder crops (Figure 9.3b) like guinea grass (*Panicum maximum*) for monsoon fodder and broom grass for winter fodder production.

FIGURE 9.3 (**a**) Broom grass for multiple benefits and (**b**) use of terrace for growing field and horticultural crops and risers for fodder crops in Eastern Himalayan zone.

FIGURE 9.4 (**a**) Agro-pastoral system in hills; (**b**) Agro-forestry of pine with turmeric.

The agri-horti-silvi-pastoral system involves agro-pastoral land use (Figure 9.4a) in the bottom one-third portions of the hill by making bench terraces as explained above, horticultural system in the middle one-third area by making half-moon terraces, grassed water ways and silviculture or agro-forestry in top portion of the hill (Figure 9.4b) by making contour bunds (Sarangi and Satapathy 2007). Under the agri-horti-silvi-pastoral system, nearly all of the rainfall (>99%) was absorbed *in-situ*, resulting in the lowest runoff (0.84 mm year^{-1}) and peak flow rate (0.60 mm hr^{-1}) and limiting soil erosion (Satapathy and Sarangi 2010). Shade tolerant intercrops such as turmeric (*Curcuma longa* L.) perform better under established trees. Highest rhizome yield (10.98 t ha^{-1}) of turmeric (variety: GL Puram) was recorded when intercropped with *Pinus kesiya* at Basar (altitude 660 m from msl), Arunachal Pradesh (Sarangi et al. 2007).

Terraced and sloping lands also provide scope for organic cultivation of cash crops such as large cardamom (*Amomum subulatum* Roxb.), ginger (*Zinggiber officinale* Rose) and turmeric belonging to family Zingiberaceae in the eastern Himalayan region (Sarangi and De 2005, 2006; Sharma and Katoch 2019). In terraces, rice- and maize-based intercropping systems such as rice + groundnut/soybean in 3:1 row ratio and maize + groundnut/soybean in 1:1 row ratio, also mixed cropping of maize with rice reduced runoff and soil loss, and resulted in higher net returns (Satapathy 2003). Cropping sequences such as rice-groundnut and rice-mustard may also be followed in the terraces (Sarangi and Lama 2013; Sarangi et al. 2010). Study conducted at Umiam, Meghalaya (980 m above mean sea level.) in rice-mustard cropping system revealed that mulching of paddy straw (@ 5 t ha^{-1}) in mustard increased the seed yield of mustard from 1.11 t ha^{-1} under control (without mulching) to 1.51 t ha^{-1} under mulching, with conservation of 2% more soil moisture in the crop root zone (Sarangi et al. 2010).

Dairy and piggery are the important livestock components of IFS for the Eastern Himalayan zone. It is a paradox that though this region inhabits about 38% population of pigs (3.95 million) of India (10.29 million) as per the 19th livestock census (2012), the productivity per animal is low and the demand for meat is increasing (Talukdar et al. 2018). However, ongoing attempts to improve pig farming in this zone resulted in the development of crossbred pigs (Rani and Asha), with better growth, production, quality of meat, larger litter size, low mortality rate and high back fat thickness compared to Indigenous pigs (Shyam et al. 2017). The crossbred pigs are developed by crossing exotic breeds (Hampshire and Duroc) with an Iindigenous breed (Ghungroo). Under the constraints of health-care infrastructure, proper nutrition and housing system for animals, low input backyard piggery has enormous potential for nutritional security and sustainable livelihood improvement of the farming community of this zone (Singh et al. 2020). Under a low input system, the local breed of pig (Ghungroo) performs better, having important characteristics like faster multiplication, higher growth rate and feed conversion ability (Boro et al. 2021). Apart from dairy and piggery, rabbitry is an attractive component of IFS for this zone as this enterprise requires less investment and space, having short generation interval, high reproductive potential, quick growth and utilization

FIGURE 9.5 (**a**) Mulberry an important multipurpose tree; (**b**) Silkworm rearing in NEH region of India.

of non-competitive feed (Das 2003). Two breeds of rabbit suitable for this zone are New Zealand White and Soviet Chinchilla; they can be successfully grown under cage, hutch (self-contained cage) or floor (deep litter) system (Bharti et al. 2011).

For better performance of livestock enterprise, strategies to generate fodder from best multipurpose trees and shrubs (MPTS) as well as inclusion of fodder crops in agroforestry system is essential. Subabul [*Leucaena leucocephala* (Lam.) de Wit] foliage with excellent level of in vitro dry matter and organic matter digestibility (65–68%) and metabolizable energy (7.95 MJ kg^{-1} dry matter) is the best MPTS followed by *Gliricidia maculate* and *Morus alba* (Datt et al. 2008). Integration of mulberry (*Morus alba*) plants (Figure 9.5a) in to the system also provides scope for silkworm (Figure 9.5b) rearing (sericulture), providing additional income to the farm family.

Pisciculture has great potential due to rich biodiversity and fisheries resources at different altitudes. Large areas of NE hill states, being a mid-altitude region, provide a major potential for aquaculture (Das 2018). Different forms of pisciculture as per the best suitability to different altitudes may be integrated to the IFS model (Figure 9.6). Integration of paddy, poultry, pig and duck with pisciculture has the best prospect both at low and medium altitudes. The common carp (*Cyprinus carpio*) is the best option for paddy-cum-fish culture with a stocking density of 6,000 numbers of fingerlings per hectare of paddy area and stocking size of around 30 g $fingerling^{-1}$, result net production of 200–300 kg fish ha^{-1} in four months of culture (Mahapatra et al. 2003).

	Low altitude (<750 m)	Medium altitude (750-1370 m)	High altitude (>1370 m)
Good prospect	• Composite fish culture • Paddy-cum-fish culture • Poultry-cum-fish culture • Duck-cum-fish culture • Pig-cum-fish culture • Air-breathing fish culture • Common carp culture	• Paddy-cum-fish culture • Poultry-cum-fish culture • Duck-cum-fish culture • Pig-cum-fish culture • Mahseer culture • Common carp culture	• Trout culture
Medium prospect		• Composite fish culture • Air-breathing fish culture • Trout culture	• Common carp culture • Mahseer culture
Poor prospect	• Trout culture • Mahseer culture		• Composite fish culture • Air-breathing fish culture • Paddy-cum-fish culture

FIGURE 9.6 Suitability of different pisciculture technologies in low, medium and high altitudes of Eastern Himalayan zone.

There is a tremendous scope of integration of horticulture with pisciculture by growing vegetables (cabbage, cauliflower, carrot, turnip, brinjal, radish, ginger), fruit crops (peach, papaya, pineapple, guava, arecanut, mango, orange, Assam lemon, banana) and flowers (rose, dahlia and gladiolus) on the pond dykes. The water and silts from pond are used for irrigation of crops and improving soil fertility respectively. The roots of the crops protect the pond embankment from erosion and add organic matter for the growth of planktons for the fish. Similarly, integration of poultry, duckery and piggery has a synergistic effect and increases net income from unit pieces of land in this zone.

9.7 IFS MODEL FOR MIDDLE GANGETIC PLAINS ZONE

This zone covers 12 districts of eastern Uttar Pradesh and 27 districts of the Bihar plains. It is spread over 17 mha of geographical area and endowed with high rainfall (1,211–1,470 mm), alluvial soil and a moist sub-humid to dry-humid climate. During the Kharif season, the predominant crops are rice, maize, pigeon pea, green gram and during the rabi season, wheat, lentil, Bengal gram, pea, sesame and groundnut are the cultivated crops. Cash crops such as potato, sugarcane, tobacco, chilies, turmeric and coriander are also grown with supplemental irrigation. Horticultural crops include mango, guava, litchi, banana and citrus as major fruit crops and onion, brinjal, tomato, cauliflower and cabbage among vegetable crops. Average operational holding size in Bihar is comparatively smaller (0.39 ha) compared to Uttar Pradesh (0.73 ha). Out of the total 146.45 million operational holdings in the country, the highest number belong to Uttar Pradesh (23.82 million) followed by Bihar (16.41 million). The productivity of field crops in this zone is below the national average, whereas the productivity of vegetables and fruit crops is higher than the national average (Mishra 2016). The per capita income in this zone (Bihar: ₹16,083; Eastern Uttar Pradesh: ₹19,512) is far below the income in the eastern region (₹26,047) and also at the national level (₹39,904). Therefore, there is an urgent need to implement IFS models to increase the income level of the farming community.

Sizable area in the middle IGP is under low land ecology. Therefore, diversifying to aquatic crops like fox nut (*Euryale ferox*), locally known as '*Makhana*' and water chestnut (*Trapa bispinosa*), popularly known as '*Singhara*' or '*Pani Phal*' with pisciculture, water lily (*Nymphaea* spp.), sweet flag (*Acorus calamus*) and kewra (*Pandanus* spp.) has a great potential to harness higher farm income. In lowland ecologies (0.3–0.6 m water depth) and pond-based systems (1.2–1.8 m water depth) integration of fox nut and water chestnut with composite fish culture (Rohu, Katla, Common carp and Mrigal) were found most remunerative for net income of ₹1.26 lakh ha^{-1} and ₹1.56 lakh ha^{-1} in pond and field-based systems, respectively (Singh et al. 2017). The seed rate for *Makhana* is 80 kg ha^{-1} sown in the month of December. In a pond-based system, mahua cake @ 2.5 t ha^{-1} is applied to eradicate the carnivorous fish. Fingerlings (@ 6000 ha^{-1}) of Rohu, Katla, Common carp and Mrigal in the ratio of 40:20:20:20 are released during March–April and again in September after the harvest of *Makhana*. Harvesting of fish is completed during December-January. In this system *Singhara* is cultivated during the Kharif season, staring from July-August and harvested during October–November. The IFS model involving Fish-*Makhana* is an eco-friendly integration as the decomposed plant parts of *Makhana* crop enhances the *in-situ* plankton population as a source of feed for fish; in turn, fish contribute to the control of *Makhana* pests and fish faecal matter acts as organic manure for the crop (Verma et al. 2008).

The zone is endowed with rich inland water resources in the form of ponds, tanks, reservoirs, rivers and lakes, offering huge potential for expanding the area under freshwater aquaculture (Pandey and Dewan 2006). Fish production in this zone was led by Uttar Pradesh with 6.99 lakh tonnes (third largest producer in India) followed by Bihar with 6.41 lakh tonnes (fifth largest producer at national level) during 2019–2020 (Handbook of Fisheries Statistics 2020). For Uttar Pradesh, duck-fish IFS model in ponds involving polyculture of carps (@ 10,000 numbers per ha) under organic farming with Khaki Campbell ducks (@ 300 numbers per ha of water area) resulted in production of about 3–5 t ha^{-1} of fish in pond, 2 t ha^{-1} of fish in trenches and 780–2,150 eggs ha^{-1} with net income increasing to the tune of ₹1.55 lakh ha^{-1} yr^{-1} (Haque et al. 2011). Polyculture of prawn-cum-fish in ponds resulted in production of 0.31–0.32 t ha^{-1} of prawn and 3.16–3.28 t ha^{-1} of fish with net income of ₹1.30–1.56 lakh ha^{-1} yr^{-1} (Haque et al. 2011).

9.8 IFS MODEL FOR UPPER GANGETIC PLAINS ZONE

This zone covers 32 districts of Uttar Pradesh covering a geographical area of about 14.34 mha, with a dry sub-humid to semi-arid climate and alluvial soil types. Irrigation sources are canals and tube wells with good groundwater potential. Large portion (70%) of the geographical area is cultivated and about 65% of the cultivated area are irrigated from canals, tube wells and wells with ratio of surface-to-groundwater irrigation is about 1:2 (Jat et al. 2016). Crop-livestock integration is the unique feature of IFS model for this region, as there is a tradition of use of crop residues (wheat straw, pulse peels and oil cakes) as the fodder for livestock and livestock provide milk, meat, dung (fuel and manure) and perform most agricultural operations (Singh 2012b). In recent years, the use of animal energy in agriculture has been declining due to an increasing number of tractors (Srivastava 2014). However, livestock, particularly cattle and buffalo, are the major component of the IFS model for this zone. Crossbred and exotic cattle in this zone constitute 22% of the national population in this group. In the upper Gangetic plains 16% of India's buffalo population are found (Table 9.4). Pigs and goats also constitute a significant population among the livestock of this zone.

Basmati rice–based cropping system is suitable for some of the districts (Saharanpur, Muzaffarnagar, Meerut, Bijnor, Ghaziabad, Moradabad, Rampur, Bulandshahr, Budaun, Bareilly, Pilibhit, Aligarh, Mathura, Agra, Mainpuri, Farrukhabad and Etawah). The notified basmati rice varieties grown are Basmati-370, Basmati-386, Type-3 (Dehraduni), Taraori, Ranbir, Pusa-1509, Pusa Basmati-1, CSR-30 and Pusa Basmati-1121 (Basmati Survey Report, APEDA 2018). Salinity, sodicity and waterlogging are some of the abiotic stresses which affect the agriculture in this region and can be managed by IFS approach. Pond-based IFS model for 1 ha land involving composite fish farming in pond (0.4 ha), rice-wheat system (0.25 ha), fruit crops (0.15 ha) on the top portion of the pond embankment and banana (G-8) on the lower embankment and vegetables (0.10 ha) and fodder crops (0.10 ha) has been recommended (Singh et al. 2008). Fruit crops included in this model are Aonla (NA-6 and NA-7) and guava (Allahabad Safeda), vegetables are spinach and garlic, fodder crops are sorghum-berseem. This pond-based IFS model for waterlogged sodic land has been reported economically feasible with a cost-benefit ratio of 1.93, 1.85, 2.50, 1.70, 2.63 and 2.21 for rice-wheat, forage, vegetables, fruits (banana only), fish and for the whole system, respectively.

TABLE 9.4
Livestock Population (×10³) in the Upper Gangetic Plains (UGP) Agroclimatic Zone, Uttar Pradesh (UP) and India

Animals/Birds	UGP	UP	India	UGP as % of UP	UGP as % of India	UP as % of India
Crossbred and Exotic cattle	8854	19557	39732	45	22	49
Indigenous cattle	8854	19557	151172	45	6	13
Buffalo	17526	30625	108702	57	16	28
Sheep	635	1354	65069	47	1	2
Goat	7302	15586	135173	47	5	12
Pig	756	1334	10294	57	7	13
Poultry	7909	18668	729209	42	1	3

Source: Government of India 2014 (19th Livestock Census)

The potential of agroforestry as a component of IFS in this zone strongly exists in the interspaces of trees such as bilayati babool (*Leucaena* spp.), babool (*Acacia nilotica*), neem (*Azadirachta indica*), shisham (*Dalbergia sissoo*) and ber (*Ziziphus jujuba*) as a sustainable land management system (Kumar et al. 2021).

9.9 IFS MODEL FOR EASTERN PLATEAU AND HILLS ZONE

This zone comprises a ew districts of eastern part of Madhya Pradesh, the western part of West Bengal (Purulia district), most of the inland districts of Odisha, Chhattisgarh, most part of Jharkhand and eastern most districts of Maharashtra. In Madhya Pradesh, the districts covered are Balaghat and Shahdol. Districts from Chhattisgarh are Durg, Rajnandgaon, Raipur, Bilaspur, Surguja, Raigarh and Bastar. Most of the inland districts of Odisha are Balangir, Sambalpur, Dhenkanal, Sundargarh, Kendujhar, Mayurbhanj, Koraput, Kalahandi, Kandhamal and Baudh. Districts from Jharkhand are Godda, Sahibganj, Dumka, Deogarh, Dhanbad, Giridih, Hazaribag, Palamu, Gumla, Lohardaga, Ranchi, Paschimi Singbhum and Purbi Singhbum. Three districts (Bhandara, Chandrapur and Gadchiroli) of Maharashtra are included in this zone. This zone covers the largest geographical area of about 39.5 mha. The soils are shallow to medium in depth and the topography is undulating with 1–10% slope. Tanks and tube wells are the main sources of irrigation water.

Forest cover is an important natural resource of this zone, particularly in Odisha, Maharashtra and Madhya Pradesh. In some of the districts of Odisha such as Sambalpur, Sundargarh, Kendujhar, Kandhamal and Baudh, the forest cover is higher compared to the state average (Table 9.5). The trend of forest cover in the Maharashtra part of this zone is alarming in all the three districts, particularly in the Gadchiroli district, the decline is to the tune of 87% compared to 2017 assessment. The forest cover is less than 10% of the geographical area in the districts of Durg (Chhattisgarh) and Deogarh (Jharkhand). Sericulture, an agro-based enterprise can be sustainably integrated with Ain (*Terminalia elliptica*) and Arjun (*Terminalia arjuna*) trees in this zone, particularly in the Bhandara, Chandrapur and Gadchiroli districts of Maharashtra.

Apart from rice, millet such as ragi (finger millet), jowar (sorghum), bajra (pearl millet) as well as small and little millet are suitable crops for many parts of this zone. Among the rabi crops, gram, jowar, green gram, black gram and moth bean are important. Cattle- and goat-rearing add significant income under the IFS model of this region. The *Manda* breed of buffalo is a sturdy animal adapted to this zone found in Koraput region of Odisha (https://nbagr.icar.gov.in). In the central and northeastern plateau of Jharkhand, rice is the main crop and monocropping is the predominant practice. The western plateau zone of Jharkhand is suitable for growing fruits crops such as mango, jackfruit, guava, banana, litchi, citrus, papaya, pears and ber. Potato and tomato are principal vegetable crops. Among the field crops oilseeds such as sesame, linseed, safflower, and pulses (green gram, black gram, pigeon pea and Bengal gram) are suitable.

The major crops in Odisha are rice, pulses (green gram, black gram and horse gram), chilies, groundnut, sugarcane, mustard and millets. Horticultural crops are coconut, cashew nut, guava, mango and papaya. In Odisha, uncertainties in monsoon rain cause scarcity in food grain production due to drought and flood (Pandey 2014b). In the central table land zone, the Hirakud reservoir is the source of irrigation water for agriculture, giving scope for double cropping of rice. However, during the non-monsoon season, the irrigation water requirement exceeds the canal water availability, necessitating replacing the rice crop by low water requiring crops (wheat, pulses, millets, oilseeds, vegetables and condiments) except sugarcane (Raul et al. 2008). Crop diversification estimates revealed that one ha of rice can be replaced by 5 ha of non-rice crop during the monsoon season and 3.3 ha of non-rice crop during the summer season. Rainwater harvesting, storage, recharge of groundwater, use of direct-seeded rice, adoption of drip and sprinkler irrigation in fruit and vegetable crops could yield more crop per drop of water in this zone.

Adoption of IFS model involving paddy, vegetables, dairy cow, buffalo, poultry, goat and pig resulted in a significant increase in the income of farmers to the tune of ₹4.5–5.0 lakh per annum

TABLE 9.5
State and District Wise Forest Cover (% of Geographical Area During 2019) in the Eastern Plateau and Hills Zone

State	District	Forest Cover (% of GA)	% Change With Respect To 2017	Source
Odisha	Balangir	36.64	4.26	https://fsi.nic.in/isfr19/vol2/isfr-2019-vol-ii-odisha.pdf
	Sambalpur	49.70	12.29	
	Dhenkanal	32.47	28.61	
	Sundargarh	44.00	9.37	
	Kendujhar	121.87	10.16	
	Mayurbhanj	39.30	14.17	
	Koraput	23.83	9.27	
	Kalahandi	30.55	1.80	
	Kandhamal	65.01	5.71	
	Baudh	51.53	1.94	
	State average	33.15	273.51	
Maharashtra	Bhandara	24.44	–7.08	https://fsi.nic.in/isfr19/vol2/isfr-2019-vol-ii-maharashtra .pdf
	Chandrapur	35.43	–32.54	
	Gadchiroli	68.81	–87.06	
	State average	16.50	95.56	
Madhya Pradesh	Balaghat	53.44	–1.94	https://fsi.nic.in/isfr19/vol2/isfr-2019-vol-ii-madhya -pradesh.pdf
	Shahdol	31.76	48.71	
	State average	25.14	68.49	
Chhattisgarh	Rajnandgaon	31.41	8.18	https://fsi.nic.in/isfr19/vol2/isfr-2019-vol-ii -chhattishgarh.pdf
	Durg	9.10	5.39	
	Raipur	29.31	10.92	
	Bastar	40.43	8.86	
	Bilaspur	29.70	0.89	
	Raigarh	36.98	9.33	
	Surguja	45.02	10.61	
	State average	41.14	63.57	

(Continued)

TABLE 9.5 (CONTINUED)
State and District Wise Forest Cover (% of Geographical Area During 2019) in the Eastern Plateau and Hills Zone

State	District	Forest Cover (% of GA)	% Change With Respect To 2017	Source
Jharkhand	Godda	18.68	2.35	https://fsi.nic.in/isfr19/vol2/isfr-2019-vol-ii-jharkhand .pdf
	Sahibganj	27.74	-0.65	
	Dumka	15.35	1.31	
	Deogarh	8.22	1.71	
	Dhanbad	10.47	9.51	
	Giridih	18.16	11.24	
	Hazaribag	38.05	1.77	
	Palamu	27.03	0.78	
	Gumla	26.91	1.26	
	Lohardaga	33.60	0.62	
	Ranchi	22.85	0.49	
	Paschimi Singhbhum	46.60	0.12	
	Purbi Singhbum	30.30	3.38	
	State average	29.62	58.41	
West Bengal	Purulia	14.63	11.88	https://fsi.nic.in/isfr19/vol2/isfr-2019-vol-ii-west-bengal .pdf
	State average	19.04	54.51	

from ₹1.5–2.0 lakh per annum under conventional agriculture in Ranchi district of Jharkhand (Burman et al. 2022). The Department of Agriculture and Farmers' Empowerment, Government of Odisha launched an IFS portal (https://ifsodisha.nic.in) for processing the IFS models to be adopted by aspiring farmers. This website has been designed by the National Informatics Centre. This website provides guidelines about IFS and step-by-step procedures for generation of an IFS model for securing assistance from the Government. For all the IFS models of Odisha, farm pond is mandatory and farmers get 40–50% (₹1,37,000–1,71,250) assistance for adoption of IFS. The IFS models include crops, compost/vermicompost pit, fishery unit, duckery, agro-forestry, livestock (crossbred cows, goat, backyard poultry) and field/hydroponics fodder crops cultivation. Other alternative options are mushroom and honey bee (apiary) units. The Government of Odisha has fixed target for providing assistance to 942, 1,884, 2,826, 3,768 and 4,710 numbers of IFS units during 2021–2022, 2022–2023, 2023–2024, 2024–2025 and 2025–2026, respectively (https://ifsodisha.nic.in/assets/IFS%20Guideline.pdf).

To sustainably increase the aquaculture production in this zone, cage culture may be an option in the water body under the IFS model by optimizing the trophic structure and function for fish/aquatic production. A cage is a confined area that allows unrestricted water exchange with the surrounding body of water used for rearing selected aquatic organisms (NFDB 2016). With the exception of the top, where the stock is fed and handled, the cages are closed on all sides. Cages may be floating type or positioned at the bottom, middle or surface of the water body. The standard size of a cage is 6 m (length) × 4 m (width) × 4 m (height). The number of cages may be 6, 12, 24 or more depending on the total area available for culture. However, as per NFDB guidelines 2016, cage culture is allowed in water bodies having a surface area of 1,000 ha or more with average water depth of 10 m. Economically viable fish species for cage culture are exotic pangasius or Sutchi Catfish (*Pangasianodon hypophthalmus*) and GIFT tilapia (a genetically improved strain of *Oreochromis niloticus*). However, GIFT tilapia is restricted to only all-male seed, obtained from authorized hatcheries. All other exotic fish species are strictly prohibited for cage culture. Cage culture can also be done with Indigenous species with high economic and nutritive values such as Bata (*Labeo bata*), Jayanti rohu (*L. rohita*), Pengba (*Osteobrama belangeri*), Pabda (*Ompok bimaculatus*), Koi (*Anabas testudineus*), Catfish (*Pangasius pangasius*), Punti (*Puntius sarana*), Bhetki (*Lates calcarifer*), Milk fish (*Chanos chanos*), Featherback (*Chitala chitala*), Pearlspot (*Etroplus suratensis*), Murrels (*Channa marulus, C. Striata*), Helicopter/Wallago catfish (*Wallago attu*) and Shellfish (*Macrobrachium rosenbergii*).

To harness the full potential of IFS in this zone, rainwater harvesting, scientific management of groundwater and surface water are essential to meet the water demand as well as fulfilling the socio-economic development aspirations of the farming community (Bera and Das 2021).

9.10 IFS MODEL FOR WESTERN PLATEAU AND HILLS ZONE

This zone covers major parts of Maharashtra (southern part of Malwa plateau and Deccan plateau), parts of Madhya Pradesh and one district of Rajasthan. This zone is the third largest in terms of geographical area (33.22 mha) after the Eastern Plateau and Hills zone and the Southern Plateau and Hills zone. This zone is a semi-arid region with annual rainfall ranging from 600 to 1,050 mm (average 900 mm) with medium to deep black soils. Agriculture is affected by floods, hailstorms, cyclones and untimely rain. The predominant cropping systems in the irrigated areas of this zone are rice-rice, rice-wheat-sugarcane, pearl millet-vegetables/potato-onion, cotton-wheat, cotton-soybean-wheat, groundnut-wheat, sorghum-wheat, sugarcane + sorghum/maize, cotton-rice and rice-sugarcane-rice. In the rainfed situation, pearl millet-wheat, mung bean-wheat and soybean-wheat are the major cropping systems. In Malwa plateau, Nimar plains and Jhabua hills, a cotton-sorghum cropping pattern is followed. Promising horticultural crops include orange (*Citrus sinensis*), grape (*Vitis ninifera*), pomegranate (*Punica granatum*) and banana (*Musa* spp.).

The sub-agroclimatic zones under this zone are western hills and plains, central plateau (Malwa plateau is plain to gently sloping, whereas Nimar plateau is undulating), Jhabua hills, submontane zone (moderate rolling topography), scarcity zone (very low rainfall of 600 mm per annum) and Vidarbha region. Rainfall in most parts of this zone, except the western ghat zone, is low and irrigated areas are less (Pandey 2014c). In Maharashtra, the overall average size of an operational holding was 1.34 ha, which was 0.44 ha, 1.33 ha, 2.59 ha, 5.59 ha and 16.66 ha in size for marginal, small, semi-medium, medium and large size categories during 2015–2016. The total number of operational holdings in Maharashtra (15.29 million) was third in India after Uttar Pradesh (23.82 million) and Bihar (16.41 million). In Madhya Pradesh there were 10 million operational holdings with average size of operational holding 1.57 ha, with sizes of 0.49 ha, 1.41 ha, 2.70 ha, 5.67 ha and 14.83 ha for marginal, small, semi-medium, medium and large categories, respectively (Agriculture Census 2015--2016).

Maharashtra is the leading state with respect to area and production of grapes in India. The cost of establishment of one-hectare grape orchard is ₹5.68 lakhs, of which major share (42%) is for installation of training system followed by plant protection cost (17%) and installation of drip system (13%). For cultivation, the cost involvement is ₹11.25 lakhs ha^{-1} with output: input ratio of 1.81, which is achieved from production of 355.98 quintals of grape ha^{-1} with gross return of ₹20.36 lakhs (Mhetre et al. 2020). Cultivation of orange crop is another attractive agro-enterprise in this zone. The per hectare establishment cost, cost of cultivation, net return and benefit cost ratio (BCR) were ₹2.08 lakhs, ₹1.08 lakhs, ₹1.34 lakhs and 2.24, respectively (Bante et al. 2015). Cultivation of pomegranate is highly economically viable option with per hectare total cost of cultivation, net return, and BCR of ₹2.15 lakhs, ₹5.21 lakhs, and 3.42, respectively (Zore et al. 2019). Banana is also a profitable horticultural crop in this zone with cost of cultivation, net return and BCR of ₹3.20–3.33 lakhs ha^{-1}, ₹2.76–3.40 lakhs ha^{-1} and 1.83–2.06, respectively (Rede et al. 2021). Studies on onion cultivation revealed that it is not profitable (Table 9.6) when cost structure includes all the costs of cultivation, expenditure on rental value of land, interest on fixed capital and family labour costs (Makhare and Korade 2018). However, to make this crop profitable, there is a need to improve marketing efficiency by introducing a cooperative and farmer producers' organization system (Makhare and Korade 2018; Shukla et al. 2019).

Pomegranate cultivation is attractive for smallholder farmers of this zone as it is a low resource demanding crop (in terms of inputs) compared to grapes and it provides scope for intercropping, thereby increasing overall income, which is well spread across the year (Phadke et al. 2022). There is a great scope for expanding organic farming both in Madhya Pradesh and in Maharashtra, being the leading states with respect to area and production of organic products in India (Table 9.7). The

TABLE 9.6
Economics of Promising Horticultural Crops in the Western Plateau and Hills Zone

Crop	Cost of Cultivation (₹ Lakhs ha^{-1})	Gross Return (₹ Lakhs ha^{-1})	Net Return (₹ Lakhs ha^{-1})	Benefit Cost Ratio	Reference
Grape	11.25	20.36	9.11	1.81	Mhetre et al. (2020)
Orange	1.08	2.42	1.34	2.24	Bante et al. (2015)
Pomegranate	2.15	7.36	5.21	3.42	Zore et al. (2019)
Banana	3.27	6.35	3.08	1.94	Rede et al. (2021)
Onion[a]	1.28	1.10	-0.18	0.86	Makhare and Korade (2018)

[a]Based on mean of cost C.

TABLE 9.7
Area and Production Under Organic Farming in Leading States and India During Last Five Years (2017–2018 to 2021–2022)

	Area (000′ ha)			Production (000′ tonnes)		
Year	MP	Maharashtra	India	MP	Maharashtra	India
2017–2018	1156.88	304.07	3566.54	575.35	377.31	1675.56
2018–2019	918.30	261.57	3428.64	738.89	860.98	2607.38
2019–2020	1161.02	293.14	3669.80	980.61	904.95	2709.12
2020–2021	1637.73	371.80	4339.20	1392.10	775.77	3469.00
2021–2022	2370.59	1133.67	9119.87	1410.89	691.42	3410.20

Source: APEDA 2020-21

livestock (cattle, buffaloes, sheep and goat) and poultry enterprises have promise as a component of IFS model for this region, as encouraged by various central- and state-sponsored schemes. The Maharashtra state has developed four central poultry hatcheries at Aurangabad, Kolhapur, Nagpur and Pune for better poultry farming (Economic Survey of Maharashtra 2021–2022). Government of Madhya Pradesh is implementing several schemes for livestock and poultry development such as a national cow and buffalo breeding program (modified central scheme), goat breed improvement program, supply of advanced breed animals to the beneficiaries, *Nandishala* scheme, Samunat Murra Pada supply scheme, unit supply scheme of five pigs, male sukar supply scheme etc. (Madhya Pradesh Economic Survey 2022–2023). '*Kadaknath*' chicken breed (locally called *Kali Masi*) found in the Jhabua district of Madhya Pradesh are well known for its resistance to hot summer and cold winter. These are suitable for backyard poultry farming having high economic value of ₹800–1000 per kg meat and ₹20–35 per egg (Swati et al. 2022).

Aquaculture in inland water resources (farm ponds, lakes, reservoirs, tanks and canals) has a great potential as a component of IFS model for this zone. There is a wide gap of actual and potential yield in aquaculture in Maharashtra, as well as it is dependent on West Bengal for fish seed (Bhendarkar et al. 2020). Therefore, there is a great scope for scientific production of fish seed to meet the demand. Production of ornamental fish seed can also be a component of an IFS model. In Madhya Pradesh out of 136 fish species, more than 58 fish species have ornamental value such as Dwarf Gourami (*Trichogaster lalius*), Honey Gourami (*Trichogaster chuna*), Banded Gourami (*Trichogaster fasciata*), Zipper Loach (*Acanthocobitis botia*), Fighting Loach (*Botia* spp.), etc. (Ansari et al. 2015).

9.11 IFS MODEL FOR CENTRAL PLATEAU AND HILLS ZONE

This zone spans over three states (Rajasthan, Madhya Pradesh and Uttar Pradesh) covering 46 districts (Five from Uttar Pradesh, 16 from Rajasthan and 25 from Madhya Pradesh). It is located at the centre of India forming plateau and hills, where some areas are barren and uncultivable (15%) and water run-off is excessive (Pandey 2014a). The rainfall varies from about 490 mm to 1570 mm in this semi-arid to dry sub-humid agro-climatic zone with mixed red and black, red and yellow and medium black soils.

An IFS model for the Kymore plateau and Satpura hills region of Madhya Pradesh involves crop production (paddy, wheat, maize, soybean, gram), vegetables (onion, cauliflower, chili), fruit crops (guava and papaya), animal husbandry (Gir breed of cattle), fisheries (common carp), backyard poultry (*Kadaknath*) and water harvesting are recommended (Gupta et al. 2018). An IFS model for one hectare area in Satpura plateau region involving crop production, horticulture and dairy component resulted net income of ₹1.36 Lakhs, consisting of ₹0.42 lakhs from crop production, ₹ 0.52 lakhs from horticulture and ₹0.42 lakhs from dairy farming (Mawase et al. 2018). Cotton, maize, groundnut, sorghum, sesame, pearl millet and rice are important *Kharif* crops. Among the rabi crops, wheat, mustard, chickpea, linseed and groundnut are important. In the Rajasthan part of this zone, rainfall is less (about 550 mm per annum). In the Bundelkhand sub-zone (Uttar Pradesh and Madhya Pradesh), rainfall is medium (about 700 mm per annum), whereas in the hill areas of Vindhya, Satpura plateau and Narmada valley areas, rainfall is comparatively high (1,120–1,570 mm). The Gird zone covering Morena, Bhind, Gwalior, Shivpuri and Guna districts of north Madhya Pradesh, there is availability of rich alluvial soils, but the rainfall in these areas is only about 670 mm per annum.

There is a need to integrate fodder cultivation in this zone, as the performance of cattle is poor due to want of green fodder supply. Cows, buffaloes, goats and sheep are the significant livestock potential to be integrated with IFS model. Agri-silvicultural system with grasses (*Cymbopogon jwarancusa, Aristida* spp., *Cenchrus ciliaris, C. setigerus*) as fodder; millets, pulses and oilseeds (pearl millet, cowpea, cluster bean, moth bean, green gram, sesame) as intercrops and trees (*Prosopis cineraria* and *Holoptelia integrifolia*) for multipurpose use have been recommended for the arid zone of Rajasthan (Bhati et al. 2017). The Bundelkhand region under this agroclimatic zone is the most degraded zone with undulating and rugged topography, scarce ground water resources, soil with poor fertility and low water-holding capacity (Singh et al. 2018). An IFS model for this region, involving agro-forestry, beekeeping, goat rearing and development of water resources has been recommended (Singh et al. 2018). The model recommends boundary plantation (*Tectona grandis, Madhuca longifolila, Azadirachta indica* and *Acacia nilotica* at 10–20 m spacing), inter-spaced trees (*Dendrocalamus strictus, Eucalyptus, Ingadulsis* and *Lawsonia alba*) and fruit crops (Aonla var. NA-7). Integration of beekeeping with a mustard-based cropping system resulted in additional net income of ₹6,000–7,500 for marginal farmers and to the tune of ₹75,000 for large farmers in the first year and the income significantly increased during the subsequent years (Singh et al. 2018). In Rajasthan, goat-based IFS model involving total cost of ₹12,744–13,620 resulted net income of ₹9443–11,183 from sale of milk, kids and manure (Singh et al. 2018).

9.12 IFS MODEL FOR SOUTHERN PLATEAU AND HILLS ZONE

This is a representative semi-arid zone covering the states of Andhra Pradesh, Karnataka and Tamil Nadu (total 35 districts). Dry farming is practiced in the major areas of this zone. This zone is second in terms of geographical area (39.41 mha) after the Eastern Plateau and Hills zone (39.5 mha). The soils are medium to deep black, red sandy and red loamy types. This zone covers south Telangana (Rangareddy, Mahbubnagar, Nalgonda and Hyderabad), north Telangana (Adilababd, Nizamabad, Karimnagar, Medak, Warangal and Khammam), north dry region of Karnataka (Bidar, Gulbarga, Bijapur, Belgaum, Dharwad, Raichur and Bellary), central region of Karnataka and Rayalseema (Chitradurga, Tumkur, Bangalore, Chitoor, Cuddapah, Kurnool and Anantapur), south region of Karnataka (Hassan, Mandya and Mysore), north region of Tamil Nadu (Dharmapuri and Salem) and central plateau of Tamil Nadu (Periyar, Coimbatore, Anna, Madurai, Tiruchirappalli and Pudukkottai). Major crops are rice, sorghum, bajra, maize, millets, groundnut, pulses, oilseeds (sunflower and safflower), coconut, sugarcane, turmeric, cotton, chilies and vegetables. In India, Karnataka ranks first in maize and sunflower production, second in nutri/coarse cereals and third in sugarcane production. Tamil Nadu ranks third in groundnut production. In Andhra Pradesh, more than 90% of the area of rice and sugarcane are irrigated. In Karnataka, 99.6% of the sugarcane area are irrigated. In Tamil Nadu, 94.9% of the rice area and 100% of the sugarcane area are irrigated (Pocket Book of Agricultural Statistics 2020).

To increase farm income, there is a need to shift from conventional rice-based mono-cropping to an IFS model involving fruit crops vegetable crops, livestock, plantation crops (arecanut, coconut, coffee and cocoa), spices (cardamom, black pepper, and cumin), fisheries, sheep, rabbit, cut flowers, poultry, vermicompost, apiary and piggery (Kumara et al. 2016). In Karnataka, maize-based cropping systems such as maize-maize (practiced in command areas), soybean-maize, maize-wheat and cotton-maize are profitable cropping systems. The livestock component includes cattle, buffaloes, sheep and goats; however, to increase the productivity of animals, there is a need to address critical gaps with respect to breed improvement, fodder availability and establishment of institutional mechanism for effective marketing.

In the southern transitional zone of Karnataka, Flue Cured Virginia (FCV) tobacco is the major crop followed by hybrid cotton, maize, finger millet, red gram and other rabi pulses. IFS model for one acre area involving agroforestry, livestock, vegetables, manure production, rainwater harvesting, border tree plantation (silver oak) and jatropha as a live fence crop is economically viable with BC ratio of 2.83 (Mahadevaswamy et al. 2014). Grape is an important commercial crop for Karnataka, particularly in the Bijapur district. To make a grape-based farming system profitable, skill on the cultivation technology, marketing and input management are highly essential (Shiralashetti and Hadapad 2016). Sugarcane is another profitable crop of Karnataka with a total cost of ₹48,752 ha^{-1} and net return of ₹66,248 ha^{-1}; however, to sustain the productivity and profitability of this system, conservation agricultural practices are required to be followed. Conservation agricultural practices help maintain soil fertility, protect the soil from salinity development and increase the use efficiency of applied irrigation water (Wali et al. 2019).

Tamil Nadu (TN) receives a major share of rainfall from the northeastern monsoon (47%) followed by the southwestern monsoon (33%). Rice-rice and rice-groundnut are the major cropping systems and rice+pigeon pea, rice+green gram and rice+black gram are major intercropping systems. Apart from these commercial crops such as sugarcane, tobacco, tea, coffee, rubber, pepper and cashew nut are promising. Oilseed crops, spices (chilies, turmeric and coriander), poultry and fishery are also components of IFS models for TN. For the lowland ecologies of TN, an IFS model with rice+*Azolla*+fish is economically profitable with 25.7% higher income over rice monoculture and 6.9% higher income over rice+fish system (Jayanthi et al. 2021). In Karnataka, an IFS model involving crops (rice-rice, maize-sunflower, vegetables and fodder crops), fish pond, goat and poultry recorded 26.3% higher productivity and 32.3% higher net return over a commercial rice-rice system (Jayanthi et al. 2021). Integration of a dairy component and poultry adds significantly to the net income (Table 9.8).

For Andhra Pradesh, an IFS model (0.56 ha) consisting of field crops (rice, maize, sorghum – grain and fodder, red gram, green gram), horticulture (banana, papaya, guava, pomegranate, apple, ber, rosewood, pomelo, citrus and vegetables), livestock (Ongole cow, Kadaknath and Aseel poultry) and fish (Rohu, Indian carp and grass carp) is recommended (ICAR Annual Report 2020).

TABLE 9.8
Recommended IFS Models for Southern Plateau and Hills Zone

IFS Model	System Production (kg ha^{-1})	Net Return (₹ ha^{-1})	Reference
Crop + Fish + Goat	37,679	1,31,118	Jayanthi et al. 2021
Crop + Fish + Poultry	29,609	97,731	
Crop + Fish + Pigeon	29,173	98,778	
Crop + Horticulture + Ongole cow + Poultry + Fish	–	1,31,000	

9.13 IFS MODEL FOR WEST COAST PLAINS AND GHATS ZONE

This zone comprises the west coast of Tamil Nadu, Kerala, Karnataka, Maharashtra and Goa. This zone is endowed with high rainfall (2,226–3,640 mm), with dry sub-humid to per-humid climate. Soils are laterite, red loamy and coastal alluvium. Irrigation facility enables double and triple cropping such as rice-rice, rice-rice-pulses, rice-rice-ragi/groundnut, rice-sugarcane, pearl millet-vegetables and maize-pulses. Under rainfed conditions, rice-maize, rice-pules, pearl millet–wheat/pulses and groundnut-maize are predominant cropping systems. Plantation crops such as coconut, arecanut, cashew nut, tea, coffee and rubber are important for this zone. Spices such as pepper, ginger, turmeric, small cardamom and orchard crops such as mango are important apart from commercial crops such as sugarcane and tapioca.

This zone includes north coastal areas of Konkan, Goa and Uttar Kannad, Kerala coast, Palakkad and Kottayam of Kerala and hill region (Shimoga, Chikmagalur and Kodagu of Karnataka; Idukki and Wayanad of Kerala and Nilgiri hills of Tamil Nadu). Climate smart IFS models involving rice-based cropping system in lowland and plantation crops for Goa, crops + horticulture + livestock (dairy + goatery + poultry) + vermicompost + boundary plantation + kitchen garden for Maharashtra, homestead farming + coconut + rice + banana for Kerala are recommended (AICRP on IFS 2021). Climate smart IFS models could generate employment for 262 man-days, address GHG emission (carbon negative), improve soil organic carbon (49.4%) and generate net income of ₹2.55 lakhs from mean area of 0.43 ha with mean cost of ₹2.63 lakhs. For coastal areas of Tamil Nadu, IFS models recommended are integrated rice + fish + poultry + mushroom + vermicompost for wetland areas, integrated goat + millet + vegetables + floriculture + vermicompost + mushroom for rainfed upland areas and sea weed culture + mushroom + vermicompost as coastal aqua farming (Final Report NAIP, Annamalai University 2014).

Coconut-based cropping system involving intercropping of annual (banana, turmeric, ginger and fodder crops) and perennial crops (pepper, arecanut and cocoa) for Konkan to Malabar coast; 'Pokkali' and 'Kaipad' rice cultivation system for Malabar region and diversified homestead farming system have been recommended for this zone (Singh 2020). Evaluation of farming system options for low-lying, flood-prone coastal lands of Karnataka (locally called '*Gajnis*') revealed highest net income (₹2.5 lakhs ha^{-1}) from semi-intensive prawn farming, followed by mixed farming of prawn and fish (₹2.4 lakhs ha^{-1}) and lowest (₹0.06 lakhs ha^{-1}) from monoculture of paddy (Keremane and Naik 2006). However, for long-term sustainability the package should include new resources, agricultural education, skill development and appropriate farm planning.

9.14 IFS MODEL FOR GUJARAT PLAINS AND HILLS ZONE

This zone covers all the districts of Gujarat, Dadra and Nagar Haveli and Daman and Diu. This is mostly an arid, semi-arid and dry sub-humid zone with low rainfall in most of the parts. The soils are deep black (Vertisols) and coastal alluvium. Annual rainfall is higher in the southern hills (about 1,800 mm) to meagre (about 500 mm) in the northwestern arid region and the north Saurashtra region. he average holding size is 1.88 ha with 68.3% of the total households are in the marginal and small category (Agriculture Census 2015–2016). The irrigated cropping systems are rice-sugarcane-rice, rice-wheat, rice-cotton + groundnut-wheat, sugarcane/maize-wheat, groundnut-wheat/cumin/isabgol, castor-wheat-cotton, pearl millet-wheat-cotton-vegetables, sorghum-cotton, pearl millet-garlic/potato-sugarcane and cotton-vegetables. Under rainfed situation, cropping systems are rice-chickpea, rice-maize/pearl millet/sorghum and pearl millet-wheat-groundnut.

The entire zone is divided into seven sub-zones, viz., southern hills (Dangs and Valsad), southern Gujarat (Surat and Bharuch), middle Gujarat (Vadodara, Panch Mahals and Kheda), north Gujarat (Banas Kantha, Mehsana, Sabar Kantha, Gandhinagar and Ahmedabad), north west arid (Kachchh), north Saurashtra (Amreli, Bhavnagar, Jamnagar, Rajkot and Surendranagar) and south Saurashtra (Junagarh). Gujarat having most salt-affected soils (50% of the coastal salt-affected soils)

TABLE 9.9
Area Under Salt-Affected Soils (Saline and Sodic) in the Gujarat Plains and Hills Zone

District	Saline Soils (ha)	Sodic Soils (ha)	Total (ha)
Kachchh	434786	51349	486135
Surendranagar	135170	98259	233429
Junagarh	39800	69358	109158
Bhavnagar	78414	21249	99663
Jamnagar	69858	13389	83247
Bharuch	62561	-	62561
Rajkot	57634	-	57634
Amreli	42952	-	42952
Kheda	37852	-	37852
Surat	26287	-	26287
Valsad	15427	-	15427
Total	1000741	253604	1254345

Source: Adapted from Mandal et al. 2018.

in all the agro-climatic zones (1.00 mha saline soils and 0.25 maha sodic soils) and saline Vertisols, pose serious constraints to crop production (Table 9.9).

Seawater ingress/inundation cause land and water salinity problems, particularly in the Kachchh and Saurashtra region with 1,125 km coast line out of the total 1,600 km coastline of Gujarat (Gururaja Rao et al. 2019). These salt-affected soils are mainly reclaimed by leaching, drainage, bio-drainage and growing salt-loving (halophytic) plants. In the areas where there is the presence of high saline groundwater and water stagnation occurs for a period of about two weeks during the wet season, short-duration rice and finger millets can be grown followed by low-water requiring high value crops such as dill (*Anethum graveolens*), mustard (*Brassica campestris*) and safflower (*Carthamus tinctorius*). The major management options for salt-affected areas of this region are selection of salt-tolerant crops and varieties (Table 9.10), rainwater harvesting, appropriate crop establishment methods, proper time of sowing/planting and agronomic operations.

The recommended IFS model for semi-arid region of Gujarat involves arable crops (rice+ maize + pigeon pea +wheat/gram), two bullocks, one cow and buffalo each, 10 goats and 20 poultry birds as backyard farming and biogas unit (Khadda et al. 2018). Livestock is the major component of the IFS model with contribution of about 26% to the total value of output (of which contribution of milk is 82% and meat 10%) from agriculture and allied sectors (Gulati et al. 2021). Gujarat is the fourth largest (12.8 million MT in 2016-17) milk producing state, contributing around 8% of total milk production in India after Uttar Pradesh, Rajasthan and Madhya Pradesh (Gulati et al. 2021). An IFS model involving field and fodder crops (0.7 ha), horticulture unit (0.25 ha), tree plantation of farm boundary, nursery for raising seedlings of vegetable crops (0.01 ha), farm pond (0.015 ha), two Mahesani breed buffaloes and vermicompost unit (0.025 ha) is recommended for north Gujarat condition (Patel et al. 2016a). This model generated net income of ₹3.61 lakhs per year in the fifth year of study, in which the average contribution from boundary plantation, cropping system, horticulture unit and livestock were 53%, 32%, 10% and 5%, respectively. The cropping system comprises field crops and fodder crops, e.g., castor + green gram, groundnut – wheat – multicut pearl millet, green gram – mustard – pearl millet and hybrid napier + cowpea – lucerne + fodder chicory. This one-hectare IFS model generated average net income of ₹990 per day round the year with creation of 306 person-days of employment. Horticulture unit comprises 40 mango trees (8 m × 8 m spacing), 80 lemon trees (in between two rows of mango at 4 m spacing), 36 custard apple

TABLE 9.10
Recommended Crops and Varieties for Saline Vertisols of Gujarat

Crop	Varieties
Rice	CSR 36, CSR 43, CSR 46, CSR 49, CSR 52, CSR 56, CSR 60
Wheat	Kharchia-65, KRL 1-4, Raj 3077, KRL 19, KRL 210, KRL 213, KRL 283
Cotton	G Cot 23, G Cot DH 7, GBav 105, GBav 106
Maize	Hybrids: DKC-8101, IJ 8214, IL 8534, JI 8212, 900 M GOLD, DKC 8081 Inbreds: DMRQPM-0903, CML 260, GM-6
Chick pea	Karnal Chana 1 (CSG 8962), ICC 9942 (salt tolerant accession)
Pigeon pea	ICPL 84074, ICP 8626, ICPL 227, PAU-881
Mustard	Pus Bold, Varuna, Kranti, CS 52, CS 54, CS 56, CS 58, CS 60
Sugarcane	Co 6806, Co 7219, Co 7717, Co 8208, Co 85004, Co 85019, Co 86032, CoC 671
Cumin	Gujarat Cumin 4
Dhaincha	CSD 123, CSD 137
Guava	Gujarat Anand Red Flesh Guava 1 (Lal Bahadur)
Sapota	Cricket Ball, Kalipatti, Pilipatti
Pomegranate	Bhagwa, Ganesh, G-137, Phule Arakta, Jalore Seedless, Mridula
Jamun	Goma Priyanka
Coriander	Gujarat Coriander-3
Fenugreek	GM-1 (Suitable for Banaskatha, Dahod, Patan, Surendranagar, Mehasan)
Dill	Gujarat Anand Dill Seed 1 (North and Middle Gujarat)
Castor	GAUCH-11 (VHB-44), GCH-9, Gujarat Anand Castor 11
Groundnut	Mallika, GJG-HPS-1, GJG-9, GJG-17, GJG-22, GJG-31, JL 501, Dharani, Phule Morna, Phule Warna
Mango	Gujarat Mango 1 (Anand Rasraj)
Lemon	Acid Lime
Custard apple*	Annona-7, Arkashan, Automoya, Chandcilli, Cherimoya
Ber	Gola, Umran, Mehrun
Karonda	Konkan Bold, CHESK-11-7, CHESK-35
Fodder crops	Marvel grass (GMG-1, GAMG-2), Dharaf grass (GAUD-1, APBN-1)

*Not suitable for heavy black soils with morum below plough layer and poor drainage

plants and seasonal vegetables. Boundary plantation consists of 10 plants each of *Ardusa* (*Ailanthus excelsa*), Eucalyptus (*Eucalyptus globulus*), Subabul (*Leucaena leucocephala*) and custard apple (*Annona reticulata*), 4 plants of Jamun (*Syzygium cumini*), 3 numbers of mulberry (*Morus rubra*) plants, 15 numbers of drum stick (*Moringa oleifera*) plants, 3 plants of Aonla (*Phyllanthus emblica*), one bamboo (*Bambusa vulgaris*) bush and 35 numbers of teak (*Tectona grandis*) trees. In the bunds, Dhaman (*Cenchrus ciliaris*) and Hybrid Napier (*Pennisetum purpureum*) are grown to supplement green fodder requirements. Farm pond (10 m × 10 m× 15 m) is used to harvest rainwater and recharge groundwater.

The IFS model for south Gujarat agroclimatic situation (Bharuch and Surat) consists of diversified field crops (cereals, pulses, sugarcane and cotton), horticultural crops (vegetable and fruits) and livestock (Patel et al. 2017). About 40% of the farming system in this region are sugarcane based, 27% are cereal crop based and 10% are cotton based systems. In the north-west and northern region (Kutchh, Patan and Mehsana) of Gujarat, pulse-oilseed based farming system is predominant (38%), followed by cotton (24%), spices (21%), cereal (14%) and horticulture based (3%) systems (Patel et al. 2016b).

9.15 IFS MODEL FOR ISLANDS ZONE

This zone covers Andaman and Nicobar (A and N) as well as Lakshadweep islands located near the equator with high annual rainfall of about 3,000 mm in a per-humid climate. This zone covers a comparatively small geographical area (0.08 mha) but forest cover is significantly high of 82% in A and N island (https://fsi.nic.in/isfr19/vol12/isfr-2019-vol-ii-andaman-nicobar-islands.pdf) and >90% in Lakshadweep Island (https://fsi.nic.in/isfr19/vol12/isfr-2019-vol-ii-lakshadweep.pdf). The average size of land holding in A & N island is 1.78 ha, whereas in Lakshadweep it is 0.27 ha, however, the majority of the farm holdings are small and marginal (Agriculture Census 2015–2016). The majority of the cultivated area is occupied by coconut (45%) followed by paddy (22%), vegetables (8%), arecanut (8%), fruit crops (7%), oil palm (3%), rubber (3%), spices (2%), cashew (1%) and others (1%). Coconut and arecanut are grown in the hill slopes, whereas oil palm and rubber are grown in the undulating terrain (Velmurugan et al. 2016). There is a great scope for organic farming in the Island zone in an IFS mode by integrating fruit and plantation crops in multi-storey cropping, with poultry, piggery and dairy enterprises. This because of existence of rich agro-biodiversity and in most of the cultivated areas by default organic inputs are used (Velmurugan et al. 2016).

IFS model for 0.75 ha land involving cropping system (rice-maize, rice-sorghum, rice-black gram, rice-vegetables, okra-radish-cucumber, Amaranthus-okra-brinjal, okra-French bean-bottle gourd, Brinjal-brinjal-bottle gourd and fodder crops), dairy, fishery and composting resulted net income of ₹2.17 lakhs per year with 50% contribution from livestock component (Swarnam et al. 2018). Soils in the A and N Islands are medium to very deep red loamy types, including marine alluvium deposits, whereas in Lakshadweep soils are highly calcareous and sandy type. In the hill terrains there is a need for soil and water conservation measures, such as contour bunding, contour trenching, bench terracing, vegetative barriers and grassed waterways. In the lowlands during the wet season due to high rainfall, there is problem of waterlogging. To address this issue and for crop diversification, various land modification options such as creation of a farm pond (FP), broad bed and furrow (BBF) system, three-tier system (original lowland, high land and medium land), paired bed system, deep furrow and high ridges are recommended. These systems create opportunity for rainwater harvesting and for integration of other enterprises such as fishery, poultry, piggery and dairy.

Freshwater fish farming comprising Indian major carps (*Catla catla, Labeo rohita* and *Cirrhinus mrigala*), red-bellied pacu (*Piaractus brachypomus*) and Sutchi catfish (*Pangasianodon hypophthalmus*) have great potentials. However, at present the Island zone depends on the mainland for aquaculture inputs (Kiruba-Sankar et al. 2022). IFS models should consider conservation of native breeds of animal/birds (Trinket cattle, Barren Island goat, Teressa goat, Nicobary pig, Andaman wild pig and Nocobary fowl), feed and fodder availability by growing fodder crops such as maize, soil and water conservation measures, livestock-based land use, marketing infrastructure and veterinary facilities (Chand et al. 2013). A coconut plantation provides scope for intercropping of pineapple, banana, elephant foot yam, groundnut, chillies, sweet potato, tapioca up to an initial 8–10 years. During the next 10–22 years, shade tolerant crops such as Colocasia, can be grown. Multistoreyed cropping of coconut, cocoa, pepper, cinnamon, clove and nutmeg and fodder crops (hybrid napier bajra, guinea grass, alfalfa, berseem) is a feasible option. Arecanut is the second most important plantation crop of the Island zone and contributes about 50% of annual income of the farmers; however, further interventions on training and offering an improved package of practices, storage and establishment of farmers producer organizations (FPOs) are needed to increase productivity and profitability (Kasinath et al. 2019).

A BBF system involving alternate broad beds and furrows (in one hectare area 10 furrows of 6 m width and 100 m length and 10 beds of 4 m width and 100 m length) creates an opportunity for growing rice, vegetables and fodder crops as well as pisciculture, giving an economically viable B:C ratio of 2.48 for the A&N islands (Sarma et al. 2010). This technology facilitates maintaining agrobiodiversity, rainwater conservation, reduction in soil salinity from ECe 6.5–10.12 dS m^{-1} to 1.89–2.0 dS m^{-1} and significantly enhances the income of the farmers in waterlogged saline soils by year-round cultivation of vegetables, fish, paddy and fodder crops (Meena et al. 2016).

Thus, for the Islands agroclimatic zone, an IFS model involving diversified crop cultivation by land modification, including field crops, particularly a rice-based cropping system, agro-forestry, fodder crops, horticultural crops, fishery and animal husbandry is recommended.

REFERENCES

Agriculture Census 2015–2016. 2019. *All India Report on Number and Area of Operational Holdings.* Agricultural Census Division, Department of Agriculture, Co-operation and Farmers Welfare, Ministry of Agriculture and Farmers Welfare, Government of India. Available online: https://agcensus.nic.in/document/agcen1516/T1_ac_2015_16.pdf. Accessed on 12 April 2023.

AICRP on IFS. 2021. *Annual Report 2020–2021.* ICAR. Indian Institute of Farming Systems Research (Indian Council of Agricultural Research), Modipuram, Meerut-250110. p 116. Available online: https://iifsr.icar.gov.in/icar-iifsr/AICRP/pdf/AICRP-IFS_ANNUAL_REPORT%202020-21.pdf. Accessed on 01 May 2023.

Annamalai, S.J.K. 2014. Long-term strategies and programmes for mechanization of agriculture in agro-climatic zone-XI: East coast plains and hills region. In: *Long-term Mechanization Strategy Papers for Different Agroclimatic Zones.* Available online: https://www.farmech.gov.in/farm.htmal. Accessed on 13 April 2023.

Ansari, M.A., Prasad, J.R., Gupta, S.K. 2015. Scope of ornamental fisheries in Madhya Pradesh. *International Journal of Scientific Research* 4(11):427–431. DOI: https://doi.org/10.361106/ijsr.

APEDA. 2020–2021. *Agricultural and Processed Food Products Export Development Authority 2020–2021.* Available online: https://apeda.gov.in/apedawebsite/organic/data.htm. Accessed on 27 April 2023.

Arora, R., Bhatia, S., Sodhi, M., Sahana, G., Mukesh, M., Ahlawat, S.P.S. 2005. *Sheep Genetic Resources of India: Garole the Prolific Microsheep of India.* Monograph 18, National Bureau of Animal Genetic Resources, Indian Council of Agricultural Research, Karnal, Haryana, India. Available online: https://www.nbagr.icar.gov.in/wp-content/uploads/2020/02/Garole-sheep.pdf. Accessed on 13 April 2023.

Ayyappan, S., Sarangi, N., Sinhababu, D.P., Das, P.C., Jena, J.K. 2004. Rice-fish farming: An economic enterprise for lowland rice farmers. In: Panda D., Sasmal S., Nayak S.K., Singh D.P., Saha S. (eds) *Recent Advances in Rice-based Farming Systems.* Proceedings of the National Symposium on Recent Advances in Rice-based Farming Systems 17–19 November 2004. Central Rice Research Institute, Cuttack, Odisha, India. pp. 190–201.

Bante, R., Pallewar, S., Shrey, R. 2015. Economics of orange production in Nagpur district of Maharashtra. *International Research Journal of Agricultural Economics and Statistics* 6(1):136–139. DOI: https://doi.org/10.15740/HAS/IRJAES/6.1/136-139.

Basmati Survey Report, APEDA. 2018. Agricultural and Processed Food Products Export Development Authority (APEDA), Basmati Export Development Foundation and GEOtrans Technologies Pvt. Ltd., New Delhi, India. pp 30. Available online: https://apeda.gov.in/apedawebsite/Announcements/Basmati_Crop_survey_Report_6_Season_2018.pdf. Accessed on 21 April 2023.

Bera, A., Das, S. 2021. Water resource management in semi-arid Purulia district of West Bengal in the context of sustainable development goals. In: Shit P.K., Bhunia G.S., Adhikary P.P., Dash C.J. (eds) *Groundwater and Society.* Springer, Cham. pp 501–519. DOI: https://doi.org/10.1007/978-3-030-64136-8_23.

Bharti, P.K., Suresh Kumar, D.S., Basumatary, R., Khan, M.H., Kadirvel, G., Bardoloi, R.K. 2011. *Scientific Management Practices for Broiler Rabbit Production in North Eastern Hill Region of India.* Published by Director, ICAR Research Complex for NEH Region, Meghalaya, India. Available online: http://www.kiran.nic.in/pdf/farmers_corner/pamphlets/Scientific_Management_Practices_Broiler_Rabbit_Productiom_NEH_Region_India.pdf. Accessed on 23 April 2023.

Bhati, T.K., Shalander, K., Amare, H., Whitbread, A.M. 2017. *Assessment of Agricultural Technologies for Dryland Systems in South Asia: A Case Study of Western Rajasthan, India.* International Crops Research Institute for the Semi-Arid Tropics (ICRISAT), Telangana, India. p 68. Available online: https://www.researchgate.net/publication/326935030_Assessment_of_Agricultural_Technologies_for_Dryland_Systems_in_South_Asia_A_Case_Study_of_Western_Rajasthan_India#fullTextFileContent. Accessed on 27 April 2023.

Bhendarkar, M.P., Brahmane, M.P., Gaikwad, B.B., Singh, N.P. 2020. The status and prospectus of fisheries and aquaculture in Maharashtra, India. *Indian Journal of Geo Marine Sciences* 49(4):567–575. Available online: https://www.researchgate.net/publication/342522520_The_status_and_prospectus_of_fisheries_and_aquaculture_in_Maharashtra_India#fullTextFileContent. Accessed on 26 April 2023.

Boro, P., Bharali, D., Sarma, M., Sonowal, M., Saharia, J., Brahma, J., Kalita, M.C., Thakuria, J. 2021. Performances of Ghungroo pigs reared under farm condition. *Journal of Entomology and Zoology Studies* 9(1):2265–2267. https://www.entomoljournal.com/archives/2021/vol9issue1/PartAF/9-1-358-918.pdf.

Burman, D., Noel, A.S., Maurya, M.K., Ramchandra, Tripath, P. 2022. Economic analysis of IFS (Integrated Farming System) in Ranchi district of Jharkhand. *The Pharma Innovation* 11(7):4306–4310. https://www.thepharmajournal.com/archives/2022/vol11issue7S/PartAX/S-11-6-322-712.pdf. Accessed on 25 April 2023.

Chand, S., Kumar, N., Roy, S.D. 2013. Livestock production challenges and strategies for tropical agro-ecosystem, Andaman and Nicobar Islands, India. *Basic Research Journal of Agricultural Science and Review* 2(10):195–201. http://www.basicresearchjournals.org.

Das, A. 2003. Prospects of rabbitry in NEH region. In: Bhatt B.P., Bujarbaruah K.M., Sharma Y.P., Patiram (eds) *Approaches for Increasing Agricultural Productivity in Hill and Mountain Ecosystem.* Director, ICAR Research Complex for NEH Region, Meghalaya, India. pp 359–361; 417.

Das, S.K. 2018. Mid hill aquaculture: Strategies for enhancing production in Northeast hill region of India. *Journal of Coldwater Fisheries* 1(1):42–47. Available online: https://dcfr.res.in/download/journalsep18/S.K._Das_Mid_hill_aquaculture_Strategies_for_enchancing_production_in_Northeast_hill_region_of_India.pdf. Accessed on 23 April 2023.

Datt, C., Datta, M., Singh, N.P. 2008. Assessment of fodder quality of leaves of multipurpose trees in subtropical humid climate of India. *Journal of Forestry Research* 19:209–214. DOI: https://doi.org/10.1007/s11676-008-0035-2.

Economic Survey of Maharashtra. 2021–2022. Directorate of Economics and Statistics, Planning Department, Government of Maharashtra, Mumbai. 136 p. Available online: http://mls.org.in/pdf2022/budget/ESM_2021_22/Economic%20Survey%20of%20%20Maharashtra%202021-22.pdf. Accessed on 27 April 2023.

Edwards, C.A. 1987. The concept of integrated systems in lower input/sustainable agriculture. *American Journal of Alternative Agriculture* 2(4):148–152. https://www.jstor.org/stable/44506997. Accessed on 12 April 2023.

Final Report NAIP. 2014. National Agricultural Innovation Project (Indian Council of Agricultural Research). Farming Systems for Livelihood Security of Small and Marginal Farmers in Disadvantaged Districts of Tamil Nadu. Annamalai University, India. Available online: https://naip.icar.gov.in/download/77735/Final Completion_Report_Annamalai_University.pdf/FinalCompletion_Report_Annamalai_University.pdf. Accessed on 01 May 2023.

Government of India. 2014. *19th Livestock Census – 2012. All India Report.* Ministry of Agriculture, Department of Animal Husbandry, Dairying and Fisheries, Krishi Bhawan, New Delhi, India. p 120. Available online: https://dahd.nic.in/sites/default/filess/Livestock%20%205_0.pdf. Accessed on 21 April 2023.

Gulati, A., Roy, R., Hussain, S. 2021. Performance of agriculture in Gujarat. In: Gulati A., Roy R., Saini S. (eds) *Revitalizing Indian Agriculture and Boosting Farmers Income.* India Studies and Business and Economics. Springer, Singapore. pp 113–144. DOI: https://doi.org/10.1007/978-981-15-9335-2_5.

Gupta, P.K., Singh, S.R.K., Mishra, R.K., Tomar, A.K. 2018. Integrated farming system for the Kymore plateau and Satpura hills region of Madhya Pradesh. *Journal of Community Mobilization and Sustainable Development* 13(3):493–499. https://www.indianjournals.com/ijor.aspx?target=ijor:jcmsd&volume=13&issue=3&article=019.

Gururaja Rao, G., Kanani, A.D., Purohit, D., Waghela, D. 2019. Coastal saline soils of Gujarat (India): Problems, reclamation measures and management strategies. In: Dagar J., Yadav R., Sharma P. (eds) *Research Developments in Saline Agriculture.* Springer, Singapore. pp 629–651. DOI: https://doi.org/10.1007/978-981-13-5832-6_21.

Handbook of Fisheries Statistics. 2020. Department of Fisheries, Ministry of Fisheries, Animal Husbandry & Dairying, Government of India, New Delhi. p 176. Available online: https://www.dof.gov.in/sites/default/files/2021-02/Final_Book.pdf. Accessed on 20 April 2023.

Haque, N., Ansari, A., Kumar, D., Singh, R., Srivastava, S.C., Singh, A.K. 2011. A case study of integrated fish farming in Uttar Pradesh, India: Profitability and sustainability. *Research in Environment and Life Sciences* 4(3):109–112. Available online: https://www.krishi.icar.gov.in/jspui/bitstream/123456789/4946/1/volume4paper24.pdf. Accessed on 20 April 2023.

https://fsi.nic.in/isfr19/vol12/isfr-2019-vol-ii-andaman-nicobar-islands.pdf. Accessed on 03 May 2023.

https://fsi.nic.in/isfr19/vol12/isfr-2019-vol-ii-lakshadweep.pdf. Accessed on 03 May 2023.

https://nbagr.icar.gov.in/en/manda-buffalo/.

ICAR Annual Report. 2020. Indian Council of Agricultural Research, Department of Agricultural Research and Education, Ministry of Agriculture and Farmers Welfare, Government of India, New Delhi. pp 228. Available online: https://icar.org.in/sites/default/files/ICAR-AR-2020-English.pdf. Accessed on 01 May 2023.

Ismail, A.M., Singh, S., Sarangi, S.K., Srivastava, A.K., Bhowmick, M.K. 2022. Agricultural system transformation for food and income security in coastal zones. In: Lama T.D., Burman D., Mandal U.K., Sarangi S.K., Sen H.S. (eds) *Transforming Coastal Zone for Sustainable Food and Income Security*. Springer Nature, Switzerland AG. pp 3–22. DOI: https://doi.org/10.1007/978-3-030-95618-9_1.

Jat, N.K., Yadav, R.S., Kumar, S., Kumar, K. 2016. Upper Gangetic plains: Uttar Pradesh. In: Gangwar B., Jat N.K. (eds) *Organic Farming: Technologies and Strategies*. Today and Tomorrow's Printers and Publishers, New Delhi, India. pp 129–147. Available online: https://krishi.icar.gov.in/jspui/bitstream/123456789/39561/1/UGPs-Organic%20FarmingTechnologies%20and%20Strategies.pdf. Accessed on 20 April 2023.

Jayanthi, C., Vennila, C., Varshini, S.V. 2021. Resilient integrated farming system models in irrigated agriculture. *Indian Farming* 71(11):20–23. Available online: https://www.researchgate.net/publication/359866139_Indian_Farming_Resilient_integrated_farming_system_models_in_irrigated_agriculture_Jayanthi_C_Vennila_C_and_Varshini_S_V#fullTextFileContent. Accessed on 04 June 2023.

Kasinath, B.L., Meena, B.L., Kumar, S., Lal, S.V., Kapoor, P., Gangaiah, B. 2019. Cultivation and marketing knowledge level of arecanut farmers in Andaman and Nicobar Islands. *Journal of Extension Education* 31(3):6343–6348. DOI: https://doi.org/10.26725/JEE.2019.3.31.6343-6348.

Keremane, G.B., Naik, B.K. 2006. Economics of Gajni farming under different farming systems in coastal floodplains of Karnataka, India. *Journal of Social and Economic Development* 8(1):1–12. Available online: https://www.researchgate.net/publication/264742165_Economics_of_Gajni_Farming_under_Different_Farming_Systems_in_Coastal_Floodplains_of_Karnataka_India#fullTextFileContent. Accessed on 01 May 2023.

Khadda, B.S., Kanak, L., Rajkumar, Jadav, J.K. 2018. Integrated farming system models for tribal farmers of semi-arid region of Gujarat. *Journal of Progressive Agriculture* 9(1):115–117.

Kiruba-Sankar, R., Saravanan, K., Haridas, H., Praveenraj, J., Biswas, U., Sarkar, R. 2022. Policy framework and development strategy for freshwater aquaculture sector in the light of COVID-19 impact in Andaman and Nicobar archipelago, India. *Aquaculture* 548(1):737596. DOI: https://doi.org/10.1016/j.aquaculture.2021.737596.

Kumar, A., Lal, S.K., Meena, R.S. 2021. Management of natural resources in Indo-Gangetic plain region of India for sustainable agricultural land use planning – a case study. *Biological Forum – An International Journal* 13(2):282–292. https://www.researchtrend.net/bfij/bfij.php.

Kumara, O., Sannathimmapa, H.G., Basavarajappa, D.N., Danaraddi, V.S., Pash, A. 2016. Predominant farming system in Karnataka, India. *Paripex – Indian Journal of Research* 5(11):145–147. Available online: https://www.worldwidejournals.com/paripex/recent_issues_pdf/2016/November/predominant-farming-systems-in-karnataka-india_November_2016_5101902260_7306182.pdf. Accessed on 01 May 2023.

Madhya Pradesh Economic Survey. 2022–2023. Directorate of Economics and Statistics, Government of Madhya Pradesh. p 276. Available online: https://mpplanningcommission.gov.in/MPES%202022-23_English.pdf. Accessed on 27 April 2023.

Mahadevaswamy, M., Shenoi, M.M., Ramesh, S., Mahadeva, C., Venkatesh, T. 2014. Sustainable livelihood through agro-forestry and livestock based integrated farming system in rainfed Alfisols of southern transitional zone (STZ) of Karnataka. *Agriculture for Sustainable Development* 2(2):97–100. http://www.sustainable agriculture.in.

Mahapatra, B.K., Vinod, K., Mandal, B.K. 2003. Fisheries in north east India and norms for sustainable development – an evaluation. In: Bhatt B.P., Bujarbaruah K.M., Sharma Y.P., Patiram (eds) *Approaches for Increasing Agricultural Productivity in Hill and Mountain Ecosystem*. Director, ICAR Research Complex for NEH Region, Meghalaya, India. pp 387–396; 417.

Makhare, P.K., Korade, V.C. 2018. Economics of onion production in Nashik district of Maharashtra. *International Research Journal of Agricultural Economics and Statistics* 9(1):97–100. DOI: https://doi.org/10.15740/HAS/IRJAES/9.1/97-100.

Mandal, A.K., Reddy, G.P.O., Ravisankar, T., Yadav, R.K. 2018. Computerized database of salt-affected soils for coastal region of India. *Journal of Soil Salinity and Water Quality* 10(1):1–13. Available online: https://krishi.icar.gov.in/jspui/bitstream/123456789/25878/1/01-AK-Mandal.pdf. Accessed on 03 May 2023.

Mawase, S., Mawase, S.K., Rai, A.K., Patidar, J. 2018. A study of economics analysis of diversification farming system in Satpura plateau region of Madhya Pradesh, India. *International Journal of Current Microbiology and Applied Sciences* 7(2):2422–2432. DOI: https://doi.org/10.20546/ijcmas.2018.702.295.

Meena, B.L., Meena, R.L., Velmurugan, A., Sharma, D.K. 2016. Impact assessment of broad bed and furrow system in *Tsunami* affected waterlogged saline areas for agrobiodiversity in island agro-ecosystem of India: A case study. *Conference Proceedings of the 1st International Agrobiodiversity Congress.* November 2016, New Delhi, India. pp 116; 6–9. DOI: https://doi.org/10.13140/RG.2.2.29910.29761.

Mhetre, A.V., Chavan, R.V., Chaudhari, S.A. 2020. Economic analysis of grape production in Sangli district of Maharashtra. *International Journal of Current Microbiology and Applied Sciences* 11:1439–1444. https://www.ijcmas.com/special/11/A.V.%20Mhetre,%20et%20al.pdf.

Mishra, J.S. 2016. *Production and Technological Gaps in Middle Indo-Gangetic Plains.* Published by the Director, ICAR-Research Complex for Eastern Region, ICAR Parisar, Patna – 800 014, Bihar, India. p 15. Available online: https://www.icarrcer.icar.gov.in/wp-content/uploads/2014/01/IGP.pdf. Accessed on 20 April 2023.

NFDB (National Fisheries Development Board). 2016. *Guidelines for Cage Culture in Inland Open Water Bodies of India.* Department of Animal Husbandry, Dairying and Fisheries, Ministry of Agriculture and Farmers Welfare, Government of India, Krishi Bhavan, New Delhi. Available online: https://nfdb.gov.in/PDF/GUIDELINES/Guidelines%20for%20Cage%20Culture%20in%20Inland%20Open%20Water%20Bodies%20of%20India.pdf. Accessed on 26 April 2023.

Pandey, M.M. 2014a. *Long-term Strategies and Programmes for Mechanization of Agriculture in Agroclimatic Zone-VIII: Central Plateau and Hills Region.* Available online: https://farmech.gov.in/06035-04-ACZ8-15052006.pdf. Accessed on 27 April 2023.

Pandey, M.M. 2014b. *Long-term Strategies and Programmes for Mechanization of Agriculture in Agroclimatic Zone-VII: Eastern Plateau and Hills Region.* Available online: https://farmech.gov.in/06035-04-ACZ7-15052006.pdf. Accessed on 21 April 2023.

Pandey, M.M. 2014c. *Long-term Strategies and Programmes for Mechanization of Agriculture in Agroclimatic Zone-IX: Western Plateau and Hills Region.* Available online: https://farmech.gov.in/06035-04-ACZ09-15052006.pdf. Accessed on 26 April 2023.

Pandey, S.K., Dewan, R. 2006. Constraints in fish farming practices in Uttar Pradesh, India – an analysis. *Journal of Indian Fisheries Association* 33:183–189. Available online: https://www.researchgate.net/publication/343181219_Constraints_in_Fish_Farming_Practices_in_Uttar_Pradesh_India_-_an_Analysis_In_J_Indian_Fish_Assoc#fullTextFileContent. Accessed on 20 April 2023.

Panwar, A.S., Ravisankar, N., Prusty, A.K., Shamim, M., Singh, R., Bhaskar, S., Malik, S.K., Tomar, R.K., Arunachalam, A., Alagusundaram, K. 2019. *Integrated Farming System for Agricultural Diversification Enhanced Income and Employment.* Indian Council of Agricultural Research, New Delhi. p 51. Available online: https://www.iifsr.icar.gov.in/icar-iifsr/prog-files/Bulletin-IFS-July-2020.pdf. Accessed on 12 April 2023.

Patel, A.M., Patel, P.K., Saini, A.K., Patel, K.M. 2016a. *Potential Integrated Farming System Model under North Gujarat Agro-Climatic Situations.* Extended Summaries Vol. 1: Fourth International Agronomy Congress, November 22–26, 2016, New Delhi, India. pp 223–224. DOI: https://doi.org/10.13140/RG.2.2.35284.83841.

Patel, R.R., Pandya, S.P., Patel, P.K. 2016b. Characterization of farming system in north west agroclimatic zone of Gujarat state. *Gujarat Journal of Extension Education* 27(2):206–208. https://www.gjoee.org/papers/194.pdf.

Patel, R.R., Patel, P.K., Patel, R.M., Patel, A.S. 2017. Characterization of farming system in south Gujarat agroclimatic zone of Gujarat. *International Research Journal of Agricultural Economics and Statistics* 8(1):117–120. DOI: https://doi.org/10.15740/HAS/IRJAES/8.1/117-120.

Patiram, Bhatt, B.P., Singh, K. 2003. Regenerative agriculture in NEH region for sustainability of jhum land. In: Bhatt B.P., Bujarbaruah K.M., Sharma Y.P., Patiram (eds) *Approaches for Increasing Agricultural Productivity in Hill and Mountain Ecosystem*. Published by Director, ICAR Research Complex for NEH Region, Meghalaya, India. pp 309–316; 417.

Paul, T.K. 2022. A study on the prospects of fish farming in West Bengal. *International Journal of Advances in Engineering and Management* 4(5):1357–1364. DOI: https://doi.org/10.35629/5252-0405135571364.

Phadke, M., Karandikar, B., Gulati, A. 2022. Grapes and pomegranate value chains. In: Gulati A., Ganguly K., Wardhan H. (eds) *Agricultural Value Chains in India*. India Studies in Business and Economics. Springer, Singapore. pp 145–193. DOI: https://doi.org/10.1007/978-981-33-4268-2_5.

Pocket Book of Agricultural Statistics. 2020. Government of India, Ministry of Agriculture and Farmers Welfare, Department of Agriculture, Cooperation and Farmers Welfare, Directorate of Economics and Statistics. Available online: https://desagri.gov.in/document-report-category/pocket-book-of-agricultural-statistics/. Accessed on 01 May 2023.

Raina, A., Sharma, V. 2021. Problems and prospects of Himalayan farmers and farming: A case study of district Kishtwar, Jammu and Kashmir. *Regional Economic Development Research* 2(1):82–95. DOI: https://doi.org/10.37256/redr.212021776.

Raul, S.K., Panda, S.N., Hollander, H.M., Billib, M.H.A. 2008. Sustainability of rice-dominated cropping system in the Hirakud canal command, Orissa, India. *Irrigation and Drainage* 57(1):93–104. DOI: https://doi.org/10.1002/ird.351.

Rede, B.H., Ratnaparkhe, A.N., Rede, G.D. 2021. Economics of banana production in Solapur district of Maharashtra, India. *Asian Journal of Agricultural Extension, Economics & Sociology* 39(11):451458. DOI: https://doi.org/10.9734/AJAEES/2021/v39i1130772.

Roy, A., Kumar, U. 2018. Emerging farming systems in Western Himalaya: A state level analysis of sustainability. *International Journal of Environmental Sciences and Natural Resource* 9(2):1–5. DOI: https://doi.org/10.19080/IJESNR.2018.09.555757.

Sarangi, S.K. 2022. Coastal homestead farming systems for enhancing income and nutritional security of smallholder farmers. In: Lama T.D., Burman D., Mandal U.K., Sarangi S.K., Sen H.S. (eds) *Transforming Coastal Zone for Sustainable Food and Income Security*. Springer Nature, Switzerland AG. pp 277–290. DOI: https://doi.org/10.1007/978-3-030-95618-9_23.

Sarangi, S.K., De, L.C. 2006. Cultivation practices of organic turmeric and ginger in NEH Region. In: Purohit S.S., Gehlot D. (eds) *Trends in Organic Farming in India*, AGROBIOS(INDIA), Agro House, Behind Nasrani Cinema, Jodhpur. pp 380–395; 438. Available online: https://www.researchgate.net/publication/335857188_Cultivation_practices_of_organic_turmeric_and_ginger_in_NEH_Region. Accessed on 23 April 2023.

Sarangi, S.K., De, L.C. 2005. Production of organic turmeric and ginger in NEH Region. *Agrobios Newsletter* IV(1):55–56.

Sarangi, S.K., Lama, T.D. 2013. Composting rice straw using earthworm (*Eudrilus eugeniae*) or fungal inoculant (*Trichoderma viridae*) and its utilization in rice (*Oryza sativa*)- groundnut (*Arachis hypogaea*) cropping system. *Indian Journal of Agronomy* 58(2):146–151.

Sarangi, S.K., Saikia, U.S., Lama, T.D. 2010. Effect of rice (*Oryza sativa*) straw mulching on the performance of rapeseed (*Brassica campestris*) varieties in rice-rapeseed cropping system. *Indian Journal of Agricultural Sciences* 80(7):603–605.

Sarangi, S.K., Satapathy, K.K. 2007. Crop production through farming system approach. In: Panwar A.S., Ngachan S.V., Munda G.C., Das A. (eds) *Sustainable Crop Production – For Food Sufficiency in NE India*. Published by Director, ICAR Research Complex for NEH Region, Meghalaya, India. pp 116–137; 478.

Sarangi, S.K., Singh, K.A., Singh, R. 2007. Performance of turmeric (*Curcuma longa*) under shade of tree species. *Range Management and Agroforestry* 28(1):44–46.

Sarma, K., Ravisankar, N., Roy, S.D., Grinson, G., Srivastava, R.C. 2010. Prospects of fish culture in the broad bed and furrow system in Andaman. *Environment & Ecology* 28(1):86–90. Available online: https://www.researchgate.net/publication/263587845_Prospect_of_fish_culture_in_the_broad_and_bed_furrow_system_in_Andaman#fullTextFileContent. Accessed on 03 May 2023.

Satapathy, K.K. 2004. Social and institutional aspects of watershed management programmes in north eastern hills region: An overview. In: Satapathy K.K., Agrawal K.N. (eds) *Watershed Management – Social and Institutional Issues in North East India*. Published by Director, ICAR Research Complex for NEH Region, Meghalaya, India. pp 1–19; 314.

Satapathy, K.K. 2003. Soil and water conservation measures for sustainable hill agriculture. In: Bhatt B.P., Bujarbaruah K.M., Sharma Y.P., Patiram (eds) *Approaches for Increasing Agricultural Productivity in Hill and Mountain Ecosystem*. Published by Director, ICAR Research Complex for NEH Region, Meghalaya, India. pp 3–13; 417.

Satapathy, K.K., Sarangi, S.K. 2010. Long-term effect of alternate land use systems on hydrology, soil properties, and economics in a fragile hill ecosystem. *Journal of the Soil and Water Conservation* 9(1):46–50.

Sharma, S., Katoch V. 2019. Status and prospects of large cardamom cultivation in India. *International Journal of Economic Plants* 6(2):093–096. DOI: https://doi.org/10.23910/IJEP/2019.6.2.0295.

Shiralashetti, A.S., Hadapad, M. 2016. Constraints of grape cultivators in Karnataka: A case study of Bijapur district. *Indian Journal of Applied Research* 6(1):44–48. Available online: https://www.worldwidejournals.com/indian-journal-of-applied-research-(IJAR)/recent_issues_pdf/2016/January/January_2016_1451651960__15.pdf. Accessed on 04 June 2023.

Shukla, K., Pandey, G.K., Kumari, M.V., Vanam, A., Singh, N. 2019. A study on marketing pattern of onion in Nashik district of Maharashtra, India. *International Journal of Current Microbiology and Applied Sciences* 8(4):151–160. DOI: https://doi.org/10.20546/ijcmas.2019.804.016.

Shukla, R., Chakraborty, A., Sachdeva, K., Joshi, P.K. 2018. Agriculture in the western Himalayas – an asset turning into a liability. *Development in Practice* 28(2):318–324. DOI: https://doi.org/10.1080/09614524.2018.1420140.

Shyam, J., Borgohain, A., Talukdar, D. 2017. Management practices and constraints perceived by piggery farmers in Kamrup district of Assam. *Indian Journal of Social Research* 58(1):65–70.

Singh, A. 2020. Coastal agriculture and future challenges. In: Singh A., Fernando R.L.S., Haran N.P. (eds) *Development in Coastal Zones and Disaster Management*. Disaster Research and Management Series on the Global South. Palgrave Macmillan, Singapore. pp 61–86. DOI: https://doi.org/10.1007/978-981-15-4294-7_5.

Singh, A.K. 2009. Probable agricultural biodiversity heritage sites in India: II. The Western Himalayan Region. *Asian Agri-History* 13(3):197–214. https://www.asianagrihistory.org/pdf/research/197-to-214.pdf.

Singh, A.K. 2012a. Probable agricultural biodiversity heritage sites in India: XIII. Lower Gangetic Plain or Delta Region. *Asian Agri-History* 16(3):237–260. https://www.asianagrihistory.org/pdf/articles/aks.pdf.

Singh, A.K. 2012b. Probable agricultural biodiversity heritage sites in India: XI. The Upper Gangetic Plains Region. *Asian Agri-History* 16(1):21–44. https://www.asianagrihistory.org/pdf/articles/AKS1.pdf.

Singh, B.N., Sinhababu, D.P., Venugopalan, M., Saha, S., Rayudu, B.T., Ghosh, A. 2001. Household food security by improving rice-based farming system in cyclone prone belt of Orissa. *Journal of the Indian Society of Coastal Agricultural Research* 19(1&2):323–332.

Singh, G., Sharma, D.K., Mishra, V.K., Singh, Y.P., Nayak, A.K., Singh, R., Kalidhonkar, M.J., Varshney, P.K. 2008. *Integrated Farming System Model for Waterlogged Sodic Soils*. Technical Bulleting No. 1/2008. Central Soil Salinity Research Institute, Regional Research Station, India. pp 12. Available online: https://www.researchgate.net/publication/289602418_Integrated_Farming_System_Model_for_Waterlogged_Sodic_Soils#fullTextFileContent. Accessed on 21 April 2023.

Singh, I.S., Kumar, L., Bhatt, B.P., Thakur, A.K., Choudhary, A.K., Kumar, A. 2017. Integrated aquaculture with fox nut – a case study from north Bihar, India. *International Journal of Current Microbiology and Applied Sciences* 6(10):4906–4912. DOI: https://doi.org/10.20546/ijcas.2017.610.461.

Singh, J., Riar, T.S., Garg, L. 2021. Prospects of integrated farming system in Punjab. *Journal of Community Mobilization and Sustainable Development* 16(1):313–315. Available online: https://www.researchgate.net/publication/357991385_Prospects_of_Integrated_Farming_System_in_Punjab. Accessed on 12 April 2023.

Singh, M., Sridhar, K.B., Kumar, D., Tewari, R.K., Dev, I., Ram, A., Uthapa, A.R., Kumar, V., Singh, R., Dwivedi, R.P. 2018. *Options and Strategies for Farmers' Income Enhancement in Bundelkhand Region of Central India*. Technical Bulletin No. 2/2018. ICAR-Central Agroforestry Research Institute (CAFRI), Jhansi, India. p 50. Available online: http://cafri.res.in/html/Technical_Bulletins/Bulletin%20Options%20and%20Strategies%20for%20Farmers%20Income.pdf. Accessed on 27 April 2023.

Singh, N.M., Medhi, K., Gogoi, S.H., Pathak, B. 2020. Reproductive growth performance and carcass traits of pigs raised under existing low input tribal backyard pig production system in Dima Hasao, Assam of North East India. *Journal of Entomology and Zoology Studies* 8(4):12–17. https://www.entomoljournal.com/archives/2020/vol8issue4/PartA/8-3-323-603.pdf.

Singh, P., Negi, G.C.S. 2014. Diversified integrated farming for sustaining small-holder farm families in Uttarakhand. *ENVIS Newsletter on Himalayan Ecology* 11(1):5. Available online: http://gbpihedenvis.nic.in/ENVIS%20Newsletter/vol11(1).pdf. Accessed on 21 April 2022.

Singh, R., Riar, T.S., Gill, J.S. 2017b. Integrated farming systems and socio-economic characteristics of Punjab Agricultural University awardee farmers. *Asian Journal of Agricultural Extension* 16(3):1–5. DOI: https://doi.org/10.9734/AJAEES/2017/30054. https://journalajaees.com/index.php/AJAEES/article/view/240.

Singh, R., Riar, T.S., Gill, J.S. 2017a. Integrated farming systems approach of PAU awardee farmers for income and employment enhancement in Punjab. *Advances in Research* 11(6):1–5. DOI: https://doi.org/10.9734/AIR/2017/33795. https://journalair.com/index.php/AIR/article/view/145.

Singh, S.K., Dige, M. 2015. *All India Coordinated Research Project on Goat Improvement.* Project Coordinator's Report 2014–2015. Central Institute for Research on Goats, Makhdoom, Uttar Pradesh, India. p 80. Available online: https://www.pcgoatcirg.icar.gov.in/pcunit/pdf-download/PC%20Report%20%202014-15.pdf. Accessed on 13 April 2023.

Sinhababu, D.P., Poonam, A. 2022. Development in rice-fish-livestock farming for higher production and income in coastal areas. In: Lama T.D., Burman D., Mandal U.K., Sarangi S.K., Sen H.S. (eds) *Transforming Coastal Zone for Sustainable Food and Income Security.* Springer Nature, Switzerland AG. pp 261–276. DOI: https://doi.org/10.1007/978-3-030-95618-9_22.

Srivastava, N.S.L. 2014. *Long-term Strategies and Programmes for Mechanization of Agriculture in Agro Climatic Zone-V: Upper Gangetic Plains Region.* Available online: https://farmech.gov.in/06035-04-ACZ5-15052006.pdf. Accessed on 21 April 2023.

Swarnam, T.P., Velmurugan, A., Subramani, T., Swain, S., Perumal, P., Jaisankar, I. 2018. Integrated farming system for enhancing farm income in rainfed lowlands of Andaman Islands. *Journal of the Andaman Science Association* 23(2):106–109. Available online: https://www.researchgate.net/publication/348080378_Integrated_Farming_System_for_enhancing_farm_income_in_rainfed_lowlands_of_Andaman_Islands#fullTextFileContent. Accessed on 03 May 2023.

Swati, Bharati, V.K., Chaurasia, O.P. 2022. Kadaknath chicken meat: Scope and source of animal origin nutrients for high-altitude regions. *Defence Life Science Journal* 7(4):288–293. https://publications.drdo.gov.in/ojs/index.php/dlsj/article/download/18015/7861/.

Syed, Z., Khan, M.S. 2017. Livestock and wild herbivores in the Western Himalaya: Competition or co-existence? *Journal of Threatened Taxa* 9(4):10084–10088. DOI: http://doi.org/10.11609/jott.3279.9.4.10084-10088.

Talukdar, P., Talukdar, D., Sarma, K., Saikia, K. 2018. Prospects and potentiality of improving pig farming in north eastern hill region of India: An overview. *International Journal of Livestock Research* 9(1):1–14. DOI: http://dx.doi.org/10.5455/ijlr.20180508053437.

Tiwari, B.K., Shukla, R.P., Lynser, M.B., Tynsong, H. 2012. Growth pattern, production and marketing of *Thysanolaena maxima* (Roxb.) Kuntzc: An important non-timber forest product of Meghalaya, India. *Forests, Trees and Livelihoods* 21(3):176–187. DOI: http://dx.doi.org/10.1080/14728028.2012.732793.

Velmurugan, A., Swarnam, T.P., Singh, D.R., Roy, S.D. 2016. Islands zone: Andaman and Nicobar Islands. In: Gangwar B., Jat N.K. (eds) *Organic Farming: Technologies and Strategies.* Today and Tomorrow's Printers and Publishers, New Delhi, India. pp 375–396. Available online: https://krishi.icar.gov.in/jspui/bitstream/123456789/52455/1/3_Chapter%20Organic%20farming_Island.pdf. Accessed on 03 May 2023.

Verma, A.K., Singh, V.B., Singh, V.P. 2003. Watershed management for sustainable hill farming system: A case study. In: Bhatt B.P., Bujarbaruah K.M., Sharma Y.P., Patiram (eds) *Approaches for Increasing Agricultural Productivity in Hill and Mountain Ecosystem.* Published by Director, ICAR Research Complex for NEH Region, Meghalaya, India. pp 87–94; 417.

Verma, A.M., Jha, V., Ahmad, S.H. 2008. Fish-makhana (*Euryale ferox* Salisb) integration – a case study of sustainable aquafarming system in north Bihar. *Journal of Indian Fish Association* 35:87–96. Available online: https://aquadocs.org/bitstream/handle/1834/33362/JIFA35_087.pdf?sequence=1&isAllowed=y. Accessed on 27 April 2023.

Wali, V.S., Kadam, K.L., Mohapatra, U., Mishra, R.K. 2019. Economics of sugarcane cultivation in Bagalkot district of Karnataka. *The Pharma Innovation Journal* 8(2):261–265. Available online: https://www.thepharmajournal.com/archives/?year=2019&vol=8&issue=2&ArticleId=3024.

Zore, R.V., Khunt, A.K., Amale, A.J. 2019. Economic feasibility of pomegranate production in Marathwada region of Maharashtra state. *International Journal of Current Advanced Research* 8(04A):18068–18071. DOI: https://dx.doi.org/10.24327/ijcar.2019.18071.

10 High-Value Aquaculture, Agribusiness Options and FPOs

Adoption of environment friendly, replicable high-value aquaculture practices as a major component of an integrated farming system (IFS) has the potential to provide various ecosystem services in terms of food security, nutritional security, nutrient recycling, biodiversity conservation, increased employment, enhanced profitability and climate resilience. This would also help in minimizing water footprint, waste load and carbon-nitrogen footprints while enhancing water use efficiency and water productivity. Besides year-round employment generation and income, high-value aquaculture practices produce less waste and lower carbon-nitrogen footprints than agriculture production sector. It reduces resource consumption and even permits increased food production, resulting in safe stocks and reduced environmental strain. This will increase the availability and affordability of wild seafood for everyone, particularly in areas where residents rely on only imported fish.

10.1 BIOFLOC AQUACULTURE

Biofloc technology (BFT), one of the advances in aquaculture, can support sustainable aquaculture objectives (Khanjani et al. 2022b). When properly applied, BFT can greatly reduce water use, mitigate environmental effects and improve farm biosafety by utilizing the organisms already present in the water (Avnimelech 2012). Three primary functions of biofloc organisms (BFOs) in BFT are (i) preserving water quality through the consumption of nitrogenous compounds; (ii) feeding the cultivated species (Wang et al. 2016); and (iii) competing with pathogens to produce probiotic qualities, which are frequently produced by heterotrophic organisms (e.g., *Bacillus* spp bacteria). By breaking down faeces and leftover food, BFOs contribute to the preservation of water quality by lowering dangerous nitrogenous chemicals like ammonia and nitrite (Suita et al. 2015; Wang et al. 2016). Total ammonia nitrogen (TAN) regulation in the environment is associated with BFO performance (Khanjani and Sharifinia 2020). Bioflocs (BFs) contains organisms such as bacteria, microalgae, yeast, rotifers, ciliates, protozoans, nematodes, copepods and crustaceans (Monroy-Dosta et al. 2013; Collazos-Lasso and Arias-Castellanos 2015; Khanjani et al. 2022a, 2022c). Increased variety in BFOs can help flocs better utilize proteins (Avnimelech 2007), mineralize waste (Martínez-Cordova et al. 2015) and manage pathogens (Emerenciano et al. 2013). The various C/N ratios (Panigrahi et al. 2018, 2019a, 2019b); environmental factors like salinity, light, oxygen levels and solids load and the type of cultivated species (Meenakshisundaram et al. 2021; Khanjani and Sharifinia 2021; Krummenauer et al. 2021; Khanjani et al. 2022b) are some of the factors that affect the diversity and density of the floc community in a BFT system. Organic carbon sources in BFT reduce ammonia-N concentration and improve the microbial community diversity (Deng et al. 2018). Furthermore, the kinds of ecological connections (commensalism, competition and predation) that exist between BFOs in a trophic network dictate the dynamics in the BFT system (Collazos-Lasso and Arias-Castellanos 2015). Proteins, lipids and vitamins can be made available to farmed species by the presence of BFOs in aquaculture, particularly in ciliates and nematodes. This promotes growth and lessens the requirement for these substances in commercial feed (Khanjani and Sharifinia 2020). Furthermore, determining the composition of BFOs can aid in optimizing the use of BFT, leading to higher output and a less environmental impact (Kumar et al. 2021; Abakari et al. 2022).

DOI: 10.1201/9781003239963-10

FIGURE 10.1 A small scale biofloc aquaculture unit.

High-density culture is permitted by BFT (Figure 10.1), which also provides the opportunity to recycle nutrients – particularly nitrogenous waste – into microbial biomass that the cultured animals can use *in-situ* while maintaining good water quality with little to no water exchange. Because nutrients may be continuously recycled and reused in the culture medium while benefiting from little to no water exchange, BFT is referred to as the upcoming 'blue revolution.' It is an environmentally benign aquaculture method based on the generation of microorganisms *in-situ*. The suspended growth known as 'biofloc' in ponds and tanks is made up of aggregates of bacteria, phytoplankton and other grazing bacteria as well as dead and living particle organic debris. It is the use of internal microbial processes in the pond or tank to serve as a water treatment solution and a source of food for cultured organisms (Wang et al. 2016; Suita et al. 2015). As a result, other names for this system include heterotrophic ponds, green soup ponds and active suspension ponds.

The principle is to maintain greater C:N ratio (Panigrahi et al. 2018, 2019a) by adding carbohydrate source, and the water quality is improved by the creation of high-quality single-cell microbial protein. Heterotrophic microbial growth takes place under such circumstances, assimilating nitrogenous waste that can be used as feed by the cultured species and functioning as a bioreactor to control the quality of the water. Because heterotrophs grow at a pace ten times faster than autotrophic nitrifying bacteria and produce more microorganisms per unit of substrate, hazardous nitrogen species are immobilized in biofloc more quickly. The foundation of this technique is the idea of flocculation inside the system.

Biofloc is a heterogeneous aggregate of suspended particles and variety of microorganisms (such as bacteria, algae, fungi, invertebrates and detritus) associated with extracellular polymeric substances (Khanjani et al. 2022a, 2022c). It is a live feed that is high in protein that is created when leftover feed and excrement in a culture system are exposed to sunlight and intense aeration, turning them into a natural food. The loose mucus matrix that binds each floc together is released by bacteria and is held together by electrostatic attraction or filamentous microorganisms. The naked eye can see large flocs, but the majority are tiny, the size range is 50–200 microns. A good nutritional value is discovered in biofloc; the dry weight protein ranges from 25% to 50%, fat ranges 0.5–15% and an excellent source of minerals and vitamins, especially phosphorus. It functions similarly to probiotics. One can use the dried biofloc as a component in the feed in place of soybean meal or fishmeal.

BFT make aquaculture activities more environmentally friendly. In contrast to conventional methods like the application of antibiotics, antifungals, probiotics and prebiotics, BFT is an inventive strategy for disease prevention and control given the emergence of new disease concerns and the rising expenses associated with energy. Aquaculture production could be made more profitable, sustainable and efficient by utilizing BFT in conjunction with bio-secure modular systems. There is reduced water usage, recycling of organic matter and *in-situ* nutrients as well as improved farm biosecurity due to disease exclusion, increased natural food production, improved feed conversion ratio and stress-free surroundings. The availability of microbial biofloc, a natural food source, makes up for aquatic organisms' increased protein requirements. Aquaculture could undergo a transformation certainly due to BFT. However, this is still in a preliminary stage and much research is necessary for its modification, calibration and execution.

10.1.1 BFT System Preparation

Before or after stocking, microbial biofloc production is a crucial phenomena for shrimp and fish culture in BFT systems. Depending on the species to be cultivated, farming intensity, buffering time for biofloc generation and system design and compatibility, there are various ways to set up the system for biofloc-based fish and shrimp farming. Additionally, the system can run in either semi-biofloc or full biofloc mode, incorporating substrates for periphyton growth or not. Ponds are prepared in accordance with standard procedures, which include drying, ploughing, replenishing the ponds with old soil (at least for four to five days) and adequately sanitizing water after it has been sieved through a 60 to 100 μ screen. All biosecurity precautions are also followed. Fertilizing with phosphorous and nitrogen fertilizers alone or in combination with any other biofertilizers promotes autotrophic bloom. Liming with hydrated lime or dolomite maintains the pH. The first step involves increasing the nitrogen content with feed or any other organic material. After that, a carbohydrate source, such as molasses, is added based on the amount of feed and the TAN level in order to maintain a high C:N ratio (15:1) and encourage the flocs (Khanjani and Sharifinia 2020; Panigrahi et al. 2019b). Usually, autotrophic algae take over ponds and, after a few days, floc forms, turning the water dark and frothy and turning the system heterotrophic with little to no algae. Following the rise of nitrite, this eventually stabilizes and can be managed by providing additional C, ammonia levels peak and then decline. Molasses can be added at intervals of one to three days over a period of two tco three weeks to encourage the transition from algae to bacteria. Based on the TAN level and feed quantity, the molasses or other carbohydrate supplements can be adjusted to maintain a high C:N ratio once the heterotrophic population has established. If producing biofloc in a shorter amount of time is the goal, the procedure may need to be altered. If the juveniles raised in nurseries are stocked, there will be adequate feed available to kick-start the bioflocs quickly with the addition of a significant amount of carbohydrates. Biofloc inoculum from other ponds or in its preserved condition can also be added to speed up the process. Before utilizing any commercial inoculum for this reason, caution should be exercised. Heavy aeration helps in keeping the floc under suspension. Pond liner allows easier maintenance of floc in suspension, avoiding dead spots, and stops accumulation of inorganic material from pond banks/walls caused by excessive water circulation.

10.1.2 Environmental Requirements for BFT

BFT operates in a biosecure environment since there is no water exchange. Bacteria in the biofloc system consume ammonia, increasing nitrite and increasing tolerance to a greater extent. The decreasing alkalinity must be corrected in order to keep it around 75–150 ppm. Meeting the high oxygen demand in the biofloc system is another crucial necessity. Additionally, an aspirator-style aerator will need to be suspended at the sludge pile. One HP is typically needed for every 350–400 kg of fish or shrimp biomass. The biofloc system requires high aeration not only to provide dissolved oxygen and remove CO_2 but also to keep biofloc in suspension and prevent production of nitrogenous metabolites, sulphides and organic acids.

10.1.3 Water Quality Management

The water chemistry in the system needs to be monitored and maintained during the cultivation time. Some indicators of water quality should be checked daily, while others should be checked only once every three days. Twice daily checks can be made for dissolved oxygen, pH, temperature and turbidity; daily checks can be made for NH_3, NO_3, NO_2 and TAN. Once a week, the algal community, salinity and hardness can be observed. A floc volume of 12–20 ml L^{-1} should be maintained. The total heterotrophic count can be checked every three days. To maintain the health of the shrimp and fish, the levels of water quality parameters should be maintained at their ideal values. Maintaining appropriate levels of all physico-chemical parameters is necessary for improved fish and shrimp production.

10.1.4 Major Cultivable Sspecies Suitable for BFT

The biofloc system functions best in species where the direct intake of floc can yield some nutritional benefits. The species that can withstand high sediment concentrations in water and generally accept low water quality are most suited for the biofloc system. Some of the species that are suitable for BFT are: air breathing fish like magur (*Clarias batrachus*), singhi (*Heteropneustes fossilis*), pabda (*Ompok pabda*), pangasius (*Pangasianodan hypophthalmus*), anabas/koi (*Anabas testudineus*); non-air-breathing fish like tilapia (*Oreochromis niloticus*), common carp (*Cyprinus carpio*), rohu (*Labeo rohita*), milkfish (*Chanos chanos*) and shellfish like tiger shrimp (*Penaeus monodon*) and Pacific white shrimp (*Litopenaeus vannamei*).

10.1.5 How to Prepare the Inoculum?

For the floc formation, 150 litres of inoculum are needed for every 15,000 litres of freshwater. Use a clean tub or can with 150 litres of water to prepare this inoculum, and then vigorously aerate it. Include 3 kg of pond soil, 1.5 g of urea or ammonium sulphate and 30 g of a carbon source (tapioca/wheat flour or jaggery). Thoroughly combine it with the water in the tub and ensure sufficient aeration. After 24 to 48 hours, the inoculum will be ready to be moved to the main tank. For floc to form, carbon sources must be added every day. For every 1 kg of feed given (with 25% of crude protein), 600 g of carbon source is to be added to the system to maintain C:N of 10:1. Once the floc volume reaches 15–20 ml further addition of carbon source is not required.

10.1.6 Advantages of BFT

- It is an eco-friendly culture system that reduces environmental impact thus higher biosecurity.
- Judicious/sensible use of land and water resources.
- It is a limited or zero water exchange system with higher productivity (It enhances survival rate, growth performance and better feed conversion in the culture systems of fish).
- BFT reduces water pollution and mitigates the risk of introduction and spread of pathogens
- It reduces utilization of protein rich feed and cost of feed.
- It reduces the pressure on capture fisheries, i.e., use of cheaper food fish and trash fish for fish feed formulation.

10.2 CARP POLYCULTURE

The primary carp species that are traditionally cultivated in ponds are the Indian major carps, or catla (*Catla catla*), rohu (*Labeo rohita*), and mrigal (*Cirrhinus mrigala*). These three species are raised together in ponds under a polyculture system because of their compatibility in terms of preferred diet and habitat. Mrigal feeds on the bottom, while catla is a surface dweller; rohu feeds in columns. In order to increase production under composite carp culture, or polyculture, exotic silver carp, grass carp and common carp with better growth potential are also included (Figure 10.2). Based on the intensity of resource utilization, standardized carp production technologies with different production levels have been developed. While pond carp polyculture typically produces

Indian major carps	Exotic carps
Catla (*Catla catla*)	Silver carp (*H. molitrix*)
Rohu (*Labeo rohita*)	Grass carp (*C. idella*)
Mrigal (*Cirrhinus mrigala*)	Common carp (*C. carpio*)

FIGURE 10.2 Indian major carp and exotic carp suitable for polyculture.

between 3–5 t ha^{-1}, some farmers in India are achieving higher production levels of 6–12 t ha^{-1} y^{-1}. Three essential inputs are the right kind of seed, feed and fertilizers; the success of the grow-out output depends on how well these are managed. In addition, fish health issues and the management of water and soil quality also play a significant role. The ideal ponds for carp grow-out production are thought to be 0.4–1.0 ha in size and have a water depth of 1.5–2.0 m. Pre-stocking, stocking and post-stocking operations are three broad categories for the environmental and biological manipulations used in carp polyculture management procedures. These activities are covered in more detail below.

10.2.1 Pre-stocking Pond Preparations

Control of aquatic weeds and removal of predatory and weed fish are among the important pre-stocking operations for achieving higher survival of the stocked seed. Drying the ponds until soil cracks followed by ploughing not only helps in eradication of undesired animals but also removes the obnoxious gases from the bottom soil. However, owing to the time requirement, non-availability of water for refilling and economic considerations, drying of ponds often is a difficult proposition. Such perennial ponds thus require necessary control measures against aquatic weeds, and predatory and weed fish prior to stocking. Even though there are many other mechanical, chemical, biological and manual techniques for controlling these weeds, the manual method is frequently recommended for weed removal. One of the most popular piscicides is mahua oil cake, which is applied at a rate of 2.0–2.5 t ha-m^{-1} three weeks before stocking. After decomposing, the oil cake not only acts as a piscicide but also functions as organic manure. Another efficient way to kill predatory and weed fish is to apply 350 kg ha-m^{-1} of commercial bleaching powder (30% chlorine) to the water. These fish can also be successfully controlled by alternate applications of urea at 100 kg ha-m^{-1} and bleaching powder at 175 kg ha-m^{-1} after 18 to 24 hours. Improvement of soil and water quality through use of lime and fertilizers is another important aspect connected with this phase of management. The ponds are usually manured with raw cow dung @ 3–4 t ha^{-1} 15 days prior to stocking as the basal dose. When mahua oil cake is used for controlling the predatory and weed fishes, the basal cow dung application, is avoided. Liming is carried out in the ponds three to four days prior to stocking

TABLE 10.1
Optimum Aquaculture Water Quality Parameters for Indian Major Carp

Water Parameter	Optimum Level
Water depth	2.0 m
Temperature	26–32°C
Dissolved oxygen	>5.0 ppm
pH	7.2–8.5
Total ammonia nitrogen	<1.0 ppm
Total nitrate nitrogen	<5.0 ppm
Nitrite nitrogen	<0.01 ppm
Sulphide	<0.03 ppm
Biological oxygen demand (BOD)	<5.0 ppm
Chemical oxygen demand (COD)	<70 ppm
Sacchi disc visibility (transparency)	25–50 cm
Ratio of phyto and zooplankton	10:1
Water colour	Light green/light brown

for correction of soil and water pH, which also helps in mineralization of the organic matter and pond disinfections. Calcium oxide (shell lime), dolomite or calcium carbonate (lime stone) is used for liming and the dosages are decided based on the prevailing soil/water pH. Usually, a pond with slightly acidic to neutral soil pH may be applied with 200–300 kg of lime. It is always important to maintain optimum water quality (Table 10.1) for grow-out operation.

10.2.2 Stocking the Ponds

The carp culture ponds are stocked with a proportionate number of surface, column and bottom feeders for minimizing the competition among them for food available in pond. A proportion of 30–40% surface feeder (silver carp and catla), 30–35% column feeders (rohu) and 30–40% bottom feeders (common carp and mrigal) are commonly recommended. In possibility of providing aquatic weeds from external sources, grass carp may be added at 5–10% level substituting other species. Usually, the grow-out ponds ate stocked with fingerlings of 80–100 mm (8–10 g) after due acclimatization. Stocking density of 5,000 fingerlings ha^{-1} is recommended for carp polyculture with a production target of 3–5 t ha^{-1} yr^{-1}. The densities can be raised to 8,000–10,000 fingerlings ha^{-1} for achieving production levels of 5–8 t ha^{-1} yr^{-1} and 15,000–25,000 juveniles ha^{-1} for 10–15 t ha^{-1} yr^{-1}. The stocking density, however, depends largely on the input use and levels of management.

10.2.3 Post-stocking Management

Fertilization schedule: Intermittent liming and fertilization, supplemental feeding, water management and health care are the main facets of post-stocking pond management. The post-stocking fertilization techniques in an average productive pond consist of applying cow dung at a rate of 0.5 t ha^{-1}, urea at a rate of 10 kg ha^{-1}, and SSP at a rate of 15 kg ha^{-1} every two weeks. However, based on the pond's plankton content and water quality, the timing and dose of the application may be postponed. Other sources of organic manure include duck droppings, pig dung, poultry litter and so on, depending on what is available. Application of *Azolla* as biofertilizer @ 40 t ha-1 yr-1, provided at weekly split dosages, has been shown to supply the complete complement of nutrients in an intense carp culture pond, i.e., 1,500 kg of organic matter, 25 kilogram of phosphorus, 90 kg of potassium, and 100 kg of nitrogen. Because biogas slurry releases nutrients more quickly and uses less oxygen, it is beneficial to use at 30–45 t ha^{-1} yr^{-1} or 80-–00 l day^{-1} ha^{-1}.

Supplemental feeding: A mixture of groundnut/mustard oil cake and rice bran at 1:1 by weight is the most commonly used supplementary feed in carp polyculture. These are provided at 5% of the stocked biomass in the first month of culture followed by sliding rate of 3–2% of the estimated biomass in subsequent months. The biomass of fish is estimated through monthly sampling and the daily feed ration is calculated as follows

Feed requirement/day = Estimated fish biomass in the pond × % feeding rate,

where, biomass = Average weight of fish × total number of fish stocked × % survival.

A survival percentage of 75–80% is usually considered for estimation of biomass. The daily ration is divided into two splits and provided in dough form preferably in feed trays or gunny bags hung at uniform distance inside the pond (Figure 10.3). Aquatic vegetation, viz, *Hydrilla, Ceratophylum, Najas*, duckweeds, etc., are additionally provided in ponds stocked with grass carp in selected corners of the pond. The nutrient-rich supplementary feed in the form of pellets or floating feed are also available for carp. Feed comprises about 60–65% of production cost. In addition to extra expense, water quality can deteriorate unnecessarily due to the addition of excess feed. Extruded, floating feeds offer the advantage of watching the feeding response as opposed to a sinking, steam-pelleted feed (Figure 10.4). Today, floating feeds are used widely due to minimal wastage of feed, minimal deterioration of water quality, low FCR, easy to digest, faster growth, higher yield and higher water productivity.

Aeration and water exchange: Normally dissolved oxygen is not a critical factor in ponds stocked at density of 5,000–10,000 fingerlings ha^{-1} with 3–5 t ha^{-1} yr^{-1} production targets. However, ponds with higher stocking density require additional aeration (Figure 10.5) to increase the dissolved oxygen content in water. Paddle wheel, aspirator and submersible types of aerators are commonly used in carp culture ponds. Usually a pond of one-hectare water area requires four aerators (1 HP each). Water exchange is required in the high-density culture ponds to avoid deterioration of water quality, which occurs due to accumulation of metabolites and decayed unutilized feed, which otherwise lead to slow growth and outbreak of diseases. Low-density culture, however, does not require any water exchange during the culture period, only except compensation for seepage and evaporation loss. In intensive culture practices replacement of certain amounts of water at regular intervals, especially during later part of the culture period is required for maintenance of water quality.

Plankton management: In addition to reducing water quality oscillations, phytoplankton are important for maintaining the equilibrium of the entire pond ecosystem. A healthy population of

FIGURE 10.3 Feeding fish using gunny bags hung at uniform distance inside the pond.

FIGURE 10.4 Different types of fish feed in use for carp polyculture.

FIGURE 10.5 Pond aeration using paddle wheel aerators.

phytoplankton reduces the concentrations of CO_2, NH_3, NO_2 and H_2S while enriching the system with oxygen during the day through photosynthesis. Because phytoplankton may tie up heavy metals and consume NH_4, a robust phytoplankton bloom can minimize dangerous chemicals. It can stop filamentous algae from growing because phytoplankton can obstruct light from penetrating the bottom. In addition to regulating water temperature and supplying the right amount of turbidity, a robust bloom also reduces temperature loss in the winter. In a pond system, a 10:1 ratio of phytoplankton and zooplankton is optimum (Figure 10.6). There is usually a lot of phytoplankton in ponds. These phytoplankton have an impact on the water quality and are crucial to the ecology of ponds.

A few species, typically fewer than ten, make up the majority of the plankton population in aquaculture ponds, which frequently contain 50 or more species of phytoplankton and three to five species of zooplankton at any given moment. In ponds, phytoplankton communities usually experience fast succession, and over the course of a few weeks, the species composition of these communities

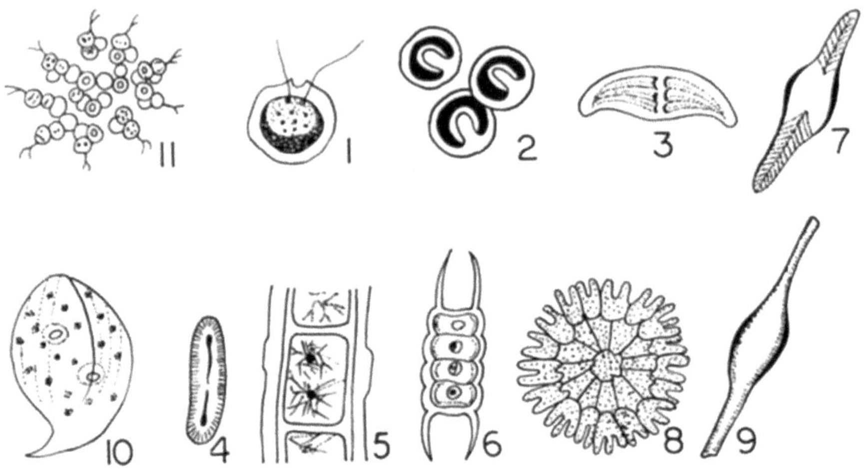

Phytoplankton: *1.Chlamydomonas, 2. Chlorella, 3.Closterium,4. Pinnularia, 5. Zygnema, 6. Scenedesmus, 7. Navicula, 8.Pediastrum, 9.Fragilaria, 10. Phacus, 11. Eudorina*

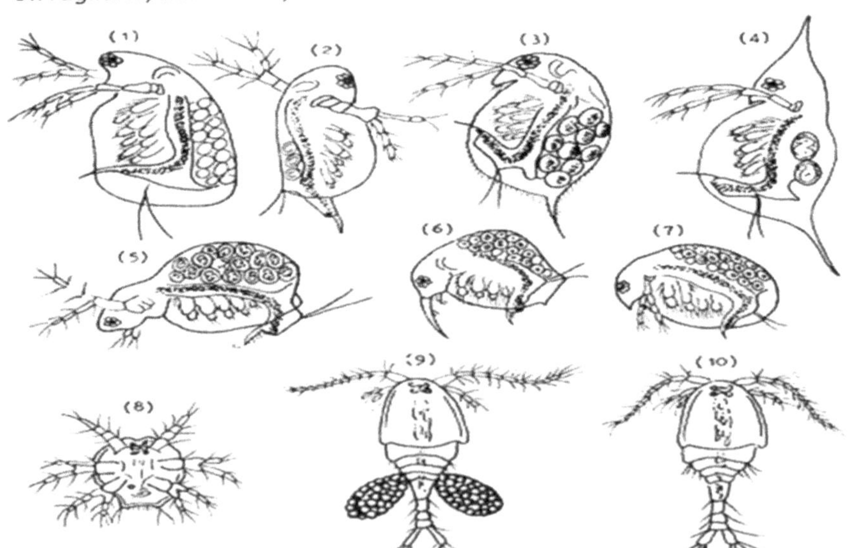

Zooplankton: *1.Sida sp., 2. Diaphanosoma, 3-4.Daphnia 5. Moina, 6-7. Bosmina, 8.Cyclops larva, 9.Cyclops with egg, 10. Cyclops without egg*

FIGURE 10.6 Common phytoplankton and zooplankton encountered in grow-out pond culture.

frequently shifts. The concentrations of nutrients, particularly those of inorganic nitrogen and phosphate, often regulate the number of phyto- and zooplankton. The term 'plankton bloom refers to the condition where these organisms give the water colour. One significant source of dissolved oxygen is phytoplankton. During daylight hours, the photosynthesis of these plants generates oxygen at a significantly higher pace than the atmospheric diffusion of oxygen into the pond water. Ammonia nitrogen is quickly removed from the water by phytoplankton, lowering the amount of this potentially harmful material in the water. Lastly, phytoplankton produces turbidity that prevents light from penetrating the bottom of ponds, and a phytoplankton bloom is an effective way to manage aquatic macrophytes that are submerged in water.

In ponds, a high concentration of some phytoplankton species can have negative consequences, most notably the generation of poisonous and unpleasant substances. Fish and prawns can absorb the malodorous substances geosmin and methyl-isoborneol (MIB), which are produced by some phytoplankton, primarily particular blue-green algae (BGA). This can give fish and prawns an off-flavour, or unpleasant smell and taste. The most common producers of this chemical are Anabaena and Aphanizomenon species, which are frequently found in large numbers in freshwater aquaculture ponds. MIB has a musty, medicinal smell that is similar to that of camphor. It seems that species of *Planktothrix*, formerly known as *Oscillatoria*, are the common cause of MIB. To reduce the possibility of off-flavour, some producers have found success in stocking planktivorous fish like tilapia and Chinese carp alongside the primary culture species. These fish consume BGA. Nevertheless, to eradicate the algae that gives ponds their off-flavour, copper sulphate treatment may occasionally be required. The usual treatment rate for copper sulphate in milligrams per litre is 0.01 times the total alkalinity concentration.

By giving no more feed or fertilizer to ponds than is necessary, efforts should be taken in pond management to prevent excessive phytoplankton blooms. However, in feed-based aquaculture, ponds usually experience extensive blooms of phytoplankton. In ponds without aeration, phytoplankton blooms frequently become excessive at feeding rates more than 30 kg ha^{-1} daily. To prevent low dissolved oxygen content at night, it is crucial to perform adequate mechanical aeration at higher feeding rates. Without a doubt, the most significant problem with water quality in feed-based pond aquaculture is low evening dissolved oxygen concentration.

10.3 AIR-BREATHING CATFISH CULTURE

Catfish are found all throughout the world and are classified into four families: the Ictaluridae, Claridae, Pangasidae and Siluridae. Southeast and South Asia have long been home to the catfish culture. In some regions, catfish are valued highly, while in others they are regarded as medium- and lower-value market species. Catfish are naturally hardy; they can go extended periods of time without water and have a secondary respiratory system. A particular market exists in certain regions of eastern India for live catfish. The most important aquaculture species of catfish is the Asian catfish, *Clarias batrachus*. It belongs to the Claridae family.

Because catfish are so popular, they are quite important commercially. Both air-breathing catfish species (Figure 10.7), *Clarias batrachus* (magur) and *Heteropneustes fossilis* (singhi), are acclimated to harsh environmental circumstances. Although the methods for producing these two significant catfish species' seeds and through induced breeding have been refined, their large-scale production has not yet been implemented. Grow-out cultures of magur, which reach 100–200 g in six to eight months, can reach production levels of 2.5–3.5 t ha^{-1} with pond management techniques that are essentially the same as those used for carp culture, including stocking 20,000–50,000 fingerlings ha^{-1}. Despite the enormous market demand for these species and the abundance of potential resources in the form of wetlands and abandoned waters that could be exploited for commercial farming, large-scale culture of these species has not yet received the attention it deserves. Due to its carnivorous eating habits, the development of a balanced supplemental feed and the availability of the required quantity of seed in the appropriate size are the two crucial issues that must be

FIGURE 10.7 Catfish top: *Clarias batrachus* (magur) and bottom: *Heteropneustes fossilis* (singhi).

addressed. Given the high consumer preference for these in various parts of the country, research into the development and standardization of induced breeding and grow-out technologies of several other non-air breathing catfishes, such as *Mystus seenghala*, *M. aor*, *Pungasius pungasius*, *Wallago attu* and *Ompak pabda*, is also being considered. These fishes can suitably be cultured both in monoculture and in polyculture systems.

During larval rearing, at least 50% water exchange is given in every alternate day for maintenance of good water quality, which otherwise leads to poor larval growth and sometimes leads to heavy mortality. Water depth of the larval lank is an important factor that needs to be given attention. Initially, a water column of 8–10 cm is kept in the larval tank to avoid heavy loss of energy spent by the larvae for making vertical trip. The larvae are reared in rearing tanks for 12–14 days during which they reach to 10–12 mm size (30–40 mg). Then the fry is reared for 30 days usually at a stocking density of 200-3 00 numbers per m^2 in outdoor cement tanks provided with 6–8 cm soil base or well-prepared earthen ponds. The tanks are fertilized with a schedule similar to that in carp rearing tanks. Floating weeds like water hyacinth or duckweeds are provided to cover a portion of the water area. These plants not only give shelter, but also harbour insects that form good food for the fry. The fish are fed with pallet feed (about 35% protein) of finely minced trash fish/molluscan meat and rice bran at 1:1 ratio @ 6–10% of stocked biomass per day during the evening hours. Fingerlings grow up to 3–4 cm (0.8–1.0 g) size during this period, which are harvested by draining the tank and are stocked in the grow-out ponds with water depth of 75–100 cm.

A 1,000–2,000 m^2 pond is the perfect size for magur grow-out culture. The ideal slope for a pond bund is between 1:2 and 1:3. The grow-out culture of magur is appropriate for the small, shallow (0.75–1.0 m water depth) earthen ponds, stone pitched ponds and cemented tanks. Magur used to crawl out of the pond during the wet season. In order to prevent them from escaping, the pond's dyke should be built in such a way during construction. The majority of farmers stock 200 fingerlings per square meter for commercial purposes, although the recommended stocking density is 50–70 m^{-2}. During grow-out culture, stocking larger-sized seed (3–5 g) demonstrates good survival and development. To guarantee full netting, the pond bottom needs to be level. As air breathers, magur typically rise to the surface of the water to take in atmospheric oxygen. This behavior draws birds that will eat them. As a result, ponds must be covered with net to keep birds from eating the fish.

Measures like (1) stocking fingerlings of equal size, (2) feeding the fish two to three times a day in at least four feeding zones of the pond and (3) separating the fastest developing individuals can be implemented to reduce the loss resulting from cannibalism.

10.4 FRESHWATER PRAWN CULTURE

Aquaculturists and fish farmers around the world have taken an interest in the culture of *Macrobrachium rosenbergii*, the giant freshwater prawn, due to its fast growth, resilience to disease, ability to adapt to artificial feed and polyculture environments and high market demand. Among freshwater prawns, the largest and quickest growing species is *Macrobrachium rosenbergii*, commonly referred to as scampi (Figure 10.8). The species' hatchery technology has been refined and standardized. In a culture period of five to six months, freshwater prawn monoculture at stocking densities of 40,000–50,000 ha^{-1} has demonstrated output values of 1.5–2.0 t ha^{-1}. Furthermore, it has been shown that polyculture of freshwater prawn and carp are an economically feasible way to increase farmers' farm income while also being a technologically sound culture method.

The grow-out monoculture of the species is carried out in earthen ponds of 0.2–0.5 ha size with a depth of 1.2–1.5 m. preferably with a sloping bottom and a sump at the lower end of the pond for easy harvesting. The grow-out ponds are often stocked with post-larvae, which usually lead to poor survival and hence low yield. Therefore, a nursery phase of 45–60 days is advocated prior to the grow-out phase for successful farming of scampi. The pre-stoking pond preparation for nursery and grow-out production involves a number of steps, including eradication of competitors and predators and liming and pond fertilization, which are similar to the grow-out carp culture.

10.4.1 Nursery Pond Management

Smaller earthen ponds or cemented tanks with an average water depth of 1 m are preferably used for nursery rearing. Floating weeds with dense root system like *Eichhornia*, *Pistia*, etc., may be provided as shelter for the post-larvae, covering approximately 10% of the water surface. Frame of bamboo or PVC pipe may be used to prevent the spreading of the weed. The nursery ponds are stocked with post-larvae (8–10 mg) at usual stocking density of 20 per m^2. Crumbled feed pellet @ 60–80% of the biomass per day for the first fortnight followed by 30–40% per day in the subsequent period may be provided as supplementary feed. In case of non-availability of formulated pellet, mixture of groundnut oil cake and rice bran may be given. The feed may be provided through broadcasting in two rations, i.e., during morning and evening hours. Aeration may be provided for 6–8 hours per day to keep the optimum dissolved oxygen level of 5–7 ppm. While water exchange is not a necessity, evaporation and seepage losses may be compensated at least at fortnight interval. During 45–60 days of rearing period the post-larvae grow to juveniles of 1–2 g with survival levels of 65–75%.

FIGURE 10.8 The giant freshwater prawn (*Macrobrachium rosenbergii*).

10.4.2 Grow-out Pond Management

Grow-out culture of *M. rosenbergii* is normally carried out for a period of four to six months. Prawn juveniles obtained from nursery pond are stocked at a density of 40,000–50,000 ha^{-1} (4–5 per m^2) under monoculture. Supplementary feed in the form of pellets (2–3 mm diameter. 25–30% protein) with a water stability of 2 hours are broadcasted around the peripheral area @ 8–10% of the biomass during the initial month, followed by 3% of biomass in the subsequent rearing period. The daily ration is provided in two to three splits with a major share provided at evening hours. The amount of daily feed is monitored through provision of four cheek trays placed at regular intervals in the pond. Since feed consumption reduces during moulting period as well as on rainy/overcast days, the daily ration may be reduced during these periods by 10–20%.

Heterogeneous individual growth (HIG), one of the many intrinsic and extrinsic characteristics related to scampi production, poses a major risk to farmers and can cause significant financial losses. It is known that *Macrobrachium rosenbergii* displays a complex social organizational hierarchy with separate subordinate, dominant and subdominant groups based on morphology. The male morphotype's predominance of a clear social hierarchy contributes to the prawns' varied growth patterns. While the female population's size distribution is rather uniform, the male population exhibits a high degree of HIG. The three developmental stages of the male maturation process are typically represented by the three male morphotypes: stunted male (SM), orange-clawed male (OCM) and blue-clawed male (BCM). These morphotypes undergo metamorphosis from SM >> OC >> BC. The developmental road begins with the SMs, which are submissive and not territorial; OCMs, on the other hand, are subdominant and signify extremely somatic growth; and BCMs, on the other hand, are the last step in the pathway of morphotypic metamorphosis. A group of prawns that have reached the terminal morphotype stage prevents other morphotypes from changing into higher stages. As a result of this the growth pattern varies greatly. A few strategies can be used to solve this issue, including: (1) periodically removing berried females, which aids in morphotypic transformation and the male maturation process; (2) periodically removing dominant BCMs to reduce the number of 'runt' individuals; and (3) harvesting fast-growing individuals in stages to increase the chances that other smaller individuals will reach their full potential for growth.

10.4.3 Water Quality Management

Constant water depth of 1.2–1.5 m is maintained in the pond through periodic compensation of the seepage and evaporation loss. Water quality (Table 10.2) in the grow-out ponds should be monitored at least every 15 days and appropriate intermittent liming and fertilization measures should be taken. In the case of intensive fanning with high stocking density, intermittent water exchange and aeration are the other additional management measures to increase the yield. Harvesting of bigger size prawn, mainly females from fourth month onwards, can increase the yield and improve the growth of smaller prawn. The major problems encountered during grow-out culture includes low dissolved oxygen content, which often occurs due to heavy plankton bloom. This problem gets precipitated during continuous overcast, which can be overcome with water exchange or aeration. Usually the prawns reach marketable size at about 50–60 g within five to six months. Harvesting need to be done only during the inter-moult period, as prawn of pre-moulting stage and just after moulting fetch a very low market price.

10.5 ORNAMENTAL FISH CULTURE

Ornamental fish form an important commercial component of the fisheries sector. The brilliant colour and exotic appearance of the fish attract one and all, greatly enhancing the aesthetic beauty of homes. The relatively minimum requirement of space or attention compared to other species is the reason for the growing interest in keeping aquaria in households. In spite of the country possessing 100 plus Indigenous species of ornamental value and about equal numbers of exotic species, the

TABLE 10.2
Optimum Aquaculture Water Quality Parameters for *M. Rosenbergii*

Water Parameter	Optimum Level
Water depth	1.2–1.5 m
Temperature	28–32°C
Dissolved oxygen	>5–7 ppm
pH	7.0–8.5
Alkalinity ($CaCO_3$)	20–60 ppm
Total hardness ($CaCO_3$)	30–150 ppm
Non-ionized ammonia (ppm NH_3-N)	<0.3 ppm
Nitrite nitrogen (ppm NO_2-N)	<0.2 ppm
Nitrate nitrogen (ppm NO_3-N)	<10 ppm
Iron (Fe)	<1.0 ppm
Hydrogen sulphide (ppm H_2S)	<0.01 ppm
Biological oxygen demand (BOD)	<3.0 ppm
Chemical oxgen demand (COD)	<70 ppm
Transparency	25–40 cm
Ratio of phytoplankton and zooplankton	10:1
Water colour	Light green/light brown

domestic ornamental fish trade mostly is confined to a few exotic live-bearers, gold fish and angel fish. Considering the increasing domestic demand and ready export market, ornamental breeding and culture offers enormous scope for adoption both at small scale and as commercial enterprise.

10.5.1 Infrastructure Facilities

The infrastructure facilities mainly include FRP/ferro-cement breeding and rearing tanks of different dimensions, smaller earthen ponds, clean water supply systems, glass aquaria of assorted sizes and proper housing for breeding and larval rearing. The ornamental fish culture unit (Figure 10.9) should have provision for a continuous supply of water and electricity. In a large-scale production unit, there must be provision for water pumps, standby generator, air blowers, live food culture units, etc. An oxygen cylinder may be kept for packing of fish during sale. The dimension of the tanks may depend on the purpose of their use, species to be bred/reared and scale of operation. Tanks of 2 × 2 × 2' to 4 × 3 × 3' or circular tanks of 4' diameter and 3' height are commonly used as breeding tanks, while the size of culture tanks normally varies from 5 × 6 × 3' to 10 × 7 × 3'.

10.5.2 Species for Culture

Ornamental fish are broadly classified as live-bearers and egg layers. Among the live-bearers, *guppies* are the simplest group to start with as the species can tolerate a wide range of water quality and accept a great variety of foods. There are many varieties of guppy, viz., Round tail, Sphere-tail, Fantail, Veil-tail, Cobra guppy, Singapore guppy, etc. The next popular live-bearer group is the molly, which includes Sail-fin molly, Balloon molly, Chocolate molly, White molly, etc. Another magnificent group of live-bearer is the Sword tail with varieties like Red swordtail, Green swordtail, Sunset swordtail, Tuxedo tail, Double-swordtail, etc. Platy is yet another group of small live-bearer with many varieties like Round tail, Veil tail, Fringe tail, Lyre tail, Pin tail, etc.

The majority of ornamental fish are egg-layers, which includes carp, barbs, rasboras, danios, tetras, catfishes, etc. Goldfish has been the most common and popular aquarium fish throughout the world for decades. Many varieties of goldfish like Lionhead, Veil-tail, Fringe-tail, Telescope-eye,

FIGURE 10.9 A small-scale ornamental fish culture unit.

Calicogold, Peacock-tail, Red-carp, Pearl-scale, etc. are common in the market. Now a days, Koi carps with different colour combinations have become popular for aquaria and garden pools. The easy seed production and bright body colouration have increased the popularity of barbs for aquarium fish keeping. Many Indigenous and exotic varieties of barbs are available, including the well-known ones like Rosy barbs, Tiger barbs, etc. Danios like Zebra danio, tetras like Black widow tetra, Neon tetra, Lemon tetra, etc. are other popular varieties of ornamental fish. The anabantid groups of fish are air breathers and lay eggs in bubble nests. The group includes Siamese fighting-fish, several varieties of gouramies like dwarf gourami, pearl gourami, moonlight gourami, kissing gourami, etc. Angelfish belonging to the chilid group is another major group for aquarium breeding. Discuss fish and its blue and brown forms are highly priced. Some Indigenous fish like Indian glassfish and loaches are also much in demand in the market.

10.5.3 Breeding of Live-Bearers

Mollies mature in 12–16 weeks, while live-bearers such as swordtails, platies, and guppies take six to eight weeks. Selected mature males and females of these species are kept in separate breeding pools designated for each species, and breeding traps consisting of perforated nylon bags are stocked within them. To aid in breeding, aquatic plants such as cobomba are added to the breeding pool. In these live-bearers, internal fertilization occurs when the male inserts its gonopodium into the female to deposit milt, fertilizing the eggs inside the female. The embryos grow within four weeks to tiny young ones and become ready for free swimming. Soon after their birth, the young ones are collected to avoid parental predation. The larvae are fed with infusoria, zooplankton, tubifex, etc. till they attain 25–30 mm in size, after which they are transferred to rearing tanks.

10.5.4 Breeding of Egg Layers

Matur males and females of each species are selected from the community tanks and stocked in breeding tanks at the ratio 1:2-3. Usually, females are introduced in the breeding tank one day in advance than the males and external fertilization takes place. In general, parents eat away their own egg and larvae. Therefore, round pebbles at the bottom and aquatic weeds and nylon threads are provided to prevent parental predation. Immediately after hatching, parents should be removed from the breeding pool. The larvae are fed with infusoria, zooplankton, tubifex, etc. till they attain 25–30 mm in size, after which they are transferred to rearing tanks. Generally, feed is provided once daily, preferably in the morning. The feeding rate depends on species, size and season. Overfeeding is more damaging than underfeeding as the excess feed worsens the water quality.

10.5.5 Water Quality Management

Because ornamental fish are more susceptible to poor water quality, ornamental fish production units required a greater level of knowledge for better water quality management. Farmers in urban areas utilize standard taps for farming. It is dechlorinated by aerating it for a few days prior to usage. In rural places, tube well water is also directly utilized. The rearing water in the region has an average temperature between 20°C and 28°C and a slightly alkaline pH. Other factors are not as important. The majority of cultivated species favor water that is soft to medium hard. Regular water exchange along with proper aeration is a must to maintain critical water quality parameters like pH (7.2–7.7), DO (6–7 ppm), nitrite (<0.75 ppm), Nitrate (<30 ppm), ammonia (almost zero) etc.

10.6 FRESHWATER PEARL CULTURE

Freshwater pearl culture using three commonly available freshwater mussel species, viz., *Lamellidens marginalis, L. corrianus* and *Parreysia corrugata*, have been standardized. Three distinct surgical techniques – mantle cavity insertion, mantle tissue implantation and gonadal implantation – have also been standardized to produce various pearl products. The results of the mantle cavity insertion technique can be shell-attached, half-round or design pearls, depending on the implanted nuclei's shape. The outcomes of the gonadal implantation operation are unattached, somewhat larger round pearls than those of the mantle tissue implantation procedure, which produces irregular to oval graft pearls or small round nucleated pearls.

10.6.1 Pond Culture

The implanted mussels are stocked in the ponds following post-operative care. The mussels are housed in nylon bags with a mesh size of 1 to 12 cm^2, with two mussels per bag. The bags are then strung from bamboo or PVC pipes and submerged in ponds at a depth of 1 m (Figure 10.10). During the sweltering summer months, the mussels can be positioned in deeper zones up to 2.0 m to prevent

FIGURE 10.10 Implanted mussels for freshwater pearl culture.

surface heating. 20,000–30,000 ha^{-1} is the stocking density at which the mussels are cultivated. A rectangular pond with appropriate inlets and outlets is the best shape for raising implanted pearl mussels. Pearl culture operations can be conducted in ponds that do not have aquatic macrophytes or algal blooms, such as those caused by *Microcystis* sp. and *Euglena* sp. P.V.C. tubing (2" dia.) platforms (16 × 8 m) are supplied with the ponds as rafts for hanging pearl mussel culture units. The implanted mussels are packed in nylon bags (1.0 cm mesh, 12 × 14 cm) with two mussels per bag at a density of 50,000–75,000 ha^{-1}. These basic culture units provide enough mesh size and pocket space to allow mussels to open and close their shell valves for feeding and movement. After that, these bags are fastened to bamboo rafts or PVC floating platform units that are kept in the cultural setting. It has been found that for optimal mussel growth and survival, the pearl mussel culture units should be suspended 1.5–2.0 m deep in the ponds. Alternatively, the mussels can also be placed in plastic crates (0.5 × 0.35 × 0.25 m) @ 20–25 mussels per crate.

10.6.2 Water Management

Ponds with a clay soil base and mildly alkaline waters, measuring between 2 and 3 m deep, are ideal for pearl growing operations. Monitoring the ponds' physical and chemical characteristics as well as their water level is necessary for the entire period of the mussels' culture. The ideal temperature range is between 25 and 30°C. The pond must be kept at its ideal pH range of 7.5–8.5, total alkalinity of 75–150 ppm, total hardness of 40–75 ppm and dissolved calcium of 25–50 ppm. The habitat used in freshwater pearl culture ponds is typically the same as that used in carp aquaculture. Since freshwater mussels make up the majority of the benthic invertebrates in pond ecosystems, their free dispersal on the bottom of the pond should be optimal for their survival and growth. Research has shown that mussels kept in bottom-set culture units had a lower survival rate (27%) compared to those kept in surface and column-set units (66%). The lower level of primary output and siltation made worse by the units' limited space could be the cause of the low survival in bottom units. The degree of macrophyte infestation and water flow in the culture environment are additional crucial factors. But there are significant issues with sampling and pearl harvesting when it comes to the pond bottom dispersion of the mussels under operation. In order to maintain plankton productivity, the ponds are periodically fed with both organic and inorganic fertilizer. Maintaining the pearl bearing mussel standing crop properly has been found to require the constant addition of green water (specifically, *Chlorococcum* sp. and *Scenedesmus* sp.) to the pearl culture ponds as direct mussel feed. An array of 200-litre Ferro-cement tanks are set up along the pond bundh to facilitate the 'open culture method' of producing green feed. The water in the tanks is fertilized as given below:

- Cow dung: 1000 kg ha^{-1} yr^{-1} (in 12 equal installments)
- Urea: 100 kg ha^{-1} yr^{-1} (in 12 equal installments)
- Single super phosphate: 100 kg ha^{-1} yr^{-1} (in 12 equal installments)

The green water appears after seven to ten days of fertilizer degradation. The pearl cultivation ponds get the water that has been enhanced with algae. Because they are mucoid filter feeders, the mussels can take in a wide range of specific organic material feed. Pearl mussels kept in captivity can be fed a diet consisting of groundnut oil cake and powdered rice bran (1:1) at 1% of the mussels' weight given every other day. Throughout the course of the 12- to 18-month culture period, periodic inspection of the mussels, including the removal of dead ones and bag cleaning, is necessary.

10.7 COASTAL SHRIMP AQUACULTURE

In Southeast Asia and India, coastal shrimp aquaculture (*Penaeus monodon*, *P. indicus*, or *Litopenaeus vannamei*) is the most popular (Figure 10.11). Hatchery-produced seeds are often

FIGURE 10.11 Coastal brackish water shrimp culture (left: shrimp seed stocking: right- grow out pond).

stocked at rates of 15 to 20 PL m^{-2} (*Penaeus monodon*) and up to 100 PL m^{-2} (*Litopenaeus vannamei/P. indicus*) in semi-intensive ponds (1–5 ha). Pumping is used in addition to the tide's usual water exchange. The shrimp are fed artificial meals in addition to naturally occurring foods that are increased by pond fertilization. For *Penaeus monodon*, the average production ranges from 1,000–4,000 kg ha^{-1} $crop^{-1}$, while for *Litopenaeus vannamei*, it ranges from 3,000–12,000 kg ha^{-1} $crop^{-1}$. The majority of intensive farms are found in non-tidal locations where ponds can be dried out and thoroughly drained before every stocking. Ponds are typically square or rectangular in shape and small (0.1 to 1.0 hectare). In India, the typical range for stocking density is 20–60 PL m^{-2}. Either electric or diesel motors must be used for heavy aeration in order to provide animals and phytoplankton with oxygen and water circulation within. Artificial diet feeding occurs four to five times a day, and it is followed by feed tray inspection. Typically, the final FCR ranges from 1.2 to 1.8:1. Reduced water exchange and closed systems have become standard practice since the white spot disease outbreak because they have a lesser risk of spreading viral illnesses through water intake. To prevent wastes from degrading the pond bottom and water quality, feed and phytoplankton blooms must be closely monitored and controlled. Since feed accounts for more than 50–60% of production costs in grow-out culture systems, effective feed management is a critical component of a successful crop. The parameters of water quality, including pH, salinity, dissolved oxygen, alkalinity, Secchi disk, H_2S and unionized ammonia must be periodically checked (Table 10.3). If a closed system culture is applied, stocking should be minimized, otherwise the pond must be harvested earlier (within 3.5 months instead of four to five months) and smaller shrimp will be yielded (20 g, instead of 30–35 g as achieved in semi-intensive and intensive systems with water exchange).

10.7.1 Water Quality Management

During the production cycle, the rearing pond's water quality deteriorates because of things like uneaten food and animal excrement. Dissolved oxygen (DO), pH, ammonia, water colour and water odor are typically used by shrimp growers as markers to assess the pond's water quality. In order to keep these parameters within the ideal range, they are routinely monitored using either scientific equipment or the farmer's expertise (Table 10.4).

10.7.2 Dissolved Oxygen

The health of the shrimp depends on the amount of dissolved oxygen in the pond water. However, plankton (20–45%) and pond substrate/sediment (50–70%) are the primary consumers of dissolved oxygen in the rearing pond. The amount of dissolved oxygen that the shrimp consumes is only 5%. There are three methods for regulating the dissolved oxygen content. First, by positioning paddle wheels correctly, the water surface area is increased. This improves the pond's water circulation and increases its oxygen content. Second, by regulating plankton density to an ideal degree, and, third, by reducing surplus organic materials, such as leftover food.

TABLE 10.3
Trouble Indicators vs. Remedial Measures in Coastal Shrimp Aquaculture

Observation	Reasoning	Remedy
Water becoming clear	Plankton crash, heavy sediment and benthic algae	Go for bottom drainage, add water, add fertilizer and go for algal inoculation
Abrupt change in water colour	Death of algae, high organic load, acid leaching	Exchange water, regulate pH
Erratic swimming at the water surface	Stress due to low O_2, high NH_3	Exchange bottom water followed by aeration
Snails climbing out the water	Low dissolved O_2 contents in the water	Exchange water/followed by aeration
Abrupt lowering of salinity	Due to heavy rain, fresh water floats on the top of salt water and forms a barrier result in low oxygen at the bottom	Drain surface water, add saline water followed by aeration
Temperature >32°C	–	Go for aeration
pH variations	Weather condition, plankton population, organic load	Application of lime/zeolite
Partial moulting	Unfavorable water quality	Water exchange: 30–50%
Increased organic load	Poor feeding management	Check feeding schedule, prefer bottom drainage
High DO in the afternoon	Over blooming	Avoid fertilization, exchange water
Turbid water	Heavy sediment and colloidal	Apply gypsum, exchange water 6–8 hours after liming
Fouling problem	Over blooming, poor bottom condition, high turbidity	Water exchange to stimulate moulting
Size variation	Unsuitable feeding procedure	Split feeding for smaller and larger groups

TABLE 10.4
Optimum Water Quality for Shrimp Pond

Major Parameters	Value
Dissolved oxygen	>5 ppm
pH	7.5–8.5
Ammonia	<0.1 ppm
H_2S	<0.01 ppm
Transparency	30–45 (cm)
Temperature	25–30°C
Salinity	15–25 ppt
Water colour	Light green/brown

10.7.3 pH Adjustment

Lime is used by shrimp farmers to keep the pH of their water within the ideal range of 7.5–8.5 and to minimize pH fluctuations during the day to less than 0.5. Every two to three days, 5.0–8.0 kg ha^{-1} of dolomite should be applied when the pH is between 7.5–7.8 and the variation between the morning and afternoon pH is less than 0.5 units. If the morning pH is less than 7.5, apply 5.0 kg ha^{-1} of dolomite each day until the pH rises to 7.5 or higher. When the morning pH is greater than 8.0 and the afternoon pH is greater than 9.0, 3.0–5.0 kg ha^{-1} of dolomite should be applied daily until the pH difference is less than 0.5. In the second half of a crop cycle, 8.0 kg ha^{-1} of dolomite should be used every day or at least every two days, depending on water colour. Every time before exchanging water, 5.0–8.0 kg ha^{-1} of dolomite should be used.

10.7.4 Water Colour Control and Adjustment

The primary source of colour in pond water is phytoplankton suspended particles. Shrimp growers need to pay attention to two management aspects: plankton species and density. To make sure there are adequate nutrients for a plankton bloom, an additional fertilizer, either organic (10–30 kg ha^{-1}) or inorganic (1–3 kg ha^{-1}), is supplied to the pond during the first two months of shrimp culture. Nutrients from uneaten meals are often at sufficient amounts after this time. In certain situations, an excessive plankton bloom may result in a plankton collapse due to an oversupply of nutrients. In an open shrimp culture system the farmers exchange pond water with natural clean seawater to reduce excess plankton density. But in a closed system, where exchanging water is not needed, shrimp farmers use algaecides such as calcium hypochloride or benzalkonium chloride (BKC) 0.1–1.0 ppm to reduce plankton density. In cases where undesirable water colour appears like the 'red tide' caused by certain types of plankton such as dinoflagellates, then the plankton can be controlled by switching-off the aerators for a period of time and applying BKC (0.1–1.0 ppm).

10.8 AGRI-BUSINESS OPTIONS AND ECONOMIC ASSESSMENT OF IFS MODELS

An integrated farming system (IFS) is considered to be a powerful tool and holds the key for ensuring income, employment, livelihood and nutritional security in a sustainable mode for small and marginal farmers, constituting 84.97% of total operational holdings and operated 44.31% of area in the country. IFS refer to agricultural systems that integrate livestock and crop production or integrate fish and livestock and may sometimes be known as integrated bio systems. Integrated farming systems seem to be the possible solution to the continuous increase of demand for food and nutrition, income stability and livelihood upliftment particularly for small and marginal farmers with few resources. Currently there are about 7374 producer companies covering over 4.3 million small producers in the country (Govil et al. 2020).

A much-discussed topic in Indian agriculture is how to integrate the farmers, especially the small farmers, with the value chain so that the net return at the farmers' end is remunerative enough for the farmers to remain interested in agriculture. India is home to more than 92 million small holdings, or approximately 21% of the 450 million small holdings worldwide – the second-largest number after China. Therefore, ensuring that small holdings are actually productive and the primary source of income for the millions of Indians who depend on them is an immense challenge.

To make agriculture commercially viable, this enterprise needs to be viewed as a company rather than a sole producer. The farmer-owned companies can come up in the rural areas and at least one in every *Panchayat* and potential agriculture production areas. Farmers must be encouraged, given authority, given education and given the capacity to establish 'farmer-owned companies' in accordance with local laws and regulations. The establishment of farmer firms can obtain assistance from civil society organizations, non-governmental organizations, corporations, government agencies, financial institutions and development agencies that support small and marginal farmers. Because farmers are essentially producers rather than experts in organizational structure and business issues, they need assistance. Under the provisions of the Companies Act 1956, companies can be formed under the following set up: (1) Private Limited Company (PLC), (2) Limited Company (LC), (3) Farmer Producer Company (FPC).

Cooperatives are registered in the Registrar of Cooperative Societies. Many cooperative institutions in a wide range of business sectors can be found in India. India has had a difficult cooperative experience since the government has mostly supported cooperatives, emphasizing welfare over profit. The Government of India (GoI) enacted the Producer Companies Act in 2002 by adding a new section IXA to the Indian Companies Act of 1956. This was done by an amendment to the earlier document. The producer companies are incorporated in with the Registrar of Company (RoC). The objective was to formulate legislation that would enable incorporation of cooperatives as companies and convert existing cooperatives into companies, while ensuring that the unique elements of cooperative business maintain a regulatory framework similar to that of private companies. The PC is formed with the equity contribution by the members and limited to them. The day-to-day operation is expected to be managed by the professionals, hired from outside, under the direction of the Board of Directors elected/selected by the general body of the PC for a specific tenure. Since farmers or the producers are the equity holders of the company, a PC as an organization provides an appropriate framework for owning the company by the producers themselves. The need to organize farmers, especially the smallholders, is a well-established fact. The basic purpose of the PC is to group small farmers of producers for a viable commercial venture.

Producer companies can help smallholder farmers participate in emerging high-value markets, such as the export market and the unfolding modern retail sector in India. As elsewhere in the developing world, in India small farmers' livelihoods are being threatened due to the liberalization and privatization of Indian agriculture and the increasing interest of private capital in the agribusiness sector. The withdrawal of the state from productive and economic functions, and changes in the organization of marketing channels, present new challenges for small-scale farmers. In this environment of greater instability and competition, organization and collective action can help to enhance farmers' competitiveness and increase their advantage in emerging market opportunities (Trebbin and Hassler 2012).

10.8.1 Background and History of FPOs

Farmers producer organizations (FPOs) are groups of small and marginal farmers who come together to improve their economic and social status through collective action. They are formed to leverage the collective bargaining power of their members to negotiate better prices for their produce, to provide access to credit and other inputs and to improve their access to markets. In India, the Producer Companies Act was enacted in 2002 with a vision to combine the principles of collective action with the structural benefits of a company (Prabhakar et al. 2012). Members are expected to have a greater sense of ownership as the investment is largely member driven. At the same time, these companies are expected to address local communities to become strong local institutions of marginalized producers. While there is little confusion over the function of an FPO as compared to a cooperative society, an FPO could actually be incorporated as a cooperative society, as a mutually aided cooperative society and as a farmer producer company (FPC) under the Companies Act. One of the main reasons for promoting producer companies is that they offer an avenue for pooling small amounts of capital from large numbers of people into greater sums for creating scalable and viable businesses.

10.8.2 Agribusiness Opportunities of Integrated Farming Systems

Integrated farming systems (IFS) are a type of agricultural production system that combines various elements such as crop cultivation, animal husbandry and aquaculture in a holistic and synergistic manner. These systems aim to improve food and nutrition security by increasing productivity and reducing environmental impact, while also promoting social and economic well-being. IFS can provide a wide range of agribusiness opportunities for farmers and their communities. These include:

- Development of value chains for different products, such as processing and packaging, to add value and increase profitability.
- Production of high-value crops and animals, such as horticultural crops, organic produce and niche meats, which can fetch higher prices in the market
- Ecotourism and agro-tourism, which can provide additional income streams for farmers and their communities
- Production of non-food products, such as biofuels, medicinal plants and fibres, which can be sold for industrial or commercial use.

10.8.3 Role of FPOs in Promoting IFS

The benefits of FPOs for farmers include increased bargaining power, access to credit and markets and improved access to technology and information. On the other hand, FPOs also face challenges such as lack of access to credit and markets and limited technical and management capacity. FPOs can play a vital role in the following ways to promote IFS.

1. FPOs can provide technical assistance to their members on various aspects of integrated farming, such as crop selection, animal rearing and waste management.
2. FPOs can help their members' access finance to invest in integrated farming practices. This can include loans for purchasing inputs, setting up infrastructure and adopting new technologies.
3. FPOs can help their members sell their produce in local and regional markets through market linkages. This can help farmers get better prices for their products and increase their income.
4. FPOs can create awareness among farmers about the benefits of integrated farming. This can help farmers understand the importance of sustainable farming practices and encourage them to adopt them
5. FPOs can facilitate knowledge sharing among their members through training programs, workshops and field visits. This can help farmers learn from each other's experiences and adopt best practices.

Farmers producer organizations (FPOs) offer a range of benefits for small and marginal farmers who are often at a disadvantage in the market due to their small landholdings and lack of access to resources and information. Some of the key benefits of FPOs include:

1. *Increased bargaining power*: By coming together in an FPO, farmers are able to negotiate better prices for their produce and secure access to credit and other inputs. They also have a stronger voice in decision-making processes related to their livelihoods.
2. *Access to markets*: FPOs can help farmers access new markets, both domestic and international, by pooling their resources and expertise.
3. *Improved access to technology and information*: FPOs can provide farmers with access to training, technical assistance and information on best practices, which can help improve the productivity and quality of their crops.
4. *Access to credit*: FPOs can help farmers access credit from banks and other financial institutions as well as provide training on financial management and record-keeping.

However, FPOs also face a number of challenges that limit their effectiveness and sustainability. Some of the key challenges include:

- *Lack of access to credit and markets*: Many FPOs struggle to access credit and markets, which can limit their ability to grow and expand.

- *Limited technical and management capacity*: FPOs often lack the technical and management skills needed to run their organizations effectively.
- *Resistance from traditional intermediaries*: FPOs can face resistance from traditional intermediaries, such as middlemen and local traders, who may be unwilling to change their established practices.
- *Poor governance and lack of transparency*: FPOs can struggle with poor governance and lack of transparency, which can lead to mistrust among members and limit their ability to attract new members.
- *Limited government support*: FPOs often face a lack of government support, which can limit their ability to access resources and expand their reach.

10.8.4 Market Analysis and Economic Viability, Bankable Projects

Producer firms' financial sustainability is influenced by a number of factors, the two most significant of which are capitalization and operating model. Producer corporations' (PC) operational models are influenced by the commodities they choose, the range of activities they engage in, the amount of value added, the variety of revenue streams they have and the level of cooperation they have with other PCs. It is crucial to remember that, in the case of producer firms, the operational models also control how frequently members transact with them, affect producer loyalty and encourage producer patronage, all of which support PCs' overall financial viability.

Generally speaking, FPOs can work with one or more agricultural commodities. Due to production seasonality, single-commodity FPCs – aside from those that deal with milk and specific vegetables – face a variety of difficulties. An FPO that focuses on paddy, for instance, can purchase food from producers only once a year, which restricts the opportunity to build lasting relationships with them. Producers may sell their produce to purchasers other than the FPO as a result, which would negatively affect their support of the PC and lower member loyalty. This, in turn, affects FPOs' business volume and profitability and also limits the benefits for their members. Therefore, many successful FPOs tend to work with multiple agricultural commodities to ensure better capacity utilization of resources and greater patronage by their members. Those PCs handling multiple commodities are typically in a better position than those handling only single commodities (with the exception of dairies). Procurement of multiple commodities ensures better capacity utilization of resources. It also enables companies to engage in more frequent transactions with members, which builds familiarity and contributes to higher volumes. This also reduces risks arising from moral hazard common in collective efforts, and improves producer loyalty by increasing PCs' relevance in the lives of producers.

Economic assessment is an important tool for evaluating the viability and sustainability of different IFS models. This includes cost-benefit analysis, which compares the costs of implementing an IFS model to the benefits it generates in terms of income, food security and environmental benefits. Break-even analysis identifies the point at which the revenues generated by an IFS model equal its costs. Net present value (NPV) analysis calculates the present value of the future benefits and costs of an IFS model, taking into account the time value of money. Economic rate of return (ERR) analysis measures the profitability of an IFS model as a percentage of the initial investment.

Economic assessment can help to identify the most viable and sustainable IFS models, and guide decision making about which models to adopt and implement. It can also help to identify key constraints and opportunities for improving the performance and profitability of different IFS models.

In order for an FPO to be economically viable and to develop bankable projects, it is important to conduct a market analysis to understand the demand for the products or services the FPO plans to offer. This analysis should consider factors such as the size of the target market, the competition and the potential for growth. The FPO should also conduct a cost-benefit analysis to determine the economic viability of the proposed project. This analysis should take into account the costs of production and marketing as well as the projected revenues. The FPO should also consider the potential risks and challenges that may arise and how they will be mitigated. Once the FPO has a clear understanding of the market and the economic viability of the proposed project, it can then develop a detailed business plan and financial projections. This will be necessary to secure funding from banks and other financial institutions. In the case of integrated farming systems (IFS), a market analysis should consider the demand for various agricultural products and the feasibility of integrating various crops and livestock into the farming system. Additionally, an analysis of the costs and benefits of using organic and sustainable farming practices should be conducted. The FPO should also consider the potential for value addition and processing to increase the market value of the produce. Furthermore, the FPO should consider the regulations and certifications required to market the IFS products and the costs and benefits of obtaining them. A well-designed and economically viable project, supported by a detailed business plan and financial projections, will increase the chances of securing funding from banks and other financial institutions. Overall, market analysis and understanding the economic viability of a project are crucial for the success of an FPO, and they are key components in developing bankable projects in IFS.

10.8.5 Role of Government and Other Stakeholders in Supporting FPOs

The government and other stakeholders can play a critical role in supporting FPOs by providing financial support, creating an enabling policy environment, providing technical assistance, promoting market linkages and creating awareness among farmers. These measures can help FPOs become more effective and contribute to the development of agriculture in India. The government can provide financial assistance to FPOs for setting up infrastructure, purchasing inputs and adopting new technologies. Other stakeholders such as non-governmental organizations, private sector companies and development agencies can also provide financial support to FPOs. The government can create an enabling policy environment for the formation and functioning of FPOs. This can include provisions for the registration of FPOs, tax exemptions and access to credit and insurance. The government and other stakeholders can provide technical assistance to FPOs on various aspects of their operations, such as marketing, quality control and value addition. The government and other stakeholders can help FPOs access markets and get better prices for their produce. This can include setting up market linkages, providing market intelligence and facilitating contracts with buyers. The government and other stakeholders can create awareness among farmers about the benefits of forming and joining FPOs. This can help farmers understand the importance of collective action and encourage them to participate in FPOs.

As of March 31, 2019, NABARD has promoted over 4,000 FPOs under two schemes: Producer Organisation Development Fund (PODF) initiated in 2011–2012 with a corpus of Rs. 50 crores and the Produce Fund of Rs. 200 crores was established in 2014–2015 for promotion of 2000 FPOs over two years (NABARD 2018; NABARD 2019). In addition, many producer companies have been promoted and funded by philanthropies such as Tata Trusts, Rabo Bank Foundation, Axis Bank Foundation, Dell Foundation, Reliance Foundation, etc. Many others have been funded by corporate social responsibility (CSR) arms of companies such as Mahyco Monsanto, Dupont, HSBC, Walmart, and others.

FPOs can act as a tool for empowering small and marginal farmers who practice IFS and for improving their livelihoods. Integrated farming systems (IFS) can provide a wide range of agribusiness opportunities for farmers and their communities, and economic assessment is an important tool for evaluating the viability and sustainability of different IFS models. By exploring the agribusiness options and conducting economic assessments, farmers and other stakeholders can make informed decisions about which IFS models to adopt and implement, and they can identify key constraints and opportunities for improving their performance and profitability. This will ultimately lead to more sustainable and profitable IFS models, which will improve food and nutrition security, increase income and livelihood opportunities and reduce environmental impact.

REFERENCES

Abakari, G., Wu, X., He, X., Fan, L., Luo, G. 2022. Bacteria in biofloc technology aquaculture systems: Roles and mediating factors. *Reviews in Aquaculture* 14(1):1–25. DOI: https://doi. org/10.1111/raq.12649.

Avnimelech, Y. 2012. *Biofloc Technology. A Practical Guide Book*. The World Aquaculture Society, Baton Rouge. p 272.

Avnimelech, Y. 2007. Feeding with microbial flocs by tilapia in minimal discharge bioflocs technology ponds. *Aquaculture* 264:140–147.

Collazos-Lasso, L.F., Arias-Castellanos, J.A. 2015. Fundamentals of bioflocs technology (BFT). An alternative for fish farming in Colombia: A review. *Orinoquia* 19(1):77–86.

Deng, M., Chen, J., Gou, J., Hou, J., Li, D., He, X. 2018. The effect of different carbon sources on water quality, microbial community and structure of biofloc systems. *Aquaculture* 482:103–110.

Emerenciano, M.G.C., Gaxiola, G., Cuzon, G. 2013. Biofloc technology (BFT): A review for aquaculture application and animal food industry. In: Matovic M.D. (ed) *Biomass Now: Cultivation and Utilization*. InTech, Belfast, Canada. pp 301–308.

Govil, R., Neti, A., Rao, M.R. 2020. *Farmer Producer Companies: Past, Present and Future*. Azim Premji University, Bangalore. p. 99.

Khanjani, M.H., Eslami, J., Ghaedi, G., Sourinejad, I. 2022a. The effects of different stocking densities on nursery performance of banana shrimp (*Fenneropenaeus merguiensis*) reared under biofloc condition. *Annals of Animal Science* 22(4):1291–1299. DOI: https://doi.org/ 10.2478/aoas-2022-0027.

Khanjani, M.H., Mozanzadeh, M.T., Foes, G.K. 2022b. Aquamimicry system: A suitable strategy for shrimp aquaculture – A review. *Annals of Animal Science* 22(4):1201–1210. DOI: https://doi.org/10.2478/aoas -2022-0044.

Khanjani, M.H., Sharifinia, M. 2020. Biofloc technology as a promising tool to improve aquaculture production. *Reviews in Aquaculture* 12(3):1836–1850. DOI: https://doi.org/10.1111/raq.12412.

Khanjani, M.H., Sharifinia, M. 2021. Production of Nile tilapia *Oreochromis niloticus* reared in a limited water exchange system: The effect of different light levels. *Aquaculture* 542:736912.

Khanjani, M.H., Sharifinia, M., Hajirezaee, S. 2022c. Recent progress towards the application of biofloc technology for tilapia farming. *Aquaculture* 552:738021.

Krummenauer, D., Reis, W.G., Abreu, P.C., Bezerra, A., Wasielesky, W., Advent, B. 2021. Effect of aeration on the development of the microbial community in shrimp biofloc technology systems. *Global Seafood Alliance*. https://www.globalseafood.org/advocate/effect-of-aeration-on-the-development-of-the-microbial-community-in-shrimp-bio floc-technology-systems/.

Kumar, V., Roy, S., Behera, B.K., Swain, H.S., Das, B.K. 2021. Biofloc microbiome with bioremediation and health benefits. *Frontiers in Microbiology* 12:741164. DOI: https://doi.org/10.3389/fmicb.2021.741164.

Martínez-Cordova, L.R., Emerenciano, M., Miranda-Baeza, A., Martínez-Porchas, M. 2015. Microbial-based systems for aquaculture of fish and shrimp: An updated review. *Reviews in Aquaculture* 7:131–148.

Meenakshisundaram, M., Sugantham, F., Muthukumar, C., Chandrasekar, M.S. 2021. Metagenomic characterization of biofloc in the grow-out culture of genetically improved farmed tilapia (GIFT). *Aquaculture Research* 52:4249–4262.

Monroy-Dosta, M.D.C., Lara-Andrade, R.D., Castro-Mejía, J., Castro-Mejía, G., Coelho-Emerenciano, M.G. 2013. Microbiology community composition and abundance associated to biofloc in tilapia aquaculture. *Revista de Biologia Marina y Oceanografia* 48(3):511–520.

NABARD. 2018. "NABARD Annual Report 2017-18. National Bank for Agriculture and Rural Development," Mumbai, India. p. 240.

NABARD. 2019. "Farmer Producers' Organizations (FPOs): Status, Issues & Suggested Policy Reforms." National Level Paper, Potential Linked Plans (PLP) 2019-20. Accessed March 28, 2023. https://www .nabard.org/auth/writereaddata/careernotices/2708183505Paper%20on%20FPOs%20-%20Status%20& %20%20Issues.pdf.

Panigrahi, A., Esakkiraj, P., Jayashree, S., Saranya, C., Das, R.R., Sundaram, M. 2019a. Colonization of enzymatic bacterial flora in biofloc grown shrimp *Penaeus vannamei* and evaluation of their beneficial effect. *Aquaculture International* 27:1835–1846.

Panigrahi, A., Saranya, C., Sundaram, M., Vinoth Kannan, S.R., Das, R.R., Satish Kumar, R., Rajesh, P., Otta, S.K. 2018. Carbon: Nitrogen (C: N) ratio level variation influences microbial community of the system and growth as well as immunity of shrimp (*Litopenaeus vannamei*) in biofloc based culture system. *Fish Shellfish Immunology* 81:329–337.

Panigrahi, A., Sundaram, M., Saranya, C., Swain, S., Dash, R., Dayal, J.S. 2019b. Carbohydrate sources deferentially influence growth performances, microbial dynamics and immunomodulation in Pacific white shrimp (*Litopenaeus vannamei*) under biofloc system. *Fish Shellfish Immunology* 86:1207–1216.

Prabhakar, I., Manjunatha, B.L., Nithyashree, M.L., Hajong, D. 2012. Farmers producer company – an innovative farmers' institution. *Environment and Ecology* 30(2):427–430.

Suita, S.M., Ballester, E.L.C., Abreu, P.C., Wasielesky Jr, W. 2015. Dextrose as carbon source in the culture of *Litopenaeus vannamei* (Boone, 1931) in a zero exchange system. *Latin American Journal of Aquatic Research* 45(3):526–533.

Trebbin, A., Hassler, M. 2012. Farmers' producer companies in India: A new concept for collective action? *Environment and Planning A* 44(2):411–427.

Wang, C., Pan, L., Zhang, K., Xu, W., Zhao, D., Mei, L. 2016. Effects of different carbon sources addition on nutrition composition and extracellular enzymes activity of bioflocs, and digestive enzymes activity and growth performance of *Litopenaeus vannamei* in zero-exchange culture tanks. *Aquaculture Research* 47(10):3307–3318.

11 Agroecology and Biodiversity

11.1 INTRODUCTION

Sustainable food production often embraces principles of agroecology, which emphasize understanding and utilizing ecological processes to design farming systems. It involves promoting biodiversity, optimizing resource use and reducing external inputs like synthetic fertilizers and pesticides. Agroecological farming systems, rooted in ecological principles and understanding and incorporating diverse crop cultivation as a core strategy, have been demonstrated as effective, sustainable options that offer alternatives to the industrial monoculture model in terms of agricultural productivity, ecological integrity and societal benefits (Francis et al. 2003). Agroecology is increasingly seen as a possible remedy for the interconnected global crises and is often defined as a revolutionary movement in agriculture and society, a scientific field and a set of practices. It opposes top-down technocratic methods and aims to challenge the power dynamics in the current exploitative and oppressive system of agriculture and food production system. It is also interconnected with social and gender equity aspects of a holistic approach to creating a fair and just food system. Social equality in food production aims to provide equitable access to nutritious food, fair treatment, and opportunity for all individuals, regardless of socioeconomic background, race, gender or geography. Integrating social equity into sustainable food production involves addressing issues of food insecurity, poverty, labour rights and community well-being. An agro-ecological approach encourages agricultural diversity and sustainable livelihoods with fewer external inputs. Implementing sustainable integrated farming systems can benefit food productivity, decrease pesticide use and improve carbon balance. An intricate relationship exists between ecological processes, biodiversity, family farming, gender equity, participation and responsibility, drudgery reduction and ecosystem services. These interrelated areas underscore the significance of sustainable agricultural practices and the necessity for inclusive and responsible approaches to integrated farming systems.

11.2 ECOLOGICAL PROCESSES AND PRINCIPLES

Ecological processes and principles form the foundation of understanding how ecosystems function and how living organisms interact with their environment. These processes and principles are crucial in ecology, which is the scientific study of how organisms interact with their environment. Ecological processes also play a fundamental role in agroecology and biodiversity conservation. Agroecology relies on the concept that agricultural systems are closely linked to ecological processes and functions.

11.2.1 Nutrient Cycling and Energy Flow

Agroecology emphasizes the importance of nutrient cycling within agricultural systems. Nutrients required for producing organic materials are cycled repeatedly within the system. Plants acquire all natural elements as gases from the air or soluble salts from the soil. Carbon (C), hydrogen (H) and oxygen (O) are absorbed from the air, while nitrogen (N), potassium (K), phosphorus (P), calcium (Ca), sulphur (S) and magnesium (Mg) are absorbed from the soil. These essential nutrients, required in significant quantities by the plants, are known as macronutrients. Among them, N, P and K are considered major nutrients, while Ca, Mg and S are secondary nutrients needed in smaller

DOI: 10.1201/9781003239963-11

amounts. Essential elements like iron (Fe), manganese (Mn), zinc (Zn), copper (Cu), chlorine (Cl), molybdenum (Mo) and boron (B) are required in trace amounts, termed as micronutrients. Plants absorb these elements to create new organic materials. Nutrient recycling occurs through the decomposition of organic matter, uptake by plants and the return of organic residues to the soil. Therefore, nutrient cycling is vital for replenishing nutrient supplies and sustaining productivity. This process minimizes the need for synthetic fertilizers and promotes soil fertility. Biodiversity, encompassing microorganisms and decomposer organisms such as earthworms, fungi, and bacteria, is crucial for the decomposition of organic matter. This process releases and alters nutrients, facilitating nutrient cycling. The Ecopath model illustrates energy flow and material cycling using the principles of nutrient dynamics. In ecosystems, energy flow is unidirectional, with light energy being continually transformed into food energy. Producers, like plants, harness solar energy through photosynthesis, transforming it into chemical energy. This chemical energy is then transferred to consumers (herbivores, carnivores, etc.) through food chains and food webs. When analysing and implementing strategies for low-carbon production, it is crucial to understand the impact of different farm management options. A sustainable production system now involves efficient energy utilization to minimize environmental impact. In ecosystems, nutrients like carbon, nitrogen and phosphorus are in a constant cycle of recycling. Decomposers release nutrients into the soil or water by breaking down dead organic matter, which is further absorbed by the plants and reused within the ecosystem.

Nutrient cycles are depicted in a model framework where each cycle consists of a reservoir pool, a large and slow-moving non-biological component and an exchange pool, a smaller and more active part where nutrients are exchanged between the biotic and abiotic components of the ecosystem. There are two main types of cycles: gaseous cycles, where the atmosphere serves as the reservoir pool, and sedimentary cycles, where the Earth's crust acts as the reservoir pool. All nutrient cycles involve interactions between the soil and the atmosphere and include various organisms that facilitate the movement of elements along food chains to decomposers. However, the rates of nutrient exchange between abiotic and biotic components vary depending on the nutrient. Generally, gaseous cycles are more complete than sedimentary ones. Both types of nutrient cycles occur naturally, but sedimentary cycles are more vulnerable to disruptions caused by human activities.

The nitrogen (N) cycle follows a gaseous type, where the Earth's atmosphere, containing 78% N, serves as the reservoir pool. The exchange pool operates between organisms (plants and animals) and the soil. Nitrogen-fixing bacteria, primarily microorganisms, convert atmospheric nitrogen into nitrates that enrich the soil. Additionally, this process occurs naturally during lightning strikes in thunderstorms. Plants absorb these nitrates, entering the food chain. When organic material decomposes, nitrates are released as ammonia. Bacteria then convert this ammonia back into nitrates. If plants do not absorb the nitrates, they are lost from the exchange pool primarily through leaching from the soil or gaseous loss to the atmosphere (denitrification and volatilization).

Unlike the nitrogen (N) cycle, the phosphorus (P) cycle is a sedimentary cycle that is easily disrupted. Phosphorus is relatively scarce in nature but essential for both plants and animals. The exchange pool involves cycling between organisms, soil, and shallow marine sediments. In ecosystems, phosphorus is absorbed from the soil by plants, which utilize it for creating proteins, and then it circulates through various food chains. When plant and animal bodies, along with their waste, decompose, P is released into the soil. From there, it can be either absorbed back into plants or washed away by rainfall into drainage systems, ultimately reaching the sea. In the sea, P becomes incorporated into marine sediments and is thus lost from the exchange pool. However, upwelling ocean currents quickly return P from the sediments. When P is in surface waters, it enters marine food chains. Seabirds that feed on fish in these areas produce P-rich droppings called guano, which serve as a P source for plants. The release of the element from P rocks in the reservoir pool during erosion and weathering processes very slowly compensates for the depletion of P from the exchange pool.

11.2.2 Competition and Predation

Organisms within an ecosystem compete for limited resources, such as food, water, and shelter. Competition can influence species distribution and abundance, leading to adaptations that reduce competition and promote coexistence. Predation is consuming one organism (the prey) by another (the predator). Predation helps control population sizes and can influence the structure of communities.

11.2.3 Symbiosis and Succession

Symbiosis is the close and long-term interaction between two distinct species. This interaction can be mutualistic (benefitting both species), commensal (benefitting one species without affecting the other) or parasitic (benefitting one species at the expense of the other). Ecological succession is the gradual change in an ecosystem's structure and composition over time. Primary succession occurs in newly formed environments (e.g., after a volcanic eruption), while secondary succession occurs in areas that have been disturbed but still retain some soil and life. Disturbances like fires, floods, or human activities can significantly affect ecosystems, with some species adapted to thrive after disturbances while others may decline.

11.3 AGRO-ECOLOGICAL ASPECTS OF IFS

11.3.1 Soil Health

Agroecology recognizes that healthy soils are the foundation of sustainable agriculture. Ecological processes, such as nutrient availability, accumulation of organic matter in soil and formation of soil structure, contribute to the overall health of the soil. Biodiversity within the soil, including beneficial microorganisms and soil-dwelling organisms, enhances soil fertility, nutrient cycling and decomposition of organic matter, further enhancing soil structure, water-holding capacity and resilience to pests and diseases.

11.3.2 Pest and Disease Regulation

Agroecology aims to reduce reliance on synthetic pesticides by promoting natural pest and disease control. Biodiversity plays a vital role in these processes. For example, diverse plant communities provide habitats and food sources for natural enemies of pests, such as predators and parasitoids, contributing to biological control. Biodiversity also supports the presence of pest-resistant crop varieties and increases ecosystem resilience to pest outbreaks.

11.3.3 Pollination

Pollination is a vital ecological process that contributes to the production of crops and the maintenance of plant biodiversity. Agro-ecological practices that promote biodiversity, such as maintaining diverse flowering plants and providing habitats for pollinators, enhance pollination services. Pollinators, such as bees, butterflies, birds, etc., transfer pollen between flowers, ensuring fruit set and seed production.

11.3.4 Genetic Diversity

Agroecology recognizes the value of genetic diversity within agricultural systems. Genetic diversity results from ecological processes such as adaptation, mutation and gene flow. By culturing a diverse range of crops and using traditional and locally adapted varieties, agroecology helps maintain genetic diversity. This diversity enhances resilience to pests, diseases and fluctuating environmental conditions.

11.3.5 Ecosystem Resilience

Agroecology seeks to establish farming systems that are resilient and capable of adjusting to environmental fluctuations. Biodiversity is essential for ecosystem resilience as it enhances the capacity to withstand and recover from disturbances. Diverse plant communities, including cover crops and agroforestry systems, contribute to soil stability, water retention, and microclimate regulation. This helps mitigate the impacts of extreme weather events, maintain productivity, and support the overall functioning of agroecosystems.

11.3.6 Biodiversity

Maintaining biological diversity and nutrient cycling are fundamental principles all agroecosystems share, making them crucial in creating sustainable agricultural systems. Traditional farming methods offer a wealth of principles and practices that can enhance modern agricultural systems' resilience to extreme weather conditions. Many of these agro-ecological strategies, such as crop diversification, preserving local genetic diversity, integrating animals, managing soil organically, conserving and harvesting water and so on, reduce vulnerabilities to climate variability.

Biodiversity refers to the diverse life forms within an ecosystem, including species, genes and ecosystems. It has a massive role in the maintenance of ecosystem stability and resilience. The carrying of an ecosystem refers to the total number of individuals of a species that the environment can sustainably support. The availability of resources and environmental conditions determine it. Organisms within an ecosystem evolve and adapt to their environment over time through natural selection, allowing them to survive and reproduce successfully in changing conditions. Each species occupies a unique ecological niche, encompassing its role, habitat, and interactions with other species. Niches help prevent direct competition between species with similar resource requirements. Ecosystems offer various benefits to humanity, such as providing clean water, purifying the air, enabling pollination, and helping to regulate the climate, which is crucial for encouraging sustainable methods.

11.4 AGROECOLOGY AND BIODIVERSITY

The concepts of agroecology and biodiversity are interlinked, playing a pivotal role towards sustainable agricultural practices and preservation of natural resources. Agroecology is a farming approach that integrates ecological principles into agricultural systems. It aims to enhance the interactions among plants, animals, humans, and the environment to develop farming systems that are sustainable and resilient. Biodiversity, however, pertains to the range and diversity of life forms within an ecosystem, including the variety of species present, genetic differences and ecosystem diversity. Biodiversity is critical for ecosystem health, offering many advantages, such as aiding in pollination, controlling pests, recycling nutrients and enhancing resilience to environmental changes.

Agroecology recognizes the importance of biodiversity and aims to promote its conservation and enhancement within agricultural landscapes. Agro-ecological practices often involve preserving or creating diverse habitats within agricultural landscapes. This can include the conservation of natural areas, establishment of hedgerows, tree planting or maintaining fallow lands. These habitats provide shelter, food and breeding grounds for varieties of flora and fauna, contributing to biodiversity conservation. Agroecology encourages cultivating diverse crops and using traditional and locally adapted varieties. Growing various crops helps maintain genetic diversity within agricultural systems, reducing the risk of crop failure due to pests, diseases or climate change. It also supports the diversity of pollinators and beneficial insects that rely on different flowering plants. Agro-ecological practices often integrate trees into agricultural systems, a practice known as agroforestry. Trees provide shade, windbreaks and soil stability while creating a favourable environment for diverse species. Agroforestry systems enhance biodiversity by providing habitats for birds, insects and other wildlife and supporting interactions between different plant species. Agroecology encourages ecological pest management strategies, such as biological control, which involves employing the natural predators of pests to maintain their populations in check. This system reduces the reliance on synthetic pesticides, which can harm beneficial organisms and negatively impact biodiversity. Agroecological practices enhance soil health through crop rotation, cover cropping and organic matter addition. Healthy soils support diverse microorganisms, fungi and soil-dwelling organisms, contributing to ecosystem biodiversity. Healthy soils also provide crucial ecosystem services, including the recycling of nutrients, purification of water and carbon sequestration.

By integrating agro-ecological principles into agricultural practices, farmers can enhance biodiversity within their farms, promote ecosystem resilience and aid in the long-term viability of food production systems. Biodiversity is often perceived as the cornerstone of healthy ecosystems and sustainable agriculture. It provides numerous opportunities and benefits, while its decline poses significant threats. Biodiversity supports pollination, natural pest control, nutrient cycling and resilience to climate change. It also offers opportunities for innovative farming practices, including agroforestry, intercropping and crop diversification.

Agroecology, as an ecological approach to agriculture, poses both challenges and opportunities in the context of modern farming and food production. It offers promising solutions to many ecological and societal challenges linked to traditional farming methods. Habitat destruction, intensive monoculture farming, pesticide use and climate change pose significant biodiversity risks. Loss of biodiversity can disrupt ecosystem services and compromise agricultural productivity. One of the critical barriers to adopting agroecological practices is the resistance to change within the agricultural industry and existing farming systems. Transitioning from conventional methods to agroecology requires education, training and financial support.

11.5 SOCIAL AND ENVIRONMENTAL ASPECTS OF AGROECOLOGY

Though agroecology is an essential component of a farming system, there still needs to be more awareness and knowledge about this concept. Many farmers and policymakers may need a more thorough understanding of agroecology's principles and benefits, which could impede the widespread adoption of these sustainable practices. Another challenge is the economic aspects. The initial investment in agro-ecological practices can be higher than conventional farming methods. Farmers may face financial challenges during the transition phase. In addition, market accessibility and demand creation are additional challenges. Accessing markets that value and pay premium prices for agroecological products can be difficult. There might be limited demand or need for infrastructure for distributing these products to consumers.

Land ownership and land utilization challenges also inhibit agroecology-based farming. In some regions, land ownership and usage policies may not support or incentivize agroecological practices, making it difficult for farmers to implement sustainable techniques on their lands. Agro-ecological practices often rely on natural pest management techniques, which, in most cases, do not provide immediate solutions to pest and disease outbreaks compared to chemical interventions. There is also a need to institutionalize the process of scaling up the technologies and adopting methodologies for achieving replicability. While many successful agro-ecological practices have been demonstrated at a small scale, scaling up these practices to large commercial farms can be challenging, and replicating them across different regions may require adaptation to local conditions. Agro-ecological practices promote biodiversity, reduce chemical inputs, conserve water and enhance soil health, resulting in farming systems that are more resilient and sustainable.

Agroecology's emphasis on carbon sequestration and climate-friendly practices contributes towards mitigating greenhouse gas emissions and enhances resilience to impacts caused by climate change. Agroecology can strengthen food security by diversifying crops, improving yields, and increasing the availability of nutrient-rich foods for local communities. By promoting small-scale farming and regional food systems, agroecology can create employment opportunities and support rural livelihoods. Agroecological practices reduce dependence on high-cost inputs like synthetic fertilizers and pesticides, making farming more economically viable in the long term. Agroecology encourages knowledge sharing and collaboration among farmers, researchers, and communities, fostering local innovation and adaptive management. Agroecology empowers smallholder farmers, women and marginalized communities by promoting inclusive decision making and recognizing traditional knowledge. Diversification and regional market access in agroecological systems can reduce farmers' vulnerability to global market fluctuations. Agroecology enhances ecosystem functions such as pollination, soil health enrichment and water regulation, contributing to overall environmental health.

11.6 SCOPE OF AGROECOLOGY

11.6.1 Family Farming

Family farming is a widely practised form of agriculture worldwide, especially in rural areas. It includes food security, poverty reduction and rural development. Family farming and agroecology are closely linked, as family farmers often play a crucial role in adopting and promoting agroecological practices. Family farming refers to agricultural activities primarily carried out by families, typically involving small to medium-sized farms managed and operated by the same family members. These farmers are often deeply connected to their land and have a long-term stake in its sustainability and productivity. Family farmers usually have a strong sense of stewardship of their land, which aligns well with the principles of agroecology. They tend to adopt sustainable practices that improve soil fertility, promote biodiversity and encourage the conservation of natural resources. Family farmers possess valuable traditional knowledge of the local environment, climate and crops. This knowledge can be harnessed and combined with agroecological principles to develop context-specific and resilient farming systems. This system of farming adopts crop diversity and rotation. Family farmers often cultivate various crops on their small plots of land. Crop diversification and rotation are essential components of agroecology, as they enhance soil fertility, reduce pests and promote overall ecosystem health. There is preservation of traditional varieties. Family farmers are more likely to preserve and exchange traditional seed varieties, contributing to agroecological resilience and maintaining agricultural biodiversity. This system also uses no or very low chemical inputs. Due to economic constraints and environmental concerns, family farmers tend to reduce the usage of fertilizers and pesticides, making agroecological practices more feasible and practical. Integration of livestock and crops is an integral aspect of this system. Family farming often involves integrating crops with livestock, which can help create more closed-loop systems where animal waste becomes valuable crop inputs.

Family farmers are deeply connected to their communities. Agroecology often encourages knowledge-sharing, cooperation, and collaboration within communities. It is resilient to climate change. Agroecology's emphasis on diversification and adaptation aligns well with the needs of family farmers particularly vulnerable to climate change impacts. It respects cultural and social values. Family farming is often associated with cultural and social values, and agroecology respects and integrates these values in its approach to sustainable agriculture. Emphasizing agroecology in family farming can differentiate their produce in the market, especially when there is increasing consumer demand for sustainably grown and locally sourced food.

11.6.2 Gender Equity, Participation and Responsibility

Gender equity and empowerment of women are essential for achieving sustainable agriculture and rural development (Seibert et al. 2019). Accelerating progress towards social and environmental objectives is critical, yet broadening agroecological practices poses considerable risks to women and other disadvantaged groups unless approached with a critical, nuanced, and feminist perspective (Zaremba et al. 2021). Agroecology is increasingly recognized as a scientific discipline, a social movement, and a set of practices aimed at transforming food systems to challenge and change the existing patterns of farmer exploitation inherent in mainstream agricultural models. To achieve its twin objectives of social and ecological well-being, agroecology must focus on empowering women and other underrepresented groups to cantered as the protagonists of the movement. The importance of gender and social considerations goes beyond the apparent social dimensions of agroecology; they are integral to all facets of agroecology.

Key considerations include:

a. Equal Opportunities: Women's equal access to land, resources, education and decision-making processes is crucial for sustainable farming and rural development.
b. Participation: Women's active participation in agricultural activities, farmer organizations and decision making ensures more inclusive and effective outcomes.
c. Responsibility: Recognizing and valuing women's contributions to agriculture and sharing the responsibilities associated with farming and household chores is important for reducing gender disparities.

11.6.3 Drudgery Reduction

Reducing drudgery in agriculture, particularly for women, is an important aspect of sustainable development. Technological Innovations are essential in developing appropriate technologies and tools that mitigate labour-intensive tasks, alleviate drudgery and improve productivity. On the other hand, improved infrastructure and services, such as clean water, energy and transportation, can significantly reduce the burden of daily agricultural tasks. Drudgery reduction in agroecology refers to the efforts made to alleviate the physical and labour-intensive tasks that farmers often face in traditional agriculture. Agroecology emphasizes sustainable and regenerative farming practices that minimize the usage of external inputs, like fertilizers and pesticides, while optimizing natural ecological processes. By incorporating agro-ecological principles, farmers can reduce the arduous and repetitive manual labour involved in farming, improving livelihoods and enhancing well-being. Here are some ways drudgery reduction is achieved in agroecology.

Mechanization: Agro-ecological practices often involve the use of appropriate technology and machinery to automate repetitive tasks, such as planting, weeding and harvesting. For example, using mechanical seeders and weeders can remarkably reduce the time and effort required for manual planting and weeding. To carry out the multifaceted activities of highly diverse agroecological farming systems, field robots may be employed to meet production objectives as well as social and ecological care (Ditzler and Driessen 2022).

Crop diversity and rotation: Agro-ecological systems promote crop diversity and rotation, which helps break pest and disease cycles and reduces the need for intensive pest management. Diverse cropping patterns also provide ecological benefits, reducing the need for constant attention to single-crop fields.

Agroforestry and perennial crops: Integrating trees and perennial crops in agro-ecological systems reduces the need for annual replanting and provides long-term yields with less labour-intensive maintenance.

Mulching and cover cropping: The use of organic mulches and cover crops in agroecology helps in weed suppression, soil moisture retention and enhancement of soil health. These practices reduce the need for frequent weeding and watering.

Composting and organic fertilizers: Agro-ecological systems rely on the preparation of compost and the application of manures to improve soil fertility. This limits the need for synthetic fertilizers, which often require substantial labour for application and can be physically demanding.

Animal integration: Integrating livestock in agro-ecological systems can help manage weeds, fertilize the soil and perform other tasks, reducing the labour required for certain agricultural activities.

Water management methods: Utilizing effective water management techniques like drip irrigation or harvesting of rainwater minimizes the labour required for manual watering and irrigation.

Farm design and layout: Proper farm design, including thoughtful placement of crops and infrastructure, can minimize the distances farmers need to travel and the time spent moving between different farm areas.

Community and collective efforts: Agro-ecological practices often emphasize community participation and knowledge-sharing, enabling farmers to work collectively on tasks such as harvesting and processing, thereby reducing individual drudgery (Val et al. 2019).

Appropriate training and support: Providing farmers with proper training and technical support on agro-ecological practices can help them implement these methods more efficiently, leading to reduced drudgery over time.

11.6.4 Ecosystem Services

Ecosystem services refer to the various advantages that humans obtain from functioning ecosystems. Recognizing and valuing these services is crucial for sustainable agriculture. Globally, agricultural systems are primarily characterized by monoculture, which prioritizes profitability but often comes at the expense of ecological integrity. In recent years, there has been growing momentum in global initiatives to promote sustainable agriculture practices that emulate natural processes, with agroecology leading the way. Besides the ecological dimension, agroecology also addresses economic and social aspects across the entire food system. Transformative agroecology is recognized as a key pathway towards achieving various Sustainable Development Goals (SDGs), given its significant potential to develop farming systems resilient to climate change while reducing biodiversity decline and enhancing ecosystem services (Sethuraman et al. 2021). Maintaining diverse pollinator populations through habitat conservation and reducing pesticide use ensures pollination services for crop production. Biodiversity conservation promotes natural pest control mechanisms, reducing the reliance on chemical pesticides, while sustainable farming practices enhance soil fertility, water infiltration and water quality, supporting long-term productivity.

REFERENCES

Ditzler, L., Driessen, C. 2022. Automating agroecology: How to design a farming robot without a monocultural mindset?. *Journal of Agricultural and Environmental Ethics* 35(1):2.

Francis, C., Lieblein, G., Gliessman, S., Breland, T.A., Creamer, N., Harwood, R., Salomonsson, L., Helenius, J., Rickerl, D., Salvador, R., Wiedenhoeft, M., Simmons, S., Allen, P., Altieri, M., Flora, C., Poincelot, R. 2003. Agroecology: The ecology of food systems. *Journal of Sustainable Agriculture* 22:99–118.

Seibert, I.G., Sayeed, A.T., Georgieva, Z., Guerra, A. 2019. *Without Feminism, There Is no Agroecology*. Civil Society Mechanism (CSM) for relations to the UN Committee on World Food Security (CFS): Rome, Italy.

Sethuraman, G., Mohd Zain, N.A., Yusoff, S., Ng, Y.M., Baisakh, N., Cheng, A. 2021. Revamping ecosystem services through agroecology—The case of cereals. *Agriculture* 11(3):204.

Val, V., Rosset, P.M., Lomelí, C.Z., Giraldo, O.F., Rocheleau, D. 2019. Agroecology and La Via Campesina I. The symbolic and material construction of agroecology through the dispositive of "peasant-to-peasant" processes. *Agroecology and Sustainable Food Systems* 43:872–894.

Zaremba, H., Elias, M., Rietveld, A., Bergamini, N. 2021. Toward a feminist agroecology. *Sustainability* 13(20):11244.

12 Rainwater Harvesting and Management for IFS

12.1 WATER HARVESTING AND CONSERVATION MEASURES

Water harvesting and conservation measures can be done adopting different measures suited to land, climate and crop conditions. Water harvesting structures (WHSs) are designed to store excess runoff water during the monsoon season for efficient use in the following season. The water harvesting devices not only store excess runoff water but also help reduce soil erosion and promote groundwater recharge.

In order to effectively utilize and conserve rainfall and manage runoff, evaporation and seepage, WHSs are designed to collect and store rainfall. The main benefits of WHSs are as follows: (1) they intercept runoff and moderate peak flow and volume (increasing water availability); (2) they improve groundwater quality and recharge the water table; (3) they prevent soil erosion and flooding; (4) they provide irrigation water to command areas in *Rabi* and *Kharif*; (5) they help reclaim additional productive land; (6) they intercept and reduce silt, which reduces sediment production in stream flows and ultimately in the reservoir; (8) they reduce surface water pollution from fertilizers, pesticides and other sediments; and (9) they enhance groundwater recharge and check water table decline. Various forms of soil and water conservation are broadly classified as *in-situ* methods and *ex-situ* methods. When *in-situ* methods are used to harvest and utilize the water on the site, *ex-situ* methods are used to harvest the water from some catchment and utilize the water in the same or in other area land using a conveyance system. Further, the land slope and rainfall amount are two important factors influencing the type of measures to be adopted for conservation of soil and water. Normally, the engineering measures are adopted for conservation on the land with slope more than 4%. The design of such measures/structures depends upon rainfall pattern or runoff availability at the site. Contour bunding, terracing, contour trenching and micro-catchment are the commonly used conservation measures in India. There are some common design steps for any mechanical measures:

(i) Survey of the site and data collection regarding soil type, land slope, land use, rainfall, evaporation, etc.
(ii) Hydrologic design under which quantification of runoff is being done.
(iii) Hydraulic design under which size/dimensions of different components of the structures are estimated. The precise knowledge on design of the structures is essential for planning its full utilization.

12.1.1 Contour Bund

Building bunds along the catchment area's contour is known as contour bunding (Figure 12.1). The surface runoff that flows between two neighbouring bunds is captured by these bunds. A contour bund typically has a height of 0.30 to 1.0 m and a length of 10 to 100 m. The bund's side slope must meet the specifications. The upstream area of the bund's storage capacity is determined by its height.

DOI: 10.1201/9781003239963-12

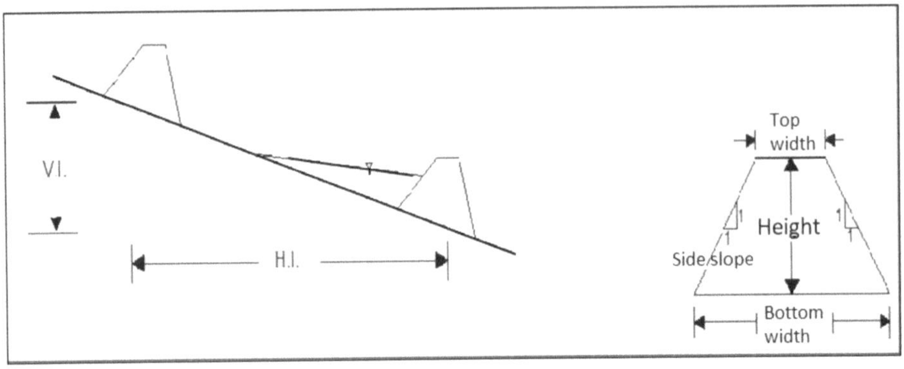

FIGURE 12.1 Cross-section of contour bunds.

For low rainfall area (<500 mm per year), construction of a *contour bund* is suggested, whereas under medium and high rainfall condition, a *graded bund* is suggested for soil and water conservation. The vertical interval (VI: vertical distance between two bunds) for different rainfall conditions are:

$$\text{VI} = 10\text{ S} + 60 \text{ (For medium \& high rainfall zones)}$$
$$\text{VI} = 15\text{ S} + 60 \text{ (For low rainfall zones)}$$

where VI (cm); S (%); Horizontal spacing = (VI/S); Height of bund (cm) = $\{(R_e \times VI)/(50)\}^{0.5}$ + Flow depth at outlet + Free board where R_e : 24 hour rainfall excess; VI (cm).

The contour bunds that are built with a grade or slope are known as graded bunds. Graded bunds are built using the same design and methodology as contour bunds. When soil conditions and rainfall intensity allow for easy capture of runoff water from the field, they are utilized as a backup plan. Through a series of channels, the extra water that is harvested or intercepted is directed to the following field. The graded bund is between 0.3 and 0.6 m tall. To catch the spilling water from the upstream bunds, the downstream bunds are made of wings. As a result, the graded bund's configuration resembles an open-ended trapezoidal bund. For this reason, it is frequently referred to as a modified trapezoidal bund. Typically, this kind of water gathering bund is utilized to irrigate the crops. The design criteria for contour/graded bunds are

- The bund's total height should not exceed 75 cm. It also requires a sufficient freeboard (10–15%).
- The bund should last three to eight years under normal circumstances.
- The graded bund's spillway capacity is typically intended to manage peak runoff, which may be anticipated once every five to ten years.
- Since the goal of a bund is to reduce runoff and soil loss, they should ideally be spaced so that the top elevation of one is equal to the bottom elevation of the bund next to it upstream.
- An outlet wide enough and long enough at the downstream end of the graded bund is a necessary component that collects water that falls over the top and to conduct it safely without scouring.

12.1.2 Contour Trench

The process of digging trenches along a level surface or contour is known as contour trenching. These are utilized for afforestation and soil and water conservation on hill slopes as well as on degraded and bare waste lands. Although these trenches have been utilized on all slopes, it is not a good idea to trench on slopes higher than 20% from a technical or financial standpoint. With different soil types and depths, contour trenches have proven profitable to utilize in both high and low rainfall scenarios. These ditches might be built sequentially or sporadically. Staggered trenches are those that are built in a staggered fashion. The area's rainfall and soil depth determine the trenches' dimensions and form. Trenches are designed to safely collect and drain excess runoff from the land above by splitting large slopes into multiple smaller slopes, which reduces erosion. When the slope reaches 50%, these measures start to become helpful.

$$VI = aS + b$$

VI: vertical interval (ft); S: slope of the land (%); a: varies from 0.4 in the arid and semiarid region to 0.8 in humid and sub-humid regions; b: 1 to 4 according to soil and cover.

VI for humid and sub-humid region also can be estimated from the following formula

$$VI = (S + 4)/\ 10, \text{ where S is in } \%$$

After calculating the VI, the horizontal distance (HI) and inclined distance (D) can be obtained as follows,

$$HI = (VI/S) \times 100$$

These trenches' design parameters are as follows: (i) Length: 100 m in humid areas and 200–300 m in arid areas (ii) Width: 1.5 or 2 m plus the filled and chopped portions (riser). (iii) Gradients: 0.5 to 1% for humid regions and 0 to 0.5% for arid or semi-arid regions for horizontal gradients. (iv) Risers: the riser slope is 1:1 for machine-made risers and 0.75:1 for hand-made risers. Riser width is calculated by multiplying riser height by either 0.75 or 1, as well as (v) the vertical intervals and ditch spacing.

12.1.3 Terrace

In regions with steep slopes, terraces are utilized to preserve soil and water. In order to control runoff and soil erosion, terracing entails building an embankment, ridge, and steps-like structure across the slope of the land. Terracing is typically used in regions with slopes greater than 10%. There are three types of terraces: bench-style, retention and diversion. To divert the overland flow from hilly slopes to an appropriate outlet, such as a grassed waterway, diversion terraces are utilized. Level terraces known as retention terraces are employed, especially in situations when it is necessary to store water on hillside surfaces. Bench terraces are terraces designed to cultivate steep hillsides; they are made of alternating rows of shelves and risers.

Bench terraces are generally constructed on the land of 6–33 % slope. These terraces are classified as level terrace, outward-sloping terrace and inward-sloping terrace based on the slope of the terrace bed. Generally speaking, level terraces are utilized in regions with medium rainfall and very permeable soil. In regions with porous soil and little rainfall, outward-sloping terraces are used. A shoulder bund is crucial for these terraces. In regions with high rainfall and permeable soils, inward-sloping terraces are preferred.

The average yearly rainfall, the types and depths of the soil, and the typical slope of the land all have a major role in determining the design elements of bench terraces. Choosing the (1) type of bench terrace, (2) terrace spacing or depth of cut, (3) terrace width and (4) terrace cross-section are the four main components of bench terrace design.

The first step of designing of bench terrace is the depth of cut, which is finalized based on the depth of productive soil and crops to be grown. After determining the depth of cut (D), the terrace width can be estimated following the formula,

$$W = 200\ (D)/S$$

where W, width of terrace (m); D, depth of cut (m); and S, slope (%).

Further, vertical interval (VI) can be determined following the formula,

$$VI = (WS)/ (100\text{-}S) \text{ for batter slope } 1:1;\ VI = (2WS)/ (200\text{-}S) \text{ for better slope } (1/2):1.$$

The design specifications of terraces are:

(i) Length: A terrace's maximum length is determined by the field's dimensions and form, the depth of its dissections and the soil's permeability and erodibility. The terraces will be more effective the longer they are. However, one should keep in mind that longer terraces increase the risk of erosion and accelerate runoff. It is advised to drain up to 100 m in a single direction under normal circumstances in a humid tropical environment. In places that are semi-arid or desert, the length can be slightly extended.
(ii) Width: The soil depth, crop requirements, cultivation methods, landowner preferences and available resources all influence the bench's breadth. The cost increases with bench width since more cut and fill are required. The ideal terrace width varies from 2.5 to 5 m for handcrafted and manually cultivated terraces, and from 3.5–8 m for machine-built and tractor-cultivated terraces.
(iii) Gradients: Depending on the soil and climate, horizontal gradients can be anywhere between 0.5 and 1%. For instance, it is safe to drain runoff in humid areas and on clay soils at 1%, but in arid or semi-arid areas horizontal gradients should not be greater than 0.5%. A terrace that slopes outwards has an outward grade of 3%, while a terrace that slopes reversely has an outward grade of 5%.
(iv) Slope limit: When the soil is deep enough, hand-made terraces should be used on slopes between 7 and 25 degrees (12% to 47%), and machine-built terraces should be used on slopes between 7 and 20 degrees (12% to 36%). In cases where the soil depths are insufficient for bench terraces, alternative rehabilitation strategies, such as hillside ditches, are to be employed. Slopes less than 7 degrees are not ideal for bench terraces. Alternatively, broad-base terraces and other low-tech conservation techniques ought to be employed.

12.1.4 Percolation Tanks (PT)/Spreading Basin

It is a farm pond-like structure constructed in relatively permeable type of soil in upper reach to facilitate groundwater recharge. The runoff from the catchment gets harvested in the structure where it gets sufficient time to slowly recharge to groundwater. It is generally constructed in the first- and second-order stream. Considering average catchment area as 50 ha. One percolation tank may be constructed in 50 ha catchment area. If the area of the tank at ground level is 30 × 30 m, the area at the bottom level is 26 × 26 m, depth as 2 m with side slope 1:1, assuming average seepage as 10 mm/day, total volume of water recharged through this structure is 1350 m^3 (approx.)

12.1.5 Check Dams

Check dams are built over tiny streams with a mild gradient; they can be built in both alluvial and hard-rock formations. The location of the check dam should have a weathered formation or a thick enough permeable substrate to allow for quick recharge of the water that has been stored. These structures typically have a height of less than 2 m and store water largely in stream courses. These are made with the breadth of the stream in mind, and extra water is permitted to overflow the wall. At the downstream of the stream, water cushions are positioned to prevent scouring from runoff. It is possible to build a sequence of these check dams to have recharging on a regional scale, allowing for the maximum runoff in the stream. To obstruct surface water flow in the stream channel and keep water on pervious soil/tock surface for longer length of time, a succession of tiny bunds or weirs are constructed across specific nala sections. When the slopes are kinder, nala bunds are built over larger nalas of second-order streams. Like a little percolation tank, a nala bund functions. A nala bund, a brickwork structure with a core wall cut off; nevertheless, cement bunds and plugs are increasingly common. For selecting a site for check dams/nala bunds (small earthen dam), the following conditions may be observed.

- Typically, the nala's entire catchment area falls between 40 and 100 hectares.
- The catchment should get less than 1000 mm of rain annually.
- The nala bed's width and depth should each be between 5 and 15 m.
- The nala bunds should be situated in an area where contouring or graded bunding has been done.
- The areas downstream of the check dam/bund should have irrigable land under well irrigation.
- For groundwater recharge to occur, the exposed rock layers in the ponded area must be porous.
- Usually built in a trapezoidal shape, the nala bund's measurements are length 10–15 m, height 2–3 m and width 1–3 m.

12.1.6 Gabion Structure

This type of check dam is frequently built across small streams with virtually no submergence outside of the stream's course in order to preserve stream flows. It consists of rocks that can be found nearby arranged in a steel weir mesh. Overflowing water runs over the structure, and some of the water is used as a recharge source. These constructions typically have a height of 0.5 m and are utilized in streams that have a width of 10 to 15 m. Eventually, the silt from the stream fills the spaces between the rocks, increasing the impermeability of the water.

12.1.7 Micro-catchment

Micro-catchment (Figure 12.2) uses a small area of ground to catch runoff yield. The runoff is channeled into a storage area where farming occurs. The land has been cleared of plants and compacted extremely effectively. This method is also known as runoff inducement. Ground catchments have also been used traditionally in Israel for the past 4000 years to cultivate cereal crops and some drought-tolerant species. Overall, micro-catchment collects runoff from non-cropped regions to utilize in farmed areas. These are often utilized in low and medium rainfall areas.

(1) Overland flow harvested from short catchment length; (2) catchment length typically between 1 and 30 m; (3) runoff stored in soil profile; (4) ratio of catchment to cultivated area typically 1:1 to 3:1; (5) normally no provision for overflow; and (6) uniform plant growth are the main characteristics of micro-catchment water harvesting. For every tree, micro-catchments typically have a size of 10–50 m^2. Bund heights vary, but they often range from 20 to 40 cm, depending on the slope that is predominant. Since bunds are frequently formed by machines, the specific form of the bund is determined by the kind of machine used, such as a motor grader or a disc plow. It is advised that the bund's height not be any lower than 25 cm. A minimum of 75 cm must be the base width. The building approach determines how the furrow is configured upslope of the bund. Bunds have to be separated by either 5 or 10 m. Cross-ties ought to be spaced 2–10 m apart and at least 2 m length. This precisely determines the size of every micro-catchment. On slopes of up to 0.5%, it is advised to give 10 m between the bunds and 5 m between them. For multipurpose trees, a typical micro-catchment size is 25 m^2. This means that 10 m of bund spacing and 2.5 m of tie spacing, or 5 m of bund spacing and 5 m of tie spacing, are equivalent. The ties are formed using soil that was excavated from the infiltration pit. At the intersection of the cross-tie and the bund, a hole is dug.

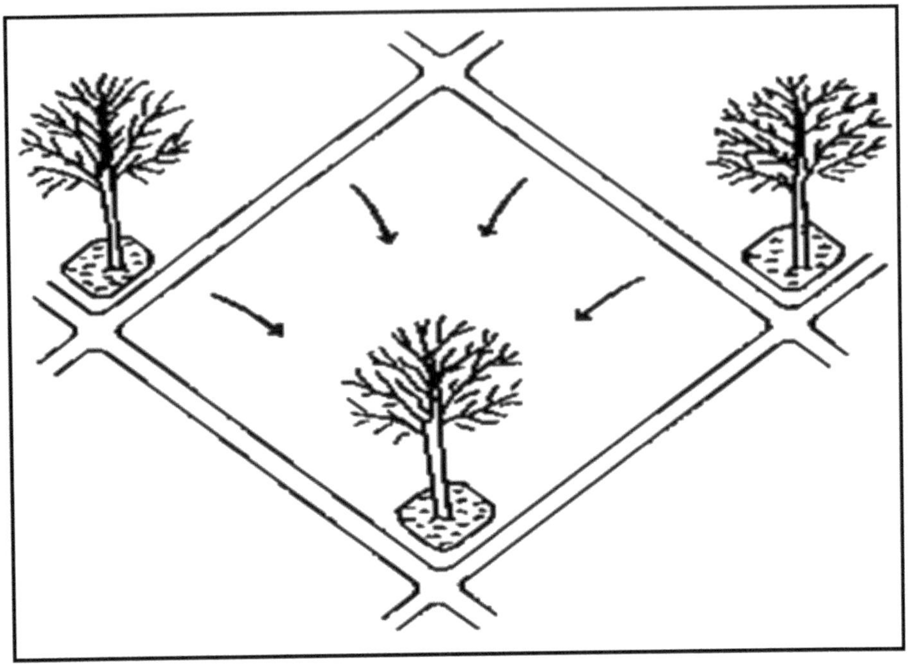

FIGURE 12.2 Micro-catchment water harvesting for trees.

Generally, a pit that is 40 cm deep and 80 cm squared is adequate. Micro-catchment design typically hinges on striking a balance between generated runoff and stored runoff.

12.1.8 Farm Pond

Farm ponds (Figure 12.3) are classified into four varieties based on the water supply and their location in relation to the land surface: dugout ponds, surface ponds, spring or creek supplied ponds and off-screen storage ponds. ***Dugout ponds***: These are created by excavating the ground, and the dirt that is left behind is used to build an embankment around the pond. Whenever aquifers are present, either surface runoff or groundwater is used to feed the pond. ***Surface ponds***: These are created by partially excavating the area and building an embankment to hold the water. Typically, a location with an existing depression is selected for the construction of the pond. Surface drainage from the catchment region feeds the pond. This kind of farm pond is among the most popular ones. The pond consists of an earthen embankment, a mechanical spillway and an emergency spillway. For irrigation purposes, it is preferable to drain the pond's water using a gravity outlet. It is crucial to choose an appropriate location for this reason. ***Spring or creek fed ponds***: These are the ponds where the water supply comes from a spring or creek. These ponds rely on the presence of natural springs or streams. ***Off-screen storage ponds***: These streams only occasionally flow, thus the ponds are built alongside them. The plan is to store the water that the streams' seasonal flow produces. Appropriate plans must be created to transfer the water from the stream to the storage ponds.

12.1.9 Rooftop Rainwater Harvesting

The lack of clean, high-quality water is becoming a major worry. Thus, one of the best ways to encourage water conservation is rooftop rainwater harvesting, which involves collecting rainwater from roof catchments and storing it in reservoirs (Figure 12.4). Rooftop rainwater collection is one of the earliest and most straightforward ways for homes to independently supply water; it has been

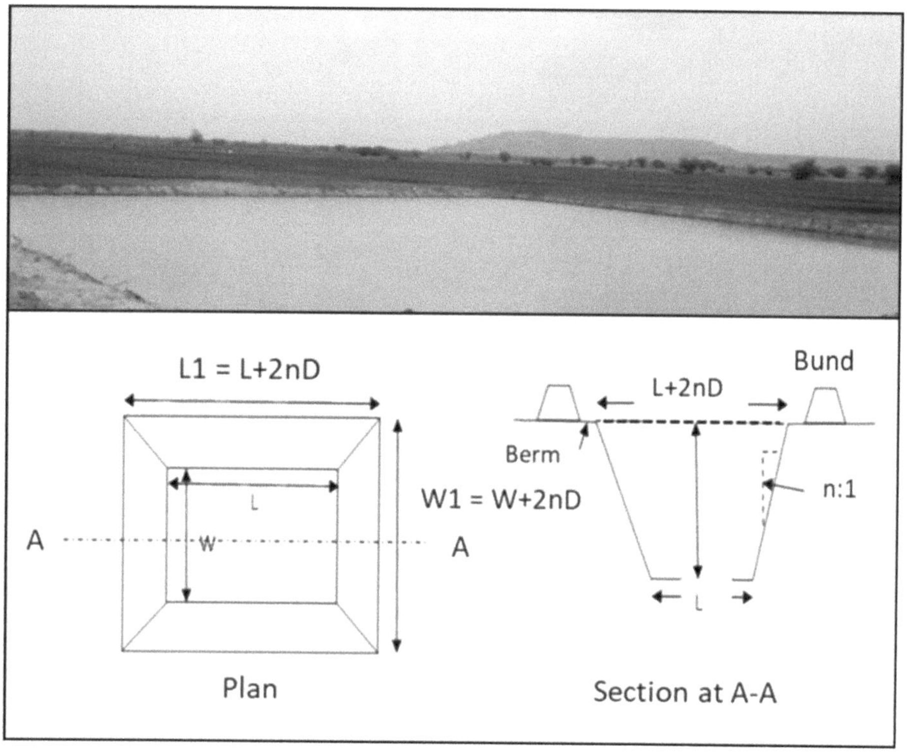

FIGURE 12.3 Cross section of dugout farm pond.

practiced for many years in South Asia and other nations. In rooftop harvesting, the building's roof serves as a collection area, and rainwater is gathered to fill a tank or be directed towards an artificial recharge system. Catchment area, conveyance, flushing and filter medium (sand gravel/charcoal) make up the majority of this system. In addition to its practical applications, rooftop rainwater collection also serves as a means of recharging groundwater supplies and supplying water for home, agricultural, animal and drinking purposes. When used correctly, this less expensive and very successful technique raises the area's groundwater level. In addition to serving organizations like schools, hospitals and other public facilities, installations can be created for a variety of scales, including homes, neighbourhoods and communities. By using artificial recharge techniques, this collected rooftop rainwater can be stored in subsurface ground water reservoirs to supply household demands through tank storage. Typically, two approaches are used for this: (i) collecting rainwater on the roof and channelling it into dry tanks, or (ii) carrying the extra water gathered from rainfall to the wells, which helps to further replenish the wells and underground water table. Tamil Nadu is the first and only state in India that calls for all homes to have a rainwater harvesting system installed on the roof. Even in Shillong, Meghalaya, rooftop rainwater collecting is the most popular technique. It is intriguing because, despite Shillong, the state capital, there is a severe water deficit in Cherrapunji and Mawsynram, which are 55 km away from Shillong and receive the most rainfall in the world.

Rooftop rainwater collecting systems have several advantages, including: (1) cheap cost; (2) lower water bills; (3) decreased need for imported water; (4) water and energy conservation; (5) better groundwater quality and quantity; (6) a filter system is not necessary for landscape irrigation; (7) the system is reasonably easy to install and maintain; (8) the water source is suitable for landscape irrigation and free of chemicals, dissolved salts and minerals.

FIGURE 12.4 Pictorial representation of rooftop water harvesting method.

12.2 WATER BUDGETING IN AQUACULTURE-BASED IFS

Since aquaculture's sustainability has been called into question on numerous occasions, international and regional organizations have put forth best management practices (BMP) to improve sustainable production and make aquaculture more ecologically conscious. Pond aquaculture's ecological sustainability is also under jeopardy due to a number of factors, including severe weather, excessive freshwater use, disease, organic pollution and chemical contamination. While aquaculture output must rise to meet the increasing demand, the scarcity of land and water resources is now limiting the amount of land that can be used for aquaculture. Furthermore, fish production uses a lot of water – much more than conventional agriculture does – and the availability of freshwater will limit its potential to flourish in the future (Verdegem and Bosma 2009). The unplanned, wasteful use of water in aquaculture is another factor impeding the industry's ability to grow. As is clear, aquaculture can have extremely high on-farm water usage, reaching up to 45 m^3 per kg of biomass produced in ponds (Verdegem et al. 2006). To make the system more water-efficient, aquaculture or intensive aquaculture production systems must be intensified in order to reduce the amount of water used on the farm per kilogram of biomass produced. Since there will be limited water available for inland aquaculture, increased water productivity and profitability will be ensured by focused efforts on input management (primarily feed), water quality monitoring and quantifying/estimating the water requirements of commercially significant high-value fish and prawn species. Therefore, it is necessary to popularize new technologies and practices that will make freshwater and brackish water aquaculture systems more resource-efficient and sustainable in the future. These include improving water use efficiency/productivity and nutritional water productivity as well as lowering water footprint.

12.2.1 Importance of Aquacultural Water Budgeting and Water Productivity

Because grow-out aquaculture has low economic output (due to rising feed costs, power supply, chemicals, aqua-drugs, etc.), it is critical to reduce operating costs by increasing water use efficiency. In fact, the aquaculture industry has been compelled to reconsider the prudent use of water in order to boost water productivity due to the unpredictability of monsoon rains, scarcity and restricted availability of freshwater resources. These days, getting water is getting harder to come by and more expensive (the cost of pumping). The freshwater reserve and supply around the world, and in India specifically, are currently in jeopardy because of rising population pressure and rising water demands in home, industrial and agricultural sectors. Because water is a scarce resource, water management should take a more comprehensive approach. Furthermore, the establishment of protocols for best water management practice (BWMP) in commercially significant grow-out aquaculture should start with water budgeting and its prudent application. The three main ways that static water ponds lose water are by evaporation, percolation and seepage. These processes cause the most nutrient loss and water quality fluctuations, which lowers the productivity of the water. Resupply/exchange of water becomes crucial to replace/maintain this water loss, pond fertility and the survival of stocked animals. Farmers used to frequently swap water in the hopes of increasing productivity, but they rarely thought through the necessity or the expense of operation, which could sometimes make the exchange unprofitable. But determining the precise amount of water needed is crucial and depends on a number of variables, including species, stocking density, growth stage, biomass, plankton and nutritional status, water loss, agro-climatic conditions, etc. The amount of water needed depends on the soil, climate and species to be supplied, culture technique and management techniques. To avoid the issue of water waste, it is vital to determine if replenishment or exchange is required, and then quantify the amount of water needed for replenishment. The water budgeting for different species and target of productions may form the practical tools for generating useful information for mitigating the challenges of water for aquatic production.

There are two types of water consumption in aquaculture: consumptive use and total use. Aquaculture uses a wide range of total amounts of water, mostly according to the culture technique employed. Water usage is lowest for cage, Biofloc and net pen cultures, and highest for intensive pond and raceway cultures. Water consumption in ponds varies depending on the intensity of production, frequency and volume of water exchange used (Boyd 2005; Boyd et al. 2007). Generally, fish production demands total water use to 4–8 m^3/kg fish in embankment ponds and 8–16 m^3/kg fish in watershed ponds. According to Verdegem et al. (2006), on-farm water use in aquaculture can currently range from 500–700 l in super-intensive recirculation systems to 45 m^3 of water per kilogram of product in vast pond systems. The degree of water exchange is a major factor in figuring out how efficiently aquaculture uses water. But for the majority of pond aquaculture, water exchange is not required (Boyd and Tucker 1998). Minimizing or doing away with water exchange conserves water and lowers pumping expenses. Additionally, the hydraulic retention time (HRT) in ponds rises with less water exchange. This lessens the amount of possible contaminants in effluent and enables natural processes to digest wastes more thoroughly (Boyd 2005). According to Boyd et al. (2007), the HRT in static ponds is often a few weeks or even months, but in ponds that exchange water, HRT is typically a week or slightly longer.

Enhancing aquacultural water productivity at various levels will be necessary by 2050 to meet growing food demands and reduce water diversion for aquaculture. Water productivity, in its broadest definition, is the goal of using less water to produce more food, revenue, improved livelihoods and ecosystem services. The ratio of the net advantages from crop, forestry, fisheries, livestock and mixed agricultural systems to the quantity of water required to generate those benefits is known as water productivity. It is also known as the net return for each unit of water utilized. Thus, 'more crop per drop' refers to physical water productivity, which is the ratio of aquaculture output to water consumption, while 'value derived per unit of water used' refers to economic water productivity. 'Increasing or improving water productivity' refers to the most efficient way to use the water that is now available to us to increase crop yields. More water is not needed when there is greater water productivity. A number of factors, including water withdrawal, water consumption and virtual water usage, must be taken into account when evaluating the sustainability of water use. Water extracted for aquaculture purposes from streams, rivers or aquifers is referred to as water withdrawal. Following withdrawal, some of the water is returned and can be recycled or reintroduced into the ecosystem. The portion that is not returned stands for water that has been consumed, that is, water that evaporates or becomes absorbed by objects and living things. The term 'virtual water use' describes the use of water in an indirect manner, such as when producing fish feed. The virtual water use should be included in the total amount of water used. Currently, the manufacturing of grains used in fish meals indirectly uses 1.7 m^3 of water per kg of production due to evaporation. The amount of water used for grains will rise in the future as fish feeds contribute more to inland aquaculture production overall, but with current technology, this consumption will level off at 3 m^3 of water per kilogram of production. Pond evaporation uses an additional 5.2 m^3 of water per kg of production on average (Verdegem and Bosma 2009). Inland aquaculture requires an average of 16.9 m^3 of freshwater removal per kilogram of production; however, if pollution is kept under control, infiltration losses (6.9 m^3) and water replacement (3.1 m^3) can be classified as green water. Infiltration, which is often calculated at 5 to 10 mm/d and depending on the type of soil and topography of ponds, is the sum of bottom percolation and lateral seepage. However infiltration losses from fish ponds will be limited to lateral seepage to nearby farms and waterways when the groundwater table is high, as it is during the monsoon season. Enhancing aquaculture can significantly lower the amount of evaporation loss per kilogram of production; therefore, studies should concentrate on raising pond water productivity while minimizing environmental effects.

12.2.2 Water Productivity and Economic Efficiency

Since water is the most valuable natural resource, conserving it, using it wisely, increasing productivity and preserving its quality are all seen as being critically important in the current environment. By 2030, more aquaculture water productivity will be needed to meet the demands for aqua-based food due to growing water diversion for the industrial sector, agriculture and other uses. According to Molden et al. (2010), aquacultural water productivity, which is calculated as the net benefits of aquaculture systems divided by the amount of water used, reflects the goals of increasing food production, revenue generation, livelihood support and ecological advantages while lowering social and environmental costs per unit of water consumed. Further, water productivity is an index of the economic value of water used (Boyd 2005), a useful indicator of efficient water management (Dasgupta et al. 2008), and it is used to define the relationship between crop produced and the amount of water involved in crop production (Ali and Talukder 2008). Higher water productivity not only reduces the need for additional water, but also minimizes the operational cost.

12.2.3 Hydrological Water Balance Study

Hydrological water balance study not only helps in estimating density-dependent water use in aquaculture operations but also helps in enhancing water use efficiency and water productivity (Figure 12.5). To make precise estimates of water use in ponds, hydrological water balance equation, inflow = outflow ± change in volume (ΔV), is used. Water use in aquaculture may be categorized as either total water use (TWU) or consumptive water use (CWU). TWU (probable inflows to ponds) = initial water filling (W_f) + management additions or regulated inflows (I) + precipitation (P) + runoff (R). CWU (possible outflows) = intentional discharge or regulated discharge (D) + overflow (O_f) + evaporation (E) + seepage (S_o) + transpiration (T) + water content in the harvested biomass (W_b). The difference between the total and consumptive water use, refers to non-consumptive water use (NWU). Commercial shrimp and fish ponds seldom receive direct input from creeks or streams. Moreover, water is hardly ever utilized for purposes other than farming, and aquatic weeds are forbidden from growing in ponds or along their margins. As a result, transpiration and creek inflow are

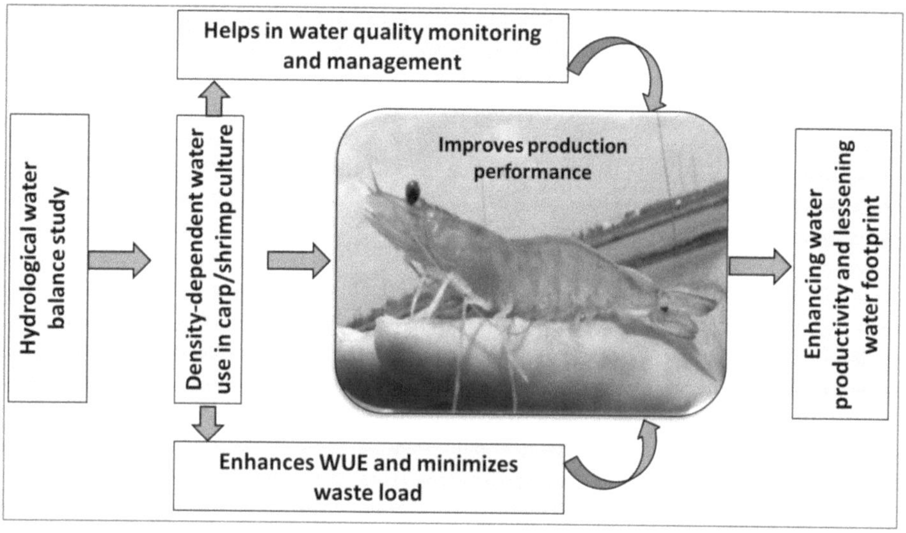

FIGURE 12.5 Advantages of hydrological water balance study.

rarely important concerns. Because embankment fish/shrimp ponds are small watersheds, groundwater inflow is rarely an issue and runoff is minimal. It is OK to overlook the harvested biomass's very small (about 0.75 m3/t) water content. Thus the appropriate water balance equation is:

$$P+I = E + S_o + O_f + D \pm \Delta V \quad (12.1)$$

A water level gauge can be installed in each pond to quantify the average water loss (evaporation + seepage) in order to estimate the CWU. When adding and withdrawing water, the outflow and inflow are also recorded using a water level gauge. Additionally, the following equation is used to estimate evaporation in order to distinguish it from regular water loss:

$$\text{Pond evaporation (mm)} = \text{Pond-pan coefficient} \times \text{Class-A pan evaporation (mm)} \quad (12.2)$$

The pond pan coefficient of 0.8, which is best suited for ponds, can be applied to the equation above. The evaporation loss is subtracted from the total loss to determine the amount of seepage from a pond. The following is an estimate of the consumptive water usage index (CWUI), which indicates the volume of water consumed per unit of fish or shrimp production:

$$\text{CWUI} = \text{CWU (m}^3\text{)} / \text{Total biomass (kg)} \quad (12.3)$$

12.2.4 Water Use and Economic Efficiency

The gross total water productivity (GTWP), net total water productivity (NTWP) and net consumptive water productivity (NCWP) can be computed while accounting for the entire volume of water utilized, as demonstrated below, to assess the effectiveness of water management:

$$\text{GTWP} = \text{Total economic value of the produce (Rs.)} / \text{TWU (m}^3\text{)} \quad (12.4)$$

$$\text{NTWP} = \text{Total economic value of the produce (Rs.)} - \text{Production cost (Rs.)} / \text{TWU (m}^3\text{)} \quad (12.5)$$

$$\text{NCWP} = \text{Total economic value of the produce (Rs.)} - \text{Production cost (Rs.)} / \text{CWU (m}^3\text{)} \quad (12.6)$$

To evaluate the efficacy of water management, total water use efficiency (WUE_t) and consumptive water use efficiency (WUE_c) are estimated as follows:

$$WUE_t \ (\text{kg m}^{-3}) = \text{Biomass production in kg ha}^{-1} / \text{TWU in m}^3 \quad (12.7)$$

$$WUE_c \ (\text{kg m}^{-3}) = \text{Biomass production in kg ha}^{-1} / \text{CWU in m}^3 \quad (12.8)$$

12.2.5 Water Footprint

The water footprint (cubic meter of consumptive water consumed per tonne of fish or shrimp produced, $m^3 \ t^{-1}$) in aquaculture is generally increased by blue and green water. The four components of water loss to the catchment – evaporation or evapotranspiration, water content in harvested biomass, contaminated water and non-return of extracted water to the same area – are taken into account when estimating the water footprint. Seepage and percolation loss, on the other hand, are not taken into account when estimating the water footprint because they do not represent a loss to the watershed and can be recovered elsewhere. Aquaculture's water footprint is often influenced by green (precipitation) and blue (surface and groundwater) water.

Thus, the appropriate equation is:

$$\text{Total water footprint } (WF_t, \ m^3 \ t^{-1}) = (D_i + O_f + E + W_b) / E_{cy} \quad (12.9)$$

where D_i = intentional or regulated discharge in m^3, O_f = overflow including other losses in m^3, E = evaporation in m^3, W_b = water content of harvested biomass in m^3 and E_{cy} = Economic crop yield in t ha^{-1}.

12.2.6 Water Use in High-Value Commercial Aquaculture: Case Studies from India

ICAR-Indian Institute of Water Management (Bhubaneswar, India), a premier central government research organization having a PAN India research mandate on water management in agriculture, aquaculture and allied sectors, conducted field experiments on water requirement, WUE, water productivity and water footprint of high-value aquaculture and aquaculture-based IFS for the period 2011–2022.

In 'Pacific white shrimp' (*Litopenaeus vannamei*) culture, with best management practice (BMP) at the optimum stocking density of 5,00,000 shrimp seed ha^{-1}, total water use could be minimized to 32,600 m^3 and water exchange minimized to 6,300 $m^3 \ ha^{-1}$, without hampering the normal growth and production. With this stocking density, shrimp productivity enhanced to a level of 10.58 t ha^{-1} in 120 days (Table 12.1). Further, with this technique, total water productivity of ₹46.6 m^{-3}, consumptive water productivity of ₹83.3 m^{-3} was generated while only 1.72 m^3 of water was required to production 1 kg of *L. vannamei* shrimp (Mohanty et al. 2018a, 2018b). In this system, the total water footprint (WF_t) amounted to 1229 $m^3 \ t^{-1}$. Various factors that affect the aquacultural water management are depicted in Figure 12.6.

In carp poly culture (Indian major carps, IMCs), with BMP at the optimum stocking density of 8,000 fingerlings ha^{-1}, total water use could be minimized to 28,200 m^3, without hampering the normal growth and production. With this stocking density, fish productivity enhanced to a level of 3.92 t ha^{-1} in 180 days (Table 12.1) with economic benefit (OV:CC) of 1.81. Further, total water productivity of ₹5.9 m^{-3}, consumptive water productivity of ₹10.3 m^{-3} was generated while only 2.9 m^3 of water was required to produce 1 kg of fish biomass (Mohanty et al. 2016; 2017a) In this system, total water footprint (WF_t) amounted to 1633 $m^3 \ t^{-1}$.

Similarly, in 'black tiger prawn (*Penaeus monodon*)' monoculture, stocking density of 2,00,000 post larvae ha^{-1} gave significantly higher yield (4.58 t ha^{-1} 125 d^{-1}), economic benefit (OV:CC, 2.47), total water productivity (₹26.6 m^{-3}) and consumptive water productivity (₹37,9 m^{-3}). At this optimum stocking density, total water use could be minimized to 28,600 m^3 and water exchange

TABLE 12.1
Technologically Validated Field Trial Result of Water Uses in High-Value Aquaculture

Water Management Parameters	Carp Polyculture	Monoculture of *P. monodon*	Monoculture of *L. vannamei*
Optimum density, ha^{-1}	8000 fingerlings	200,000 PL	500,000 PL
Culture duration	180 days	125 days	120 days
Water depth (m) maintained	1.2 up to 90 DOC 1.75 up to 180 DOC	1.0 up to 30 DOC 1.2 up to 125 DOC	1.2 up to 30 DOC 1.5 up to 120 DOC
Evaporation losses ($\times 10^4$, m^3)	0.58	0.63	0.62
Seepage losses ($\times 10^4$, m^3)	0.49	0.55	0.52
Regulated outflow ($\times 10^4$, m^3)	-	0.70	0.63
Other losses ($\times 10^4$, m^3)	0.06	0.08	0.05
Total loss (CWU) ($\times 10^4$, m^3)	1.13	1.96	1.82
Initial water level ($\times 10^4$, m^3)	1.20	1.00	1.20
Precipitation ($\times 10^4$, m^3)	0.6	0.46	0.51
Regulated inflow ($\times 10^4$, m^3)	1.02	1.40	1.55
Total water use ($\times 10^4$, m^3)	2.82	2.86	3.26
Consumptive water use index, $m^3\,kg^{-1}$	2.88	4.28	1.72
Productivity, $t\,ha^{-1}$	3.92	4.58	10.58
Feed conversion ratio, FCR	1.74	1.40	1.63
Sediment load, $m^3\,t^{-1}$ biomass	53.0	42.1	46.3
OV-CC ratio	1.81	2.47	2.11
Total water productivity, ₹ m^{-3}	5.9	26.6	46.6
Consumptive water productivity, ₹ m^{-3}	10.3	37.9	83.3
Total water footprint (WF_t, $m^3\,t^{-1}$)	1633	3079	1229

Note: PL, post-larvae; DOC, days of culture; CWU, consumptive water use; OV-CC, output value-cost of cultivation.

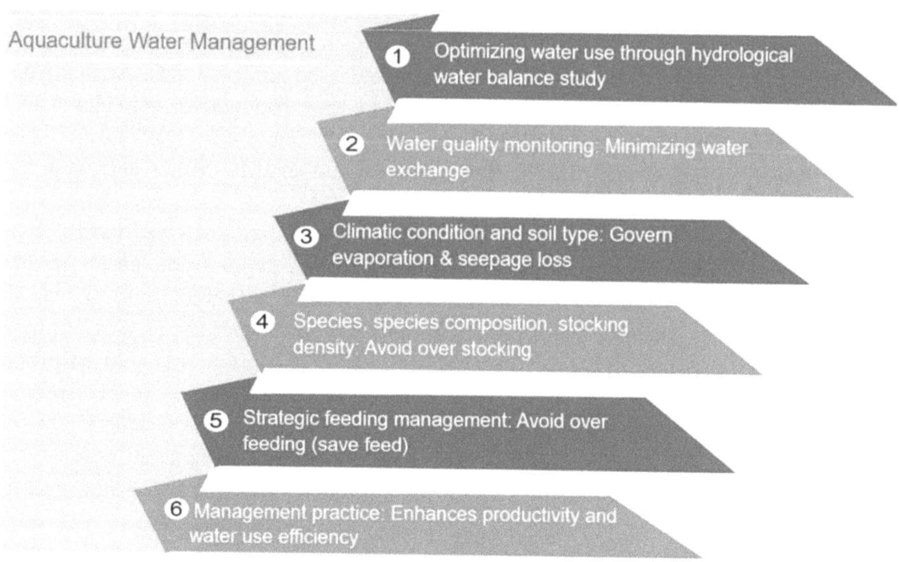

FIGURE 12.6 Factors affecting aquacultural water management.

TABLE 12.2
Water use in different freshwater carp polyculture systems

Water Use Parameters	Single Stock-Single Harvest	Single Stock-Multi Harvest	Multi Stock-Multi Harvest	Intercropping of IMC-Minor Carp
Stocking density, ha^{-1}	10,000 IMC fingerlings	10,000 IMC fingerlings	4,000 IMC fingerlings[a]	8,000 fingerlings Including 50% IMC
Culture duration	360 days	360 days	360 days	360 days
Water depth (m)	1.2 up to 180 DOC 1.5 up to 360 DOC	1.2 up to 180 DOC 1.5 up to 360 DOC	1.2 up to 180 DOC 1.5 up to 360 DOC	1.2 up to 180 DOC 1.5 up to 360 DOC
TWU ($\times 10^4$, m^3)	2.40	2.53	2.5	2.43
CWU ($\times 10^4$, m^3)	0.73	0.84	0.82	0.76
CWUI, $m^3 kg^{-1}$	1.97	1.95	1.78	1.77
Productivity, t ha^{-1}	3.7	4.3	4.6	4.3
WUE_c (kg m^{-3})	0.50	0.51	0.57	0.56
FCR	1.76	1.63	1.51	1.72
NCWP, ₹ m^{-3}	7.7	8.1	9.2	8.3
WF_t, $m^3 t^{-1}$	1122	1093	998	1026

Note: CWU, consumptive water use; CWUI, consumptive water use index; DOC, days of culture; FCR, feed conversion ratio; NCWP, net consumptive water productivity; WF_t, total water footprint; WUE, water use efficiency; WUE_c, consumptive water use efficiency; TWU, total water use.

[a]Partial harvesting was carried out at sixth and ninth months followed by restocking @1.25 times of the harvested numbers.

minimized to 7,000 $m^3 ha^{-1}$, without hampering the normal growth and production while, only 4.28 m^3 of water was required to production 1 kg of *Penaeus monodon* biomass (Mohanty et al. 2014; 2015; 2017b). In this system, total water footprint (WF_t) amounted to 3079 $m^3 t^{-1}$ (Table 12.1).

Water requirement of different freshwater carp polyculture systems was also carried out to optimize water requirement. Out of four different production systems such as (1) IMC grow-out culture: *Single Stock-Single Harvest system*, (2) IMC grow-out culture: *Single Stock-Multi Harvest system*, (3) IMC grow-out culture: *Multi Stock-Multi Harvest system*, (4) intercropping of IMC-Minor carp; the Multi Stock-Multi Harvest system (MSMH) was found to be more efficient, productive and profitable (Table 12.2) in terms of water use efficiency (0.57 kg m^{-3}), productivity (4.6 t ha^{-1}), FCR (1.51), total water footprint (998 $m^3 t^{-1}$) and net consumptive water productivity (INR 9.2 m^{-3}). Next to the Multi Stock-Multi Harvest system, the intercropping system of IMC-Minor carp was considered efficient, productive and profitable.

Water budgeting was done in the 'Aquaculture-based Integrated Farming System,' which integrated a number of different components, including agriculture (*kharif* rice, green gram and vegetables with drip irrigation and sprinklers), on-dyke horticulture (banana and papaya with drip irrigation) and poultry (*Vanaraja*). Out of the system's total crop water use (3.14×10^4 m^3 240 d^{-1}), the estimated TWU, consumptive water use index (CWUI) and productivity in carp polyculture alone was 2.3×10^4 m^3 240 d^{-1}, 2.6 $m^3 kg^{-1}$ fish production and 2.86 t ha^{-1}, respectively. The estimated evaporation and seepage losses were 2.8 and 2.1 m^3 water kg^{-1} fish production, respectively, and contributed significantly to CWU. The system as a whole resulted in a net profit of Rs.1,77,117 ha^{-1} with an output value–cost of cultivation ratio of 3.1 and net consumptive water productivity of Rs.19.2 m^{-3}.

CONCLUSION

There is increasing demand on the aquaculture sector to increase resource efficiency in production. Aquaculture's 'Blue Revolution,' which refers to its explosive growth, has to 'go green' by incorporating the concepts of 'ecological aquaculture,' which incorporates aquaculture water management. Finding the perfect amount of water required for aquaculture success is essential given the growing shortage of water resources and the risks of environmental damage. Responsible practices are necessary for aquaculture's future growth in order to increase operational effectiveness, aid stop water waste and maintain the quality of pond water. Enhancing the performance of aquaculture necessitates a hydrological water balance research to ascertain density-dependent water demand. The success of aquaculture thus depends on three key factors: water budgeting, density-dependent water consumption and water quality monitoring. Furthermore, demand-driven aquaculture with low to moderate water exchange helps minimize the water footprint, sediment load and effluent outputs while simultaneously improving water productivity and maintaining a quality of water appropriate for growth.

REFERENCES

Ali, M.H., Talukder, M.S.U. 2008. Increasing water productivity in crop production- A synthesis. *Agricultural Water Management* 95:1201–1213.

Boyd, C.E. 2005. Water use in aquaculture. *World Aquaculture* 36(3):12–15; 70.

Boyd, C.E., Tucker, C.S. 1998. *Pond Aquaculture Water Quality Management.* Kluwer Academic Publishers, Boston, MA.

Boyd, C.E., Tucker, C.S., Mcnevin, A., Bostick, K., Clay, J. 2007. Indicators of resource use efficiency and environmental performance in fish and crustacean aquaculture. *Reviews in Fisheries Science* 15:327–360.

Dasgupta, S., Pandey, B.K., Sarangi, N., Mukhopadhyay, P.K. 2008. Evaluation of water productivity and fish yield in sewage-fed vis-a`-vis fertilized based carp culture. *Bioresource Technology* 99:3499–3506.

Mohanty, R.K., Ambast, S.K., Panda, D.K., Thakur, A.K., Mohanty, S. 2017a. Density-dependent water use in carp polyculture: Impacts on production performance and water productivity. *Aquaculture* 470:32–39.

Mohanty, R.K., Ambast, S.K., Panigrahi, P., Mandal, K.G. 2018a. Water quality suitability and water use indices: Useful management tools in coastal aquaculture of *Litopenaeus vannamei*. *Aquaculture* 485:210–219.

Mohanty, R.K., Ambast, S.K., Panigrahi, P., Thakur, A.K., Mandal, K.G. 2018b. Enhancing water use efficiency in monoculture of *Litopenaeus vannamei*: Impacts on pond water quality, waste production, water footprint and production performance. *Aquacultural Engineering* 82:46–55.

Mohanty, R.K., Mishra, A., Ambast, S.K., Mandal, K.G., Panda, D.K., Panigrahi, P. 2017b. Effects of various shrimp (*Penaeus monodon*) densities on their growth, water and sediment quality, and water budget. *Aquaculture International* 25:2161–2176.

Mohanty, R.K., Mishra, A., Panda, D.K., Patil, D.U. 2015. Effects of water exchange protocols on water quality, sedimentation rate and production performance of *Penaeus monodon* in earthen ponds. *Aquaculture Research* 46:2457–2468.

Mohanty, R.K., Mishra, A., Panda, D.K., Patil, D.U. 2016. Water budgeting in a carp-prawn polyculture system: Impacts on production performance, water productivity and sediment stack. *Aquaculture Research* 47:2050–2060.

Mohanty, R.K., Mishra, A., Patil, D.U. 2014. Water budgeting in black tiger shrimp *Penaeus monodon* culture using different water and feed management systems. *Turkish Journal of Fisheries and Aquatic Sciences* 14:487–496.

Molden, D., Oweis, T., Steduto, P., Bindraban, P., Hanjra, M.A., Kijne, J. 2010. Improving agricultural water productivity: Between optimism and caution. *Agricultural Water Management* 97:528–535.

Verdegem, M.C.J., Bosma, R.H. 2009. Water withdrawal for brackish and inland aquaculture, and options to produce more fish in ponds with present water use. *Water Policy* 11(Supplement 1):52–68.

Verdegem, M.C.J., Bosma, R.H., Verreth, J.A.J. 2006. Reducing water use for animal production through aquaculture. *Water Resources Development* 22(1):101–113.

13 Socioeconomic Impact of Integrated Farming Systems (IFS)

13.1 INTRODUCTION

An integrated farming system (IFS) integrates multiple units, such as crop cultivation, animal husbandry and aquaculture, in a holistic and synergistic way. By boosting production and lowering environmental impact, these systems seek to increase food and nutrition security while advancing social and economic well-being. This chapter features an overview of the potential influences of IFS on the human, physical, social, financial and natural assets of farmers. It highlights how IFS can improve food and nutrition security, community development, income, technical and management skills, health and well-being and conservation of natural resources. Additionally, it provides real-life examples to illustrate the impact of IFS.

13.2 SOCIOECONOMIC IMPACT ON FARMERS

Integrated farming systems (IFS) can improve farmers' incomes and livelihoods by reducing their dependency on external inputs like pesticides and fertilizers, resulting in reduced production costs and increased profitability. This is facilitated by higher productivity and yields resulting from interactions between various system components. Farmers can diversify their sources of revenue through different components of IFS. This leads to improved access to markets, as farmers can produce a broader range of products that can be sold at higher prices. Farmers can create extra revenue streams by merging several agricultural activities like crop production, animal care and agroforestry. For instance, combining livestock cultivation with crops allows farmers to profit from selling meat, milk and other livestock products alongside their crops, thus raising their overall income. IFS provides an opportunity for enhanced farm productivity. IFS techniques frequently increase farm productivity and efficiency by utilizing resources efficiently and fostering ecological balance. These are two benefits of combining various agricultural activities. Incorporating nitrogen-fixing legumes into crop rotations, for instance, enhances soil fertility and decreases the demand for synthetic fertilizers, increasing crop yields. Similarly, using livestock dung as organic manure can enhance soil fertility and availability of nutrients. IFS has great potential to reduce the risks involved in agriculture. As farmers can produce various crops and animals, diversifying income streams helps minimize risk and increase resilience. Farmers can minimize market swings and climatic unpredictability risks by diversifying their operations using IFS practices. Farmers are less prone to crop failure or changes in market prices when they plant various crops and engage in several farming activities. This lowers overall risk and guarantees a more consistent income all year.

Adoption of IFS ensures improved food security and nutrition in the farming community. At the farm and household levels, IFS practices help to increase food security and nutrition. Crop diversification means farmers and their families can access a greater variety of foods and a more wholesome diet. Combining cattle and fish production can also offer a reliable source of protein and other crucial elements. This variety in food production lowers the possibility of food shortages and expands access to wholesome food. IFS provides scope for improved resource management. Sustainable

DOI: 10.1201/9781003239963-13

resource management is emphasized in IFS practices. In hilly or mountainous areas, terrace farming, for instance, reduces soil erosion by minimizing water runoff and preserving the fertile topsoil. Agroforestry practices contribute to the conservation of water resources, the enhancement of biodiversity and the provision of cover and windbreaks, all of which favour crop growth and overall farm production. Several environmental benefits are associated with this system of farming. IFS practices encourage environmentally friendly farming techniques. Farmers support ecosystem health by minimizing the use of synthetic inputs, conserving soil and maintaining biodiversity. For instance, agroforestry systems improve water quality, trap carbon and preserve wildlife habitat. Farmers who adopt IFS are empowered by knowledge and skill development. Farmers frequently need to develop new knowledge and skills to implement IFS practices. Through training programmes and capacity-building projects, farmers can learn about diverse agricultural systems, animal husbandry, crop management and ecosystem services. This increases their capacity for farming and their general flexibility and resilience. However, implementing IFS can also be challenging for farmers, particularly in the face of competing demands for land and resources. Additionally, farmers may require training and technical assistance to adopt new practices and technologies.

13.3 SOCIOECONOMIC IMPACT ON COMMUNITIES

IFS can also positively impact communities by improving food and nutritional security and increasing access to income and employment opportunities. This can be accomplished through an increased availability of nutritious and varied food as a result of the diversification of crops and animals produced by farmers. Due to the diversification of income streams for farmers and the establishment of value chains for various products, there is an increase in income and employment opportunities. Moreover, integrated farming systems (IFS) often exhibit enhanced resilience to climate change, enabling them to withstand extreme weather events better. Enhanced ecosystem services, such as pollination, control of pests and soil conservation, can benefit the entire community. However, the socioeconomic impact of IFS on communities may differ depending on factors such as population density, land availability and market access. Additionally, communities may require education and awareness-raising to understand the benefits of IFS and how to participate in developing and implementing these systems.

13.3.1 Food and Nutrition Security

Food and nutrition security is a serious issue for many small and marginal farmers, who often lack access to diverse and nutritious foods. Integrated farming systems (IFS) can help improve food and nutrition security by improving the variety and yield of crops and livestock and promoting sustainable and organic farming practices. The 'Back to Village' concept, which aims to bring development to rural areas by focusing on local resources and traditional knowledge, can be a helpful starting point for IFS adoption. By leveraging Indigenous technical knowledge (ITK) and cultural heritage, farmers can develop more resilient and sustainable farming systems better suited to local conditions.

A SWOT (strengths, weaknesses, opportunities and threats) analysis can be valuable for identifying the potential advantages and challenges of IFS adoption. For example, strengths may include increased diversity and productivity of crops and livestock, while weaknesses may consist of limited access to markets and credit. Opportunities may include access to government support and training, while threats may consist of resistance from traditional intermediaries. IFS can positively impact farmers' social, physical, economic, human and natural assets. For example, by increasing crop diversity and productivity, farmers can improve their food and nutrition security while also increasing their income. Additionally, integrated farming systems (IFS) can play a role in safeguarding and preserving essential natural resources like water and soil by encouraging sustainable and organic farming practices.

Overall, IFS can help improve food and nutrition security for small and marginal farmers and can have a beneficial impact on physical, social, economic, human and natural assets. The 'Back to Village; concept, leveraging ITK and cultural heritage, SWOT analysis and case studies, can offer a valuable framework for the effective adoption and implementation of IFS. Integrated farming systems (IFS) can significantly impact farmers' physical, social, financial, economic and natural assets.

Physical assets: IFS can positively impact farmers' physical assets by increasing crop diversity and productivity. This can improve food and nutritional security for farmers and their families, as they can access various nutritious foods. Additionally, IFS can help protect and conserve natural resources like soil and water by promoting sustainable and organic farming practices.

Social assets: IFS can also positively impact farmers' social assets. For example, by joining a farmers producer organization (FPO), farmers can increase their bargaining power and improve their access to markets, credit and other inputs. IFS can also promote community development by creating opportunities for farmers to collaborate and share knowledge and resources.

Economic assets: IFS can positively impact farmers' economic assets by increasing their income. For example, by increasing crop diversity and productivity, farmers can sell more produce and receive higher product prices. Additionally, IFS can help farmers access credit and other financial services that can be used to invest in their farms and businesses.

Human assets: IFS can also positively impact farmers' human assets. For example, IFS can help farmers improve their technical and management skills by providing training and technical assistance. This can enhance productivity and efficiency on their farms. Additionally, IFS can promote health and well-being by promoting sustainable and organic farming practices that protect farmers and their families from harmful chemicals.

Natural assets: IFS can positively impact farmers' natural assets by promoting sustainable and organic farming practices. For example, by using agroforestry and intercropping practices, farmers can improve soil health and biodiversity while reducing the need for fertilizers and pesticides. Additionally, IFS can help protect and conserve natural resources like water and soil by promoting sustainable and organic farming practices.

Case studies of successful IFS adoption can provide critical insights into the benefits and challenges of these systems (Paris 2002). For example, a case study of an FPO in India that adopted IFS practices, such as crop rotation and intercropping, reported increased crop yields and improved food and nutrition security for its members. Another case study of an FPO in Africa that adopted agroforestry practices reported increased income and improved soil health for its members.

Ponnusamy et al. 2017 explored the impact of integrated farming systems on improving the income of farmers and securing their sustainable livelihoods across two regions in Tamil Nadu and four in Haryana. It scrutinized how different enterprises, such as poultry, fisheries, sheep and goat rearing and horticulture, when combined with staple crop cultivation and dairy farming, influence the total earnings of farmers. The economic advantages of these diverse combinations of enterprises were assessed using partial budgeting, revealing an increase in profits ranging from ₹7880 to ₹57,530 per hectare. Highly significant positive correlations were reported between overall income and a range of socioeconomic elements such as the size of land owned, the establishment of long-term assets, the assurance of food and nutritional security, the creation of job opportunities, patterns in marketing activities, educational attainment and the possession of livestock. Additionally, a trend analysis from 1994 to 2014 in Haryana revealed a shift in farmers' preferences towards demand and profit-oriented activities like agroforestry, mushroom culture, aquaculture and cattle and poultry rearing. However, significant constraints in adopting integrated farming systems were noted, primarily due to heavy initial investments and labour shortages.

A study reported that combining crop cultivation, livestock rearing, poultry farming and residential gardening was the most favoured approach among both integrated and mixed farming systems in Bangladesh. It concluded that adopting integrated farming could lead to enhanced income for farmers and create more job opportunities compared to mixed and conventional farming methods in the regions examined.

Furthermore, it was also found that integrated farm households had a significantly better food security situation and reduced poverty compared to households adopting mixed farms. Notable upscaling was observed across various capitals (namely, human capital, physical capital, natural capital, social capital and financial capital) among farm households adopting integrated farming compared to those practising mixed farming (Uddin et al. 2015).

13.3.2 Environmental Impact

While examining a secure method for weighting within a life cycle impact assessment (LCA) that considers factors such as climate change, biodiversity decline rates, nitrogen and phosphorus cycles, depletion of stratospheric ozone, utilization of global freshwater resources and shifts in land use trends, a study was performed to evaluate the effects of various farming systems including organic, traditional and integrated farming practices with alternative land uses. The integrated farming practices, which included a portion of the land allocated for natural forest, had a minimum overall impact score. Conversely, traditional farming practices utilizing silvergrass and managed forests showed maximum overall impact scores. The main issue contributing to the uncertainty in the results was the diverse evaluations concerning the acceptable limit for the impact category related to the loss of biodiversity (Tuomisto et al. 2012).

Climate change influences agricultural yield and income of farmers; however, integrated farming systems (IFS) present strategies to alleviate these influences. A research study delved into the viewpoints of farmers on climate change, the obstacles they face in adapting and the tactics they employ for adaptation, focusing on a group of 2,160 farmers practicing integrated farming systems (IFS). In the last 38 years, a pattern of rising rainfall has been noted in regions with high humidity (4.18 mm year^{-1}) and semi-aridity (0.35 mm year^{-1}), whereas areas that are sub-humid and arid have experienced a decline in rainfall, measuring −2.02 mm year^{-1} and −0.20 mm year^{-1}, respectively. Annual temperature rises varied between 0.011–0.014°C across different agro-climatic zones (ACZs). Approximately 79% of IFS farmers reported perceiving changes such as increased temperature, reduced rainfall, unpredictable commencement of the monsoon season, intense rains toward the end of the season, periods of drought during the season and more regular occurrences of flooding. Farmers employed various adaptation measures in different regions with adaptation indexes as arid (50.2%), semi-arid (66.6%), sub-humid (83.3%) and humid (91.6%), respectively. Most farmers practicing integrated farming systems (IFS) faced difficulties in executing strategies to combat climate change, which encompassed meta barriers, limitations in capabilities and constraints related to water resources. Consequently, it is concluded that knowledgeable farmers engaged in varied and lucrative farming with small to medium-sized landholdings are more focused on issues related to climate change and are more likely to adopt adaptive strategies to mitigate its environmental impact (Paramesh et al. 2022).

13.3.3 Income and Employment Generation

The decreasing scale of landholdings presents a significant challenge in achieving profitable and sustainable agriculture. Considering the decline in land availability per person, it is essential to develop farming approaches and innovations that ensure adequate job opportunities and income for small-scale farmers (those owning under 2.0 hectares), representing the highest farming population in developing nations. It is unlikely that a single farming enterprise, such as traditional monocropping, can sustain small-scale farmers. When managed efficiently, integrated farming systems (IFS) present a lower risk due to the collaborative benefits among various agricultural activities, the variety of produce and the promotion of ecological preservation. Consequently, IFS is advocated for enhancing small and marginal farms throughout Asia, with researchers formulating plans that offer additional revenue sources and job prospects while reducing the risks for farmers with smaller landholdings (Behera and France 2016).

13.3.4 Poverty Reduction

Following in the footsteps of South and Southeast Asian countries, various national and international organizations are actively implementing integrated rice and fish farming systems (IRFFS) to combat poverty and insecurities related to food and nutrition. However, there is a need for more thorough empirical research on the factors impacting the implementation of such technologies, especially within underprivileged communities.

The Indigenous small-scale IRFFS in different Asian countries continues to support the livelihood of farmers with small and marginal landholdings. However, continuous research and extension efforts are needed to up-scale innovations related to this IFS. Key factors influencing adoption include the gender of the family head, availability of irrigation facilities, educational attainment, and disputes with fellow community members. The size of the farm and access to credit also significantly impact the adoption process, indicating limitations related to land ownership and financial resources. Therefore, it can be challenging for farmers with constraints in landholdings and finances to adopt the technology. A study reveal that IRFFS can significantly and positively contribute to the well-being of farm households as indicated by the increase in annual household income, income earned from farms and fish consumption rate (Islam et al. 2015).

13.4 CONCLUSION

Integrated farming systems (IFS) promise to improve food and nutritional security and boost the income and livelihood prospects of farmers and their communities. However, implementing IFS can be challenging and requires a multi-stakeholder approach, including partnerships among farmers, researchers and policymakers. Investing in research and development and promoting policies and incentives that support the adoption of IFS will be critical to their success. Additionally, understanding the specific socioeconomic impact of IFS in different agro-climatic zones and communities will be essential for improving their effectiveness and sustainability.

REFERENCES

Behera, U.K., France, J. 2016. Integrated farming systems and the livelihood security of small and marginal farmers in India and other developing countries. *Advances in Agronomy* 138:235–282.

Islam, A.H.M.S., Barman, B.K., Murshed-e-Jahan, K. 2015. Adoption and impact of integrated rice-fish farming system in Bangladesh. *Aquaculture* 447:76–85.

Paramesh, V., Kumar, P., Shamim, M., Ravisankar, N., Arunachalam, V., Nath, A.J., Mayekar, T., Singh, R., Prusty, A.K., Rajkumar, R.S., Panwar, A.S. 2022. Integrated farming systems as an adaptation strategy to climate change: Case studies from diverse agro-climatic zones of India. *Sustainability* 14(18):11629.

Paris, T.R. 2002. Crop–animal systems in Asia: Socio-economic benefits and impacts on rural livelihoods. *Agricultural Systems* 71(1–2):147–168.

Ponnusamy, K., Devi, M.K. 2017. Impact of integrated farming system approach on doubling farmers' income. *Agricultural Economics Research Review* 30(347-2017-2750):233–240.

Tuomisto, H.L., Hodge, I.D., Riordan, P., Macdonald, D.W. 2012. Exploring a safe operating approach to weighting in life cycle impact assessment–a case study of organic, conventional and integrated farming systems. *Journal of Cleaner Production* 37:147–153.

Uddin, M.T., Khan, M.A., Islam, M.M. 2015. Integrated farming and its impact on farmers livelihood in Bangladesh. *SAARC Journal of Agriculture* 13(2):61–79.

14 Road Map for the Future *Doubling Farmers' Incomes*

14.1 NEW IFS CONCEPTS

The future integrated farming system needs to focus on vertical and multiple land use, as land is the critical constraint for farming due to urbanization, land degradation, industrialization and other developments that take away land from farming. Farming as a profession or hobby for both rural and urban populaces now requires new technologies for the use of residential areas as an avenue for growing diverse crops. Water bodies (both saline and non-saline), such as ponds, canals, rivers, seas and oceans, can be venues for multiple productions of human food requirements. Requirements for alternative farming systems are for increasing off-shore production to ensure food security, innovations for restorative farming and achieving a blue-green economy. Some of the emerging concepts and practices in moving towards a new integrated farming system are described in this chapter.

14.2 MODELS OF NEW FARMING SYSTEMS

14.2.1 Three-Dimensional Farming (3D Farming)

The open ocean, coastal waters and even tanks can be used to grow marine plants such as seaweeds and marine aquaculture to produce algae, fish, shellfish, oysters, mussels, etc. at various depths of water under the sea surface without relying on external feed and fertilizers (NOAA 2018). This use of the vertical space of water bodies is called 3D farming (Photo 14.1). This new farming method for producing more seafood is becoming popular in many countries, like China, Italy, Japan and the United States of America (Buschmann et al. 2017). There are two systems of 3D farming. An open ocean farming system involves horizontal ropes attached to sturdy floats, resistant to hurricanes, which are linked to underwater lines supporting seaweed crops. These ropes also host hanging net enclosures where scallops and mussels are cultivated. Cages for clams and oysters are also attached to the surface ropes and placed on the seafloor. This type of farm is hardly noticeable from the shore (Gertz 2017). This system does not require additional freshwater, nutrients, synthetic chemicals, feed or soil for growth, as it obtains everything it requires from sunlight and sea. Another system is the closed 3D system, where production takes place within biospheres anchored to the seabed at approximately a 10-m depth. These biospheres, filled with air, function as diving bells, and multiple shelves are installed inside the dome to accommodate plant pots. This represents a polyculture vertical farming system that operates without any inputs because the sea plants filter and store carbon, rendering it a sustainable food production method. The cultivated produce includes red cabbage, lettuce, beans, basil, strawberries, assorted herbs and medicinal/cosmetic plants. Some notable benefits comprise a stable temperature within the pod, natural evaporation maintaining humidity, elevated carbon dioxide levels fostering high photosynthesis and the absence of pests or diseases, eliminating the necessity for pesticides.

14.2.2 Aquapod Farming

Globally, the consumption of fish by humans is expected to rise by 16.3%, adding an additional 25 million tonnes to a total of 180 million tonnes by the year 2029 (OECD/FAO 2020). An increasing demand for fish leads to overfishing, particularly in the marine system, which puts pressure on fish populations to recover. This situation necessitates planned aquaculture as an approach to meet

DOI: 10.1201/9781003239963-14

PHOTO 14.1 3D farming to use vertical space of water bodies.

the increasing demand for seafood. Aquapod farming is a non-conventional form or alternative to near-shore aquaculture, also known as open ocean aquaculture. It involves cultivating marine fish, invertebrates or algae in open ocean sites, situated at a distance from land, usually 1 km or more away; in deep waters, typically deeper than 20 m; and exposed to wind, waves and ocean currents without shelter from nearby landmasses or islands (Page 2013). This is a free-floating fish habitat in the deep ocean resembling a massive submerged sphere. They are large, open-ocean fish farms that float beneath the surface. Aquapods enable farmers to cultivate fish in the open ocean, away from densely populated coastlines. This sphere-shaped (geodesic domes) structure (with triangular panels made of fibreglass and reinforced plastic) is installed inside the water body in a mooring system with anchors. It uses a system that mixes feed with seawater and then pumps it in as a slurry (Butterman 2015). Under controlled conditions, it is used for farming fish, crustaceans, molluscs, aquatic plants, algae and other organisms. It is a revolutionary concept which could improve livelihoods and alleviate food shortages for small-scale farmers.

14.2.3 Cage Farming

Cage farming is a modern aquaculture form that suspends netted enclosures in an aquatic environment such as a sea or lake. This structure allows for the culture of fish and other aquatic animals. This farming method utilizes existing water bodies like lakes, reservoirs, ponds and rivers but confines the fish within a cage or enclosure. This allows water to flow freely between the fish and the environment, promoting water circulation and the removal of waste into the surrounding water. The type of fish or aquatic animals to be cultured depends on the type of water – fresh, brackish or marine. It may be used for the production of fry, fingerlings or food fish.

This farming system has advantages and disadvantages as well. It employs simple technology, making it easy to install and manage the system. It efficiently utilizes fish feeds and reduces the need for manual labour. The system enables better control over fish populations, and, in case of natural disasters, it can be relocated to a safer location. It also facilitates better disease treatment and needs less investment, as it utilizes the existing water bodies. Since it is in an enclosed system, handling and harvesting the fish are easy. Cage farming protects cultured aquatic species from outside predators. This technology can also be used to clean up eutrophicated waters by cultivating caged planktivorous species like silver carp (Soltan 2016). However, this system needs scientific feeding practices. The external water quality issues, such as low oxygen levels, must be addressed. Maintaining an optimum population is required to avoid crowding, stress and disease spread. Sometimes, diseases can be transmitted by wild fish outside the cage. Also, caged fish are debarred from getting the natural aquatic food, leading to nutritional deficiency. This system in public waters should address the nutritional enrichment issues leading to algal growth and other downstream impacts.

14.2.4 Seaweed Farming

Seaweeds are macroalgae with multiple human uses; therefore, they are called sea vegetables and have nutritional (iodine, tyrosine, vitamins, minerals and antioxidants) and health benefits. Seaweeds are different from seagrasses as the latter are vascular plants with leaves, stems and roots, whereas the former are multi-cellular algae, generally lacking vascular tissues. The principal genera of seaweeds are *Eucheuma, Gracilaria, Kappaphycus, Porphyra, Saccharina* and *Undaria*. There are also toxic and harmful seaweeds which cause severe illness and death, particularly *Gracilaria, Caulerpa* and *Acanthophora* (Cheney 2016). However, there have been no reported cases of poisonings caused by the most commonly consumed and commercially valuable seaweeds, such as *Porphyra, Laminaria* and *Undaria* (Cheney 2016). Wild harvest of seaweeds is an age-old practice, and its scientific farming has recently evolved. Seaweeds have been utilized as animal feed, fertilizers, construction materials, medicinal substances and industrial resources (Delaney et al. 2016). Seaweeds are used in cosmetic industries, have pharmaceutical/medicinal use, animal feed additives and fertilizers, and they are considered safe and natural (Kılınç et al. 2013; Polat et al. 2021). Various abiotic factors (such as water movement, light, temperature, salinity and dessication) and biotic factors (life stages and age groups) control the uptake of nutrients (nitrogen, phosphorus, carbon) by seaweeds (Roleda and Hurd 2019). The fundamental concept of floating agriculture technology could be applied to seaweed farming; however, the yield of seaweeds could be enhanced by introducing artificial nutrients through chemical fertilization of surface water near seaweed raft farms (Notoya 2010).

Blue carbon (BC) refers to the absorption of CO_2 by seaweed beds, which can serve as an effective method for capturing and sequestering carbon as a sink (Sultana et al. 2023). Seaweeds also have the potential for bioremediation of the aquaculture environment, salinity control and eutrophication (Tanaka et al. 2020). Seaweeds (*Halimeda microloba*) as ameliorants for salt-affected (sodic) soils have been reported with a significant decline of the soil pH from 9.8 to 8.0 (Ravichandran et al. 2023). At present, the global production of seaweeds is meagre (30×10^6 fresh weight tonnes) compared to terrestrial crops, grasses and forests (16×10^{11} tonnes). However, given the ongoing trend of diminishing resources such as arable land, fertilizers and water for irrigation, seaweeds should be able to efficiently provide food, feed and other high-value products at a lower cost, as well as offer ecosystem services.

14.2.5 Vertical Farming

Constraints of land availability and climate change in agriculture have opened up the avenue of vertical farming. This concept operates in several countries: the United States, the United Kingdom, Germany, France, Netherlands, Japan, China, Singapore, South Korea, Taiwan, Thailand, Kuwait and UAE. Vertical farming offers numerous advantages, such as environmental benefits, higher production, enhanced resource utilization efficiency and better quality produce. Nonetheless, it demands significant technological investment and carries a high initial cost. It should be taken up as a business with a clear marketing strategy. In India, to popularize vertical farming, ICAR-Indian Institute of Horticultural Research (IIHR) has developed a low-cost (₹22,000 per unit) structure called Arka vertical garden structure (https://www.iihr.res.in/arka-vertical-garden-structure). This structure is used for growing vegetables (tomatoes, chilis, brinjal, beans, peas), medicinal (brahmi, pudina, pepper mint, ashwagandha, shatavari, etc.) and ornamental (marigold) plants in a limited sunlit utility area, balcony and terrace by utilizing the vertical space. The government of Kerala state is popularizing this vertical farming structure under Rashtriya Krishi Vikas Yojana 2022–2023 by giving a subsidy of ₹10,525 per unit (http://shm.kerala.gov.in/arka-vertical-garden/).

14.2.6 Integrated Multi-trophic Aquaculture (IMTA)

Integrated multi-trophic aquaculture (IMTA) involves cultivating multiple species with different trophic levels together to create a symbiotic relationship, where the waste or byproducts from one species are reused as inputs (such as fertilizer, food and energy) for another species (Bueno 2021). IMTA employs an ecological engineering approach to manage ecosystems, aiming to replicate a simplified natural trophic network. This system is applicable both for freshwater and for brackish water ecosystems. This differs from polyculture in selecting species to utilize different trophic levels for nutrition. Integration of seaweeds and aquatic animals is a common IMTA model. However, integrating three species such as fish, shellfish and seaweeds can increase farm profits, reduce environmental costs and generate environmental and social benefits (Knowler et al. 2020). Invertebrates (bivalves) and plants (seaweeds) reutilize the nutrients lost from the fish feeding and excreta of fish. This results in environmental benefits as the net loss of nutrients to the environment is significantly reduced. In India, IMTA farming involving Cobia (*Rachycentron canadum*) produced 50% higher *Kappaphycus alvarezii* (Eucheuma seaweed); seaweed rafts integrated with Cobia cages yielded 225 kg per raft compared to 150 kg per raft when not integrated (Sasikumar and Viji 2016).

14.3 ROOF TOP FARMING

This system utilizes the roof top spaces of pucca as well as *kachha* (thatched) houses. In *kachha* houses, vining vegetables such as pumpkin, gourd (ridge gourd, pointed gourd, smooth gourd, bitter gourd, ash gourd), and Indian spinach are allowed to grow over the thatched spaces. In pucca houses, the horizontal floor of the rooftop of the building is used to place containers filled with soil and manure for growing a variety of crops, including vegetables, medicinal plants, ornamental plants, etc. On average, this IFS needs 71.6 litres of irrigation water, 5.5 litres of compost and 0.53 m^2 of land per kilogram of harvested food, all while preserving local biodiversity with approximately 20 different crops per farm (Dorr et al. 2023). A hydroponics system (Photo 14.2) may also be installed along with a protective structure to grow seasonal as well as off seasonal high value crops to make the farming profitable on a commercial venture. This system serves the purpose of supplying fresh green vegetables to city dwellers as well as derive environmental benefits in preventing pollution, sequestering carbon dioxide and covering the concrete jungles with green vegetation. The current issues of land scarcity, food insecurity and climate change can be addressed to some extent, and this farming system could revive ecosystem services.

14.3.1 Roof Top Protective Farming

This is a form of controlled environmental agriculture (CEA). This system is commercial in nature with installation of controlling system of a greenhouse. Here, the crop-growing conditions of carbon dioxide, moisture, temperature, humidity and fertigation are controlled per the requirement. It needs to maximize photosynthetically active radiation. Vertical farming principles are also applied to install different layers of growing platforms to harvest the maximum crop per unit area. Nevertheless, this system should employ energy-efficient control operation methods and utilize state-of-the-art energy simulation for greenhouses (Iddio et al. 2020). It requires a significant investment in production as well as marketing. The structure needs engineering designs; production

PHOTO 14.2 Rooftop hydroponics system for cultivation of vegetables.

operations need a multi-disciplinary team with expertise in various aspects such as crop agronomy, plant nutrition, irrigation experts, insects and disease control.

14.3.2 Roof Top Gardening

This form of roof top agriculture utilizes the floor of the rooftop by placing growing beds of different sizes. The growing beds may be made of wooden, HDPE or UPVC stands. Structures could also be placed here to utilize the vertical space. Different types of vegetables, as per the food habit, are grown to meet the family requirements, and if the space is enough, surplus production may be marketed. This is a low-cost system compared to protective roof top farming. This system utilizes the harvested rainwater, reduces the emission of CO_2, protects the building from heat during summer and becomes a source of oxygen for the owner. Organic farming principles may also be applied for roof top farming to produce chemical free vegetables. Compost, vermicompost, oil cake, neem oil and cakes are some of the inputs that are essential requirements for roof top farming. If there is an issue of shading, some shade-loving plants may be placed. Many small tools, such as trowels, soil scoops, khurpi, pruners, watering canes, handheld sprayers, etc., facilitate the operations of this farming process.

14.4 FLOATING AGRICULTURE

Floating agriculture is an innovative option in flood-prone areas, where land-based agriculture becomes impossible due to the high water depth. This is also called raft farming. A raft is a structure made up of wood or any other material that can float on water. The phenomenon of buoyancy makes the raft float. Different types of materials are used to prepare the raft bed. Bamboo is the commonly used material; however, other plant/weed biomass such as banana pseudo stem and biomass of water hyacinth (*Eichhornia crassipes*) and water lettuce (*Pistia stratiotes*) can be used for preparing the floating bed. In many waterlogged areas, water hyacinth is a menace that can be effectively utilized for floating agriculture. However, cost is an essential factor for this system. The investment should be less than enough to make this system profitable. Different vegetables are grown on the soil, and a manure/compost mixed bed is prepared on the raft. Other nutrient-rich materials such as rice husk, rice husk ash, pond bottom soil and oil cake may be added to this bed. Some rafts may also have structures for climbing types of vegetables to utilize the space efficiently.

14.4.1 Float Farming for Vegetables

Short duration vegetables particularly the green leafy crops such as spinach, coriander, water spinach and amaranths may be grown. Crops such as tomatoes, cabbage and cauliflower are suitable during winter. If there is provision for vertical spread, crops such as bitter gourd, cucumber and pointed gourd can be grown. If there is a chance that the floodwaters recede soon, seedlings for vegetable crops such as brinjal, chili, onion and tomato can be grown for use in the land after the recession of the flood. The weed biomass balls are prepared to insert the germinated seeds of vegetable crops and arranged over the floating bed for germination and growth.

14.4.2 Floating Aquaculture

This is an integrated system in which a floating house is made over the water body. The house is used for habitation as well as for growing vegetables, poultry birds, ducks, etc. The underground of the house has a floating enclosure for growing fish. There is provision for feeding the fish from an outlet on the floating bed. This system effectively utilizes the natural water body for fish and other aquatic/agricultural production. Pond-based floating agriculture can adopt an innovative approach known as the integrated floating cage aquageoponics system (IFCAS), allowing for the simultaneous production of fish and vegetables (Haque et al. 2015). The underground cage space is used to culture the GIFT tilapia strain (*Orechromis niloticus*) @ 100 numbers per m^{-3} space. In the rest area of the pond, *Catla catla, Labeo rohita, Cyprinus carpio* and *Cirrhinus cirrhosus* are cultured @ 14,820 ha^{-1} in the ratio of 1:2:2:1. Tilapia are fed floating feed while the carp are given supplementary feed. Vegetables are cultivated on the IFCAS scaffold, and the tilapia are raised in the net cage built underneath. On the sides of the floating bed, dried pond mud and manure mixed medium are used for growing vegetables [Indian spinach (*Basella alba/rubra*), tomato (*Solanum lycopersicum*), beans (*Phaseolus vulgaris*), etc.]. The vertical space above the water body is used for growing climbing types of vegetables [cucumber (*Cucumis sativus*), snake gourd (*Trichosanthes cucumerina*), bottle gourd (*Lagenaria siceraria*), bitter gourd (*Momordica charantia*), etc.]. The space is created by constructing a scaffold on the top of the structure using locally available materials such as bamboo and net. This system may be adopted for the shaded ponds, where growing vegetables on the pond dyke is not possible.

14.5 AQUAPONICS AND HYDROPONICS

Aquaponics refers to a symbiotic integrated farming method that combines conventional aquaculture (raising aquatic animals like fish, crabs, snails, crayfish or prawns in tanks) with hydroponics (growing plants in a water-based nutrient solution). It is also a form of soil-less farming. Plants, as well as aquatic animals, get nutrients from the water. This system can also combine the concepts of protected cultivation, vertical farming and aeroponics. However, this farming system needs access to good quality water, electricity and a market. The sophisticated system may use the polyhouse or greenhouse concepts. These systems have temperature and humidity adjustments as per crop requirements.

14.6 POLICY INITIATIVES AND INSTITUTIONAL MECHANISM FOR PROMOTING IFS

Integrated farming systems (IFS) is a method that integrates various agricultural activities and practices to develop a more sustainable and efficient farming system. Governments often recognize the potential benefits of IFS and may implement various efforts, schemes, projects, policy initiatives and institutional mechanisms to support and promote IFS. The specific initiatives and mechanisms can vary depending on the country's agricultural makeup, the environmental challenges and the level of commitment to sustainable farming practices. Some of these governmental schemes may fall into Research and Development (R&D) Funding, Training and Extension Services, Subsidies and Financial Incentives, Demonstration Farms, Market Support, Policy Reforms, Facilitating Access to Resources, Climate Resilience and Adaptation, Promotion of Collaboration and Partnerships, Monitoring and Evaluation, Certification and Standards and Promotion of Farmer Cooperatives and Organizations. Governments may allocate funds to support research and development activities related to integrated farming systems. This funding can be used to study the feasibility, benefits and challenges of implementing IFS practices. Government agencies may offer training programs and extension services to farmers, imparting knowledge about the principles and techniques of integrated farming systems. These programmes can enhance farmers' knowledge and skills, encouraging them to adopt IFS practices. Governments may also provide financial incentives and subsidies to farmers who adopt integrated farming systems. These incentives can include grants, low-interest loans or tax breaks, making it more attractive for farmers to transition to IFS. Establishing demonstration farms that showcase successful IFS practices can raise awareness and inspire other farmers to adopt similar approaches. Governments can help promote the marketing and sale of produce from IFS farms by creating market linkages, improving market infrastructure and supporting value-addition initiatives. Policymakers may review existing agricultural policies and regulations to ensure they are conducive to IFS adoption. They may modify policies to encourage sustainable practices and address any barriers that hinder the transition to IFS. Governments may also strive to guarantee that farmers have access to essential resources like water, land, seeds and organic inputs, all crucial for effectively adopting IFS. They may integrate IFS practices into climate resilience and adaptation strategies, recognizing the potential of IFS to mitigate the effect of climate change on agriculture. Additionally, governments can foster partnerships between research institutions, agricultural organizations, private sector stakeholders and NGOs to promote and support IFS initiatives collectively. Developing certification programmes and standards for integrated farming systems can help ensure that the adoption of IFS practices meets sustainability and environmental stewardship criteria. Governments may establish mechanisms to monitor the progress and impact of IFS adoption, helping to identify successful practices and areas for improvement. Encouraging the formation of farmer cooperatives can facilitate knowledge sharing, collective marketing and resource pooling among farmers practising IFS.

14.6.1 Doubling Farmers Income

Keeping in view the economic situation of the farming community, the unattractiveness of farming as a profession for the new generation due to low income and involving risk and uncertainties, the Government of India (O.M. No. 5-3/2016-FW dated 13 April 2016) formed a committee for giving recommendations on doubling farmers income (DFI). The committee led by Dr Ashok Dalwai gave a detailed report on this, comprising 14 volumes in September 2018 for carrying out an income revolution of Indian farmers (https://agricoop.nic.in/en/Doubling). Farmers encounter vulnerabilities of both risks and uncertainties in the farming environment, along with the unpredictability of market forces (Report of the Committee on DFI Vol. I, 2017). The major strategic approach to be adopted for DFI may be a combination both by the government side and by the farming community. The farmers may adopt measures for reduction of the cost of cultivation, improved resource

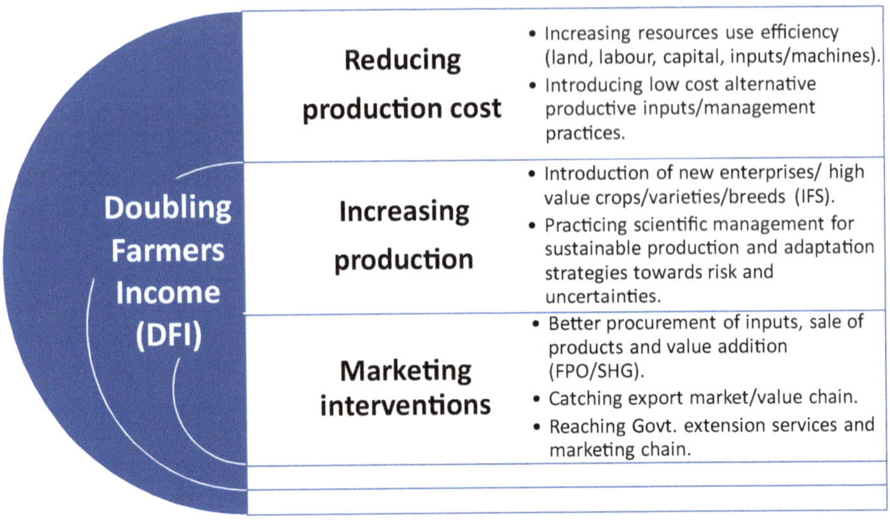

FIGURE 14.1 Major strategies to be adopted by the farming community for doubling income.

use efficiency, increasing and sustaining production through science-led approaches, better price realization for farm produce through marketing interventions and getting connected to government extension services (Figure 14.1). Government should implement policies and increase investment for better infrastructure in the areas of research, input availability, marketing, post-harvest facilities and risk reduction.

Farmers should focus on increasing cropping intensity, diversifying towards high-value crops, and boosting crop, livestock and fish productivity (Singh 2020). The consumption patterns and expected demand for major food commodities in India indicate that priority should be given to producing fruits, milk, fish, edible oils, vegetables, eggs and pulses (Kumar et al. 2016). This needs strategies for diversification towards high-value produce from horticulture, livestock and fisheries with facilities provided by integrated cold-chain and agri-logistics systems (Report of the Committee on DFI Vol. II 2017). Horticulture and produce from fisheries, dairy and livestock constitute about 70% of agriculture's contribution to the national gross domestic product. Therefore, these enterprises are the prime drivers for DFI (Report of the Committee on DFI Vol. III, 2017). Promotion of contract farming for crops (e.g., barley, basmati rice, cotton, maize, soybean, fruits and vegetables), poultry (e.g., eggs, broilers) and food processing can integrate small and marginal farmers into organized market channels (Report of the Committee on DFI Vol. IV 2017). Contract farming has begun in Punjab for malting barley; in Haryana for barley, basmati rice and potato; in Chhattisgarh for bitter gourd, beans, green chili, tomato, ladyfinger, bottle gourd and cucumber; in Gujarat for cotton; in Madhya Pradesh for cotton; in Maharashtra for sponge gourd, ridge gourd, ladyfinger, chili, tomato, white onion, potato, baby corn, soybean, beans, pomegranate and sweet corn; and in Karnataka for organic cotton.

To attain DFI, India needs to boost the annual usage of quality seeds, fertilizers and agricultural power supply by 12.8%, 4.4% and 7.6%, respectively. In the case of livestock, enhancing herd quality, providing better feed, increasing artificial insemination, reducing calving intervals and lowering the age at first calving are potential avenues for growth. Research organizations should develop technological innovations to enhance production capabilities and resource utilization efficiency. Improved agronomic practices for precision farming, use of laser land levelling, precision seeder/planter and various technologies such as a system of rice intensification (SRI), zero tillage (ZT) planting, direct seeded rice (DSR), and raised bed planting need to be upscaled (Sarangi et al. 2020a, 2020b). IFS models are to be developed and standardized for various bio-physical and socioeconomic settings with a combination of compatible technologies. The models should be in the form of a package with a major focus on increasing net farm income, sustainability and environmental benefits. The package involves technologies for production, protection, harvesting, post-harvesting and value addition with an interdisciplinary approach.

Farmers can augment their income through secondary agricultural activities. Broadly, secondary agriculture is classified into three types: Type A encompasses adding value to primary agricultural production systems, Type B involves alternative enterprises, and Type C comprises enterprises dependent on residues and waste from primary agriculture (Report of the Committee on DFI Vol. IX 2018). Type A secondary agriculture is of two types: (1) input-centric (nursery for horticultural crops, forest species, field crops, flowers, production of vermicompost, bio-fertilizers, biopesticides, animal/fish feed, soil and water testing, etc.) and (2) harvest/post-harvest centric (grading, pre-conditioning/pre-cooling, packaging of fruits, vegetables, flowers, spices, micro and small enterprises to process primary produce, e.g., garlic extracts/paste, turmeric powder, pickles, jam, custom hiring centres, agri-tourism, etc.) activities. Some type B secondary agricultural activities are beekeeping, mushroom cultivation, lac cultivation, sericulture, etc. Type C secondary agriculture activities may include using rice straw to produce fibre boards, urea-enriched fodder blocks, cotton stalk products, etc. Micro-enterprises may be promoted to augment women farmers' income, such as making pickles, juice, jam, jelly and papad, which can generate income by employing fewer than ten persons, including the owner and family members.

Marketing the produce is a real challenge for farmers in the present scenario. Promotion of farmers producer organizations (FPO), village produces organizations (VPO), farmers producer companies (FPC) and self-help groups (SGH) for efficient production, procurement and marketing is essential for achieving DFI of small-holder farmers. In this regard, attracting adequate private/public investment in agriculture, modernizing infrastructure, and placing emphasis on the energy and irrigation sector through strong coordination by government agencies are required. The committee on DFI recommends the implementation of the Model Agricultural Produce and Livestock Marketing (Promotion and Facilitation) Act, 2017, named APLM Act, 2017, which removes entry and trade barriers and promotes a unified market for agricultural produce (Report of the Committee on DFI Vol. IV 2017).

14.6.2 Policy Support, Schemes and Institutional Mechanism

Government efforts, schemes and projects to support IFS vary across countries and regions. Here are some examples of policy initiatives and institutional mechanisms worldwide and in India. World Agroforestry Centre (ICRAF) and the International Fund for Agricultural Development (IFAD) promote agroforestry systems and IFS globally through research, capacity building and knowledge sharing. They provide technical support to governments and farmers to adopt IFS practices and develop policies that support agroforestry and sustainable agriculture. ICRAF and IFAD have been actively involved in promoting IFS globally. Both organizations recognize the importance of IFS in achieving sustainable agricultural development, improving rural livelihoods and conserving natural resources. ICRAF is a global research institution focusing on agroforestry and sustainable land-use systems. It plays a crucial role in promoting IFS through research, capacity building and knowledge sharing. Some of ICRAF's key activities include:

a. Research and Development: ICRAF conducts research on agroforestry practices and their integration into farming systems. This research generates evidence of the benefits of IFS, such as improved soil fertility, enhanced biodiversity, climate change resilience and increased farm incomes.
b. Technical Support and Training: ICRAF offers technical assistance and training to farmers, extension workers and policymakers in implementing agroforestry and integrated

farming systems (IFS) practices. This includes guidance on selecting appropriate tree species, designing farm layouts and integrating livestock, crops and trees for optimal benefits.
c. Policy Advocacy: ICRAF actively engages with policymakers, advocating for policies that support agroforestry and IFS. ICRAF helps shape policies that promote their adoption and integration into national agricultural strategies by emphasizing the social, economic and environmental benefits of these systems.

IFAD, a United Nations specialized agency, concentrates on rural development and alleviating poverty. It recognizes the role of IFS in improving agricultural productivity, enhancing food security and reducing poverty in rural areas. IFAD's efforts to promote IFS include:

a. *Financial support*: IFAD provides financial support to projects and programs that promote sustainable agriculture, including IFS. IFAD helps farmers and rural communities adopt and implement IFS practices by providing loans, grants and technical assistance.
b. *Knowledge sharing*: IFAD aids in the establishment of knowledge-sharing and learning platforms to facilitate the exchange of experiences, best practices and lessons acquired in the implementation of integrated farming systems (IFS). This includes organizing workshops, conferences, and networks where farmers, experts and policymakers can share their knowledge and experiences.
c. *Capacity building*: IFAD invests in capacity-building activities to promote the skill and knowledge development of farmers, extension workers and other stakeholders engaged in promoting integrated farming systems (IFS). This includes training programs on agroforestry, sustainable land management and integrated approaches to farming.

Through their combined efforts, ICRAF and IFAD contribute to promoting and adopting IFS practices globally. By providing research, technical support, financial assistance and policy advocacy, these organizations play a major role in advancing sustainable agriculture, conserving biodiversity and improving livelihoods in rural communities.

The European Union's Common Agricultural Policy (CAP) promotes sustainable agriculture practices, including IFS, through financial incentives, subsidies and rural development programmes. It encourages farmers to adopt agro-ecological approaches, diversify their production systems and promote biodiversity conservation. The United States Department of Agriculture (USDA) provides technical and financial assistance to farmers who adopt sustainable agricultural practices, including IFS. Initiatives like the Conservation Stewardship Program (CSP) and Environmental Quality Incentives Program (EQIP) provide financial aid to farmers adopting conservation practices, including agroforestry and integrated farming systems.

The Government of India is promoting IFS directly/indirectly through more than 20 schemes covering agriculture, horticulture, livestock, fisheries, insurance, etc. (Balkrishna et al. 2020). Government schemes in India introduced after 2014 to promote sustainable agriculture include supporting climate smart farming, building soil health, cover insurance and reducing the impact of disasters on agriculture (Khan et al. 2022). India's Sustainable Agriculture Mission promotes adopting sustainable agricultural practices, including IFS. It emphasizes the integration of crops, livestock, horticulture and agroforestry to enhance productivity, conserve natural resources and improve rural livelihoods.

The National Mission for Sustainable Agriculture (NMSA) is among the eight missions outlined in the National Action Plan on Climate Change (NAPCC), which was developed in the fiscal year 2014–2015. The introduction of NMSA aimed to boost agricultural yield, especially in rain-fed regions, by prioritizing Integrated Farming Systems (IFS) to enhance productivity and mitigate risks linked to climate variability (Report of the Committee on DFI Vol. V 2017). The main highlights of this programme include improving livelihood opportunities and food security while reducing crop failures. This scheme encourages and supports both on-farm and off-farm conservation techniques, efficient management of water resources, promotion of integrated farming systems (IFS) and improved agronomic practices such as enhancing water retention, better soil management, judicious use of chemicals and energy and increased soil carbon storage. It also involves creating a database on soil resources through land use surveys, soil profile studies and soil analysis on GIS platforms, as well as crop-specific integrated nutrient management (INM) practices. Additionally, it focuses on developing strategies for climate change adaptation and mitigation by involving knowledge institutions and professionals, disseminating and adopting rain-fed technologies in disadvantaged areas and preparing location-specific plans. This scheme also promotes coordination with other schemes/missions such as MGNREGS, IWMP, RKVY, NFSM, MIDH and NMAE&T, etc. NMSA prioritizes improving water usage efficiency, integrated farming, soil health management and coordinating conservation efforts, particularly in rain-fed regions. It consists of four main components: Rain-fed Area Development (RAD), On Farm Water Management (OFWM), Soil Health Management (SHM) and Climate Change and Sustainable Agriculture: Monitoring, Modelling, and Networking (CCSAMMN) (NMSA Online).

14.6.2.1 Sub-mission on Agroforestry (SMAF)

This is a sub-mission operating within the framework of NMSA. Its key features include expanding tree plantation on arable land using trees suitable to local agro-climatic and land use conditions, promoting the establishment of new small and high-tech large nurseries and encouraging various agroforestry practices/models such as sustainable agri-silvicultural systems, silvi-pastoral systems, agri-silvopastoral systems and other forms like apiculture with trees and aqua-forestry. It also promotes peripheral and boundary plantation, compiling a database on the area under agroforestry and cultivating endemic and medicinal plants. Agroforestry development, climate change mitigation, providing livelihoods to the rural population, increasing income levels, and biodiversity conservation are the key features of this programme (https://vikaspedia.in).

14.6.2.2 National Agroforestry Policy, India

India has a National Agroforestry Policy that aims to promote agroforestry systems, including IFS, across the country. The policy focuses on increasing tree cover on agricultural lands, promoting agroforestry-based livelihoods and conserving biodiversity.

14.6.2.3 Rashtriya Krishi Vikas Yojana (RKVY), India

RKVY is a centrally sponsored scheme in India that aims to promote the holistic development of agriculture and allied sectors. It plays a crucial role in supporting the promotion of IFS in the country. It also provides financial assistance to states and union territories for implementing projects and activities related to agricultural development, including IFS. This financial support helps farmers adopt and integrate diverse farming practices and technologies, such as agroforestry, crop-livestock integration and organic farming. Moreover, it supports capacity-building efforts and training programmes designed to improve the knowledge and skills of farmers, extension workers and other stakeholders in IFS practices. This involves conducting workshops, seminars and demonstrations to raise awareness and promote understanding of integrated farming approaches. The RKVY provides funds to develop essential infrastructure and facilities to support the implementation of IFS. This can involve activities such as establishing demonstration farms, creating seed banks and fodder banks and building farm ponds or water harvesting structures to improve water availability for IFS. It facilitates the dissemination of appropriate technologies and best practices related to IFS, which includes the promotion of improved crop varieties, livestock breeds and agroforestry models suitable for different agro-climatic regions. The scheme encourages the adoption of sustainable practices and technologies that improve farm productivity while conserving natural resources. It promotes research and innovation in the field of integrated farming systems, which includes funding research projects focused on developing and validating IFS models, assessing their economic viability and studying their impact on productivity, income generation and environmental sustainability. The RKVY incorporates monitoring and evaluation mechanisms to evaluate the progress and impact of promoting IFS. Regular monitoring helps in identifying successful interventions, addressing implementation challenges and making necessary modifications to enhance the effectiveness of IFS initiatives.

The Indian government's objective through the RKVY is to encourage the adoption of integrated farming systems as a sustainable agricultural approach. Through offering financial aid, capacity building, infrastructure enhancement, technology distribution and research backing, it significantly encourages farmers to adopt and implement integrated farming practices in various states in India.

14.6.2.4 State-Level Initiatives

Many states in India have their own schemes and programs to support IFS and sustainable agriculture. Here are some examples of state-specific schemes that promote IFS in different states in India:

14.6.2.4.1 Odisha

Millet Mission: The Odisha Millet Mission promotes millet cultivation and integrated farming practices. It promotes the integration of crops, livestock and agroforestry to boost soil health and biodiversity in offering sustainable livelihoods to farmers.

14.6.2.4.2 Kerala

Kerala Agriculture Value Chain Development Programme promotes sustainable agriculture practices, including IFS. It supports the adoption of organic farming, agroforestry and livestock integration to enhance productivity, conserve natural resources and improve farmer incomes.

14.6.2.4.3 Tamil Nadu

Sustainable Agriculture Mission: The Sustainable Agriculture Mission in Tamil Nadu promotes sustainable agricultural practices, including IFS. It promotes organic farming, agroforestry and livestock integration to enhance farm productivity, soil fertility and environmental sustainability.

14.6.2.4.4 Himachal Pradesh

Himachal Pradesh Horticulture Development Project: This project supports the adoption of integrated farming practices, including horticulture-based IFS. It encourages the cultivation of high-value horticultural crops alongside livestock rearing, agroforestry and water management methods, aiming to enhance rural livelihoods and preserve natural resources.

14.6.2.4.5 Punjab

Punjab State Farmers' and Farm Workers' Commission (PSFWC): The PSFWC encourages the adoption of IFS practices in Punjab. It offers farmers technical guidance, training and financial aid to adopt practices like crop diversification, agroforestry and integrated livestock management.

14.6.2.4.6 Madhya Pradesh

Mukhya Mantri Krishi Vahini Yojana: This scheme in Madhya Pradesh promotes IFS by offering financial aid for purchasing machinery and equipment required for integrated farming practices. It aims to enhance farm productivity, income generation and resource conservation through the adoption of integrated farming systems.

These are a few examples of state-specific schemes that promote integrated farming systems in different states in India. These schemes provide farmers financial assistance, technical support, capacity building and infrastructure development to encourage sustainable agricultural practices and enhance farm productivity while conserving natural resources. It is important to recognize that certain initiatives and programmes can change, so consulting the most recent government policies and programmes is recommended for the most accurate information.

14.6.2.5 National Agriculture Policy (NAP) (2000)

The reduction in the agriculture sector's contribution to the GDP has a significant impact on the majority of farmers, particularly those in the marginal and small farmer categories. To resolve these problems of livelihood and to improve the agricultural sector of India, the first National Agricultural Policy (NAP) was framed on 28 July 2000 with the objective of technologically, environmentally and economically sustainable agricultural growth. The main features and focus area of this policy are the macro-management approach; promotion of sustainable agriculture by efficient usage of natural resources like soil, water, environment and biodiversity, etc.; providing sufficient and timely inputs to farmers; women empowerment; promoting private sector participation in agriculture and tenancy reforms.

14.6.2.6 National Policy for Farmers (2007)

In 2006, the National Commission on Farmers (NCF) submitted the final report on the National Policy for Farmers (NPF) to the Government of India (GOI) based on a draft report compiled by NCF in consultation with state governments. Based on Prof. M. S. Swaminathan's, chairman of NCFs report, the government accepted the National Policy of Farmers in 2007. NPF-2007 aimed to enhance the economic sustainability of farming and boost the net income of farmers. The policy objectives include asset reforms concerning water, land, livestock, fisheries and biodiversity; providing appropriate price and trade policy mechanisms; health and trade security of farmer families; social security system for farmers; supply of disease-free planting material and quality seeds; mainstreaming the human and gender dimension; support services for women; proper use of information and communication technology; revamping the agricultural curriculum and teaching methods to empower every graduate in farm and home science to become an entrepreneur and to ensure that agricultural education is gender-inclusive; development of agriculture market infrastructure; effective execution of the minimum support price (MSP); establishing community food grain banks; implementation of timely and efficiently reachable institutional credit; and offering suitable avenues for non-farm employment opportunities for rural households.

14.6.2.7 Remunerative Approach for Agriculture and Allied Sector Rejuvenation (RKVY-RAFTAAR)

In 2007, the Government of India launched this programme. Previously, it was known as Rashtriya Krishi Vikas Yojana (RKVY). Until 2013–2014, the scheme was carried out as an additional central support to the state plan scheme, fully funded by the central government. However, starting in 2014–2015, it was converted into a centrally sponsored scheme with full central assistance. Since 2015–2016, the funding ratio has been changed to 60:40 between the central and state governments. However, in the northeastern states and Himalayan states, the ratio is 90:10, while for Union Territories, the funding pattern is 100% central grant. This scheme aims to bring together agriculture and related sectors like crop and horticulture management, livestock rearing, dairy production and fisheries activities along with agricultural research and education, marketing, food storage, soil and water conservation, financial services for agriculture, other related programmes and cooperation endeavours. This scheme also encourages convergence with other programmes. In this yojana, the formulation of district and state agriculture plans is obligatory. Based on feedback during the implementation of RKVY in the 12th five year plan, RKVY guidelines have been refurbished as RKVY-RAFTAAR to enhance the effectiveness, efficiency and inclusiveness of the programme for the outstanding period of the 14th finance commission.

14.6.2.8 National Food Security Mission (NFSM)

In 2007, the National Development Council (NDC) introduced the National Food Security Mission (NFSM) with the aim of boosting rice production by 10 million tons, pulses by 2 million tons and wheat by 8 million tons by the conclusion of the eleventh plan. The primary goal of this mission is to elevate cereal crop production, revive soil fertility and productivity at the individual farm level and improve the economic status of farms.

REFERENCES

Balkrishna, A., Mittal, R., Kumar, A., Singh, D., Arya, V. 2020. Analysis of policy interventions in agriculture and ICT based mechanistic approach towards sustainability: An Indian perspective. *Asian Journal of Agriculture and Rural Development* 10(1):194. DOI: https://doi.org/10.18488/journal.1005/2020.10.1/1005.1.194.213.

Bueno, P.B. 2021. Widening the horizon of Asian mariculture with IMTA. *Journal of the Indian Society of Coastal Agricultural Research* 39(2):239–248. DOI: https://doi.org/10.54894/JISCAR.39.2.2021.110550.

Buschmann, A.H., Camus, C., Infante, J., Neori, A., Israel, Á., Hernández-González, M.C., Pereda, S.V., Gomez-Pinchetti, J.L., Golberg, A., Tadmor-Shalev, N., Critchley, A.T. 2017. Seaweed production: Overview of the global state of exploitation, farming and emerging research activity. *European Journal of Phycology* 52(4):391–406. DOI: http://dx.doi.org/10.1080/09670262.2017.1365175.

Butterman, E. 2015. *A Fishy Engineering Innovation*. Available online: https://www.asme.org/topics-resources/content/a-fishy-engineering-innovation. Accessed on 28 June 2023.

Cheney, D. 2016. Toxic and harmful seaweeds. In: Fleurence J., Levine I. (eds) *Seaweed in Health and Disease Prevention*. Academic Press. pp 407–421. ISBN 9780128027721, DOI: https://doi.org/10.1016/B978-0-12-802772-1.00013-0.

Delaney, A., Frangoudes, K., Li, S.A. 2016. Society and seaweed: Understanding the past and present. In: Fleurence J., Levine I. (eds) *Seaweed in Health and Disease Prevention*. Elsevier Academic Press, London. pp 7–40. DOI: https://doi.org/10.1016/B978-0-12-802772-1.00002-6.

Dorr, E., Hawes, J.K., Goldstein, B., Fargue-Lelièvre, A., Fox-Kämper, R., Specht, K., Fedeńczak, K., Caputo, S., Cohen, N., Poniży, L., Schoen, V., Górecki, T., Newell, J.P., Jean-Soro, L., Grard, B. 2023. Food production and resource use of urban farms and gardens: A five-country study. *Agronomy for Sustainable Development* 43(1):18. DOI: https://doi.org/10.1007/s13593-022-00859-4.

Gertz, E. 2017. *Vertical Ocean Farms that Can Feed Us and Help Our Seas*. Available online: https://ideas.ted.com/vertical-ocean-farms-that-can-feed-us-and-help-our-seas/. Accessed on 28 June 2023.

Haque, M.M., Alam, M.R., Alam, M.M., Basak, B., Sumi, K.R., Belton, B., Murshed-E- Jahan, K. 2015. Integrated floating cage aquageoponics system (IFCAS): An innovation in fish and vegetable production for shaded ponds in Bangladesh. *Aquaculture Reports* 2:1–9. DOI: https://doi.org/10.1016/j.aqrep.2015.04.002.

https://agricoop.nic.in/en/Doubling. Accessed on 21 July 2023.

https://vikaspedia.in/agriculture/policies-and-schemes/crops-related/krishi-unnati-yojana/national-mission-for-sustainable-agriculture-1/sub-mission-on-agroforestry-smaf. Accessed on 13 November 2024

Iddio, E., Wang, L., Thomas, Y., McMorrow, G., Denzer, A. (2020). Energy efficient operation and modeling for greenhouses: A literature review. *Renewable and Sustainable Energy Reviews* 117:109480. DOI: https://doi.org/10.1016/j.rser.2019.109480.

Khan, I.Y., Goswami, M., Nautiyal, S., Gupta, A.K., Premkumar, A., Baidya, S., Prakash, S. (2022). Promoting climate smart agriculture through policies and schemes in india: Temporal and strategic overview. *International Journal of Ecology and Environmental Sciences* 48(5):547–569.

Kılınç, B., Cirik, S., Turan, G., Tekogul, H., Koru, E. 2013. Seaweeds for food and industrial applications. In: Muzzalupo I. (ed) *Food Industry, 1st Edition*. IntechOpen. pp 735–748. DOI: http://dx.doi.org/10.5772/53172.

Knowler, D., Chopin, T., Martínez-Espiñeira, R., Neori, A., Nobre, A., Noce, A., Reid, G. 2020. The economics of Integrated Multi-Trophic Aquaculture: Where are we now and where do we need to go? *Reviews in Aquaculture* 12(3):1579–1594. DOI: https://doi.org/10.1111/raq.12399.

Kumar, P., Joshi, P.K., Mittal, S. 2016. Demand vs supply of food in India – Futuristic projection. *Proceedings of the Indian National Science Academy* 82(5):1579–1586. DOI: http://dx.doi.org/10.16943/ptinsa/2016/48889.

National Oceanic and Atmospheric Administration (NOAA). 2018. *Story Map: Farming in Water*. Available online: http://www.noaa.gov/stories/story-map-farming-in-water. Accessed on 28 June 2023.

Notoya, M. 2010. Production of biofuel by macroalgae with preservation of marine resources and environment. In: Seckbach J., Einav R., Israel A. (eds) *Seaweeds and their Role in Globally Changing Environments*. Springer, Dordrecht. pp 217–228. DOI: https://doi.org/10.1007/978-90-481-8569-6_13.

OECD-FAO Agricultural Outlook, OECD Agriculture statistics (database). 2020. DOI: http://dx.doi.org/10.1787/agr-outl-data-en.

Page, S.H. 2013. Aquapod systems for sustainable ocean aquaculture. In: Christou P., Savin R., Costa-Pierce B.A., Misztal I., Whitelaw C.B.A. (eds) *Sustainable Food Production*. Springer, New York, NY. pp 223–235. DOI: https://doi.org/10.1007/978-1-4614-5797-8_187.

Polat, S., Trif, M., Rusu, A., Šimat, V., Čagalj, M., Alak, G., Meral, R., Özogul, Y., Polat, A., Özogul, F. 2021. Recent advances in industrial applications of seaweeds. *Critical Reviews in Food Science and Nutrition* 63(21):4979–5008. DOI: https://doi.org/10.1080/10408398.2021.2010646.

Ravichandran, M., Chinnadurai, S., Subha, B., Muthulaxmi, V., Sivakumar, S.R., Subbiah, S., Dineshkumar, R. 2023. Ameliorated reclamation potential of *Halimeda microloba* on sodic soil and its impact on the *Vigna radiata*. *Arabian Journal of Geosciences* 16:404. DOI: https://doi.org/10.1007/s12517-023-11509-8.

Report of the Committee on Doubling Farmers' Income Volume I. 2017. *March of Agriculture since Independence and Growth Trends*. Department of Agriculture, Cooperation and Farmers Welfare, Ministry of Agriculture and Farmers Welfare, Government of India. p 119. Available online: https://agricoop.nic.in/en/Doubling. Accessed on 21 July 2023.

Report of the Committee on Doubling Farmers' Income Volume II. 2017. *Status of Farmers' Income; Strategies for Accelerated Growth*. Department of Agriculture, Cooperation and Farmers Welfare, Ministry of Agriculture and Farmers Welfare, Government of India. p 191. Available online: https://agricoop.nic.in/en/Doubling. Accessed on 21 July 2023.

Report of the Committee on Doubling Farmers' Income Volume III. 2017. *Post Production Agri-Logistics: Maximizing Gains for Farmers*. Department of Agriculture, Cooperation and Farmers Welfare, Ministry of Agriculture and Farmers Welfare, Government of India. p 176. Available online: https://agricoop.nic.in/en/Doubling. Accessed on 21 July 2023.

Report of the Committee on Doubling Farmers' Income Volume IV. 2017. *Post Production Interventions: Agricultural Marketing*. Department of Agriculture, Cooperation and Farmers Welfare, Ministry of Agriculture and Farmers Welfare, Government of India. p 260. Available online: https://agricoop.nic.in/en/Doubling. Accessed on 21 July 2023.

Report of the Committee on Doubling Farmers' Income Volume V. 2017. *Sustainability Concerns in Agriculture*. Department of Agriculture, Cooperation and Farmers Welfare, Ministry of Agriculture and Farmers Welfare, Government of India. p 106. Available online: https://agricoop.nic.in/en/Doubling. Accessed on 21 July 2023.

Report of the Committee on Doubling Farmers' Income Volume IX. 2018. *Farm Linked Activities and Secondary Agriculture*. Department of Agriculture, Cooperation and Farmers Welfare, Ministry of Agriculture and Farmers Welfare, Government of India. p 154. Available online: https://agricoop.nic.in/en/Doubling. Accessed on 21 July 2023.

Roleda, M.Y., Hurd, C.L. 2019. Seaweed nutrient physiology: Application of concepts to aquaculture and bio-remediation. *Phycologia* 58(5):552–562. DOI: https://doi.org/10.1080/00318884.2019.1622920.

Sarangi, S.K., Maji, B., Sharma, P.C., Digar, S., Mahanta, K.K., Burman, D., Mandal, U.K., Mandal, S., Mainuddin, M. 2020b. Potato (*Solanum tuberosum* L.) cultivation by zero tillage and paddy straw mulching in the saline soils of the Ganges Delta. Potato Research 64(2):277–305. https://link.springer.com/article/10.1007%2Fs11540-020-09478-6.

Sarangi, S.K., Singh, S., Kumar, V., Srivastava, A.K., Sharma, P.C., Johnson, D.E. (2020a). Tillage and crop establishment options for enhancing the productivity, profitability, and resource use efficiency of rice-rabi systems of the salt-affected coastal lowlands of eastern India. *Field Crops Research* 247:1–15. DOI: http://doi.org/10.1016/j.fcr.2019.03.016.

Sasikumar, G., Viji, C.S. 2016. Integrated multitrophic aquaculture systems. In: Joseph I., Ignatius B. (eds) *Course Manual Winter School of Technological Advances in Mariculture for Production Enhancement and Sustainability*. ICAR-Central Marine Fisheries Research Institute (CMFRI), Kochi, Kerala, India. pp 47–55. Available online: http://eprints.cmfri.org.in/10600/1/PDF-Manual.pdf. Accessed on 04 July 2023.

Singh, S. 2020. Doubling farmers' incomes: A critical review of policy. *Review of Agrarian Studies* 10(2):138–146.

Soltan, M. 2016. *Cage Culture of Freshwater Fish*. Available online: https://www.researchgate.net/publication/294087263_Cage_culture_of_freshwater_fish?channel=doi&linkId=56bdc8e308aedb6510d18f81&showFulltext=true. https://doi.org/10.13140/RG.2.1.4802.2803. Accessed 29 June 2023.

Sultana, F., Wahab, M.A., Nahiduzzaman, M., Mohiuddin, M., Iqbal, M.Z., Shakil, A., Mamun, A., Khan M.S.R., Wong L., Asaduzzaman M. 2023. Seaweed farming for food and nutritional security, climate change mitigation and adaptation, and women empowerment: A review. *Aquaculture and Fisheries* 8(5):463–480. DOI: https://doi.org/10.1016/j.aaf.2022.09.001.

Tanaka, Y., Ashaari, A., Mohamad, F.S., Lamit, N. 2020. Bioremediation potential of tropical seaweeds in aquaculture: Low-salinity tolerance, phosphorus content, and production of UV-absorbing compounds. *Aquaculture* 518:734853. DOI: https://doi.org/10.1016/j.aquaculture.2019.734853.

15 Policy Framework Requirements for the Success of IFS

15.1 INTRODUCTION

The success and sustainable growth of developmental activities hinge on establishing suitable policies that support their implementation and expansion. A policy in agriculture refers to a set of guidelines, regulations or laws that govern the development and management of the sector. These policies are created by governments, international organizations or other entities to provide a framework for achieving specific objectives related to agriculture and its allied activities. These policies can cover many areas, such as production, marketing, trade, research and development, land management and environmental protection. Some common objectives of policies include improving food security, increasing agricultural productivity, promoting sustainable farming practices and supporting rural development. Agricultural policies can take different forms, depending on the context and objectives. For example, policies can be designed to support smallholder farmers, encourage private-sector investment, promote agricultural trade or protect natural resources. They can be implemented through different mechanisms, such as subsidies, price support, regulations or incentives.

Integrated farming systems (IFS) combine various elements, such as crop cultivation, animal husbandry and aquaculture, holistically and synergistically. These systems aim to improve food and nutrition security by increasing productivity, reducing environmental impact and promoting social and economic well-being. For IFS to be successful, a supportive policy framework is essential. Robust and participatory governance frameworks and institutions and the availability of technology, data and innovation are key drivers in the extensive range of policies, funding and legal frameworks essential for revolutionizing food systems (FAO et al. 2021). Systems approaches are required to develop cohesive sets of guidelines, investments and legislation to facilitate mutually beneficial resolutions while addressing trade-offs. These approaches include territorial strategies, ecosystem-based approaches, Indigenous Peoples' food systems approaches and interventions that will tackle prolonged crises. Establishing a supportive setting for farmers, producers and other stakeholders is crucial for achieving sustainable development in the agricultural sector.

The Integrated Farming System (IFS) can play a crucial role in the economic advancement of numerous nations. For IFS to succeed, a supportive policy framework for resource management and land use must be provided. Incentives, subsidies and tax breaks must be provided to help farmers pay the expenses of implementing innovative techniques and technology. Regulations are necessary to ensure the responsible utilisation of natural resources such as soil, water and biodiversity. These policies should encourage the creation of value chains for various IFS goods to increase farmers' access to markets. The following are the key policy frameworks that can help create an enabling environment for rapid dissemination of IFS.

DOI: 10.1201/9781003239963-15

15.2 ENABLING POLICIES FOR RAPID DISSEMINATION OF IFS TECHNOLOGIES

15.2.1 Land Management Policy

A policy is a framework that outlines the government's goals, objectives and strategies to promote agricultural growth and development. This policy should focus on improving farmers' access to land, water, inputs and markets. Land management policies are critical in promoting responsible land use practices and ensuring the responsible use of natural resources (Hajimirzajan et al. 2021). Integrated farming systems are a way to maximize the productive use of land while also promoting sustainable practices. Land management policies can support integrated farming systems by providing clear zoning and land use regulations. This can ensure that appropriate areas are designated for farming, forestry and other land uses that are compatible with each other. It can also help reduce conflicts between land uses and ensure that agricultural activities do not negatively impact neighbouring communities. It should also address climate change, food security and rural development issues. The policy must ensure that resources are conserved and support sustainable land management techniques like contour farming, terracing and agroforestry.

15.2.2 Investment Policy

An investment policy can help create an enabling environment by attracting private investment into the agriculture sector. This policy should provide incentives for private investment, such as tax breaks and subsidies, and encourage public-private partnerships. Land management guidelines can support integrated farming systems through investment in rural infrastructure, like roads, bridges and irrigation facilities. This can improve accessibility to markets, reduce transportation costs and increase the profitability of farming. It can also increase the resilience of farming communities to natural disasters and climate change. The policy also should support small-scale farmers. This can include access to credit, insurance, and marketing support. It can also involve providing social protection measures, such as safety nets, to protect smallholder farmers from income shocks and economic instability. Investing in R&D will help in better understanding IFS and create the best practices and technology possible.

15.2.3 Trade Policy

A trade policy can help create an enabling environment by ensuring access to regional and international markets. This policy should focus on reducing trade barriers and improving market access for agricultural products.

15.2.4 Research and Development Policy

A research and development policy can help create an enabling environment by promoting innovation and improving productivity in the agriculture sector. This policy should fund research and development programmes and encourage collaboration between researchers and stakeholders. IFS are intricate systems requiring constant study and development to improve performance and adjust to shifting circumstances.

15.2.5 Environmental Policy

An environmental policy can help create an enabling environment by promoting sustainable agriculture practices. This policy should focus on minimizing the adverse effects of agriculture on the environment, promoting the conservation of natural resources and encouraging the adoption of sustainable farming practices. These practices, such as using cover crops, intercropping and crop rotation, can help reduce soil erosion, enhance soil fertility and increase the productivity of the land. Policies can provide incentives and technical support to farmers to adopt these practices.

15.2.6 Agricultural Extension Services Policy

Land management policies can support integrated farming systems by investing in agricultural extension services. These services provide technical assistance and training to farmers on the latest farming techniques, technologies, and best practices. This can help farmers adopt integrated farming systems and improve their productivity. It should encourage conducting training programmes and provide extension services to give farmers the information and expertise they need to adopt and maintain IFS. A comprehensive package of institutional tools including localized technical extension and educational servies would be helpful in the expansion of IFS in low-income countries (Suh 2015).

15.2.7 Regulatory Policy

A regulatory policy can help create an enabling environment by ensuring that regulations are consistent and transparent. This policy should provide clear guidelines and standards for agricultural practices and ensure they are enforced fairly and efficiently. The policy should foster cooperation between farmers, researchers and decision-makers to recognize and exploit significant possibilities and problems.

15.3 SOCIAL AND ECONOMIC DEVELOPMENT

IFS can have a good effect on social and economic development, but this effect needs to be realized in a climate of supportive policies. This should include policies that support the creation and implementation of IFS by women and other excluded groups. The policy initiatives should promote education and public awareness of the benefits of IFS and how to participate in their development and use. Rural communities can flourish and become more accessible to basics like transportation, health care and education thanks to policies and initiatives. An enabling environment is essential for small and marginal farmers to embrace and execute IFS successfully. The policies should support small and marginal farmers and promote sustainable and organic farming practices. Ensuring small and marginal farmers have secure land tenure rights, which can be used as collateral for credit, is an essential aspect of the policy. This can be done by implementing land registration and certification systems and by recognizing the rights of small and marginal farmers. Overall, an enabling environment that supports small and marginal farmers promotes sustainable farming practices and recognizes the importance of ITK and cultural heritage, which is essential for the successful adoption and implementation of IFS.

15.4 CASE STUDIES FOR POLICY DEVELOPMENT

This section gives an account of different case studies for policymakers to take notice of and take appropriate actions for policy development. Latest developments in agriculture, technology and policy should be utilized and scaled for transforming conventional land use to productive ecosystems (ICRISAT 2023). Case studies can be an effective way to understand the potential benefits and challenges of initiating, adapting and advancing integrated farming systems (IFS). Examples of case studies on Farmers' Producer Organizations (FPOs) may be useful for the purpose of policy framework development and execution (NABARD 2022).

An FPO in India that adopted IFS practices such as crop rotation and intercropping reported increased crop yields and improved food and nutrition security for its members. This case study highlights the potential benefits of IFS for small and marginal farmers, such as increased crop diversity and productivity and improved food and nutrition security. One such FPO is Jaivik SRI Farmers' Producer Company Ltd. Koraput, Odisha, registered on 18 May 2016, facilitated by Pragati, Koraput and supported by NABARD, India. The members comprise 551 farmer shareholders, of which 63 are women, covering 37 villages. The amount of procurement of different products from FPO members was INR 649,000, which involved produce like paddy, rice, millets and pulses. Quality seeds of Indigenous and high-yielding seeds of paddy, millets, pulses and vegetables are timely made available to the farmers at a fair price. Around 90% of the farmer members of the Producer company can now get quality seeds in time and at fair prices compared to the market. The farmer members practice integrated farming with crops like paddy, millets, pulses and vegetables, along with horticulture and livestock rearing like goatery and poultry. A total of 143 farmer members are adopting aquaculture, which fetches them additional income and meets their consumption needs.

The company has been carrying out a special programme to promote millet in tribal regions of Odisha, with the support of the Odisha Millet Mission, in Nandapur Block of Koraput District since June 2017. Due to the implementation of the programme, 3525 farmers of 12 Gram Panchayats in Nandapur block have adopted the System of Millet Intensification. The farmers are getting incentives and selling the surplus at a minimum support price fixed by the government. During Kharif 2019–2020, the company established a linkage with DSMS to market 4,000 quintals of finger millet, which will fetch a minimum support price of Rs3150 per quintal. The company has established a linkage with the National Horticulture Mission and received Rs 500,000 for excavating a pond for fish culture. The company has established convergence with the horticulture department for drip irrigation for 114 farmers and seedling support for 450 farmers. The company, in convergence with ICAR-CIFA, Bhubaneswar, promotes fish farming in four blocks of Koraput District, namely Borigumma, Kotpad and Koraput blocks. During 2019, carp fish fingerlings were stocked for 131 fish farmers. Three carp fish breeding portable hatchery units were set up by the Jaivik SRI Farmers Producer Company in Borigumma (2 no) and Koraput (1 no) block with support from Pragati and ICAR-CIFA, Bhubaneswar. Each hatchery can produce 10 lakh spawn in one production cycle. It can take up to four to five productions in a year, which will be marketed by the company. The company, in convergence with ICAR-Indian Institute of Soil and Water Conservation, promotes off-farm livelihood for its members through training and support in poultry farming, quail farming and mushroom cultivation. So far, 120 members are trained on these off-farm livelihoods. Thus, the synergy between different sectors of agriculture and allied areas could be harnessed to bring about a significant change to society in particular and the state in general.

A case study of ICRISAT-led integrated crop-livestock systems (ICLS) in the drylands of Africa helped farmers in a drought-prone area to adopt IFS practices such as rainwater harvesting, soil conservation and crop diversification. As a result, farmers were able to increase crop yields and reduce their dependency on rainfall. It has been seen that crop-livestock systems' water productivity rose, enhancing their resilience against drought conditions and climate change. They assist soil fertility, food and energy production, water and carbon cycle regulation, ecological services and cultural advancement. Additionally, integrated systems' inclusiveness promotes greater equity in production processes. In particular, goats and groundnuts are seen as women's traditional realm in southern Africa. A greater emphasis on these products and their supply chains will undoubtedly improve the situation for women, for instance, goat milk and meat support Zimbabwe's food security, economic well-being and income. Goat money is used to purchase food, cover educational expenditures and pay for health care for people. Enhancing the feed systems through the integration of agricultural and animal production promotes revenue growth, education and food and nutritional security. Enhanced systems for animal feed and marketplaces present significant opportunities for adding value to agricultural products, as demonstrated by the work done by the International Crops Research Institute for the Semi-Arid Tropics (ICRISAT) and the International Livestock Research Institute (ILRI). Small and medium-sized businesses created around poultry, cattle and their related feed systems can improve nutrition, lower poverty and enhance the income of all actors in the value chain, not just farmers.

The system employs various technologies, including water harvesting methods like micro-catchments, planting pits, half-moons, trenches and animal manure. It also involves planting high-value fruit trees like Moringa and Pomme du Sahel (jujube) and annual Indigenous vegetables such as okra, roselle and sicklepod. This blend of technologies forms an integrated climate-smart and gender-sensitive system, enhancing soil health and productivity by capturing rainfall and producing biomass. This case study illustrates how integrated farming systems (IFS) can assist farmers in adapting to changing climate conditions and enhancing their resilience.

Another example of a farmers' producer organization (FPO) in Asia focusing on integrated farming systems (IFS) is Thailand's Maejo Organic Agriculture Project in Chiang Mai Province. The project has been an effective FPO that has been applying IFS practices. The project aims to advance sustainable agriculture and raise local small farmers' living standards. To accomplish their objectives, they have embraced various IFS practices, such as terrace farming, agroforestry and crop diversification. The Maejo Organic Agriculture Project has documented various advantageous effects of implementing IFS practices, including enhanced food and nutrition security, decreased soil erosion and higher farmer revenue. These methods have made it feasible for farmers to grow various crops, including high-value crops, thereby diversifying their income sources. The project's terrace farming methods reduced soil erosion by reducing surface runoff, while agroforestry systems provided extra goods like fruits and nuts in addition to helping to save soil.

These case studies can help policymakers and farmers understand the potential benefits and challenges of initiating, adapting and advancing IFS. They can also inspire and guide policymakers and farmers looking to implement similar systems in their communities.

15.5 CONCLUSION

Integrated farming systems can improve food and nutrition security and increase income and livelihood opportunities for farmers and their communities. However, implementing IFS can be challenging and requires a supportive policy framework that includes investment in research and development, sustainable land use and resource management as well as social and economic development. By addressing these policy framework requirements, governments and other stakeholders can create an environment conducive to the success of IFS and ensure that farmers and communities realize their benefits. IFS policies can be essential tools for promoting sustainable and inclusive agricultural development. They can assist in establishing a conducive environment for farmers and other stakeholders, ensure the effective utilization of available natural resources and contribute to achieving broader development objectives.

REFERENCES

FAO, IFAD, UNICEF, WFP, WHO. 2021. *The State of Food Security and Nutrition in the World 2021. Transforming Food Systems for Food Security, Improved Nutrition and Affordable Healthy Diets for All.* FAO, Rome. DOI: https://doi.org/10.4060/cb4474en.

Hajimirzajan, A., Vahdat, M., Sadegheih, A., Shadkam, E., El Bilali, H. 2021. An integrated strategic framework for large-scale crop planning: Sustainable climate-smart crop planning and agri-food supply chain management. *Sustainable Production and Consumption* 26:709–732.

ICRISAT. 2023. Innovations to transform drylands towards inclusive, resilient and sustainable agri-food systems. In: Kumar A., Kumar S., Mandapati J., Peachey R. (eds) *International Crops Research Institute for the Semi-Arid Tropics*. Telangana, India. p 80.

NABARD. 2022. *Case Studies of FPOs in India 2019–2021.* NABARD Research Study 25.

Suh, J. 2015. An institutional and policy framework to foster integrated rice–duck farming in Asian developing countries. *International Journal of Agricultural Sustainability* 13(4):294–307.

Index